Tl

Ser

Fat
Mi
Sin
Ge
An
Ka
Ka
Ma
Fab
Vít
Chr
Giu
Igo
Jere

Aims and Scope

The Urban Book Series is a resource for urban studies and geography research worldwide. It provides a unique and innovative resource for the latest developments in the field, nurturing a comprehensive and encompassing publication venue for urban studies, urban geography, planning and regional development.

The series publishes peer-reviewed volumes related to urbanization, sustainability, urban environments, sustainable urbanism, governance, globalization, urban and sustainable development, spatial and area studies, urban management, urban infrastructure, urban dynamics, green cities and urban landscapes. It also invites research which documents urbanization processes and urban dynamics on a national, regional and local level, welcoming case studies, as well as comparative and applied research.

The series will appeal to urbanists, geographers, planners, engineers, architects, policy makers, and to all of those interested in a wide-ranging overview of contemporary urban studies and innovations in the field. It accepts monographs, edited volumes and textbooks.

More information about this series at http://www.springer.com/series/14773

Mladen Obad Šćitaroci ·
Bojana Bojanić Obad Šćitaroci ·
Ana Mrđa
Editors

Cultural Urban Heritage

Development, Learning and Landscape Strategies

Springer

Editors
Mladen Obad Šćitaroci
Department of Urban Planning,
Spatial Planning, Landscape Architecture
University of Zagreb
Zagreb, Croatia

Bojana Bojanić Obad Šćitaroci
Department of Urban Planning,
Spatial Planning, Landscape Architecture
University of Zagreb
Zagreb, Croatia

Ana Mrđa
Department of Urban Planning,
Spatial Planning, Landscape Architecture
University of Zagreb
Zagreb, Croatia

ISSN 2365-757X ISSN 2365-7588 (electronic)
The Urban Book Series
ISBN 978-3-030-10611-9 ISBN 978-3-030-10612-6 (eBook)
https://doi.org/10.1007/978-3-030-10612-6

Library of Congress Control Number: 2018965445

This Springer imprint is published by the registered company Springer Nature Switzerland AG
The registered company address is: Gewerbestrasse 11, 6330 Cham, Switzerland

Preface: The Heritage Urbanism Approach and Method

Introduction

The research papers published in this volume are the result of four years work on the Urban and Spatial Models for the Revival and Enhancement of Cultural Heritage research project. The project and the researchers faced the question of how to preserve and present cultural heritage and how to tackle the challenges arising from modern needs and new interventions on historic buildings and in the sphere of cultural heritage. Various types of heritage have been explored from different periods—ranging from cultural landscapes to individual buildings, from ancient archaeological sites to modern architecture, from the tangible to the intangible heritage—from different perspectives—interdisciplinary, multidisciplinary and transdisciplinary. To achieve this, the project was open to everyone who could contribute to its development, including researchers and experts of different professions and interests. Everyone attempted to answer the question posed in the title of this research project: what the potential models for the revival and enhancement of cultural heritage are?

Much is being said and written about the need for the revival and enhancement of heritage, and many declarations and charters have been adopted. Simultaneously, many heritage sites are deteriorating and disappearing, primarily due to the lack of a clear, long-term vision, high costs, the lack of resourcefulness of the local authorities and potential investors, complex administrative procedures, the prejudice that revival is difficult and complicated, etc. In such circumstances, cultural heritage is rarely a priority, except in notable examples. This research does not focus on these examples of exceptional value, but on the "ordinary" heritage which is prevalent in all towns, settlements and landscapes and does not necessarily have universal global or national value but is important at the regional or local level. This is the heritage that deteriorates the most, although the heritage of the greatest value is not exempt from decay either. Judging by the state of preservation, we can recognise heritage that is not alive and needs revitalisation, and heritage that is alive, but needs enhancement.

Stimuli for Research

The motivation for this research came from the numerous cultural heritage sites in Croatia. The Adriatic coast of Croatia incorporates more than 25 centuries of urban culture—from the first Greek cities in the fourth century BC and even older proto-urban Illyrian settlements which developed simultaneously with the neighbouring Etruscan settlements on the Apennine Peninsula. In the Croatian part of the Pannonian Plain, 8000 years of continuity of settlements are visible through the archaeological remains in eastern Croatia (Vinkovci and Vukovar). From those past millennia to the present day, multiple traces of towns, settlements and cultural landscapes are layered over one another as visible and recognisable traces of humankind.

Another reason for this research was the poor state of the cultural heritage, not only in Croatia, but globally, which is an issue that cannot be resolved on account of the myriad difficulties and challenges of the modern world, where there is no money for culture, where people die of hunger, and global pollution is so critical that it causes various problems throughout the world, among other factors.

The third reason for the research was the need to objectify the revitalisation of heritage so that subjective decisions can be avoided for each case of heritage revival and enhancement. This is a frequent consequence of the absence of clear and elaborate criteria for the assessment of heritage and criteria for new heritage interventions, while models for cultural heritage management are being sought and tested on a small number of cases. Conservation methods of renewal and legal protection are well known, but they do not give satisfactory results because heritage cannot live based on formal protection only.

Therefore, it is necessary to establish an adapted modern method and identify acceptable renewal models to objectify the implementation of heritage revitalisation. By comparing other countries, a similar situation has been discovered in many of them. Their cultural (built) heritage is also deteriorating, many historical buildings do not have a permanent purpose or any purpose at all, and if the heritage is not used, it deteriorates and disappears completely.

The Main Research Problem and Research Question

The research question arose from the state of cultural heritage—how can heritage which is not in use become a winner instead of a loser, how can it transform from a burden to the community to a sustainable resource, and how can dying, inactive heritage become active heritage integrated into the life of a town or a settlement?

Heritage is a non-renewable resource. We can build new heritage and make it look old, but it would have no authenticity or originality; it would be untrue and would not be proof of the continuity of life and the art of heritage. For this reason, there is no alternative for the renewal and enhancement of cultural heritage. The question is not whether to preserve heritage, but how to renew it and how to preserve its historic properties (its specific features and identity) and adapt it to modern needs at the same time. This research question was the foundation of the research in an attempt to find potential solutions for the revival and enhancement of cultural heritage. Answers were primarily sought within urban and spatial planning models, although different solutions were also found and identified in other scientific and professional fields.

Relevance and Timeliness of the Research Topic

Exploring the criteria and models for cultural heritage enhancement involves internationally relevant issues given that cultural heritage is often mentioned in many charters, conventions and declarations which emphasise the importance of the preservation, enhancement and revitalisation of cultural heritage as a driver of economic and social development. The research papers in this book, just as those in the whole of the research project, follow international guidelines and principles. The timeliness of this topic is seen in the fact that it has not been explored yet, although there is a vast amount of literature on heritage, individual heritage topics, and numerous projects on the renewal and modernisation of the built heritage.

Plenty of papers on cultural heritage have been published offering primarily a cultural or historical perspective. A smaller number of papers offer a conservator's point of view, and even fewer papers are concerned with methodology. Rare scientific papers explore the methods and models of heritage revitalisation, especially in the field of architecture, urbanism and spatial planning, and attempt to consider them in a holistic manner. We hope that this volume, as well as the other papers published as a part of the research project, will contribute to the enhancement of the methodology of the revitalisation of cultural heritage. This research will improve the current situation because it offers models for the revitalisation of abandoned and neglected cultural (built) heritage in the context of economic sustainability and recognition of the ambiance.

Research Starting Points

Examining examples of good and poor practices and researchers' personal experiences (most them are architects and urban planners) showed that formal heritage protection is not sufficient and that heritage needs to be restored by active use. To understand the approach used in the research and presented in this book, several important starting points heed to be underlined:

- Apart from justified exceptions, cultural heritage cannot be preserved solely in the form of museum exhibits, but needs to be actively used;
- The survival of built heritage is possible only if we integrate it into modern life;
- It is hard to justify new construction while our historic buildings deteriorate and have no purpose;
- Although all cultural heritage demands an individual approach, the methods of renewal and revitalisation can, and sometimes must, be the same;
- New interventions in heritage are allowed and justified, but under the condition of preserving the authenticity and integrity of the heritage.

Numerous charters and conventions also served as a starting point for research:

- Recommendation on the Historic Urban Landscape (2011) by UNESCO encourages the active protection of cultural heritage and sustainable management as a precondition for development. It also encourages heritage to be embedded into planning policy and practice and argues that planning should enable mechanisms for balancing protection and sustainability.
- The Charter on Sustainable European Cities (2007) and the European Urban Charter II (2008) emphasise that the effective integration of cultural heritage depends on multidisciplinary and interdisciplinary research and considers that research is an outcome.
- The Valletta Principles for the Safeguarding and Management of Historic Cities, Towns and Urban Areas (2011) underline the need to protect and integrate historic and traditional areas into modern community life and to recognise this need as a foundation for urban planning and spatial development.
- The Paris Declaration on Heritage as a Driver of Development (2011) recommends integrating heritage into sustainable regional development plans, tourism and the economy.

Research Hypotheses and Goals

By examining urban and spatial planning models for the revival and enhancement of cultural heritage, we sought to confirm and support the following two hypotheses:

- Sustainable and creative planning protects heritage values, and, at the same time, these values become a primary driver (resource) of sustainable economic development;
- Preserving heritage values leads to enhancing people's environment, and enhancing heritage contributes to the quality of life and rehabilitation of spaces (towns, settlements and landscapes).

The research seeks to improve the urban and spatial planning methods and indicate potential models to contribute to the revitalisation and enhancement of cultural heritage. This could help increase the speed and effectiveness of the development of the local community to enable it to use its built and other cultural heritage as a resource for economic recovery and growth.

Several goals are underlined for the enhancement of the built heritage and other types of heritage:

- To determine the conditions of the heritage, the causes of deterioration and the opportunities for improvement;
- To identify the value and potential of the heritage which can and must become a driving force and an active participant in social, cultural, economic and tourist development;
- To affirm the art of heritage, to enable the continuity of heritage and create the prerequisites for excellence in modern interventions in heritage so that these interventions can someday become a new heritage;
- To suggest professionally relevant and scientifically founded criteria and models for the creative use of heritage;
- To consider heritage as an active space integrated into contemporary life.

Heritage Urbanism as a New Paradigm

In today's world, we can identify about seventy, perhaps even a hundred, different views and directions to consider and interpret urbanism. The starting point for all is a town or settlement as a constant, a built structure comprising a series of artefacts and public and private spaces. In a world of differences, different priorities and interests, various theories and practices appear which aim to contribute to the realisation of the ideals of life in urban and rural communities. These ideals always have their urban constants that make the town a town, or the settlement a settlement, and these are movement systems, a system of built structures, a system of public and landscape space, and infrastructure. These urban structures are upgraded or reinterpreted by different views on specific topics, so we can talk about landscape urbanism, ecological urbanism, environmental urbanism, new urbanism, sustainable urbanism, traditional urbanism and many others (Fig. 1).

Through our examination of models for the revival and enhancement of cultural heritage, we developed a spatial and urban planning view on the use of modern heritage that we named Heritage Urbanism (acronym HERU). This is a new paradigm. Heritage Urbanism considers the revitalisation and enhancement of heritage through the context of spatial and urban planning and landscape and finds models for its integration into modern living. The context, i.e. the environment, exerts an impact on the heritage and its revival, and reusing and finding new uses for the heritage are stimulating and affect the development of the environment.

Agrarian Urbanism, Anti-Urbanism, Archeological Urbanism, Beautiful Urbanism, Behavioral Urbanism, Big Urbanism, Blue Urbanism, Border/Trans-Border Urbanism, Bricole Urbanism, Brutal Urbanism, Bypass Urbanism, Clean Urbanism, Consumer-Based Urbanism, Contest Urbanism, Denied Urbanism, Dialectical Urbanism, Digital Urbanism, Disconnected Urbanism, Ecological Urbanism, Emergent Urbanism, Environmental Urbanism, Everyday Urbanism, Exotic Urbanism, Futuristic Urbanism, Green Urbanism, Guerilla Urbanism, Gypsy Urbanism, **Heritage Urbanism**, Holy Urbanism, Hybrid Urbanism, Informal Urbanism, Infrastructural Urbanism, Instant Urbanism, Intelligent Urbanism, Irresponsible Urbanism, Landscape Urbanism, Lite Urbanism, Magical Urbanism, Market Urbanism, Micro Urbanism, Middle Class Urbanism, Mobile Urbanism, Neoliberal Urbanism, Networked Urbanism, New Urbanism, New Suburbanism, Nuclear Urbanism, Open Source Urbanism, Opportunistic Urbanism, Paid Urbanism, Parametric Urbanism, Participatory Urbanism, Performative Urbanism, Political Urbanism, Pop-Up Urbanism, Postmodern Urbanism, Post-Traumatic Urbanism, Propagative Urbanism, Radical Urbanism, Real Urbanism, Recombinant Urbanism, Retrofuture Urbanism, Secon-Rate Urbanism, Slum Urbanism, Stereoscopic Urbanism, Sustainable Urbanism, Tactical Urbanism, Temporary Urbanism, Traditional Urbanism, Unitary Urbanism, Walkable Urbanism, ...

Fig. 1 Heritage Urbanism as one of the variety approaches within urbanism

Heritage Urbanism as a Viewpoint

The actions and reasoning of urbanism and spatial planning are interdisciplinary by nature since they include different aspects of thinking about space. Therefore, the urbanistic view on heritage implies an integral approach, which includes many different perspectives on heritage: the spatial and urban planning perspective; the landscape perspective; the architectural perspective; the cultural and historical perspective; the technical and infrastructural perspective; the legal perspective; the economic perspective; and the ecological perspective. Heritage is most often seen from the perspective of conservation, but this perspective alone cannot solve numerous issues related to the sustainability of heritage.

The spatial and urban planning perspective enables one to perceive heritage in the context of a town, a settlement or a region and to notice how heritage builds the town and how it can contribute to the development of a town or a settlement. The landscape perspective explores the possibilities of preserving and enhancing landscape values and ambience and visual experience in the cultural landscape where cultural heritage is located. The architectural perspective focuses on accomplishing excellence in new (modern) construction in inherited spaces. The cultural and historical perspective explores and respects inherited identity values. The technical and infrastructural perspective focuses on the transport system and communal infrastructure which affect the heritage areas and may contribute to heritage enhancement. The legal perspective directs its attention to ownership and procedures that can alleviate or aggravate, i.e. accelerate or slow down, the procedure of heritage restoration. The economic perspective provides insight into different scenarios of the financial sustainability of heritage renewal and the further use of heritage after renewal. The ecological perspective points to the potentially positive or negative impacts of the environment on heritage or of the repurposed heritage on its immediate and extended surroundings.

From the Heritage Urbanism viewpoint, cultural heritage must be an active subject in the contemporary cultural, social, economic and spatial development of local communities and it should enhance the excellence of space and of human life. Heritage can be adapted to modern needs, and it can be a driving force of local community development. Therefore, we must give heritage a chance to be reused and to start life anew. New interventions in heritage must affirm both the heritage and its surrounding environment, and they must be creative and demonstrate architectural brilliance. Heritage also demands appropriate cultural and economic management so that it may survive in the long run. It is also crucial to coordinate all state departments whose decisions and laws can have a positive or negative impact on heritage preservation and renewal. It is particularly important to coordinate the work of the authorities responsible for culture, spatial planning, regional development, the economy, tourism, fiscal policy and the judiciary.

The Heritage Urbanism Method

Originally, only a building or a building complex was considered as cultural heritage, but later this definition was extended to encompass the whole ambience, while today cultural heritage is an interconnected system and the coexistence of anthropogenic and natural landscape which is known by the name of cultural landscape. To solve the issues relating to the revitalisation of neglected and abandoned cultural heritage by integrating it into contemporary life, we must look for solutions by means of spatial and urban planning methods. These methods will establish the basic spatial, functional, infrastructural and cultural determinants for the revitalisation of cultural heritage on which all other aspects (architectural, economic, tourism, social, etc.) of heritage rehabilitation depend.

The views on the integral perception of heritage based on international documents connect cultural heritage protection with urban and spatial planning by using the idea of sustainable development. This includes affirming heritage as an active space and a spatial resource integrated into contemporary life and allowing it to become a driving force and an active participant in sustainable spatial, social, cultural, economic and tourism development.

Heritage Urbanism complements the methods used in spatial and urban planning and conservation. Heritage is considered together with its environment and all the effects this environment has on heritage and vice versa. At the same time, the impact of the revitalised heritage on its surroundings is also explored. The Heritage Urbanism method is applied when considering the revitalisation of cultural heritage.

It comprises three main components: factors, criteria and models for heritage rehabilitation. The factors include those of identity, influence and value, while the criteria and models help to enhance and revitalise the heritage. Besides concrete examples of heritage, comparative examples are also selected for examination (case studies) through which we can identify the criteria and models used in these chosen examples. Such a comparison helps to determine new criteria and models for a specific case of heritage enhancement or revitalisation.

Factors of identity include determining the basic characteristics of heritage, its features of identity which make it recognisable, authentic and original. This primarily consists of exploring the spatial, urban, architectural, landscape, cultural and historical features, complemented with other features depending on the type and character of the heritage. It is essential to identify the current and the historical context.

Factors of influence should be separated from the research context. These factors show how and to what extent the context affects the state and the character of the heritage, what changes have occurred in the past regarding reuse and adaptation, in what way the current spatial and social context affects the status and opportunities for heritage enhancement, and whether future positive or negative influences can be expected.

Factors of value are determined for every example of cultural heritage individually, depending on its identity and the influence of context. Valorisation of heritage enables scientifically founded and a professional determination of criteria for new interventions in heritage, regardless of whether these interventions refer to the enhancement of heritage that is in good condition or the revitalisation of abandoned and ruined heritage. Valorisation is conducted from the conservational, cultural, historical, spatial, ambient, architectural, aesthetic and functional aspect.

Criteria for new interventions in heritage, both those for enhancement and revitalisation, are based on the character of the heritage (its identity features) and its value and enrich heritage with a new contemporary contribution. There is a difference between general criteria used in numerous cases and specific criteria that originate from the factors of identity, influence and value. By changing the criteria, objectivity increases, and subjectivity predominantly decreases when deciding on new interventions in heritage areas.

Criteria for heritage enhancement are determined when the heritage in question is in good condition, is living and is being used, but needs improvement. These criteria must ensure the preservation of the most valuable identity features of heritage and prevent any sort of cultural heritage devaluation or destruction. Enhancement refers to conservation and architectural enhancement, as well as the technological enhancement of buildings and their adaptation to modern standards and needs. During these procedures, the usual technical standards must be adjusted to respect the features and values of the cultural heritage and to find a balance between the requirements and possibilities.

Revitalisation criteria are determined when the heritage in question is in poor condition, has no purpose and is not in use, so it needs to be repurposed and given new life. Preferably, the renewal should ensure that the level of authenticity is kept as high as possible, i.e. that historic structures and other remains are preserved, just like essential identity features, and that modern interventions provide the revival and reinterpretation of the past environment.

Models of heritage revival and enhancement are both determined in accordance with the factors of identity, influence and value, and in harmony with the criteria for new interventions. To establish potentially applicable models, it is advisable to use a comparative analysis of the selected relevant examples (case study) to determine the historical and modern models in similar situations, which will guarantee the selection of the best model for a specific case. The models offered will present a creative approach to solving the revitalisation problem because all heritage cases are different due to their factors of identify, influence and value that make each case unique. Whenever possible, it is recommended to apply the models of integrative planning which allow for sustainable development in the long run. The central interest is the interaction between the preservation and enhancement of the cultural heritage and the development strategy of the space.

By using the Heritage Urbanism method, we enhance the previous methods of spatial renewal and conservation and cultural heritage revitalisation. The Heritage Urbanism method focuses particularly on abandoned and neglected heritage in need of revitalisation and of integration into the daily life of the local community, thus contributing to social, economic and spatial development. The method also helps to make the implementation of cultural heritage renewal and rehabilitation more objective. It stimulates thinking about various heritage revitalisation scenarios derived from the viewpoints of spatial and urban planning, architecture, landscape and ambiance, the economy and other aspects, along with the essential conservation viewpoint that needs to be in harmony with all the other viewpoints.

The Heritage Urbanism method consists of interdisciplinary, multidisciplinary and transdisciplinary views on cultural heritage revival. It includes the harmonisation of different starting points in the fields of culture, spatial planning, regional development, the economy, tourism, etc. Selecting the optimal model of heritage revival will result from exploring factors, criteria and models from different standpoints, among which the following are important:

- spatial and urban planning—heritage involved in the daily life of urban and rural areas;
- architectural—achieving excellence in new contemporary buildings on heritage sites;
- landscape—conservation and enhancement of landscape values, and environmental and visual experiences;
- cultural and historic—acknowledgement of heritage identity values;
- conservation—conservation and affirmation of the historical tissue and the character of heritage;
- tourism—interaction between tourism and heritage promotes the enhancement of destination quality as a place for living, working and investing;

- marketing—spatial branding in the field of tangible and intangible cultural heritage and the designed natural environment, related to the renewal of heritage and/or the implementation of urban contemporaneity;
- economic—economic management of cultural heritage, including transparency in financial management, which is an important prerequisite for the survival and life of heritage;
- strategic—strategic planning and management involving the design and implementation of the main goals and initiatives based on rethinking resources and assessing the surroundings in which cultural heritage is located by securing its survival.

This is an applied method. Its application in the field of urban and spatial planning, and in the preservation and renewal of the cultural/built heritage, especially in conservation, is of the utmost importance. It can also be used indirectly in other areas related to spatial planning, such as construction, tourism, spatial economy, environmental protection and art history.

This method can readily contribute to the culture of buildings and to the culture of space, to cultural heritage enhancement and its integration into life, to the preservation of values and the cultural identity of an area, to the development of the economy and tourism, to sustainability and the quality of space and life in it, as well as to the enhancement of legislation and regulations in the field of spatial planning.

The papers in this volume follow the approach and the method of Heritage Urbanism. They strive to stimulate new views and possibilities for the new life of cultural heritage. We hope that the research results presented in this book will inspire new thoughts on the urbanism of heritage/Heritage Urbanism, in other words that they will create respect for cultural heritage, the contemporaneity of the historical, the potential applications of modern technologies and new functions in cultural heritage environments.

Zagreb, Croatia Mladen Obad Šćitaroci

Acknowledgements

We would like to thank all those who contributed to the publication of this book, which is the result of the "Urban and Spatial Models for the Revival and Enhancement of Cultural Heritage" research project, or, in short, Heritage Urbanism (HERU). The project was conducted from 2014 to 2018 at the Faculty of Architecture, University of Zagreb, and was funded by the Croatian Science Foundation (research project HRZZ-2032). The head researcher was Prof. Mladen Obad Šćitaroci. Thematic research by 36 researchers contributed to the project development and to the results published in this volume. The main task of the research is to identify historical and current models of cultural heritage revival and enhancement as a starting point for establishing future models to respect inherited, contemporary and sustainable future development.

The following project researchers were divided into groups by theme:

Module 1—Heritage in Landscape/Cultural Landscape as Heritage: Biserka Dumbović Bilušić, Boris Dundović, Sanja Gašparović, Irma Huić, Jasenka Kranjčević, Mirna Meštrović, Ana Mrđa, Ksenija Petrić, Marko Rukavina, Ana Sopina, Flavio Stimilli;
Module 2—Heritage in the City/The City as Cultural Heritage: Nikša Božić, Alenka Delić, Jesenko Horvat, Vedran Ivanković, Tihomir Jukić, Zlatko Karač, Damir Krajnik, Zehra Laznibat, Mara Marić, Ivan Mlinar, Boško Opalić, Marina Pavković, Ksenija Radić Knežević, Dario Sironić, Marijana Sironić, Zorana Sokol Gojnik, Alen Žunić;
Module 3—City Experience/The City as Experience: Bojana Bojanić Obad Šćitaroci, Igor Gojnik, Tin Oberman, Tamara Zaninović;
Module 4—The Normative and Legal Aspect of Cultural Heritage: Nenad Lipovac, Lea Petrović Krajnik, Koraljka Vahtar-Jurković.

A total of 17 topics (some of which contain sub-topics) are explored: cultural landscape, archaeological heritage, fortification architecture, cultural heritage of the countryside (castles and villas), rural heritage, island and coastal cultural heritage as the foundation of tourism development, the historic urban tissue (the heritage of continuity and transformation), small-town heritage, the heritage of twentieth-century modernism, the heritage of twentieth-century residential construction, garden heritage, cultural heritage as a starting point for place branding, public spaces interconnecting the town, soundscapes as cultural heritage, the brightness of space as heritage, legal and conceptual aspects of heritage, and cultural heritage in urban planning documentation.

As part of the project, research was conducted for 17 doctoral theses. Most theses have already been completed, while several are still in progress following the completion of the project. Postdoctoral and graduate students studying architecture and urban planning at the Faculty of Architecture, University of Zagreb, were also involved in the research.

Some of the research published in this volume is the result of international collaboration with researchers from other universities around the world who participated and provided their own interpretations of the project's starting points. The research network involved in the preparation of this book consisted of 36 researchers from different universities in Austria, Bosnia and Herzegovina, Croatia, Great Britain, Hungary, Italy, Israel, Montenegro, Portugal, Slovenia, Spain, Turkey and USA. We are grateful to all authors for their contributions to this book and for the affirmation of the models for cultural heritage enhancement.

We are grateful to the following for organising and preparing the research papers for publication in this book: Tin Oberman, Marko Rukavina and Ana Sopina.

Many experts and researchers who took part in two conferences organised while the research was in progress were also indirectly included in this project. There were 280 researchers from 22 countries and 42 universities in the world. We are grateful for their contribution to our research on models of heritage revival and enhancement in the areas where they live and work. The two conferences mentioned were:

- Cultural Heritage—Possibilities for Spatial and Economic Development, Zagreb, 22–23 October 2015, Faculty of Architecture, University of Zagreb, Conference Proceeding: ISBN 978-953-8042-10-2 (print), ISBN 978-953-8042-11-9 (eBook);
- Models of Revitalisation and Enhancement of Cultural Heritage—Multidisciplinary Dialogue, Zagreb, 24 May 2017, Faculty of Architecture, University of Zagreb, Conference Proceeding: ISBN 978-953-8042-29-4 (print), ISBN 978-953-8042-30-0 (eBook).

Contents

Introduction: The Revitalisation and Enhancement of Cultural Heritage in the Context of the Heritage Urbanism Approach

The themes of this book are strategies and models for cultural heritage enhancement from the multidisciplinary point of view. The main research problem is to identify historical, current and possible future models for the revival and enhancement of cultural heritage, taking into consideration three factors: respect for the inherited, for the contemporary and for sustainable future development.

The research approach relies on consideration of spatial-urban planning and a reflection on cultural heritage, from which the expression Heritage Urbanism developed. The Heritage Urbanism approach affirms heritage integrated into its surrounding environment and with all the influences that the immediate and wider environment has on it. The context, i.e. the environment, affects the revitalisation of the heritage, just as new use and reuse of the heritage have a stimulating and developmental impact on its environment. Heritage is therefore not considered in isolation but as part of the wider living space and the local community. In this way, heritage is considered on different scales—from the cultural landscape, cities and settlements to individual buildings, structures and intangible heritage.

This book was created through a multidisciplinary dialogue between many authors of different professions and different research and professional experience. It is a link between the global and the local, or the specific and the general. The contents of the book are compatible with the European Cultural Heritage Strategy for the twenty-first century (Strategy 21). This Strategy is built into the Namur Declaration, adopted at the European Cultural Convention in Namur (Belgium) on 23–24 April 2015. The Declaration Strategy 21 is based on the existing Council of Europe conventions in the field of heritage, in particular, the Faro Convention on the Value of Cultural Heritage for Society. "This Convention is based on the idea that knowledge and use of heritage form part of the citizen's right to participate in cultural life as defined in the Universal Declaration of Human Rights. The text presents heritage both as a resource for human development, the enhancement of cultural diversity and the promotion of intercultural dialogue, and as part of an economic development model based on the principles of sustainable resource use".[1]

[1] https://www.coe.int/en/web/conventions/full-list/-/conventions/treaty/199 (28.01.2018).

Some of the key components of Strategy 21 are the subject of consideration and research presented in this book, such as cultural heritage—a key component of European identity; identities and cultural diversity; quality of life and the living environment; management of cultural heritage; heritage and citizenship; heritage and societies; heritage and the economy; heritage and knowledge; governance in the field of heritage; and heritage and sustainable development.

The goal is to contribute to the enhancement of previous methods of cultural heritage renovation and enhancement, to improve the methods of the spatial protection of heritage and to contribute to the development of the local community through the use of cultural and, in particular, architectural heritage. Cultural heritage is perceived primarily through conservation, but that comes with limitations. If heritage is perceived and experienced solely through conservation, it becomes a static object. The goal is to make it an active subject. This primarily implies life in heritage, as well as a new purpose and a new life for abandoned heritage. Heritage could be considered as a resource that generates revenue for its own and for the sustainability of the local community. To achieve this, it should be developed in accordance with contemporary needs and technological achievements, but on scientifically based and professional criteria and on sustainable models. This research, consequently, is based on the interaction of experiments (case studies) and theory.

The book is divided into five thematic units with examples primarily from Europe, but also from other parts of the world.

The first part—Heritage in Development—includes the following topics: heritage management; cultural tourism; archaeological heritage; cultural landscape; the transformation and sustainable development of small historic towns; urban heritage regeneration; and the application of the Space Syntax method in developing the Heritage Urbanism approach.

The second part—Heritage Experience Learning—consists of the following topics: identity of place; heritage branding; sustainable planning and valorisation of heritage tourism; associative landscape; city identity; reusing city areas; sacred architecture; and walkspace and soundscape.

The third part—Landscape Heritage—includes heritage in the coastal hinterland; landscape regeneration; natural heritage in an urban environment; strategies of conservation and urban design; brownfields in protected areas; bastion fortifications; historic gardens; country houses; and historic road corridors.

The fourth part—Concluding Words on Heritage Urbanism Models—presents our final thoughts on the various models that can be used in the restoration, revitalisation and enhancement of cultural heritage and which are the result of the analyses of historical and contemporary models.

The book may serve as an inspiration for a different view of cultural heritage, among other things, to discover the contemporaneity of the historical and to search for the contemporary in historical spaces. It encourages reflection on the application of modern technologies and functions in heritage ambiences. A contemporary approach is promoted because the current models of heritage enhancement yield slow progress, and heritage decays faster than we can preserve, renovate and integrate it in everyday life.

This book will contribute to the culture of building and to the culture of space, to cultural heritage enhancement, to the preservation of the value of space and cultural identity, to the development of the economy and tourism, to the sustainability and excellence of space and life in it and to the improvement of legislation and regulations in the field of spatial planning. It can be used as a handbook for university courses, as a manual for designers, planners and other professionals and as a stimulus for new and ongoing research.

Zagreb, Croatia
March 2018

Mladen Obad Šćitaroci
Bojana Bojanić Obad Šćitaroci
Ana Mrđa

Chapter 1
Heritage Literacy: A Model to Engage Citizens in Heritage Management

Darko Babić, Meltem Vatan Kaptan and Clara Masriera Esquerra

Abstract Heritage management is a complex and demanding task; when successful, it will always show (either in the forefront or in the background) the compound multidisciplinary approach standing behind it. Heritage interpretation, not surprisingly, very often plays a key role in this process. Contemporary understanding of heritage management, by professionals as well as by laymen (since they are interconnected), must consider all the relevant (site/county/country/region) specific factors, foremost the social or economic ones beside essential preservation. Heritage management tends towards ensuring tangible (as well as intangible, or more precisely indirect) benefits for local communities and in this way towards the development of society in general. Critical heritage studies over the last few years have significantly influenced the perception of heritage, and consequently the essence of heritage management and heritage interpretation. Stress on the participative and inclusive approach has become crucial, where multi/polyvocality is (almost) self-evident. While the aforementioned words are regular buzzwords today, this article looks for their origins. Surprisingly, this practice could be easily tracked to the early 1970s and the eco-museums movement which is quite a revealing experience. We attempt to demonstrate how the practices of eco-museums could be interlinked with contemporary demands, the need for participative and inclusive heritage interpretation, and management approaches/practices. Finally, the paper will point forward the need for socially responsible heritage management which could indeed be recognised as a demand for heritage literacy and as a model/tool to mitigate diverse interests where contemporary heritage management is concerned.

D. Babić (✉)
Faculty of Humanities and Social Sciences, University of Zagreb, Zagreb, Croatia
e-mail: dbabic@ffzg.hr

M. Vatan Kaptan
Faculty of Architecture and Design, Bahcesehir University, Istanbul, Turkey
e-mail: meltem.vatankaptan@arc.bau.edu.tr

C. Masriera Esquerra
Facultat de Lletres, Universitat Autònoma de Barcelona, Barcelona, Spain
e-mail: clara.masriera@uab.cat

M. Obad Šćitaroci et al. (eds.), *Cultural Urban Heritage*,
The Urban Book Series, https://doi.org/10.1007/978-3-030-10612-6_1

Keywords Heritage literacy · Eco-museum · Heritage interpretation · Polyvocality · Participative approach

1.1 Introduction

Today's still dominant understanding of heritage is a relatively recent innovation.[1] Humans have always had a certain relationship to what we can generically label as heritage (Carman and Stig-Sørensen 2009; Harvey 2008) where today's approach to the relationships and interactions which exist between us and the material and immaterial culture, structures and sites that surround us and in which we exist, nature and the environment in which we live and act, etc., represents only one of the possible approaches to heritage. If we extend this thesis to the global level and (especially) to different cultures, we may note that the experience of a human's relation to heritage as it is today could be defined by two important characteristics:

1. This relationship is dominantly *Eurocentric* (Smith 2006); in other words, affiliation is almost exclusively marked with the perspective(s) of *Western civilisation* and its understanding of the heritage, i.e. by the norms it has established, including only sporadic oscillations;
2. Concerning the former, the perspective and inclusion of different concepts of heritage (at the universal level) have happened only recently, at the beginning of this century.[2]

These two reasons clearly indicate that the notion of heritage is extremely complex and that understanding the same and consequently how we approach it and foremost manage it can be properly understood only through the systematically built and mutually respected interdisciplinary methodology of various scientific disciplines called upon to offer ideas. In this sense, our contribution here, which primarily brings a socio-humanistic understanding (of the idea) of heritage,[3] must be read as lobbying for even more of an interdisciplinary approach and for research at all levels.

Participating in or indeed consuming heritage in our own or our neighbouring regions, in Europe or anywhere else in today's globalised world, is very simple, which

---

[1] From the nineteenth century.

[2] Compare here, for example, the moment when the UNESCO Convention on Intangible Heritage (2003) were adopted, or the tendency (achieved through the Association for Critical Heritage Studies and regular conferences) concerning the need to critically examine ideas and concepts of heritage that have been intensively developing at the global level only since 2012, when the first conference of the Association for Critical Heritage Studies took place (in Gothenburg, Sweden).

[3] It is worth noting here that there is no single socio-humanistic perspective for understanding heritage. More precisely, there is no unique understanding within social and humanistic sciences of what heritage is, where perspectives differ significantly depending on tradition, previous experiences and preferences of the researchers in certain scientific disciplines. Nevertheless, recent trends indicate a common focus towards a scientific discipline (of heritage studies) directed at the idea of heritage per se, heritage as a phenomenon, no matter if this is sometimes expressed by means of contradictory attitudes.

makes the quality of experience that is directly dependent on the offered heritage service or management critically important. The world in which we live may look similar but it has drastically changed over the last few decades, i.e. in only one and a half generations. The lifestyle and culture (and heritage!) of various nations or communities on our and especially on other continents were until recently reachable only through specialised magazines, while now they have become part of the standard offer displayed in travel brochures that we find in our email inboxes. The opportunities to learn about natural and cultural values of a(ny) region and the experiences of visiting cradles of civilisations are daily within our reach. Distances are reduced, but, at the same time, as it seems, so are the possibilities to face the desired difference of the other and the different. While we want to learn something new, something different, what we usually get is much closer to uniformity and monoculture than the originally desired experience facilitated by (the idea of) heritage.

Based on many years of experience related to heritage, our advocacy for a far higher quality of heritage experience must resort to redefining the idea of heritage management, or, more particularly, to asserting the importance and necessity of high-quality heritage interpretation and what we define as heritage literacy. If we want to avoid uninteresting monoculture (both as heritage professionals and as casual or regular cultural/heritage consumers), what we have said above represents the only way to proceed to ensure quality for all stakeholders. Redefined heritage management, through heritage interpretation and literacy thereof, is the "key to success", but provided that we know what heritage actually denotes and what consequences it implies in today's world.

1.2 The Present(s) of Heritage

According to Howard (2003), a common way for certain places or phenomena to become legitimate (as heritage) is to compare them with those whose values and significance are already recognised and proven.[4] When such a practice is used by the user/visitor/heritage consumer, we do not have the right to hold anything against this practice. Reasoning about the user, directly or indirectly, is always linked to the idea of (heritage) management. Our Western, European way of understanding the world, and thus heritage which (ironically) we selflessly offer (and impose on) all other cultures, was until very recently[5] utterly puzzling, and often today remains confused, based on the thesis that heritage in its basics is completely intangible. And while we cope, more or less successfully, with the idea of the existence of a category rather new to us, designated as intangible heritage, and (occasionally) recognise it in our own cultures, some researchers in the field of heritage, e.g. Laurajane Smith,

[4]The author cites a number of territories in Europe which are designated as a "little Switzerland", or, for instance, "Venice of the North".

[5]See note 1, the UNESCO *Convention for the Safeguarding of the Intangible Cultural Heritage* was adopted only in 2003 and officially entered into force in 2006, only 12 years ago.

further confuse us with conclusions such as "There is, really, no such thing as heritage" (Smith 2006). Or Howard (2003) who, in his book *Heritage: Management, Interpretation, Identity*, takes an only seemingly contradictory stance, noting that heritage can actually be whatever we desire it to be and that the existence of desire is crucial in the process: "…things actually inherited do not become heritage until they are recognised as such. Identification is all". Finally, let us mention in this short overview of the possible challenges to heritage (mis)understandings the frequently quoted authors Graham et al. (2000) who, in their book *A Geography of Heritage: Power, Culture and Economy*, straightforwardly claim that "heritage can be visualized as a duality - a resource of economic and cultural capital", in other words that heritage is actually "a commodity, moreover one that is simultaneously multi-sold in many segmented market places".

Numerous authors (West and Ansell 2010; Carman and Stig-Sørensen 2009; Harvey 2008, 2001; Hernández 2008; Davison 2008; Smith 2006; Howard 2003; Graham et al. 2000) who directed their research towards the study of the history of heritage agree with the idea that its beginning is impossible to determine unambiguously and, accordingly, that a detailed chronology cannot be made. In conference proceedings, i.e. in a very comprehensive manual for researchers in the field of heritage, David C. Harvey in the preamble states that the main reason for the above is that the history of heritage, unlike the history of an individual institution or a significant movement or document, does not have a formal starting date and hence any further chronological systematisation is utterly arbitrary and directly dependent on the preferences of individual authors. Even more, he stresses in particular that (the idea of) heritage does not equate to an individual object and so does not exist by itself, nor does it imply any movement or project separately. This statement is important because, as he explains, "(heritage is about) the process by which people use the past" (Harvey 2008), and we will add the present too. Since every "society has had a relationship with its past, even those that have chosen to ignore it…" (Harvey 2001), the history of heritage in fact tends to explain the relationships in a society, in particular, the relevant historical development of the efforts of exercising the privilege of using heritage within a particular society, or humanity as a whole if we are talking about the idea of universal heritage. The specificity of a historical review always remains characterised by the fact that every era defines its own criteria; that is, the way heritage is used is always determined by the present, and hence, the history of heritage is always a kind of history of the present, or more precisely "a historical narrative of an endless succession of presents, a heritage of heritage that can have no terminal point" (Harvey 2008).

Almost without exception, all authors researching the historicity of heritage agree that it was the modern era that brought significant change; a certain democratisation of rights to the creation and control of heritage consumption happened at this moment (Harvey 2008) and "heritage becomes a public concern and its care an expression of the interests and responsibilities of civic societies" (Carman and Stig-Sørensen 2009), a concept that we can unreservedly support in its basic principle even today.

In the twenty-first century, we recognise that heritage may involve infinitely different shapes and culturally different expressions, and it participates in our lives in quite particular ways (Carman and Stig-Sørensen 2009) by defining relations in a society (in smaller communities, but also regarding humanity as a whole), where each day we discover new and/or differently interpreted patterns and relations. During the past two decades, we have witnessed at the same time a conceptual expansion of the concept of heritage (e.g. its immateriality) and the parallel steep rise of interest in a comprehensive and complex understanding of the meaning of heritage and the role heritage plays in a society.

If we accept the idea that heritage is a subset of culture, it is clear that the idea of representation becomes crucial for its understanding. The fact is that the meanings of certain objects, or events, or phenomena have varied over time, or, to put it more precisely, the same do not exist unless credited by us, humans (in the present time). This seriously challenges the traditional notion of heritage, which has always been dominated by only two dimensions: the past (the time from which the object or sites we keep or take care of dates) and the future (for whom we keep it), and brings the present to the foreground. In other words:

> The concept of time has remained central: heritage is a view from the present, either backward to a past or forward to a future. In both cases, the viewpoint cannot be other than now, the perspective is blurred and indistinct and shaped by current concerns and predispositions, while the field of vision is restricted to a highly selective view of a small fraction of possible pasts or envisaged futures (Graham et al. 2000).

According to the changed perspective, heritage does not actually exist until some very specific elements inherited from the past, but also some others incurred in the present according to our current preferences, are identified and labelled as such. This idea, taken from Graham et al. (2000), leads us to point out, in discussing heritage, that we therefore have access to a specific resource where someone, in line with certain needs, selects elements and then turns them into a particular kind of resource whose purpose is to meet some very certain (but always modern) needs. Since resources are later converted into products intended for consumption, it is evident that heritage is a form of commodification. Indeed, in the case of heritage there is at least a double form of usage and consumption—on the one hand at the cultural (or socio-political) level, and on the other at the economic level, where in both cases marked heritage possesses a quite specified market value (Graham et al. 2000; Ashworth et al. 2007). The most common forms of heritage use as economic resources are relatively well known[6] and easily identifiable, associated primarily with the creation of development strategies (i.e. development plans for rural or urban areas) quite often connected with tourism. Besides presenting an economic resource convertible into a more or less successful commercial product, another characteristic of heritage is that it is a potential and very rich cultural or socio-political resource. Through the use of heritage, it is possible to create and influence a range of important relations within society, which gives heritage an extremely powerful socio-political function. This process takes place in such a way that the selected elements are converted, by manipulated interpretation, into heritage which

[6]See more in Ashworth and Howard (1999), Graham et al. (2000), Howard (2003), Rypkema (2005), Ashworth et al. (2007), Boccardi (2015), Kohl and McCool (2016), etc.

becomes a specific medium, communicating complex layers and characteristics of identity (Graham et al. 2000). In other words, a series of attributes that have an extremely important role in human existence, such as language or religion, ethnicity, class, nationality and many other categories, acquire certain values, and consequently, their confirmation (or denial) is precisely defined through heritage construction. Since we realise our identities by identification and affiliation to a wider generally accepted concept—especially one that gives us a sense of closeness and direction, enrichment or sometimes escape when we need it (Lowenthal 1985, 1996)—it does not come as a surprise that the concept of heritage, regardless of the way it has been set, combines all the listed levels/layers and plays a significant role in our lives. By consuming and participating in heritage, we so commonly and (un)consciously accept it, without thinking too much about who created it and why or how it was created.

1.3 Heritage, Management and Literacy

To better understand the idea of heritage and all relationships (and possibilities) that emerge from it, it is necessary to accept sometimes confronting views and to act in a dialectic situation (e.g. to deal with heritage claiming at the same time that heritage does exist and/or arguing that heritage is always and exclusively intangible). As far as heritage management is concerned, the situation is confusing but at the same time revealing since "…heritage is heritage only because it is subjected to the management and preservation/conservation process, not because it simply is" (Smith 2006).

The last claim points out how heritage is formed, the fact that any definition of heritage is basically a process of heritage management, or precisely a socially and culturally (and also economically) managed process by which certain meanings and values are attributed to certain selected elements[7] after which they are identified as heritage. Over the last three decades, use of the term heritage management has experienced exponential growth, extending to virtually all areas of human activity, whereas the meaning of it still dominantly implies an (extensive) set of practical methods and techniques (from planning, through various forms of realisation, to evaluation) that can be applied to a particular segment of recognised heritage. Most of the quite extensive literature in the field of heritage management often do not question the basic phenomenon (and ways of its forming) and only apply certain techniques of technocratic management. Even more, potential users/visitors, who are the actual owners of the addressed heritage, are often forgotten; i.e. their role in the heritage management process is seen only through the lens of a passive consumer.

As we have already explained, heritage is primarily defined by representation and attributing values and meanings happening in the present time, which is thus (a sort of) heritage management (Smith 2006). Accordingly, further understanding of the term management, i.e. the concept of management, is required. Management is most commonly identified with the practice of running companies (business entities) or

[7]Objects, buildings, phenomena, ideas, etc.

organisations. As a form of organisational activity, it designates the formation and maintenance of an environment in which individuals through their work achieve the wanted goals in a desirable, successful and efficient way. Thus, management can be viewed as the effective employment of human and material resources which is realised in line with specific rules known in advance, while adhering to verified principles and applying appropriate techniques in the context of which four basic elements exist: planning, organising, guiding/managing and finally controlling or evaluating. It is certainly worth mentioning that one of the pioneers in the field of organisational theory, the American Mary Parker Follett (1868–1933), at the beginning of the twentieth century defined management as the philosophy or the art of resolving matters through mediation and/or in cooperation with people (Daft and Marcic 2009). In the aforementioned approach, understanding of management is particularly interesting here since management obviously goes beyond merely technocratic discourse and implies the inclusion of multifaceted social and humanistic layers and relationships which play a crucial role in any management and, we will say, especially in (any) heritage management, since both heritage and management are strongly defined by societal inputs and contexts.

Heritage has been a frequently used economic resource in the last 30 years or so. However, and this is our major point, heritage as a special form of knowledge started to be discussed only very recently, and even more importantly, heritage as knowledge is practically non-existent within (traditionally understood) heritage management. As we have already explained, heritage as a form of knowledge, the way it is constructed and how it is used have the potential to determine our values and impact our attitudes. In other words, the (conscious) selection and *management* of particular elements of our surrounding heritage become not only an economic resource but a resource with a very powerful and significant social and political dimension. The nature of such knowledge always depends on certain time-determined and socially conditioned circumstances in which it is created because there is always somebody who decides about heritage as well as what it is (Ashworth et al. 2007) with all its appending features, values and indeed possible use.

The presented ideas direct us towards the crucial question—not how, but who defines heritage, and for whom? Namely, the perceived, wider understanding of heritage management that assists us in recognising heritage can actually be anything identified and managed. More precisely, heritage is a specific form of managing values and meanings through the process of identification, representation and finally participation, and thus, heritage and heritage management are inseparable categories. And all this brings us closer to the idea of heritage literacy as a possible model to engage citizens, the real owners of heritage, in all/any heritage management process where the term literacy needs to be understood beyond standard definitions of reading and writing skills (thus of being able to read and to write) but as an ability to *navigate in a specific social context* defined by specific characteristics of a group/local community. In other words, literacy is directly connected with the (local) heritage. This indirectly derives from the Council of Europe Faro Convention[8] which introduced

[8]The Faro Convention on the Value of Cultural Heritage for Society.

the idea of heritage communities and their rights concerning heritage as far as the European context is concerned.

In brief, heritage literacy stands for the idea that every community has the undisputed right to shape and define its heritage and to manage it in the way that the (local) community understands and recognises that it will best shape its progress and development as well as contribute to the common good. Reaching a level of heritage literacy is demanding but necessary if we want to democratise heritage and to put heritage practically on a pedestal concerning human rights. To be able to do so and to be in a position to enable citizens to act as savvy heritage managers, (at least) three interconnected models are essential:

1. a critical approach to heritage;
2. eco-museological experiences;
3. heritage interpretation.

The first model, although the order could be changed since all three are equally important, was already discussed in the previous two sections, so the arguments there will only be summarised here. The critical approach addresses all stakeholders involved in heritage. It points out that heritage and heritage management are far more complex than they appear. Understanding how heritage is constructed, as well as how it is used, must be more transparent, where different voices are heard and respected, and where it is necessary to take into account all relations and dependencies (socio-cultural, economic, etc.) which exists in a local community. At its (basic) level, in accordance with contemporary heritage theories, this will minimally include reflections on and overcoming challenges of heritage discourse imposed by politicians, developers, heritage experts and others who have power (or a privileged position) over the possible interests of the locals. This segment of heritage literacy advocates that the (local) community must not only be informed but also be involved in all heritage management processes. Or from the opposite perspective, those to whom heritage management is currently committed (usually heritage experts) must all the time be aware that heritage management is a democratic, participatory and inclusive process where expertise must be directed to benefit the addressed local community.

1.4 Eco-museological Experiences

The second model we are addressing is eco-museums, i.e. eco-museological practices and connected theory. Eco-museums are relevant since they started, in a way, with the approach we are now discussing almost five decades ago. Use of the term eco-museum is quite frequent today,[9] where often misunderstandings about the fundamental qualities and the visionary approach of eco-museums exist. Eco-museums, created originally in France in the early 1970s, are special kinds of materialisation of a new way of thinking and understanding of relationships between society/the

[9]Including within projects funded by the European Commission.

local community and the environment. Focusing on local development, based on heritage, is the immanent determinant of eco-museums. Although in their organisational form and activities it is possible to identify some previous institutional experiences,[10] eco-museums possess distinctive characteristics which encourage the idea of an institution which serves the entire community. Even more, from the very beginning, directly or indirectly, they introduced innovative considerations about the relation between man/community and the related heritage, concepts we are discussing intensively today. This includes ideas of immateriality, the participative approach, polyvocality, inclusion, sustainable development, pro-activity, etc.

The first eco-museum, the Museum of Man and Industry, was set up in central France in 1971 near the towns of Le Creusot and Montceau-les-Mines[11] as a reaction to the rapid decline of the once successful regional economy based on industry, which reflected negatively on the 150,000 inhabitants living there. This was the set context, a kind of zero point, at which the museologists Hugues de Varine, Georges Henri Rivière and Marcel Evrard, in collusion with the existing regional development policy at that time, offered a solution by suggesting the creation of something seemingly ordinary (a museum) but extremely unusual in terms of its goals. The intentions of the Museum of Man and Industry were to restart the local economy, but in fact much more than that, as it tended to help local people to find meaning, to rediscover their identities misplaced due to rapid changes and to create new development opportunities throughout the region. Schneider's Castle,[12] dating from the eighteenth century, was designated to be the centre of the Museum where the historical development and the main features of the region, the daily life of the population and their industrial and artistic products were presented in an appropriate museological way. But more important for the Castle/Museum was the assigned role of the symbolic point of departure concerning interpretation of the territory. This was the starting point for the identification, introduction and exploration of the entire region which the eco-museum decided to embrace. Across the rest of the covered area, a special form of scattered, fragmented museum was developed, which actually means that the diverse tangible and intangible heritage in a defined landscape was identified, logically linked, museologically processed and interpreted in situ, that is, in places where they had originally been located without relocating them in the main building. In this way, bearing in mind that we are talking here about the beginning of the 1970s, there was an extremely important theoretical and practical innovation—the focus was shifted from a museum institution realised as strictly defined and concrete museum buildings to the interpretation of the totality of a territory conceptually covered by an eco-museum. In addition, the primary tasks of a few hired professionals-experts in this project were to launch the museum, to act foremost as sorts of catalysts of

[10]For example, open-air museums or *heimatmuseums*.

[11]In an area of approximately 500 km^2.

[12]The Schneider family, which possessed large industrial complexes in the area on which the local economy was dependent, had been accused of collaboration with the Nazi regime after the Second World War and accordingly the management of its industries had been transferred to Paris. The dislocated management entailed a lack of interest, which, together with the changing trends in industry, resulted in the neglect and deterioration of industrial plants, and therefore in a loss of jobs.

into the cultural and natural, conceptually introduced by the idea of landscape, or more precisely the concept that is perhaps best conveyed by the French word *terroir*.

Summarising everything presented above, we may say that eco-museum practice does not tend to be an applicable prototype that can be easily copy-pasted but, if properly understood, the theory/model of our (re)thinking of heritage and its potentials can be applied. Therefore, we are inclined to believe that eco-museums have the capacity to adapt to and embrace the essence of diverse territories and diverse heritage. In other words, they represent a model that is conceptually sufficient and adaptable to cultural diversity and to different perceptions of values and meanings implied by the idea of heritage (within different cultures), thus capable of overcoming not only the lack of "traditional" museums but also the existing practices of heritage management. Eco-museums are a bottom-up approach and thus give a different meaning to the idea of heritage management.

1.5 Heritage Interpretation

Third, an important part of the heritage literacy concept for it to be successful is heritage interpretation which may be directly connected to ideas presented in the previous section addressing eco-museums. In his attempt to explain how eco-museums were different, Pierre Mayrand, one of the main developers of eco-museums and related new museology ideas used the model of the *creativity triangle* in which heritage interpretation, more concretely the concept of participatory heritage interpretation, plays a key role. In other words, the community (in a three-year cycle) by creating an eco-museum and by developing and managing it becomes able to interpret itself and in doing so *comes into power* to influence its own development. According to Mayrand, the process of development of an eco-museum starts by the initiative coming from the local community and it is realised through the assistance of experts in the form of interpretative activities within a geographically defined/bordered area. The existence of (initial) interpretation further on sensitises (other) members of the local community and significantly increases their interest in the territory where they live and their awareness of a connection with what is interpreted, and consequently their pride in a local *sense of place*. As a direct consequence, the members of a community become more active and involved in heritage actions, including requesting the creation of an eco-museum. Finally, over time, through eco-museum experience and owing to the related feedback, the population is able to interpret itself and define the guidelines of its development (Davis 1999, 2011; Mayrand and Mairesse 2000). Dominique Poulot would say eco-museums have transformed the social use of heritage as they have contributed to the development of interpretation in the constant questioning of possession within society, more concretely how an addressed local community becomes aware of what it actually is (Poulot 1994). Similarly and also instructive here, André Desvalles in his analysis of the predecessors of eco-museums indicated that the innovative approach of heritage interpretation defined by the American writer and great lover of nature Freeman Tilden[17] influenced the ideas of the eco-museum.

One of the most common challenges we are (still) facing today in regard to heritage is a lack of (qualitative) communication. Heritage interpretation is the key since it could successfully provide an understanding of the activities we/local communities are pursuing or want to implement. The main idea or *raison d'être* of heritage interpretation is simple, yet powerful: "Through interpretation, understanding; through understanding, appreciation; through appreciation, protection" (Tilden 1977).

The word interpretation comes from the Latin *interpretari* and implies an explanation of the meaning (especially of a text), exposition and understanding.[18] However, the phrase heritage interpretation can be understood at least at three, apparently similar, but substantially very different, levels. At the first level, it involves a set of practical methods and techniques of presenting heritage. This set of purposeful and very pragmatic advice addresses primarily the question of how and where this approach still dominates in scientific[19] literature. At the next level, it implies a specific form of one-way communication in which experts (i.e. those who know) transmit messages (values and meanings) of the heritage to recipients, or, more precisely, to those who are willing to listen. This approach is most often (directly or indirectly) used by authors who wish to go beyond the exclusively utilitarian level. Finally, the third level implies an understanding of heritage interpretation that relies on the original meaning of the term interpretation and accordingly sees heritage interpretation as a concept depending on subjective interpretation by an individual and his/her individual capabilities, taking into account also the issue of social, cultural and any other differences existing between individuals or groups/communities. Although the first and second levels are still dominant today, we are here primarily interested in the last level where the main focus is on the idea of the active creation of meaning which is interpreted by us, humans, instead of the one-way communication of information. Specifically, we believe that in order to understand the values and significance of heritage for us as individuals, as well as for us as members of a particular group (community/society), it is necessary to interpret it. And how, when and at what niveau—this falls within the exclusive domain of personalised experience dependent on a social, cultural, economic and politically conditioned context. Only heritage interpretation, which respects the last approach, and attitudes can lead us towards desirable heritage literacy.

Heritage interpretation is as old as homo sapiens, man as a living being with mental capacities, since we have always interpreted the surroundings in which we live, be it the nature around us or what we have created with our hands or minds (i.e. culture). The idea of the importance of targeted and consciously delivered heritage interpretation can be primarily linked to the development of national parks in the

[17] According to Davis (1999, 2011).

[18] Interpretation also means acting, demonstration, performance.

[19] Better to say vocational.

USA at the end of the nineteenth century[20] (Beck and Cable 2002; Merriman and Brochu 2006; Kohl and McCool 2016; Beck et al. 2018). While E. Mills's ideas mark the beginning of the formation of standards for nature guiding approximately 40 years later, the main goals of heritage interpretation were elaborated in six very clearly structured principles formed by Freeman Tilden, an American writer, nature lover and supporter of the idea of national parks. Tilden's work marked the true beginning of heritage interpretation as a specific field, as a specific profession[21] where his six principles are still relevant today and where they fundamentally point to the quality of meaningful heritage interpretation. Tilden's six basic principles of heritage interpretation are:

- Any interpretation that does not somehow relate what is being displayed or described to something within the personality or experience of the visitor will be sterile.
- Information, as such, is not interpretation. Interpretation is a revelation based on information. But they are entirely different things. However, all interpretation includes information.
- Interpretation is an art, which combines many arts, whether the materials presented are scientific, historical or architectural. Any art is to some degree teachable.
- The chief aim of interpretation is not instruction, but provocation.
- Interpretation should aim to present a whole rather than a part and must address itself to the whole person rather than any phase.
- Interpretation addressed to children (say, up to the age of twelve) should not be a dilution of the presentation to adults, but should follow a fundamentally different approach. To be at its best, it will require a separate programme (Tilden 1977).

It is interesting to note here that we have only recently (re)discovered Tilden's ideas which not only explain what heritage interpretation is but (indirectly) what heritage is, as well as some misunderstanding we had about it. One is the long-standing idea (which is still often present in practice) of the artificially created distinction between natural and cultural heritage which Tilden erased in his principles. For him, there is only one single heritage. In addition, in his principles, it is clear that he insists on the active inclusion of visitors in any form of heritage interpretation, or, in other words, that it would basically be wrong to see a visitor as a pupil who just happens not to be sitting in school; i.e. it would be wrong to see heritage interpretation as just another lesson in the school curriculum (West and McKellar 2010). He points to the necessary interaction between the interpreter and the user, who only by means of joint

[20] Among the first who left a mark in the field, and entitled to bear the name "father of heritage interpretation", was the American naturalist, occasional mountain guide and writer Enos Mills, who among other things deserves credit for the foundation of the Rocky Mountain National Park. In the future area of the Park, Mills worked as an independent and self-proclaimed guide-interpreter from the 1880s.

[21] In his book *Interpreting Our Heritage* F. Tilden defines interpretation as: "An educational activity which aims to reveal meanings and relationships through the use of original objects, by first hand experience, and by illustrative media, rather than simply to communicate factual information" (Tilden 1977).

participation form the meanings and co-create interpretation—consciously or unconsciously—something which is still too often forgotten today. As we have already mentioned within the field of heritage interpretation, discussions still often focus on thinking how to best and most efficiently use interpretation to control visitors[22] and incomparably less on the needs, values or opinions of the visitors (Uzzell 1998). Heritage interpretation is always determined by personality and individual experiences brought in by (every) visitor. In other words, stress is on the need to respect the opinions and familiarities of the visitor(s) and consequently there is no single, uniform understanding of heritage. Accordingly, we firmly believe that in Tilden's principles we may read a request for polyvocality and a multicultural approach,[23] and thereby different modalities on which heritage per se is formed and perceived in different cultural environments. Tilden can further be credited for insisting on the necessity to observe and respect the totality of life, here related to the totality of heritage as part of our life, as stated in his fifth principle. As Pierssené (1999) reminds us, the main intention of true, holistic heritage interpretation is unthinkably simple—with it we simply want to explain the world we live into ourselves and to others. Heritage interpretation, when implying the ideas set forth herein, must without any doubt be understood as more complex than merely a temporary/occasional task, or yet another school class. Its methods should always be glocal, while the area over which it may extend should in no way be limited. This means that within heritage interpretation there cannot be two identical sites, as there cannot be general users, and accordingly no pre-set tool for interpretation to instigate the required inspiration and discovery. The latter recalls the ideas set out in relation to eco-museums, certainly with reason. The ideas of respecting the diversities of visitors/users, the specific features of the local site/context, the necessity of a holistic approach and putting focus on ensuring inspiration (provocation, not instruction!) are the preconditions to achieve the basic aim of heritage interpretation, mentioned at the beginning of this paragraph.[24]

1.6 Heritage Literacy—Conclusion

The main purpose of this article was not ultimately to answer the question about what heritage is, or to define how we should manage it. The main idea and goal here, in accordance with the presented principle of heritage interpretation, is to offer desirable provocation instead of instruction: provocation that calls for a re-thinking of the term heritage, the role(s) of heritage in contemporary society, as well as to challenge existing heritage management models as (possibly) inadequate. We understand that the presented model of heritage management, represented in the idea

[22]While heritage interpretation could be used to assist this goal, it cannot start and end with it.

[23]No matter that F. Tilden did not use the aforementioned terms.

[24]"Through interpretation, understanding; through understanding, appreciation; through appreciation, protection" (Tilden 1977).

of heritage literacy, is not the only way we may continue. However, and foremost from the socio-humanistic perspective, it indeed looks the most desirable.

What we have argued is that heritage is actually a specific form of heritage management (of managing values and meanings) and that heritage and heritage management are inseparable categories. Further, to be able to manage heritage in this sense, people, those who own heritage (no matter if at the local, regional, national or international level), must be aware and understand how heritage is constructed/created and how it is interpreted. Introducing literacy, heritage literacy, in this context seems extremely important. Heritage literacy is a form of heritage management, which is simultaneously knowledge management as well as a practical skill to manage (local) heritage. In other words, people/local communities, the real owners of heritage, must be able to understand the process(es) of heritage construction to be able to navigate within them (a critical approach to heritage). Only after ensuring this kind of awareness and participation in knowledge-related heritage management could they extend their rights of ownership over heritage, foremost via (the crucial step of) heritage interpretation to reach the desirable position of heritage managers. The presented eco-museum practices and experience are an intermediate model, a desirable but not necessarily the only way to reach desirable heritage literacy.

Heritage literacy embraces the idea of systematic, global, lifelong and holistic methods where each individual (or group) must have an inalienable and guaranteed right to participate, benefit and use (the locally recognised) heritage on the path towards self-development, self-esteem and the creation of the local community's own experiences deriving from the endless collective heritage of humanity. Implemented by individuals or a collective (i.e. local communities), it represents a commitment towards the universal values (via heritage) that humanity possesses and it is a basic prerequisite to ensure our overall development in the direction of a common prosperity. Reaching a desirable level of heritage literacy, valid for diverse communities around the world, is no easy task but (must be) a desirable goal where the three addressed models (a critical approach to heritage; eco-museological practices; heritage interpretation) are ideal (while not necessarily the only) pointers. Any other way, the heritage of humanity (represented by/contributed to by the heritage of local communities) will remain nothing more than a mirror of existing power relations which shamelessly do not respect and do not use the richness of human diversity, with all its beauty and endless creativeness, to achieve the common good. To attain it, heritage literacy, as a sort of *socially responsible heritage management*, is indeed necessary.

References

Ashworth G, Graham B, Tunbridge J (2007) Pluralising past: heritage, identity and place in multicultural societies. Pluto Press, London

Ashworth G, Howard P (1999) European heritage planning and marketing. Intellect Books, Bristol

Beck L, Cable T (2002) Interpretation for the 21st century: fifteen guiding principles for interpreting nature and culture. Sagamore Publishing, Champaign
Beck L, Cable T, Knudson DM (2018) Interpreting cultural and natural heritage: for a better world. Sagamore-Venture Publishing, Urbana
Boccardi G (2015) From mitigation to adaptation: a new heritage paradigm for the anthropocene. In: Albert MT (ed) Perceptions of sustainability in heritage studies. De Gruyter, Berlin, pp 87–98
Carman J, Stig-Sørensen ML (2009) Heritage studies: an outline. In: Stig-Sørensen ML, Carman J (eds) Heritage studies methods and approaches. Routledge, London, pp 11–28
Corsane G et al (2007) Ecomuseum evaluation: experiences in Piemonte and Liguria, Italy. Int J Heritage Stud 13(2):101–116
Daft RL, Marcic D (2009) Understanding management, 6th edn. South-Western College Pub, Mason
Davis P (1999) Ecomuseums: a sense of place. Leicester University Press, London
Davis P (2008) New museologies and the ecomuseum. In: Graham B, Howard P (eds) The Ashgate research companion to heritage and identity. Ashgate, Aldershot, pp 397–414
Davis P (2011) Ecomuseums: a sense of place, 2nd edn. Continuum, London
Davison G (2008) Heritage: from patrimony to pastiche. In: Fairclough G, Harrison R, Jameson JH, Schofield S (eds) The heritage reader. Routledge, London, pp 31–41
Graham B, Ashworth G, Tunbridge J (2000) A geography of heritage: power, culture and economy. Arnold, London
Harvey DC (2001) Heritage pasts and heritage presents: temporality, meaning and the scope of heritage studies. Int J Heritage Stud 7(4):319–338
Harvey DC (2008) The history of heritage. In: Graham B, Howard P (eds) The Ashgate research companion to heritage and identity. Ashgate, Aldershot, pp 19–36
Hernández MA (2008) Conservation and restoration in built heritage: a western European perspective. In: Graham B, Howard P (eds) The Ashgate research companion to heritage and identity. Ashgate, Aldershot, pp 245–265
Howard P (2003) Heritage: management, interpretation, identity. Continuum, London
Hubert F (1989) Historique des ecomusees. In: Hubert F (ed) La museologie selon Georges Henri Riviere. Dunod, Paris, pp 142–154
Kohl JM, McCool SF (2016) The future has other plans: planning holistically to conserve natural and cultural heritage. Fulcrum, Golden
Lowenthal D (1985) The past is a foreign country. Cambridge University Press, Cambridge
Lowenthal D (1996) The heritage crusade and the spoils of history. Cambridge University Press, Cambridge
Mayrand P, Mairesse F (2000) Entretien avec Pierre Mayrand. Publics et Musées 17–18:223–231
Merriman T, Brochu L (2006) The history of heritage interpretation in the United States. National Association for Interpretation, Fort Collins
Murtas D, Davis P (2009) The role of the ecomuseo dei terrazzamenti e della vite (Cortemilia, Italy) in community development. Mus Soc 7(3):150–186
Pierssené A (1999) Explaining our world: an approach to the art of environmental interpretation. E & FN Spon, London
Poulot D (1994) Identity as self-discovery: the ecomuseums in France. In: Sherman DJ and Rogoff I (eds) Museum culture: histories, discourses, spectacles. University of Minnesota Press, Minneapolis, pp 66–84
Rypkema DD (2005) The economics of historic preservation: a community leader's guide, 2nd edn. National Trust for Historic Preservation, Washington
Smith L (2006) Uses of heritage. Routledge, London
Tilden F (1977) Interpreting our heritage, 3rd edn. The University of North Carolina Press, Chapel Hill
Uzzell DL (1998) Interpreting our heritage: a theoretical interpretation. In: Uzzell DL and Ballantyne R (eds) Contemporary issues in heritage and environmental interpretation: problems and prospects. The Stationery Office, London, pp 11–25. Available via Academia. http://academia.edu.documents.s3.amazonaws.com/1438249/CIHEI-UZZPRIN015.pdf. Accessed 12 Apr 2018

Varine-Bohan H (1973) Un musee 'eclate': le Musee de l'homme et de l'industrie. Museum 25(4):242–249. https://doi.org/10.1111/j.1755-5825.1973.tb01432.x
West S, Ansell J (2010) A history of heritage. In: West S (ed) Understanding heritage in practice. Manchester University Press, Manchester & The Open University, New York, pp 7–46
West S, McKellar E (2010) Interpretation of heritage. In: West S (ed) Understanding heritage in practice. Manchester University Press, Manchester & The Open University, New York, pp 166–205

Chapter 2
The Uses of Space Syntax Historical Research for Policy Development in Heritage Urbanism

Garyfalia Palaiologou and Sam Griffiths

Abstract The application of space syntax methods to heritage-related questions has a long track record both in the field of space syntax research and beyond, for example in archaeology. These studies deploy the theories and methods of space syntax to explore the sociocultural dimension embedded in spatial systems of historic and archaeological significance. Space syntax analysis provides a link between the material and immaterial aspects of 'spatial' culture. It offers a critique of built environment typologies defined in terms of stylistic periodisation by advancing an understanding of the role of *spatial configuration* in the production and reproduction of space–time events. In the context of urban heritage studies, this means looking beyond the value of buildings as individual objects to buildings as elements in emergent arrangements of social space. Building on the comprehensive review of the disciplinary interface between urban history and space syntax historical studies provided by Griffiths (The use of space syntax in historical research: current practice and future possibilities, 2012), this chapter advances 'heritage urbanism syntax' with the aim of orientating this body of historical research towards contemporary issues of urban heritage. It identifies three kinds of *heritage urbanism syntax*: (1) conservation areas; (2) street scales and (3) spatial cultures in order to assist critical reflection on the application of this perspective to urban heritage contexts. The chapter highlights how a diachronic understanding of spatial cultures enables an integrative approach to heritage urbanism that situates heritage within both historical and contemporary urban landscapes. It describes the potential contribution of space syntax to inclusive bottom-up definitions of heritage and resilient heritage futures.

Keywords Space syntax · Heritage · History · Historical research · Spatial cultures

---

G. Palaiologou (✉)
School of Architecture, Civil and Building Engineering, Loughborough University, Loughborough, UK
e-mail: g.palaiologou@lboro.ac.uk

S. Griffiths
Space Syntax Laboratory, The Bartlett School of Architecture, University College London, London, UK
e-mail: s.griffiths@ucl.ac.uk

M. Obad Šćitaroci et al. (eds.), *Cultural Urban Heritage*,
The Urban Book Series, https://doi.org/10.1007/978-3-030-10612-6_2

2.1 Introduction

Space syntax research has more to offer to the urban heritage sector than is currently acknowledged in the field of heritage studies. Despite the long-standing engagement of the field with heritage-related topics and case studies, as well as the recent development of space syntax research in relation to the field of urban history, heritage studies is not a well-defined field in space syntax research.[1] As Griffiths (2012), in highlighting the strong tradition of historical case studies produced by space syntax researchers points out, where a research theme goes largely unacknowledged *within* a research community it acts as an impediment to opening up constructive interdisciplinary dialogue *beyond* that community. A lack of articulation of urban heritage, a similarly recessive (and related) theme in space syntax research, also has implications for the possibility of achieving a wider application of syntactic approaches to the study of urbanism. Bearing this in mind, the aim of this chapter is to probe the interface of spatial urban history using space syntax methods, with urban heritage studies and policy development, in order to identify new research trajectories for future practice.

Defining built cultural heritage outside limited conceptions of architectural and historical value is a highly pertinent topic in sustainable heritage management theory and practice (Tweed and Sutherland 2007). This means looking beyond the value of buildings as individual objects to buildings viewed as components in emergent arrangements of social space. Space syntax theory focuses on spatial systems and patterns of open space which are configured by building aggregates. This implies a critique of traditional approaches to the historic built environment in terms of stylistic periodisation in preference for an understanding of the role of *spatial configuration* in the production and reproduction of space–time events. Space syntax theory proposes a dynamic link between urban life and urban space—a ‘synchronic’ (because configurational) description of space–time events (Hillier and Hanson 1984, p. 94) with social connotations. Space–time events are defined in relation to qualities of visual and physical accessibility (Penn 2003), distance and direction (Dalton and Bafna 2003) and the potential for generating social encounters (Hillier 1996). Beginning with an explanation of this link between *material* and *immaterial* urban spatial cultures, the discussion develops to consider how the study of historical spatial syntaxes can make a contribution to alternative assessments of urban ‘spatial’ heritage.

At the core of space syntax theory is the proposition that the structure of space is itself a cultural artefact that defines how social groupings manifest their spatio-temporal existence. As such, space syntax theory renders space relevant to cultural and historical (and historic) attributes, and by extension, to assets, values and definitions of heritage. By understanding spatial cultures as the product of local topological and geometrical conditions, space syntax can give quantitative description to place-specific aspects of cultural value. An increasing amount of research on urban

[1]See also Griffiths, S. and Vaughan, L. (2019). Mapping spatial cultures: contributions of space syntax to research in the urban history of the 19th Century city. *Urban History*. Special edition on History and Mapping (forthcoming).

histories using space syntax methods discusses historically specific formative processes of urban change and the historical value of urban structures. Building on the comprehensive review of the disciplinary interface between urban history and space syntax historical studies provided by Griffiths (2012), this chapter advances 'heritage urbanism syntax' as a complementary development that extends this body of historical work towards contemporary issues in urban heritage studies and practices. To this end, the chapter focuses on the study of histories of spatial cultures as a bridge between space syntax and heritage studies. The intention is to help space syntax researchers to reflect critically on how historical thinking can be usefully applied in heritage contexts. It argues in favour of the role space syntax plays in analysing and identifying multiple facets of urban heritage, and in the potential to produce hermeneutic (Griffiths 2011, 2012) as much as prescriptive applications of space syntax methods.

2.2 Space Syntax and Urbanism

The theoretical and disciplinary affiliation between space syntax and urbanism is direct and straightforward by definition and area of study. Urbanism may simply be understood as a specific mode of life which is typically—but not exclusively (Wirth 1938, p. 7)—manifested in cities and metropolises. Talen's (2005, p. 2) more elaborate definition considers [American] urbanism as '*the vision and the quest to achieve the best possible human settlement [in America], operating within the context of certain established principles.*' Space syntax is a theory and methodology for the study of the formation of human settlements and architectural complexes (Hillier et al. 1976). It emerged to address the disjunction between the 'designed' and the 'lived' (Westin 2011, p. 232) in response to debates about architecture and the city which followed criticisms on the difficulties associated with the modernist phase (Hillier 1986; Hanson 2000). Its conceptual basis seeks to address the space–society relationship as a dynamic and open-ended formative process of spatial and social structures in mutual co-dependence (Hillier and Vaughan 2007); space is not just a background to human activity—it is a system with inherent spatial and social agency (Harvey 1973; Hillier and Leaman 1973; Lefebvre 1991).

The role of physical space in enabling social life is the main focus of space syntax, and a vast body of research has interrogated this topic in the context of urbanities (as well as buildings). Space syntax research uses mathematical modelling tools and measures that provide topological–geometric (describing changes in path direction and orientation) and metric (based on units distance) attributes of spatial systems conceived as *networks* and analysed as *graphs* (Hillier and Iida 2005). Empirically derived categories of social data can be mapped onto these configurational descriptions (typically in a GIS) to create a situated understanding of the functional patterns of space use in relation to their spatio-temporal arrangement (Hillier 1996, p. 194). Space syntax does not propose a deterministic relationship between space and specific social outcomes, rather it seeks to understand how the network effects of distance and connectivity relate to statistically probabilistic distributions of movement flows and human co-presence across different areas of urban space (Hillier 1999).

2.3 Space Syntax and Heritage

The application of space syntax methods to heritage-related questions has a long track record both in studies originating from within the space syntax discipline and outside it, for example in archaeology[2] (Ferguson 1996, p. 149; Smith 2011, p. 176; Letesson 2013; Stöger 2015, p. 63). These studies deployed space syntax to explore the sociocultural dimension embedded in spatial systems of historic or archaeological significance. By adding a spatial dimension to Laurajane Smith's (2016) distinction between 'authorised' and 'alternative' heritage discourses, we may identify the following categories of space syntax work on heritage: firstly, studies of *designed* urban heritage refer to research on monumentality in cities where the outcomes of architecture and planning curate urban spaces loaded with cultural symbolism.[3] Secondly, studies on *assigned* urban heritage include research on historic urban areas, such as the old city core, which are not necessarily monumental but are designated as heritage by planning law and protected for preservation.[4] The third category, *lived/emergent* urban heritage, represents a long-standing prominent trend in space syntax efforts to interpret the collective cultural value and identity of everyday urban spaces.[5]

Overall, this body of research is concerned with advancing knowledge of the operational role and remit of spatial agency in the formation of urban cultures—in other words to understand more about what space *does* in heritage terms. Currently, there are only few exceptions (specifically in preservation, urban design, and urban regeneration) that make a case for using syntactic research to influence practice in the historic urban landscape. For space syntax to be acknowledged in heritage studies and for heritage to gain traction as a topic to engage space syntax researchers, both communities need to see the value in interdisciplinary projects that can overcome the disciplinary barriers that define areas of practice. It also needs to communicate the usefulness of syntactic analysis to relevant practitioners by making results accessible and translating them into constructive insights for policy development. The next

---

[2]There are over thirty publications which use space syntax tools in the study of multifarious archaeological periods and regions: Old and New World New archaeology including the Canadian Arctic and South America (see Spence-Morrow 2009, p. 1 for an extensive listing of publications).

[3]Hillier (1989) on the spatial cultures of Teotihuacan in Mexico, Versailles in France, and Brasília, Brazil; Fong (1999) and Mari and Karadima (2005) on monuments; Peponis et al. (1988), Koseoglu and Onder (2009) and Dalton and Bafna (2003) on landmarks; de Holanda (2010) on the uniqueness of Brasília, Brazil; Psarra (2018) on city design production.

[4]Karimi (2000, 2002) on urban conservation; Önder and Gigi (2010), Kubat et al. (2012) and Chiang and Deng (2017) on urban heritage enhancement strategies.

[5]Aazam (2005) and Clark (2007) on rituals; studies on industrial heritage (c.f. Griffiths 2005, 2008, 2017; Davis and Dino 2015); Griffiths (2016) analyses the spatiality of 'processional culture'.

section takes a first step towards the integration of the two fields in theory, research and practice for the development of *heritage urbanism syntax.*

2.4 The Study of Spatial History

This section highlights emerging research trajectories that aim to develop a constructive liaison between spatial history and heritage. It draws on the critique by Griffiths (2012) who assesses the methodological approaches and future prospects for the application of syntactic analysis in historical research. Here we expand the conversation by Griffiths to argue that historical understanding of spatial urban processes is in line with current shifts in heritage management policy and practice. We propose three areas of research for the development of heritage urbanism syntax: (1) conservation areas, (2) street scales and (3) spatial cultures. These suggest possible areas in which the study of spatial history can contribute to policy-making for inclusive, sustainable and resilient heritage futures.

The study of spatial history enables the decoding of diachronic properties of spatial systems. There is a series of 'syntactical growth processes' studies (for summary, see p. 5 in Griffiths 2012) that examine the spatial evolution of urban and rural networks through 'time-series comparisons' (Griffiths 2009). These studies analyse and compare the synchronic structure of historical spatial systems between given time-periods using historical cartography (see Pinho and Oliveira 2009a, b; and Serra and Pinho 2011). When combined with morphological study ('syntactical morphological histories', see Griffiths 2012, p. 6), syntactical observations extend to consider social context and aim to identify the historical interdependencies of urban development in relation to the configurational properties of urban landscapes. This is what Griffiths calls a 'chronology' of functional space (2009, p. 660) in his study of the industrial heritage of Sheffield. Functions of spaces can be any socio-economic or cultural variable. Not only can we use space syntax to understand the historical role of a specific street/area, at a given historical time-period, within the spatial structure of a city as whole (synchronic), but also trace historical (diachronic) changes in the spatial structure of urban systems themselves (ibid.). Space syntax historical research has substantial contributions to offer to what the heritage sector calls a landscape-integrated approach (Harvey and Waterton 2015) to heritage management and practice. The syntactical study of spatial history examines space, time and function as interrelated layers of formative processes of the urban landscape, which is seen as complex system; it interrogates how these layers and processes initiate change in the urban landscape and how they are, in turn, affected by change.

To identify the diachronic characteristics of space, the study of spatial history emphasises the process of formation and transformation of urban space. In heritage studies, the call to focus on heritage as emergent process comes from the need to understand the ephemeral dimensions of perceived heritage (Harvey 2001). Harvey (ibid., p. 327) adopts a definition of heritage as '*a contemporary product shaped from history […] a value-laden concept, related to processes of commodification, but intrinsically reflective of a relationship with the past, however that 'past' is perceived and defined.'* The subjectivity of heritage definitions is gradually promoted in heritage practice through the shift from category-driven to process-driven heritage value assessments and guidelines for management (Veldpaus 2015). On global heritage practice fronts, the UNESCO 2011 Recommendation on the Historic Urban Landscape is such an effort, still a work in progress, to escape static concepts of heritage (Bandarin and van Oers 2012, 2014). The passage from theory to practice remains a challenge (Veldpaus et al. 2013) and for updated tools to be formulated, there is a requirement for the redefinition of heritage values (Poulios 2010) and their assessment. The following sections suggest ways in which space syntax study of spatial history may contribute to accomplishment of this oeuvre.

2.4.1 Conservation Areas: Lateral Versus Confined

Ambiguity in characterisation methods of tangible heritage in the built environment is an unresolved problem in conservation practices and planning law (Palaiologou 2017). Whitehand (2009) makes a remark on the methodological gap between academic research and delimitation practices. He overviews a series of case studies (including World Heritage sites, such as central Sibiu, Romania and St Petersburg, Russia) to point out the potential benefits of the Conzenian method of 'morphological regionalisation' in planning. As Whitehand explains (ibid., p. 6) morphological regionalisation is the process of mapping and identifying 'urban landscape units'. The method derives from the Conzenian historico-geographical approach for studying urban form (Conzen 1960, 1966). It proposes the study of history of urban form change to discern 'morphological periods' and the historical pace and types of urban development. Historical understanding is integral in this method, followed by an emphasis on two keywords: *process* and *landscape*. By considering morphological rather than stylistic periodisation, the historico-geographical method escapes from materialistic-oriented norms of assessing the urban landscape. Instead, it considers density, footprint/built ratio, volume and land use. These are important metrics for understanding urban development processes, but it could be argued that their spatialisation or area delimitation across the urban landscape is to great extent predefined by the planning legacy of zoning (Taylor 2002; Bogaert et al. 2005). It is then anticipated that historical remains of zoning practices will create units or patches (McDonnell and Pickett 1990) in the urban landscape with similar morphologies. In other words, the problem with established delimitation methods in urban conservation is that it recycles delimitation (or land use) practices of the past.[6] The pertinent question here is: *How relevant and up to date are urban conservation zoning practices in facing today's challenges for sustainable urban heritage management, and sustainable*

[6]The first author has conducted EPSRC-funded research on 'Delimitating UNESCO Historic Urban Landscapes', presented in the 23rd International Seminar on Urban Form, in Nanjing, China, 2016 (Palaiologou et al. 2016).

urban development in general? We argue that this is a key inquiry to which the historical syntactic study of urban configurations can provide alternative answers.

The use of zoning tools in urban conservation practice, more so than in other planning practices, is to define rigid regulatory boundaries for the protection of historic areas in urban systems (Elliot 2008). The aim of creating an 'impenetrable' compartmentalisation of the historic urban landscape in terms of planning laws and management (ibid.) is evidenced in the definition of 'buffer zone' areas as 'added layers of protection' (Peterson 2005). These areas, which are in proximity to the historic core but not of historic value themselves, are often considered as being in need of even more forceful conservation rules than the official heritage sites themselves (Turner 2009, p. 17). It follows that it is not clear how the conservation of regulatory boundaries plays a role in enabling a landscape approach—namely, an approach that aspires to treat urban landscapes in 'holistic' management frameworks (Rodwell 2003, p. 67). By delimitating the perimeters of a historic area and of the areas of influence, heritage management practices focus on the *tangible* values without necessarily addressing *intangible* agencies of change and socio-economic impacts (Mürner 2009, p. 12). In contrast to fragmented analysis and assessment of urban space characteristics, the space syntax approach proposes a study of spatial networks, of connected spaces, where streets are considered to have a role as structural city layer by performing as 'backbones' of urban life (Hillier et al. 1993). Syntactic analysis of spatial histories is a space–time integrative approach (Griffiths 2012), in that it treats the wider landscape as inseparable and relational system, whilst it enables the consideration of sociological contexts over time.

Looking at the spatial histories of eighteenth and nineteenth Sheffield, Griffiths (2009) builds a methodology for the 'bottom-up characterisation' of urban transformation and demonstrates how diachronic knowledge can shed light on otherwise obscure historical assumptions about the emergence and spatial identification of local places (hinterland villages) against or alongside global forces (urbanisation). To develop further syntactic work in this area, future curricula for using space syntax in urban conservation could aim in addressing the ways in which the study of spatial history can inform, or even redefine, urban conservation delimitation practices from a polygon-based concept, to nodal-based systems, and explore the potentials of street-wise,[7] rather than area-wise, character assessment. Also, historical syntactic analysis may be applied to any spatial system that involves a human–environment interaction component, such as green infrastructure. Finally, the study of spatial history considers a space–time continuum that is crucial for tangible heritage to build strong and uninterrupted links with present and future urbanities. Heritage-led urban regeneration requires a balance between heritage values and sustainable development (Stubbs 2004; Pendlebury 2013; Alsalloum 2018), tested and evaluated in terms of its integration in this historical space–time continuum of urban phenomena (Guzmán et al. 2017, p. 200).

[7]See Palaiologou (2016) for an application of street-wise character assessment of Radlett, UK, which was conducted for local policy development.

2.4.2 Street Scales: Adaptable Versus Static

The previous section suggested that the study of spatial history enables methodological developments in urban conservation for bottom-up (i.e. related to spatial structure) assessment of the diachronic character of urban places, and that it is worth exploring the benefits of street-wise assessment of the historical urban fabric. Here the argument is further developed to discuss how knowledge of historical spatial processes acknowledges streets as dynamic urban heritage places (see Griffiths 2005). It questions heritage management practices that are permeated by static and materialistic views on the role of tangible heritage in urban life. It reviews space syntax studies of spatial histories that reveal the multiple and shifting roles of streets (local, citywide, regional, etc.) within constantly changing spatial systems over time. These studies come from Griffiths's (2012, p. 6) category of 'syntactical morphological histories' and develop an open-ended dialogue between physical space and social context (ibid.).

The work of Vaughan et al. (2013) on the spatial histories of two 'adaptable suburbs' of outer London is characteristic of this line of research which looks at the diachronic spatial resonance of local places. Following the processes of suburban transformation for 150 years in terms of infrastructure and land use, the study makes a case for spatial 'mechanisms of adaptability' which enabled the resilient identity of the suburban high-street. Whilst being assigned new roles in relation to the wider urban landscape, the authors studied how the high-streets of Surbiton and South Norwood maintained their spatial significance locally as legacy of their particular roles within the historical suburban street network. We can now see that this kind of work also establishes connections between spatial history and urban heritage. In the case of the London suburbs, the study of spatial history reveals a heritage urbanism syntax which overturns normative arguments about urban expansion and suggests that change is not necessarily associated with loss of character (ibid.). Instead, high-street 'changeability' (Törmä et al. 2017) fosters a dynamic place identity formation—what Vaughan et al. (2015) call 'hedgerow ecology'. Palaiologou (2015) picks up how the historical transactions of the parts–whole relationship is still evident in the case of currently fully assimilated high-streets in the London urban core, such as Upper Street in Islington. It is the dynamic continuity of its structural role within the wider system, which distinguishes the high-street compared to other streets (see Griffiths 2015). This consideration of continuity as a dynamic process gives way to adaptable rather than static definitions of heritage. The position of a street within the wider urban landscape is not considered as being a static, fixed property—it is studied as relational feature of the urban landscape that responds differently to the temporal and spatial scales at which heritage can be defined.

Space syntax research contributes to the study of the historical negotiation between continuity and change in spatial systems, where streets form the core spatial structure. It therefore offers a way of identifying the 'heritage syntax' of urban places. Spatial configuration, in contrast to visual appearance or land use, proposes a much more dynamic view of place formation: spatial configuration assumes a relational dependency between localities and the wider landscape. Conservation practices that lack flexibility and responsive measures rupture the identity of urban places (c.f. Lees 1994; Bianchi and Boniface 2002; Wang 2012) by disconnecting urban heritage from its urban context (Ripp and Rodwell 2015), by displacing communities (Martínez 2016), by exposing heritage to disaster risk (Ravankhah et al. 2017) and by politicising heritage values (Fraser 2008). All these issues are directly related to global sustainability indicators (Guzmán et al. 2017, p. 194) and pose the question: *How adaptable and integrative are heritage management practices in today's challenges for sustainable urban heritage management, and sustainable urban development in general?* For urban heritage, and culture in general, to work as a driver for urban sustainability, static and exclusive approaches to heritage values need to be revisited (Pendlebury 2013; de la Torre 2013; Poulios 2010). From a space syntax point of view, understanding the historical role of streets helps planning-related agency in the urban landscape to take a synergetic view between continuity and change, between heritage and development (Zaninović et al. 2018).

2.4.3 Spatial Cultures: An Inclusive Approach to Heritage Urbanism

'Spatial culture' is a recurring concept in space syntax theory and a fundamental one. It is advanced by Hillier (1989) in 'The architecture of the urban object' to conceptualise the space–society relationship within a theoretical framework which acknowledges the existence of generative social meaning from spatial arrangements. This section articulates the ways in which the historical study of spatial cultures fertilises new theoretical grounds and implications for defining alternative heritage values. The discussion unfolds the potential contribution of historical understanding of spatial cultures in proposing more 'inclusive assessments' (Veldpaus et al. 2013) of the heritages of spatial cultures. Building on Griffiths's (2011, 2012) observation about the 'hermeneutic possibilities' that quantitative time–space descriptions produced by space syntax analysis generate in the study of history, the argument here extends this to the field of heritage studies and the assessment of cultural meanings shifting over time. In heritage studies, Harvey (2001, p. 320) proposes a relevant argument with regard to the study of 'heritagisation as a process', to bridge the study of history and heritage from an interdisciplinary theoretical viewpoint. Griffiths in this respect follows Harvey in identifying collaborative and inclusive liaisons across history, heritage and spatial studies.

A syntactic reading of spatial cultures is preoccupied with understanding the link between spatial configuration, spatial function and cultural meaning (Hillier 1989; Griffiths and von Lünen 2016). For Hillier and Hanson (1984, pp. 42–45), the mechanism through which this link becomes embodied for humans in space is a 'description retrieval' process: the abstraction of information relating to generic social meaning via the ways the social 'encounter field' is organised by space–time relations. In other words, the syntax of spatial descriptions relates to syntax of cultural meaning. Whereas these syntaxes have been examined in terms of synchronic relationships, they are yet underexplored in terms of their ‘temporal ontology’ (Griffiths 2011, p. 79). Synchronic understanding deals with the instantaneous relational aspects of a fixed spatial system at a fixed time, an ahistorical spatial configuration. By focusing on the temporality of spatial descriptions, Griffiths (2011) emphasises the diachronic aspects of spatial configurations and brings to the fore a core argument for spatial cultures: the spatiality of cultural phenomena cannot be fully understood if distanced from historical context and processes of cultural formation. These processes involve tracing continuities, disruptions and changes on the ways societies are organised in space; but also interpreting the means by which immaterial meaning becomes mediated in and by the built world (ibid., p. 76). The key question here is: *What does spatial history tells us in terms of assessment and definition of heritage values?*

The ‘interpretative act’ plays a key role in ‘description retrieval’ (Griffiths 2011, 2012) and bears constructive relevance to discussions of alternative heritage definitions which consider non-authorised heritage values (Stephenson 2008; Waterton and Smith 2010; Emerick 2014). Interpretation acknowledges an inherent subjectivity[8] in the ways we identify spatially mediated cultural meaning—namely, it implies the *sociality* of the process of identity formation (Waterton and Smith 2010, p. 8). The ways of experiencing spatial cultures as everyday rituals (Lefebvre 1992), as normative spaces of ordinary life, depend on the integrated understanding of spatial, social and temporal contexts. In other words, to that extent that space remains an ‘invariant’ over time (a theoretical rather than an historical proposition), people and cultural habits will not. This means that the ‘uses of spatial cultures’—to borrow Smith’s (2006) argument on the uses of heritage—are constantly shifting as result of temporal processes. This is similar to what Harvey (2001, p. 320) argues for heritage, suggesting that ‘*heritage has always been with us and has always been produced by people according to their contemporary concerns and experiences*’.

Space syntax historical research on spatial cultures means understanding the continuity and change of spatial descriptions, and not only of space as physical object (Griffiths 2012). The focus in this approach is a study of the historical space–society relationship to develop knowledge on how space is lived. As Griffiths clarifies (ibid.), the intention is not only to achieve a prescriptive description of material aspects of spatial cultures, but also to develop an informed interpretation of historical evidence on the ways the material and immaterial interact at a place over time. Rather than taking on a reductionist approach to defining assets of conserved urban landscapes which is promoted by AHD (Gospodini 2004, p. 228), we may argue that the hermeneutic potential which is afforded by syntactic analysis of spatial cultures can generate an integrative and inclusive framework for assessing urban heritage. Heritage urbanism syntax can serve as a conceptual framework and tool for studying

[8]Discussing research practice in the filed of comparative urbanism, Nijman (2007, p. 5) argues that ‘*we must acknowledge the shortcomings of previous comparativist schemes and the emergence of an “interpretative social science,” one that emphasizes understanding rather than law-like explanation.*’

the process of time–space generated cultural meaning by incorporating a multileveled understanding of both tangible and intangible cultural meanings.

From a heritage management perspective, the significance of understanding historical spatial cultures becomes most evident in the cases of (a) ownership of heritage and (b) heritage reconstruction. Questions of *heritage ownership and belonging* become renegotiated when merging value of the present with the value to the past and possible value to the future (Harvey 2001). The spatialisation and adaptability of migrant urban cultures are an example where space syntax has actively contributed in overturning existing misconceptions of loss of cultural identity (Kershen and Vaughan 2013; Vaughan 2015; Vaughan and Sailer 2017). Also, the work of Nadia Charalambous (Charalambous and Peristianis 2002; Charalambous 2004) on urban spaces of political conflict suggests how the integration of social and spatial concepts, 'humanistic and technical' viewpoints, can provide new tools for the interpretation and conservation of cultural heritage (Charalambous 2007). The second strand of *reconstructing heritage* refers to studies which could deploy the study of historical spatial cultures to inform space reconstruction projects after the occurrence of disasters. The application of space syntax tools in Disaster Risk Management (DRM) practice is a heavily underexplored area. There exist a small number of studies which use syntactic analysis as disaster risk preventive tool (c.f. Gil and Steinbach 2008; Esposito and Di Pinto 2015; Maureira and Karimi 2017), but the prospects of using space syntax as reconstructive framework for *resilient* spatial cultures have not yet been tested. The potential role of cultural heritage in the delivery of a turquoise agenda (Lizarralde et al. 2015) is at the forefront of heritage management discourse (Jigyasu et al. 2013) and the study of spatial heritage against sustainability goals is a promising research area for space syntax researchers of spatial history.

2.5 Conclusion

This chapter argued that there is substantial scope in heritage studies for the application of syntactic analysis. Building on Griffiths's account of studies of history within space syntax (2012), the chapter frames a starting point for the integration of space syntax, history and heritage studies by identifying three core areas of future research activity: conservation areas, street scales and spatial cultures. Throughout the chapter, the discussion made explicit links to contemporary policy priorities (such as urban regeneration, sustainability, resilience and DRM).

As a concluding note, the development of 'heritage urbanism syntax' requires multi-disciplinary affiliation. From a space syntax perspective, the challenge is to identify analytical methods tailored to address the history–heritage relationship in order to systematise, without compromising, the hermeneutic potential of syntactic description. Overall, what the space syntax field may add to heritage urbanism is the facility to address the question of the historical role of space in the formation of present and future spatial cultures which are sustainable, inclusive and resilient.

References

Aazam Z (2005) To, around and from the centre: how the Kabah continues to integrate the society and its rituals. In: van Nes A (ed) Proceedings of the 5th international space syntax symposium, vol 1. TU Delft, Faculty of Architecture, Delft, pp 310–22

Alsalloum A (2018) Development in world heritage sites. Cathedral Communications/The Building Conservation Directory, pp 25–28. Available via Liverpool University. https://livrepository.liverpool.ac.uk/3016622/1/BCD18F6_WorldHeritage.pdf. Accessed 23 May 2018

Bandarin F, van Oers R (2012) The historic urban landscape: managing heritage in an urban century. Wiley, West Sussex

Bandarin F, Van Oers R (eds) (2014) Reconnecting the city: the historic urban landscape approach and the future of urban heritage. Wiley, West Sussex

Bianchi R, Boniface P (2002) Editorial: the politics of world heritage. Int J Herit Stud 8(2):79–80

Bogaert J, Farina A, Ceulemans R (2005) Entropy increase of fragmented habitats: a sign of human impact? Ecol. Indicators 5:207–212

Charalambous N (2004) Spatial patterns of ethnic identity in a post global world. Traditional dwellings and settlements Working Paper Series 173, IASTE, Berkeley, University of California

Charalambous N (2007) Social and spatial patterns of cultural heritage. Paper presented at the XXI international CIPA symposium, Athens, Greece, 1–6 October 2007

Charalambous N, Peristianis N (2002) "Cypriot boundaries" in traditional dwellings and settlements. Working Paper Series 159, IASTE, Berkeley, University of California

Chiang Y-C, Deng Y (2017) City gate as key towards sustainable urban redevelopment: a case study of ancient Gungnae City within the modern city of Ji'an. Habitat Int 67:1–12

Clark DLC (2007) Viewing the liturgy: a space syntax study of changing visibility and accessibility in the development of the Byzantine church in Jordan. World Archaeol 39(1):84–104. https://doi.org/10.1080/00438240601148798

Conzen MRG (1960) Alnwick, Northumberland: a study in town-plan analysis. Institute of British Geographers Publication 27 (2nd edn), George Philip, London

Conzen MRG (1966) Historical townscapes in Britain: a problem in applied geography. In: House JW (ed) Northern geographical essays in honour of G. H. J. Daysh. University of Newcastle upon Tyne, Newcastle upon Tyne, pp 56–78

Dalton RC, Bafna S (2003) The syntactical image of the city: a reciprocal definition of spatial elements and spatial syntaxes. In: Proceedings of the 4th international space syntax symposium, University College London, London

Davis H, Dino B (2015) Industrial morphologies: the historic London furniture industry and the new city of production. Paper presented in 22nd ISU International Seminar on Urban Form, Sapienza University of Rome, Faculty of Architecture, Rome, 22–26 September 2015

de Holanda F (2010) Brasília: Cidade Moderna, Cidade Eternal. FAU UnB, Brasília

de la Torre M (2013) Values and heritage conservation. Herit Soc 6(2):155–166. https://doi.org/10.1179/2159032X13Z.00000000011

Elliot DL (2008) A better way to zone: ten principles to create more livable cities. Island Press, Washington, D.C.

Emerick K (2014) Conserving and managing ancient monuments: heritage, democracy, and inclusion. Heritage Matters Series 14. Boydell Press, Woodbridge

Esposito A, Di Pinto V (2015) Calm after the storm: the configurational approach to manage flood risk in rivercities. In: Karimi K et al (eds) Proceedings of the 10th international space syntax symposium, University College London, London, pp 70:1–70:16

Ferguson TJ (1996) Historic Zuni Architecture and society: an archaeological application of space syntax. Anthropological Papers of the University of Arizona, vol 60. University of Arizona Press, Tucson

Fong P (1999) The role of the monument. In: Major MD, Amorim L (eds) Proceedings of the 2nd international space syntax symposium, Universidade de Brasilia, Brazilia, pp 1–24

Fraser N (2008) Rethinking recognition: overcoming displacement and reification in cultural politics. In: Olson K (ed) Adding insult to injury: Nancy Fraser debates her critics. Verso, London, pp 129–141
Gil J, Steinbach P (2008) From flood risk to indirect flood impact: evaluation of street network performance for effective management, response and repair. WIT Trans Ecol Environ 118:335–344
Gospodini A (2004) Urban morphology and place identity in European cities: built heritage and innovative design. J Urban Des 9(2):225–248. https://doi.org/10.1080/1357480042000227834
Griffiths S (2005) Historical space and the practice of "spatial history": the spatio-functional transformation of Sheffield 1770–1850. In: van Nes A (ed) Proceedings of the 5th international space syntax symposium, vol 2. TU Delft, Faculty of Architecture, Delft, pp 655–668
Griffiths S (2008) Historical space and the interpretation of urban transformation: the spatiality of social and cultural change in Sheffield c1770–1910. Dissertation, University of London
Griffiths S (2009) Persistence and change in the spatio-temporal description of Sheffield parish 1770–1910. In: Koch D, Marcus L, Steen J (eds) Proceedings of the 7th international space syntax symposium. TRITA-ARK- Forskningspublikation, Stockholm, pp 37:1–37:15
Griffiths S (2011) Temporality in Hillier and Hanson's theory of spatial description: some implications of historical research for space syntax. J Space Syntax 2(1):73–96
Griffiths S (2012) The use of space syntax in historical research: current practice and future possibilities. In: Greene M, Reyes J, Castro A (eds) Proceedings of the 8th international space syntax symposium, PUC, Santiago, Chile, pp 8193:1–8193:26
Griffiths S (2015) The high street as morphological event. In: Vaughan L (ed) Suburban urbanities: suburbs and the life of the high street. UCL Press, London, pp 32–50. https://doi.org/10.14324/111.9781910634134
Griffiths S (2016) Spatial culture, processional culture and the materialities of social memory in nineteenth-century Sheffield. Distinktion: J Soc Theory 17(3):254–275. https://doi.org/10.1080/1600910X.2016.1183219
Griffiths S (2017) Manufacturing innovation as spatial culture: Sheffield's cutlery industry c.1750-1900. In: van Damme I, Blondé B, Miles A (eds) Cities and creativity from the Renaissance to the present. Routledge, London, pp 127–154
Griffiths S, Vaughan LS (2019) Mapping spatial cultures: contributions of space syntax to research in the urban history of the 19th Century city. Urban History, Special edition on History and Mapping (in press)
Griffiths S, von Lünen A (eds) (2016) Spatial cultures: towards a new social morphology of cities past and present. Routledge, London
Guzmán PC, Pereira Roders AR, Colenbrander BJF (2017) Measuring links between cultural heritage management and sustainable urban development: an overview of global monitoring tools. Cities 60:192–201. https://doi.org/10.1016/j.cities.2016.09.005
Hanson J (2000) Urban transformations: a history of design ideas. Urban Des Int 5(2):97–122
Harvey D (1973) Social justice and the city. Edward Arnold, London
Harvey DC (2001) Heritage pasts and heritage presents: temporality, meaning and the scope of heritage studies. Int J Heritage Stud 7(4):319–338. https://doi.org/10.1080/13581650120105534
Harvey DC, Waterton E (2015) Editorial: landscapes of heritage and heritage landscapes. Landsc Res 40(8):905–910. https://doi.org/10.1080/01426397.2015.1086563
Hillier B (1986) City of Alice's dreams. Architects' J (July):39–41
Hillier B (1989) The architecture of the urban object. Ekistics 56(334/33):5–21
Hillier B (1996) Space is the machine: a configurational theory of architecture. Cambridge University Press, Cambridge
Hillier B (1999) Centrality as a process: accounting for attraction inequalities in deformed grids. Urban Des Int 4:107–127
Hillier B, Hanson J (1984) The social logic of space. Cambridge University Press, Cambridge
Hillier B, Iida S (2005) Network effects and psychological effects: a theory of urban movement. In: van Nes A (ed) Proceedings of the 5th international space syntax symposium, vol 1. TU Delft University of Technology, Delft, pp 553–564

Hillier B, Leaman A (1973) The man-environment paradigm and its paradoxes. Arch Des 78(8):507–511
Hillier B, Vaughan LS (2007) The city as one thing. Prog Plan 67(3):205–230
Hillier B et al (1976) Space syntax. Environ Plan 3(2):147–185
Hillier B et al (1993) Natural movement: or configuration and attraction in urban pedestrian movement. Environ Plan B 20(1):29–66
Jigyasu R et al (2013) Heritage and resilience: issues and opportunities for reducing disaster risks. International Centre for the Study of the Preservation and Restoration of Cultural Property (ICCROM), Marsh, United Kingdom. Available via issuu. https://issuu.com/emilioiannarelli/docs/heritage_and_resilience_book_for_gp. Accessed 23 May 2018
Karimi K (2000) Urban conservation and spatial transformation: preserving the fragments or maintaining the 'spatial spirit'. Urban Des Int 5:221–231
Karimi K (2002) Iranian organic cities demystified; a unique urban experience or an organic city like others. Built Environ 28(3):187–202
Kershen AJ, Vaughan LS (2013) There was a Priest, a Rabbi and an Imam…: an analysis of urban space and religious practice in London's East End, 1685–2010. Mater Religion: J Objects Art Belief 9(1):10–35
Koseoglu E, Onder DE (2009) Defining salient elements of
environment and memory
subjective and objective landmarks in Ayvalik, Turkey. In: Koch D, Marcus L, Steen J (eds) Proceedings of the 7th international space syntax symposium. TRITA-ARK- Forskningspublikation, Stockholm, pp 059:1–059:11
Kubat et al (2012) Application of space syntax in developing—a regeneration framework for Sharjah's heritage area. Greene M, Reyes J, Castro A (eds) Proceedings of the 8th international space syntax symposium, PUC, Santiago, Chile, pp 8040:1–8040:17
Lees L (1994) Gentrification in London and New York: an Atlantic gap? Hous Stud 9(2):199–217. https://doi.org/10.1080/02673039408720783
Lefebvre H (1991 {original: 1974}) The production of space (trans: Smith DN). Blackwell, Malden MA; Oxford; Carlton
Lefebvre H (1992) Éléments de rythmanalyse. Éditions Syllepse, Paris (English trans: S Elden, G Moore 2004). Continuum, New York
Letesson Q (2013) Minoan halls: a syntactical genealogy. Am J Archaeol 117:303–351
Lizarralde G et al (2015) Sustainability and resilience in the built environment: the challenges of establishing a turquoise agenda in the UK. Sustain Cities Soc 15:96–104. https://doi.org/10.1016/j.scs.2014.12.004
Mari I, Karadima C (2005) The perceptual organisation of an axial monument—from passing to facing. In: van Nes A (ed) Proceedings of the 5th international space syntax symposium, vol 1. TU Delft University of Technology, Delft, pp 581–594
Martínez PG (2016) Authenticity as a challenge in the transformation of Beijing's urban heritage: the commercial gentrification of the Guozijian historic area. Cities 59:48–56. https://doi.org/10.1016/j.cities.2016.05.026
Maureira VM, Karimi K (2017) The everyday and the post-disaster urban systems as one thing: a configurational approach to enhance the recovery and resilience of cities affected by tsunamis. In: Proceedings of the 11th international space syntax symposium, Lisbon, Portugal, 3–7 July 2017, pp 91:1–91:17
McDonnell MJ, Pickett STA (1990) Ecosystem structure and function along urban–rural gradients: an unexplored opportunity for ecology. Ecology 71:1232–1237
Mürner J (2009) Preface. In: World heritage and buffer zones. World Heritage Papers 25. UNESCO, pp 11–13
Nijman J (2007) Introduction—comparative urbanism. Urban Geogr 28(1):1–6. https://doi.org/10.2747/0272-3638.28.1.1
Önder DE, Gigi Y (2010) Reading urban spaces by the space-syntax method: a proposal for the South Haliç Region. Cities 27:260–271

Palaiologou G (2015) High street transactions and interactions. In: Vaughan L (ed) Suburban urbanities: suburbs and the life of the high street. UCL Press, London, pp 175–203. https://doi.org/10.14324/111.9781910634134

Palaiologou G (2016) Character assessment: Radlett neighbourhood plan. Aldenham Parish Council, Radlett Hertfordshire. Available via Aldenham Parish Council. https://www.radlettplan.org/copy-of-why-a-plan. Accessed 23 May 2018

Palaiologou G (2017) Viewpoint—urban morphology and world heritage practice. Urban Morphol 21(1):3–6

Palaiologou G, Karimi K, Vaughan L (2016) The use of spatial network analysis in area characterisation—towards an assessment process of UNESCO historic urban landscapes. The case of Liverpool, UK. In: Paper presented in the 23rd international seminar on Urban Form, Nanjing, China, 8–10 July 2016

Pendlebury J (2013) Conservation values, the authorised heritage discourse and the conservation-planning assemblage. Int J Heritage Stud 19(7):709–727. https://doi.org/10.1080/13527258.2012.700282

Penn A (2003) Space syntax and spatial cognition: or, why the axial line? Environ Behav 35:30–65

Peponis J et al (1988) Social structure of space and relations of community in six Greek cities. Secretariat of Research and Technology, Ministry of Energy Industry, Research and technology, Athens

Peterson A (2005) Buffer planning: historical overview. In: Kozlowski J, Peterson A (eds) Integrated buffer planning: towards sustainable development. Ashgate, Aldershot, pp 79–116

Pinho P, Oliveira V (2009a) Combining different methodological approaches to analyze the Oporto metropolitan area. In: Koch D, Marcus L, Steen J (eds) Proceedings of the 7th international space syntax symposium. TRITA-ARK- Forskningspublikation, Stockholm, pp 88:1–88:13

Pinho P, Oliveira V (2009b) Cartographic analysis in urban morphology. Environ Plan 36:107–127

Poulios I (2010) Moving beyond a values-based approach to heritage conservation. Conserv Manag Archaeol Sites 12(2):170–185. https://doi.org/10.1179/175355210X12792909186539

Psarra S (2018) The role of spatial networks in the historic urban landscape: learning from Venice in the fifteenth and sixteenth centuries. Hist Environ: Policy & Pract 9(3–4):249–273. https://doi.org/10.1080/17567505.2018.1516380

Ravankhah M et al (2017) Integration of cultural heritage into disaster risk management: challenges and opportunities for increased disaster resilience. In: Albert MT, Bandarin F, Pereira Roders A (eds) Going beyond. Heritage studies. Springer, Cham, pp 307–321

Ripp M, Rodwell R (2015) The geography of urban heritage. Hist Environ: Policy & Pract 6(3):240–276. https://doi.org/10.1080/17567505.2015.1100362

Rodwell D (2003) Sustainability and the holistic approach to the conservation of historic cities. J Arch Conserv 9(1):58–73. https://doi.org/10.1080/13556207.2003.10785335

Serra M, Pinho P (2011) Dynamics of periurban spatial structures: investigating differentiated patterns of change on Oporto's urban fringe. Environ Plan 38:359–382

Smith L (2006) Uses of Heritage. Routledge, Abingdon

Smith ME (2011) Empirical urban theory for archaeologists. J Archaeol Method Theory 18(3):167–192. https://doi.org/10.1007/s10816-010-9097-5

Spence-Morrow G (2009) Analyzing the invisible—syntactic interpretation of archaeological remains through geophysical prospection. In: Koch D, Marcus L, Steen J (eds) Proceedings of the 7th international space syntax symposium. TRITA-ARK- Forskningspublikation, Stockholm, pp 106:1–106:10

Stephenson J (2008) The cultural values model: an integrated approach to values in landscapes. Landsc Urban Plan 84(2):127–139

Stöger H (2015) Roman neighbourhoods by the numbers: a space syntax view on ancient city quarters and their social life. J Space Syntax 6(1):61–80

Stubbs MJ (2004) Heritage-sustainability: developing a methodology for the sustainable appraisal of the historic environment. Plan Pract Res 19(3):285–305. https://doi.org/10.1080/0269745042000323229

Talen E (2005) New urbanism and American planning: the conflict of cultures. Routledge, New York
Taylor PD (2002) Fragmentation and cultural landscapes: tightening the relationship between human beings and the environment. Landsc Urban Plan 58(2–4):93–99. https://doi.org/10.1016/S0169-2046(01)00212-2
Törmä I, Griffiths S, Vaughan LS (2017) High street changeability: the effect of urban form on demolition, modification and use change in two South London suburbs. Urban Morphol 21(1):5–28
Turner M (2009) Introduction—on buffs and buffers. World Herit Pap 25:15–18
Tweed C, Sutherland M (2007) Built cultural heritage and sustainable urban development. Landsc Urban Plan 83(1):62–69. https://doi.org/10.1016/j.landurbplan.2007.05.008
Vaughan LS (2015a) The Ethnic Marketplace as Point of Transition. In: Kershen AJ (ed) London the promised land revisited the changing face of the London migrant landscape in the early 21st century. Ashgate, Farnham, pp 35–54
Vaughan LS et al (2015) An ecology of the suburban hedgerow, or: how high streets foster diversity over time. In: Karimi K et al (eds) Proceedings of the 10th international space syntax symposium, University College London, London, pp 99:1–99:19
Vaughan LS, Sailer K (2017) The metropolitan rhythm of a "majestic religion": an analysis of the socio-spatial configuration of synagogues in nineteenth century Whitechapel. In: Holmes C, Kershen AJ (eds) An east end legacy: essays in memory of William J. Fishman. Routledge, London, pp 188–210
Vaughan LS, Dhanani A, Griffiths S (2013) Beyond the suburban high street cliché—a study of adaptation to change in London's street network: 1880–2013. J Space Syntax 4(2):221–241
Veldpaus L (2015) Historic urban landscapes: framing the integration of urban and heritage planning in multilevel governance. Technische Universiteit Eindhoven, Eindhoven
Veldpaus L et al (2013) Urban heritage: putting the past into the future. Hist Environ: Policy Pract 4(1):3–18. https://doi.org/10.1179/1756750513Z.00000000022
Wang S-Y (2012) From a living city to a world heritage city: authorised heritage conservation and development and its impact on the local community. Int Dev Plan Rev 34(1). https://doi.org/10.3828/idpr.2012.2
Waterton E, Smith L (2010) The recognition and misrecognition of community heritage. Int J Heritage Stud 16(1–2):4–15. https://doi.org/10.1080/13527250903441671
Westin S (2011) The life and form of the city: an interview with Bill Hillier. Space Culture 14(2):227–237
Whitehand JWR (2009) The structure of urban landscapes: strengthening research and practice. Urban Morphology 13:5–27
Wirth L (1938) Urbanism as a way of life. Am J Sociol 46(1):1–24
Zaninović T et al (2018) Urban landscape and spatial heritage: the case of gateway-pathways in Zagreb, Croatia. Hist Environ: Policy & Pract 9(3–4):274–305. https://doi.org/10.1080/17567505.2018.1514568

Chapter 3
Applying Cultural Tourism in the Revitalisation and Enhancement of Cultural Heritage: An Integrative Approach

Daniela Angelina Jelinčić and Yoel Mansfeld

Abstract This chapter focuses on cultural tourism as a means of cultural heritage revitalisation representing specific *soft* models, which can help revive heritage sites. The visible trends of increasing cultural tourism together with the new profile of tourists interested in cultural heritage go hand in hand with the revitalisation needs of the cultural heritage sector. Still, not all heritage sites are equally attractive nor are they all successful in cultural tourism development. Current research has pointed out the leading principles of heritage revitalisation based on cultural tourism development. These can be applied in order to increase the attractiveness of a given site while maintaining its physical, economic and socio-cultural sustainability. The chapter discusses an integrated approach to revitalisation which entails social, territorial, economic as well as the knowledge/education component directly visible in participatory governance, public management of heritage and the gradual integration of sustainability aspects into heritage management. The strong involvement of the community in the revitalisation process is the key principle. The accumulated body of knowledge has further detected several success factors linked to modern storytelling in heritage interpretation for the cultural tourism market; these are the use of creative industries as bearers of symbolic cultural values, engaged activities through participatory experience tourism, and the creation of a tourism offer based on experience economy design principles. The need to involve all five senses in experience creation and to stir visitors' emotions is emphasised.

Keywords Cultural tourism · Integrated revitalisation · Participatory approach · Creative industries · Experience economy

D. A. Jelinčić (✉)
Institute for Development and International Relations, Zagreb, Croatia
e-mail: daniela@irmo.hr

Y. Mansfeld
Department of Geography & Environmental Studies, Center for Tourism, Pilgrimage & Recreation Research, University of Haifa, Haifa, Israel
e-mail: yoel@geo.haifa.ac.il

M. Obad Šćitaroci et al. (eds.), *Cultural Urban Heritage*,
The Urban Book Series, https://doi.org/10.1007/978-3-030-10612-6_3

3.1 Introduction

The lack of financial resources of most governments remains one of the burning issues in the protection and maintenance of cultural heritage. This issue is of particular importance for countries in transition, where the turbulent history and the geopolitical reconfiguration have resulted in a large number of neglected or abandoned historic buildings, which until today have suffered from steady decay. These buildings are often connected to degraded areas with economic and social problems, which require immediate intervention (Jelinčić 2017).

The lack of finances has encouraged the search for alternative financing models which has resulted in a number of relatively recent scientific articles and studies (e.g. Čopić et al. 2011; Ventura et al. 2016; Oppio and Torrieri 2016; European Commission 2016). These models range from different options of public, private or public–private support which are applied differently across countries. Lately, tourism has been put forward as a great ally to cultural heritage due to its rising trend and the growing interest of tourists in cultural heritage sites. Although successful cases of physical renovation of cultural heritage by way of tourism investments/income exist, this chapter focuses on *soft* sustainable models of cultural heritage revitalisation that make the sites more attractive both for the local community and tourists.

3.2 Cultural Tourism and Heritage

Statistics on cultural tourism document great interest in cultural heritage attractions. The European Commission (2018) estimates that 'cultural tourism accounts for 40% of all European tourism; 4 out of 10 tourists choose their destination based on its cultural offering'. Besides, 'an increase in interest in the past … has also taken place. This dynamic has been clearly reflected in a rise in cultural activities in recent decades and social longing for culture and heritage' (UNWTO 2015). Estimates suggest that approximately 85% of the general population can be considered current or potential heritage tourists (Timothy 2011). In parallel, there is greater awareness of the need to preserve and protect cultural heritage which is supported by the growing number of UNESCO-listed heritage sites (Jelinčić 2017).

Cultural heritage sightseeing activities still note success in relation to tourism revenue statistics in a number of destinations (e.g. Florence, Dubrovnik, Venice) but, at the same time, the great number of visitors represents a threat to heritage attractions. The cultural tourism impact must be viewed from three perspectives: (1) tourist crowds at heritage sites are not only detrimental for the sites but, at the same time, (2) decrease the level of destination experience for the visitors, as well as (3) the quality of life for the local community. 'A new profile of tourist-seeking experiences focused on relaxation, discovery, enjoyment and knowledge' (UNWTO 2015) is associated with today's tourists and, if wanting to differ from a number of similar attractions they compete with, heritage sites are expected to offer memorable experiences. Thus, tourists' stay in the destination must be pleasurable and deprived of negative cues. At the same time, the local population must maintain or even enhance their quality of life, not only regarding economic revenues, but also concerning their everyday lives in general. The prerequisites for the reconciliation of local and tourist experiences while preserving and protecting cultural heritage sites entail careful tourism planning.

3.3 An Integrated Development Approach in Heritage Revitalisation

The main principle in heritage revitalisation which has been advocated lately concerns the so-called integrated approach to heritage revitalisation and management. The recently adopted Council of Europe's 'European Cultural Heritage Strategy for the 21st century' envisions the promotion of a 'shared and unifying approach to cultural heritage management, based on an effective legal framework for the integrated conservation of heritage' (Council of Europe 2017a).

Apart from the previous work of the Council of Europe, this Strategy also takes into account the work carried out by UNESCO and the European Union, as well as the work of INGOs and networks focusing on heritage. The reference texts[1] stress the need for greater citizen participation in decision-making and management related to their heritage. The participatory approach has also been promoted lately at the level of the European Union through the so-called open method of coordination (OMC). It is a relatively new means of governance in the European Union, based on voluntary cooperation. In the field of culture, this method 'was conceived as a flexible, non-binding and voluntary framework for structuring cooperation around the objectives of the Cultural Agenda and for fostering the exchange of best practices' (Psychogiopoulou 2015), whereas all relevant actors, including civil society organisations were invited to define key strategic cultural policy goals.

In the field of cultural heritage, by employing the participation principle, the long-term goal of greater knowledge, respect, preservation, safeguarding and enrichment of that heritage can be achieved. Thus, the European Cultural Heritage Strategy for the 21st century promotes

> good governance based on participatory management involving primarily the national, regional and local levels. For it is the local authorities that have a direct relationship with their heritage and manage it on a day-to-day basis. It is therefore at this level that citizens must play a greater role in the implementation of this Strategy, within the context of public action and in closer co-operation with the work carried out by the professionals and the public agencies concerned (Council of Europe 2017b).

Heritage is, therefore, conceived as a shared responsibility.

---

[1]The UNESCO Convention on the safeguard of intangible heritage (2003), the UNESCO Convention on the diversity of cultural expressions (2005) and the Framework convention on the value of cultural heritage for society (Council of Europe 2005).

The Strategy is based on three main components which are in mutual interaction:

1. The *social* (S) component which sees heritage as the key to promoting diversity and empowers communities for participatory governance.
2. The *territorial and economic development* (D) component stresses the contribution of heritage to sustainable development, based on local resources, tourism and employment.
3. The *knowledge and education* (K) component focuses, through heritage, on education, research and training issues.

Integrated cultural heritage management is rooted in the sustainable urban development approach which comprises

> a system of interlinked actions which seeks to bring about a lasting improvement in the economic, physical, social and environmental conditions of a city or an area within the city. The key to the process is *integration*, meaning that all policies, projects and proposals are considered in relation to one another. In this regard, the synergies between the elements of the plan should be such that the impact of the plan as a whole adds up to more than would the sum of the individual parts if implemented in isolation (European Investment Bank 2008).

Progressive recent policy-making and policy-implementation focus on integrated plans. This is a new although still not widely accepted governance framework, but calls for a paradigm shift. The approach is desirable since it ensures sustainability and possibly fosters innovations.

The integrated approach is threefold and entails:

1. *participatory governance* ensuring citizen participation in decision-making and management related to the heritage field;
2. *public management of heritage* which includes horizontal integration across various sectors and departments, and vertical integration by addressing local, regional and national spheres of government;
3. gradual *integration of sustainability aspects* into the heritage management which allows managing social and economic dimensions. This includes spatial planning, education, science, tourism, entrepreneurship and employment (Radej et al. 2017).

According to this approach, cultural heritage policies should be repositioned and placed at the heart of an integrated approach focusing on the conservation, protection and promotion of heritage by society as a whole, by both national authorities and local and regional communities.

Heritage revitalisation is directly related to tourism, especially in the last condition but the two previous ones are equally important for achieving success (Jelinčić 2017).

3.4 Community Involvement in Heritage Revitalisation

The ongoing discussion of over-tourism brings to the fore the relative role of host communities in heritage revitalisation. Concepts such as 'community-based' heritage tourism or 'community-centred' heritage tourism emphasise the need for heritage revitalisation to be based on and to take into consideration locals' needs, expectations and sensitivity towards tourism impacts (Zapata et al. 2011; Mgonja et al. 2015).

The centrality of host communities' involvement in shaping cultural tourism development and heritage revitalisation largely stems from the sustainability paradigm (Taylor and Murph 2017). Over the years, there has been a common understanding that community involvement in planning, decision-making and actual operation of heritage properties is a prerequisite for its sustainability and success. However, the predominant 'bon ton' in this respect has been a 'top-down' approach (Salazar 2012). The top-down approach is largely based on a centralised heritage revitalisation process with hardly any consultation with the local communities. This usually inflicts external cultural policies taken outside the community domain and which reflect out-of-community needs and interests. In the name of sustainability (a socio-cultural one), international organisations such as UNESCO, UNWTO and others try to impose standardised measures that are assumed to be globally efficient and at the same time protect the socio-cultural interests of locals.

Many case studies around the world have concluded that heritage development promoted and implemented centrally has not yielded adequate results in terms of addressing locals' interests and expectations. Thus, the expected outcomes, such as community empowerment, community resilience, community involvement and improvement in communities' standard of living and quality of life, have not been attained. This has led many scholars to suggest an alternative approach based on bottom-up processes. These are generated by host communities, and much better reflect the communities' development interests, on the one hand, and their limits for acceptable change, on the other hand (Taylor and Murph 2017).

Based on the sustainability ideology, the bottom-up heritage development process puts the community in the centre. It thoroughly examines the interests, needs, expectations and limitations of a given host community (Mtapuri and Giampiccoli 2016). Only after these are processed and documented does it proceed to the policy-making phase, the planning phase, and subsequently to the actual development of revitalisation processes. Furthermore, in many cases, communities are not homogenous. They tend to be comprised of various stakeholders representing different interests related to heritage preservation and development. Such circumstances require 'mapping' the various interests and seeking some kind of consensus in the community before an agreed attitude and policies towards heritage revitalisation are formed (Su and Wall 2014; Theerapappisit 2012).

The bottom-up approach is, therefore, the main principle of integrated revitalisation planning and management since it entails participatory decision-making which can accommodate not only the issues related to heritage but also those which relate to other aspects of citizens' concerns (such as tourism development, urban planning, entrepreneurship, employment and education).

3.5 Soft Factors in Heritage Revitalisation

Still, an integrated revitalisation approach mainly concerns the physical aspects of heritage planning. On the other hand, there are a number of *soft* models which may have strong impacts on heritage revitalisation in terms of making it more vibrant and appealing for tourism consumption, thus ultimately also influencing the revenues. The proposed models are as follows:

1. Creative industries model (CIM);
2. Creative participatory model (CPM);
3. Experience economy model (EEM).

As previously mentioned, the contemporary tourist seeks experiences one can easily identify with. Since the creative industries are 'expressions of contemporary, and what is often also popular culture' (Jelinčić and Senkić 2017), consumers are easily drawn to tourist attractions which use them either in the physical design of their displays or in storytelling. An average tourist does not necessarily have a great body of cultural knowledge of a certain attraction or historical period; the contemporary presentation of heritage by way of the creative industries (e.g. design, film, music, interactive devices) enhances the visitor's understanding and immersion in the heritage attraction. Besides, they are a powerful tool for cultural branding, possibly being the attractor for drawing new audiences; for the development of cultural entrepreneurship which is related to additional tourism supply (e.g. souvenirs), as well as for innovations which may lead to unique tourism attractions. The creative industries model (CIM) is, therefore, focused on the use of the creative industries in the presentation and interpretation of cultural heritage. Although tourists are likely to identify with the creative industries' products/services, this, however, does not guarantee a shift from sightseeing activity. The shift is possible if the CPM is applied.

Experiences are best absorbed if a tourist is involved in an engaged action. This has influenced the development of creative tourism and participatory experience tourism (PET). The first 'offers visitors the opportunity to develop their creative potential through active participation in courses, and learning experiences that are characteristic for the holiday destination where they are undertaken' (Richards and Raymond 2000), whereas the second entails 'a process of adding value to an experience through active participation by the tourist' (de Bruin and Jelinčić 2016). PET, however, does not necessarily involve creativity. Both types can have a positive impact on visitors' length of stay, satisfaction, new knowledge acquired, entertainment, perception of the heritage site and the destination and may offer escapist experiences, as defined by Pine and Gilmore (1998), which are labelled as the highest level of engagement. Engaging the tourist in a creative and/or participatory activity related to the heritage site adds to the cultural tourism offer which switches from merely sightseeing to engaged activity, thus ensuring not only an additional activity and length of stay at the site but also the deeper relationship of the tourist with it. The creative participatory model (CPM), therefore, consists of engaged tourist activities through creative and/or participatory experience tourism and represents a higher level than the previous model. The combination of CIM and CPM results in better outcomes.

The experience economy 'studies ways of effective and/or increased consumption of products/services by creating experiences around them' (Jelinčić and Senkić 2017). Although it was first coined in business and marketing studies, it has lately taken root also in tourism since experiences are naturally inherent to this industry; in the very act of travel, tourists are in search of experiences. Heritage attractions which apply experience economy design principles are more competitive, ultimately attract more tourists, and have higher satisfaction rates. Pine and Gilmore (1998) detected five experience design principles: (1) attributing a theme to each experience; (2) harmonising impressions with positive cues; (3) eliminating negative ones; (4) supplying memorabilia as additional experience enhancers; and (5) engaging all five senses in experience creation. If applied correctly, they add to the heritage revival and, at the same time, have the potential for engaging the tourist emotionally, which creates a strong and long-lasting linkage with the destination. Thus, the EEM model can be labelled as the highest level a heritage site can offer since it engages the tourist emotionally.

All three models mutually interact: the use of the creative industries may be used in designing the experience and should preferably engage the tourist in a participatory activity. The application of one or more models contributes to the revival of heritage sites and enhances the opportunities for increasing the number of visitors to the attraction.

3.6 Conclusions

Both sectors of culture and tourism promote a participatory, bottom-up approach to planning and to the management of their resources. Tourism, being the sector which deals with the immediate effects of its planning and management policies since it involves numerous direct users of its resources, however, may have experienced the bad effects of top-down approaches sooner than the cultural sector. This is, especially, visible in the destruction and decay of heritage resources due to the lack of integrated tourism policy planning. The cultural sector, on the other hand, has only recently begun to reflect on its policies and strategic plans in an integrative way. Although the integrative approach has existed in the urban planning sector, it has only recently been advocated throughout the whole cultural sector. Thus, its effects will certainly be visible first in the planning documents for the built heritage.

While it is certainly useful, participatory integrative planning which focuses only on the physical aspects of the built heritage hardly warrant tourism visitors to the heritage site. It adds to the heritage appeal but does not necessarily draw tourists. In order to achieve this goal, it is necessary to apply some of the proposed *soft* models of heritage revitalisation or a combination of them. They provide the content for the heritage 'shell', possibly achieving a unique selling point and eventually increasing visitor numbers. Integrated heritage revitalisation planning thus entails another dimension of integration: putting together the body and soul, the infrastructure and stories, the tangible as well as intangible aspects of the community life. Only the application of this approach may further enhance community engagement in tourism planning, at the same time proudly promoting its own heritage.

Another concern which needs to be taken into account if heritage revitalisation wants to result in sustainable planning, development and operational outcomes is to measure host communities' limits for acceptable change. As one of the pillars of sustainable tourism is the socio-cultural perspective—any heritage revitalisation must take into consideration the local community's 'red lines', preferences and sensitivities. Thus, tourism growth remains a legitimate aim. However, it needs to be calibrated, reflecting the community's limits for acceptable change both in terms of number of visitors, tourist behaviour and the kind of heritage portrayed to the tourist. At the same time, there is a need to cater to the cultural tourists' needs and expectations and to guarantee them a meaningful experience. Satisfying these two important stakeholders is a complex challenge that undoubtedly requires the implementation of integrated strategic planning, development and operational tools.

References

Council of Europe (2005) Framework convention on the value of cultural heritage for society. Available via Council of Europe https://urldefense.proofpoint.com/v2/url?u=https-3A__www.coe.int_en_web_conventions_full-2Dlist_-2D_conventions_treaty_199&d=DwIFaQ&c=vh6FgFnduejNhPPD0fl_yRaSfZy8CWbWnIf4XJhSqx8&r=sFws8nDuhsr0k5z8PUT3eTIAuFv56KNfV33VTIXnVzPyyDcHfpjq7d-k5Ph9iBiD&m=6rljj-OmoqD5SLXI1aK6ANfHnOwgnwz4lKFPVZyYjeo&s=YrFwW4r0ZzJvss0-6dscXCP-ls-JMgNX3t8aw7n9VMA&e=. Accessed 18 February 2018

Council of Europe (2017a) European cultural heritage strategy for the 21st century. Available via Direção-General de Património Cultural. http://www.patrimoniocultural.gov.pt/static/data/noticias/strategy21.pdf. Accessed 25 Jan 2018

Council of Europe (2017b) Recommendation of the committee of ministers to member states on the European cultural heritage strategy for the 21st century. Available via Council of Europe. https://rm.coe.int/16806f6a03. Accessed 27 Feb 2018

Čopič V, Uzelac A, Primorac J et al (2011) Encouraging private investment in the cultural sector. European Parliament, Brussels

de Bruin A, Jelinčić DA (2016) Toward extending creative tourism: participatory experience tourism. Tour Rev 71(1):57–66

European Commission (2016) Towards more efficient financial ecosystems: innovative instruments to facilitate access to finance for the cultural and creative sectors (CCS): good practice report. European Union, Directorate-General for Education, Youth, Sport and Culture, Luxembourg

European Commission (2018) Cultural tourism https://ec.europa.eu/growth/sectors/tourism/offer/cultural_en. Accessed 25 Jan 2018

European Investment Bank (2008) JESSICA: A new way of using EU funding to promote sustainable investments and growth in urban areas. Available via European Investment Bank. http://www.eib.org/attachments/thematic/jessica_2008_en.pdf. Accessed: 25 Jan 2018

Jelinčić DA (2017) Modeli kulturnog turizma u funkciji revitalizacije i unaprjeđenja kulturne baštine. In: Obad Šćitaroci M, Bojanić Obad Šćitaroci B (eds) Modeli revitalizacije i unaprjeđenja kulturnog naslijeđa. Arhitektonski fakultet Sveučilišta u Zagrebu, Zagreb, pp 82–83

Jelinčić DA, Senkić M (2017) Creating a heritage tourism experience: the power of senses. Etnološka tribina: J Croat Ethnol Soc 47(40):109–126. https://doi.org/10.15378/1848-9540.2017.40.03

Mgonja JT, Sirima A, Backman KF et al (2015) Cultural community-based tourism in Tanzania: lessons learned and way forward. Dev South Afr 32(3):377–391
Mtapuri O, Giampiccoli A (2016) Towards a comprehensive model of community-based tourism development. S Afr Geogr J 98(1):154–168
Oppio A, Torrieri F (2016) Supporting public-private partnership for economic and financial feasibility of urban development. Procedia—Soc Behav Sci 223:62–68
Pine BJ II, Gilmore JH (1998) Welcome to the experience economy. Harvard Bus Rev 76(4):97–105
Psychogiopoulou E (2015) The cultural open method of coordination. In: Psychogiopoulou E (ed) Cultural governance and the European Union: protecting and promoting cultural diversity in Europe. Palgrave Macmillan, London, pp 37–48. https://doi.org/10.1057/9781137453754_4
Radej B, Pirkovič J, Paquet P (2017) Smart heritage policy. In: Working paper 10(1). Slovenian Evaluation Society. https://www.sdeval.si/attachments/article/628/DZ%201-2017%20-%20Smart%20Cultural%20Heritage.pdf. Accessed 2 Jun 2018
Richards G, Raymond C (2000) Creative tourism. ATLAS News 23:16–20
Salazar NB (2012) Community-based cultural tourism: issues, threats and opportunities. J Sustain Tour 20(1):9–22
Su MM, Wall G (2014) Community participation in tourism at a world heritage site: Mutianyu Great Wall, Beijing, China. Int J Tour Res 16(2):146–156
Taylor M, Murph C (2017) An analysis on community based heritage tourism: a resource for a small community in rural County Clare, Ireland. World Acad Sci, Eng Technol, Int J Soc Tour Sci 4(4):4–9
Theerapappisit P (2012) The bottom-up approach of community-based ethnic tourism: a case study in Chiang Rai. In: Kasimoglu M, Aydin H (eds) Strategies for tourism industry—micro and macro perspectives. InTech, Rijeka, pp 267–294
Timothy DJ (2011) Cultural heritage and tourism: an introduction. Channel View Publications, Bristol
UNWTO (2015) Affiliate members global reports, volume twelve—cultural routes and itineraries. UNWTO, Madrid
Ventura C, Cassalia G, Della Spina L (2016) New models of public-private partnership in cultural heritage sector: sponsorships between models and traps. Procedia—Soc Behav Sci 223:257–264
Zapata MJ, Hall CM, Lindo P et al (2011) Can community-based tourism contribute to development and poverty alleviation? Lessons from Nicaragua. Curr Issues Tour 14(8):725–749

Chapter 4
Conceptualising the Relationship Between Tangible Cultural Heritage and the Tourism Market

Ingeborg Matečić and Oliver Kesar

Abstract The conceptualization model of the relationship between tangible cultural heritage (TCH) and the tourism market is largely determined by the complexity of the tourism market itself, the number and the complexity of the relations between the stakeholders involved in the cultural heritage management (CHM) and tourism destination management, the legislative framework and the government policies which both the stimulation of tourism activities and preservation and management of cultural heritage relies upon and the characteristics and value of tangible cultural heritage. Government policies and legislation determine the balance in the allocation of power within such relationships. Furthermore, the relationship between tourism supply-side stakeholders and CMH take on different forms and move between dichotomies of co-operation and conflict. The stakeholders in such relationships are numerous and diverse, which adds to the dynamics of interaction and, consequently, to its complexity. The relations between stakeholders are made additionally complex by the characteristics of TCH as a pure or mixed public good where competition in consumption may or may not exist. Based on different forms of cultural heritage, there are different levels of excludability. The model is applicable in the domain of tourist attraction management as well as tourism destination planning.

Keywords Cultural heritage · Tourism market · Stakeholders · Relationships

4.1 Introduction

The relationships between culture and tourism are causal and can be studied through various outcomes of their interactions. The tourism studies, when analysing the impact of culture on tourism, mostly refer to its economic effects. The Organisation

I. Matečić (✉) · O. Kesar
Department of Tourism, Faculty of Economics and Business,
University of Zagreb, Zagreb, Croatia
e-mail: imatecic@efzg.hr

O. Kesar
e-mail: okesar@efzg.hr

M. Obad Šćitaroci et al. (eds.), *Cultural Urban Heritage*,
The Urban Book Series, https://doi.org/10.1007/978-3-030-10612-6_4

for Economic Co-operation and Development (OECD) study analyses the relationship between tourism and culture in the context of the attractiveness and competitiveness of tourism destinations (OECD 2009). However, many case studies have shown that culture is not only the competitiveness determinant of a tourism destination (Porter 1990), but also the factor of its attractiveness (Ritchie and Zins 1978). During the twentieth century, tourism and culture were observed as largely separate aspects of tourism destinations. Cultural resources were perceived as a part of local cultural heritage, predominantly related to the education of the local population and the establishment of cultural identities, while tourism was broadly considered as a leisure activity separate from everyday life and local community culture (OECD 2009). Since the brink of the new millennium, the goals, mission and management of culture and tourism have started to converge, overlap and mutually evolve.

The role of cultural assets, such as tangible and intangible cultural heritage, have changed during the years and eventually become pull factors for temporary visitors, but also a factor of differentiation between tourism destinations. Many studies in this interdisciplinary area have proved that tourism plays an important role in the emergence and affirmation of the concept of heritage (Gravari-Barbas 2018) to the extent of identifying it as a "heritage producing machine" (Gravari-Barbas 2012). In the survey on relations between tourism and culture, member states of the World Tourism Organization (UNWTO) have jointly estimated that the size of cultural tourism market across participant countries was 35.8% of total inbound tourism (UNWTO 2018). For that reason, it can be regarded that the investments in cultural events and cultural heritage are "excellent means to revitalize the economy of cities" (Galla 2012, 4) in terms of image improvement, urban development and increase of destination attractiveness. The impact of culture and cultural tourism upon urban economies is meaningful (D'Auria 2009). As a result of a tourism activity, cultural heritage ceased to be perceived as a static object and experienced solely through conservation practices, but rather appears to be an active component of the tourism market. Nowadays, cultural heritage and tourism are greatly intertwined having a highly dynamic relationship.

4.2 The Complexity of Tourism Market

Given that tourism is regarded as one of the largest and most powerful economic structures that serves as an important source of benefits and development opportunities for communities worldwide, the understanding of relations and interactions between constituting entities of tourism market is essential and requires a comprehensive approach. According to Goeldner and Ritchie (2012), tourism can be defined as "the processes, activities, and outcomes arising from the relationships and the interactions among tourists, tourism suppliers, host governments, host communities, and surrounding environments that are involved in the attracting and hosting of visitors". Given that each tourism destination has its own setting and its own market players, the relations and interactions among market stakeholders cannot be considered as stereotype and simple, but rather as complex and multi-causal, generating different outcomes and prompting different concerns.

In the context of cultural tourism market, the analysis of the roles of various groups that participate in and are affected by tourism development reveals seven key destination stakeholders whose missions, goals and activities determine the way tourism is developed:

1. *Tourism demand*—a heterogeneous group of tourists (and excursionists) who travel away from home or stop by while on tour for a variety of cultural purposes, e.g. to enjoy architectural heritage and visual arts, visit historical sites or famous museums, attend some traditional dance performance, participate in a creative workshop and buy locally produced handicrafts and souvenirs;
2. *Tourism supply*—a variety of businesses providing goods and services for visitors attracted by cultural attractions, such as museums, theatres, art galleries, cultural performance venues, cultural routes, archaeological sites and theme parks, but also standard tourist services like transportation, accommodation, restaurants, pubs, shops, tour guides and communication services;
3. *Public sector*—represented by the government structures at local, regional or national level that are in charge for tourism development planning and policy making, provision of the legal framework for cultural heritage protection and management, regulations for tourism service providing, promotion of tourism destinations and attractions, provision of stable business environment, etc.;
4. *Host community*—local residents in urban or rural settings affected by and involved in tourism development, whose local cultural values and local distinctiveness serve as tourist attractions that bring expected benefits to host community in terms of increase of employment, income, local taxes and entrepreneurial activities, encourage community spirit and local pride, etc.;
5. *Non-profit sector*—a subset of civil society that includes various non-profit associations (NPAs) and non-governmental organizations (NGOs), like local and regional cultural associations and cultural societies, performing arts institutions, local action groups, cultural heritage-related voluntary groups and cultural activists;
6. *Media*—newspapers, magazines, billboards, virtual platforms, television and radio channels and other medias for advertising and distribution of news and other information valuable to local residents, businesses, policy makers and visitors, which have powerful influence on creating public opinion, launching changes in the community, protecting public goods, increasing public awareness, etc.; and
7. *Academe*—represented by the prominent universities, faculties, institutes and schools, including their scientists, researchers, professors and lecturers, who use scientific approach and methods for shaping the future of tourism development by providing vision, plans, strategies, education, audit and policy recommendations for different tourism market players.

Previously mentioned destination stakeholders create a complex network of relations that can be observed as competitive or co-operative, depending on their interests, goals and activities. When it comes to the cultural tourism market, it must be emphasized that such a multifaceted tourism product attracts visitors whose motivation, interests, previous experiences and cultural backgrounds create a myriad of narrow market segments looking for specialized tourism products, developed and optimized to meet their demands. Each of these segments has different profile and expectations, and therefore implies specially designed cultural tourisms product that must be visible, recognizable, accessible and attractive. Once a visitors' interest for a particular cultural tourism product is recognized, it activates the supply-side, regulatory and non-profit stakeholders to participate in product design, development and implementation, including heritage protection and authenticity verification activities. It seems important to point out that cultural tourism products, particularly heritage-based, must fit into the framework of local cultural identity, destination development concept and marketing strategy, following the strict guidelines and permissions provided by local authorities, and under the supervision of heritage protection institutions.

When focused on tourism destination setting, the relations between local tourism market players are regularly influenced by the social, cultural, political, technological, legal and natural environments. Isolated observation of any of these environments is misleading due to interconnectedness and interdependence between them, so an interdisciplinary and systematic approach to analysis of relations between various environments and tourism market is more than necessary.

Another important aspect of analysing relations between tourism market stakeholders is the international context of tourism development and business operations, which is heavily influenced by international organizations (e.g. United Nations World Tourism Organization), international professional associations (e.g. World Travel and Tourism Council) and multinational corporations (e.g. TUI AG), which use international agreements, standards and principles to manage and adjust relations among market players at the local level.

For more than four decades, tourism market is no longer driven solely by traditional (direct) interactions between demand, supply and intermediaries, but has shifted a significant part of its business operations to the virtual (digital) environment. The Internet and information technologies have radically changed the way market players operate, allowing them to increase interactivity and visibility on the market. According to Buhalis and Law (2008), the Internet has changed tourism market by altering barriers to entry, minimising switching costs, revolutionising distribution channels, facilitating price transparency and competition, while enhancing production efficiency. It has also radically changed the way that tourism businesses are conducted in the marketplace, as well as how visitors interact with service providers and organizations.

The growing complexity of relations on tourism market implies the introduction of a destination management, as multi-part and independent body that coordinates destination stakeholders' activities and develops a common market strategy. No matter whether established as an organization (DMO) or as a company (DMC), destination management should support synergic efforts from all stakeholders in order to improve diversity and quality of destination tourism offer. According to Čorak and Boranić Živoder (2017), unlike a decade or two ago, contemporary destination management is far more complex and driven by an acute need to ensure sustainable destination development through optimal use of resources and welfare of the local community.

4.3 The Characteristics of Tangible Cultural Heritage

As cultural industries produce cultural products, cultural heritage represents an adequate tool in doing so. Cultural heritage involved in a cultural product serves as a background for special interest tourism development, namely cultural tourism or heritage tourism. According to Riganti and Nijkamp (2007), cultural heritage is considered to be a property of historical, cultural and socio-economic importance in modern society and, as such, cannot be considered as an ordinary capital good.

The characteristics of TCH are manifold, having both economic and cultural values incorporated in its essence. The cultural value is responsible for differentiation of these assets from others, while in practice, regarding the management of heritage sites, "assigning an appropriate value to heritage assets and to the services they provide is an all-pervading problem, whether the value sought is economic, cultural or a mix of the two" (Throsby 2012). Analogously, TCH is often perceived as a property of historical, socio-cultural and economic importance for a tourism destination. Examining TCH from the cultural heritage economics perspective and its behaviour on the market, TCH gives rise to two types of values: (1) *economic* (market values) and (2) *non-economic values* (cultural, aesthetic, symbolic, spiritual and historical), which both subsequently produce impacts within a tourism destination. The economic value of cultural heritage is often referred to as the *use value* or *non-use value*, or *market* and *non-market value* (Navrud and Ready 2002). The cultural values of TCH are numerous; therefore, measurements of solely economic value cannot represent the result of the entire valuation process. Each type of heritage value has its own methodology and techniques used to estimate it. They can be divided into two basic groups, economic and non-economic valuation methods (Matečić 2016).

Moreover, heritage value and its commercialization for tourism purposes has created favourable conditions "for investment and income production, and generated debate about the economic and non-economic impacts of the process [occurring] on sites… and on the people who live and work around heritage sites. The debate over commercialization raises ethical, practical, and technical questions…" (Gould and Burtenshaw 2014). Tourism related economic activities developed around cultural heritage site may generate: (a) *direct economic effects*, such as increase of income from entry fees or food service on-site, and (b) *indirect economic effects*, such as job creation and new business opportunities off-site, in the nearby communities. These effects can be considered as positive commercialisation impacts. However, the estimation of the cultural heritage value and potential spillovers in the local economy from investing into cultural heritage is rather difficult to carry out. Moreover, the effects of investing into cultural heritage are "diverse and vary by site, project and over time" (Bowitz and Ibenholt 2009).

Apart from its positive effects, the commercialization of cultural heritage can have detrimental ones too. For example, overcrowding of the site and potential degradation of the physical environment due to foot traffic or pollution; disrupted aesthetic value of the site due to construction of facilities designed to serve visitors and staff or presence of massive numbers of stands with souvenirs on site; commoditization of heritage—disassociate the meaning of traditional artefacts, practices, and people from their original context and presents them for display and enjoyment by visitors, who may not appreciate their cultural significance (Timothy and Nyaupane 2009); displacement of long-standing residents from areas demarked to become heritage sites, etc. (Gould and Burtenshaw 2014).

Non-economic positive effects of cultural heritage valorisation could be seen in the enforcement of the identity of local community and fostering the pride of local people. Such non-economic effects generated within societies on national and regional level are observed as social cohesion and integration of marginalised groups (Bouchard 2009); building of a new value system (Ingelhart 2000); affirmation of creativity, talents and excellence; development of cultural diversity, national identity and the identity of different cultural groups (Throsby 2001); facilitating creativity and innovation (Potts and Cunningham 2008). The mentioned impacts imply that cultural heritage fosters affirmation of creativity, talents and excellence, facilitates innovation, enhances social cohesion and enforces local identity. There are many perceptions regarding cultural heritage and roles it plays within the society. The context of TCH within the CHM framework will be elaborated in greater detail under the following title.

4.4 The Concept of Tangible Cultural Heritage

Prior to discussion on the concept of cultural heritage, it is relevant to point out the approach to cultural heritage as a public good. "Cultural heritage goods are similar to environmental goods in that they are typically public goods" (Navrud and Ready 2002). Regarding the economic theory of "public goods", economists define public goods as being "non-rival" in consumption and having "non-excludable" benefits (Hardin 2013). The pure public good has to have both properties. The problem associated with the concept of public goods is the so-called free-rider problem and it closely linked to the properties of public goods. Samuelson (1954) noted that "some goods, once they are made available to one person, can be consumed by others at no additional marginal cost; this condition is commonly called jointness of supply or non-rivalness of consumption. Therefore, in standard price theory, in which price tends to equate to marginal cost, such goods should have a zero price. But if they are priced at zero, they will generally not be provided. In essence, price theory commends free riding on the provision of such goods" (Hardin 2013). In other words, *non-rivalrous* good is a good that several people can consume, without diminishing its value and formally at no additional cost (Landriani and Pozzoli 2014). In terms of cultural heritage, it means that two different people can enjoy an object of cultural heritage at the same time without interfering with each other's enjoyment. However, the consuming value begins to diminish at the point when there are too many people trying to do the same thing at the same time. Crowds are sometimes so huge they obstruct the view on the object. In practice, many cultural heritage goods have intermediate levels of rivalness, which means they are unable to accept limitless numbers of visitors (Navrud and Ready 2002). Same authors refer to that kind of goods as *congestible* public goods and distinguish them from pure public goods.

The *excludability* is another important aspect of cultural heritage preservation (Navrud and Ready 2002). Private providers and owners see no interest in preserving cultural heritage unless they receive some financial benefits from it. The profit-driven markets usually have insufficient interest to produce non-excludable goods. Therefore, the conclusion presents itself and can be rather simplified. If society relies entirely upon private owners of heritage sites or any other type of cultural heritage to preserve it, only the cultural heritage with high market demand will be protected (Navrud and Ready 2002). In this respect, heritage tourism or cultural tourism creates tremendous commercial opportunities for heritage sites producing numerous activities and generating noteworthy financial gains.

Another feature of Samuelson's view on the public goods that would make them problematic in practice is the impossibility of their exclusion. Once supplied, it is supposedly impossible to exclude anyone from the consumption of a public good. It is often noted that this feature is analytically interesting, but empirically often beside the point (Hardin 2013). Namely, *non-excludable* public goods are the ones that each person can benefit from and cannot be prevented from consuming it. Empirically, as regards cultural heritage there are various degrees of excludability (Navrud and Ready 2002). An artefact in museum can be excludable if a person does not buy an entry ticket. Therefore, the museum excludes potential visitors who do not pay entrance fees. On the other hand, sightseeing in a historic fortified town is not excludable activity since there is no admission charge for visitors to enter heritage site.

The economic theory is interested in the nature of cultural heritage resources and the need for public intervention for financing and regulating it (Landriani and Pozzoli 2014). If cultural heritage resources are perceived as generators of public services, then their public nature would not be dependent on the legal status (public or private) of the provider, but on the fact that it can satisfy community needs (Anselmi 1995).

With regard to this approach Landriani and Pozzoli (2014) proposed an advanced classification of public goods::

- *common goods*—Inspired by intergenerational solidarity, such as natural resources and cultural heritage;
- *goods belonging necessarily to public*—Essential utilities can only be benefited by the state or local authorities, including infrastructure networks that may be outsourced to private companies;
- *social public goods*—Satisfy fundamental rights, such as health, housing and local public services;
- *profitable public goods*—That is residual category, referring to the private goods that are property of public entities.

Provided classification is based on the utility of goods regardless of their ownership. Cultural heritage assets can be regarded as common goods or common pool of resources. The idea behind perceiving heritage assets as common goods assumes that the commons could be used and enjoyed, but not exclusively owned by an individual (Lucarelli 2010) or a public entity. The exclusivity of the ownership can sometimes lead to an abuse of rights. The abuse of the property rights represents the threat to those organizational principles set within Public International Law and the Constitution of each state. The solution could be sought in creation of the "universal ownership" to avoid unilateral ownership and property rights abuse at legal level. Nevertheless, public institutions should serve as providers of the public welfare and therefore cultural heritage assets should be perceived as assets of its citizens (Lucarelli 2010). However, the category of common assets requires a new way of management, different from those ruled by market regulations and the laws of the State (Landriani and Pozzoli 2014).

Regarding the conceptualization of TCH, based on the literature review on CHM provided by du Cros and Lee (2007), could be perceived that scientists tend to approach cultural heritage issues from different points of view. They associate cultural heritage assets with the notions of:

1. *Power*—Power tensions in CHM are known to arise in regard to issues of control over heritage activities (du Cros and Lee 2007) and public–private partnerships especially in the area of local participation in CHM (Stenseke 2009). Issues over heritage assets can be found in other domains of politics, such as colonialism and repatriation of cultural property (AusAnthrop 2006), human remains (Pardoe 1991), and indigenous autonomy and land rights (Lilley 2000).
2. *Quality of life*—Attitudes on cultural heritage preservation and conservation are different. They can be either negative seeing heritage as a burden of the past (Tunbridge and Ashworth 1996) or positive perceiving it to be important for community or individual reaffirmation (Lowenthal 1988), for creativity and economic development and eventually for better quality of life (Hall 2002). Debates about tourism impacts within CHM discourse in relation to sustainable tourism development are very common among researchers (e.g. Hall and Lew 2009).
3. *Specialist and community care*—Relations between stakeholders can sometimes be undergoing a lot of pressure, and collision of opinions is most likely to occur. The different and conflicting interests can sometimes arise among stakeholder groups about "who knows best regarding what criteria and principles should be followed" (Stone 1992). Cultural heritage specialists are considered to be more experienced and skilful than other destination stakeholders, but without the help of community they cannot become experts in heritage management (Hall 2002). Moreover, public, private and third-sector partnerships for the conservation of the historic urban environment are increasingly playing a pivotal role in these processes (Macdonald 2011).
4. *Commodity*—Commercialization processes have led to the commodification of cultural heritage in order to satisfy the needs of different users, usually tourists. Comprehending cultural heritage as an asset or commodity has its roots in the

capital theory. Many scientists have researched the commodification of cultural heritage assets and its impacts (Cohen 1988), and the issues of authenticity of cultural heritage (e.g. McIntosh and Prentice 1999).

Therefore, the issues of the allocation of power, public–private cooperation, quality of life and sustainable community development, processes of commodification, authenticity and dichotomies between specialist and community knowledge in managing cultural heritage represent key issues regarding the concept of TCH and its management practices. Additionally, the understanding of not only the concepts of TCH, but also of its key features or properties, its behaviour on profit-driven markets, its values and valuation processes as well as economic and non-economic impacts it generates, enables further research on relationship between TCH and the tourism market and thus establishment of the model and its various input and output variables.

4.5 The Interrelation Between Tangible Cultural Heritage and Tourism

The relations between tourism stakeholders and CHM have many forms and have been studied so far by many scientists. Commonly, tourism stakeholders are responsible for the promotion and development of tourism product and CHM for the provision and conservation of cultural heritage assets (McKercher et al. 2005). Same authors consider this relationship to be somehow "awkward" and set between the conflict/co-operation dichotomies. CHM and tourism relationship is claimed to be conflicting (Jansen-Verbeke 1998). The cultural heritage authorities often blames tourism for commodification and the compromise of cultural values for the sake of earnings (Urry 1990; ICOMOS 1999), but at the same time, tourism stakeholders argue that the tourism values are seriously weakened when that kind of attitude prevails in CHM (Fyall and Garrod 1996). The part of dichotomy regarding co-operation states that partnership in management of cultural heritage assets can bring many opportunities to all stakeholders. It has been argued that heritage tourism revives people's cultural roots (Donert and Light 1996) and culture as a tourist attraction is a powerful tool for conservation of region's historic, cultural, religious and industrial past (Frew and Shaw 1995). Furthermore, the relationships are affected by many different circumstances occurring within the tourism destination. McKercher et al. (2005) found, while analysing relationship between tourism and CHM in the established urban destination of Hong Kong, that there are five conditions which influence this relationship: (1) the independent evolution of tourism and CHM; (2) the politically imposed power balance between stakeholders; (3) the diversity of stakeholders with different levels of knowledge; (4) the diversity of heritage assets under consideration; and (5) the different ways in which assets can be consumed. Some of the tourism destinations have been faced with the independent evolution of tourism and CHM meaning that tourism and CHM operate separately "in parallel, but independent worlds" (McKercher et al. 2005) where each role has been clearly defined and

OECD (2009) The impacts of culture on tourism. Available via OECD. http://www.oecd.org/document/53/0,3746,en_2649_33956792_42040117_1_1_1_1,00.html. Accessed 10 Jan 2017
Pardoe C (1991) Competing paradigms and ancient human remains the state of the discipline. Archaeol Ocean 26(2):79–85. https://doi.org/10.1002/j.1834-4453.1991.tb00267.x
Porter ME (1990) The competitive advantage of nations. The Free Press, New York
Potts J, Cunningham S (2008) Four models of the creative industries. Int J Cult Policy 14(3):217–232. https://doi.org/10.1080/10286630802281780
Riganti P, Nijkamp P (2007) Benefit transfer of cultural values: lessons from environmental economics. J Environ Policy Law 2:135–148
Ritchie JRB, Zins M (1978) Culture as determinant of the attractiveness of a tourism region. Ann Tour Res 5(2):252–267. https://doi.org/10.1016/0160-7383(78)90223-2
Samuelson PA (1954) The pure theory of public expenditure. Rev Econ Stat 36(4):387–389. https://doi.org/10.2307/1925895
Stenseke M (2009) Local participation in cultural landscape maintenance: lessons from Sweden. Land Use Policy 26(2):214–223. https://doi.org/10.1016/j.landusepol.2008.01.005
Stone J (1992) The ownership of culture: reconciling our common and separate heritages. Archaeol Ocean 27(3):161–167. https://doi.org/10.1002/j.1834-4453.1992.tb00300.x
Throsby D (2001) Economics and culture. Cambridge University Press, Cambridge
Throsby D (2012) Heritage economics: a conceptual framework. In: Licciardi G, Amirtahmasebi R (eds) The economics of uniqueness: investing in historic city cores and cultural heritage assets for sustainable development, the urban development series. The World Bank, Washington, DC. https://doi.org/10.1596/978-0-8213-9650-6
Timothy DJ, Nyaupane GP (2009) Cultural heritage and tourism in the developing world: a regional perspective. Routledge, London
Tunbridge JE, Ashworth GJ (1996) Dissonant heritage: the management of the past as a resource in conflict, 1st edn. Belhaven Press, London
UNWTO (2018) Tourism and culture synergies. United Nations World Tourism Organization, Madrid
Urry J (1990) The tourist gaze. Sage, London

Chapter 5
Heritage Protection Policies from the Perspective of the Social Sciences: The Case of Croatia and Non-EU South-East European Democracies

Saša Poljanec-Borić

Abstract Heritage protection is a social invention established in proto-democratic countries. Through two centuries of modernisation, heritage protection has become an important public-sector activity in all Western European democracies. Transition in parallel with accession to the EU enabled the earlier integrated countries to benefit from the transfer of best practices of heritage protection from the core EU countries. Such a transfer has not taken place in belatedly integrated Croatia or in other non-integrated South-East European democracies. The mentioned deficit fundamentally threatens sane heritage protection schemes in South-East Europe. Sane heritage protection policies could be reintroduced in South-East Europe by integrating heritage urbanism principles into urban, rural and space development policies in order to couple the protection, revival and enhancement of cultural heritage at the same time. Therefore, heritage urbanism could help bridge the asymmetries in heritage protection policies on the European continent.

Keywords Heritage urbanism · Protection · Policy · Modern · Postmodern

---

S. Poljanec-Borić (✉)
Institute of Social Sciences Ivo Pilar, Zagreb, Croatia
e-mail: sasa.boric@pilar.hr

M. Obad Šćitaroci et al. (eds.), *Cultural Urban Heritage*,
The Urban Book Series, https://doi.org/10.1007/978-3-030-10612-6_5

5.1 Heritage Protection in the Perspective of the Social Sciences

This discussion is based on the insight that the concept of "heritage protection" is a "social invention" (Conger 1973) related to modernisation. The very concept of "heritage protection" as well as subsequent practices of the preservation of heritage is a reflection of the collective intention of modern societies to cherish the artistic expressions of past and/or rare and beautiful natural phenomena for the benefit of the present and/or next generations. Modern, post-World War II European democracies considered heritage protection as a public-sector activity genuine to the welfare state (Titmuss 1958). Thus, they saw preserved heritage as a sign of an effective national cultural policy (Malraux 1959). Underpinned by postmodernisation (Inglehart 1997), the core EU democracies thought of preserved heritage as of tool for creating a positive image of a (national) territory and an important market driver for constantly growing tourism demand (O'Shaughnessy and O'Shaughnessy 2000). For this reason, by the end of the twentieth century most of the core European democracies allowed activities of exchange that bring economic benefits based on the attractiveness of historical monuments (Hewison 1987) and/or rare and beautiful natural phenomena in various protected places and areas. The climate of postmodernised fragmentation and uncertainty (Bauman 1992) and the growing interest of tourism for the preservation of cultural heritage (Ritchie and Crouch 2003) led to the development of a large number of heritage protection schemes within the realm of mixed economies (Stiglitz and Rosengard 2015). This process was pervasive in the core EU democracies, and it helped maintain and rendered accessible to the public a cornucopia of artefacts of commemorative and/or aesthetic value, disregarding the distinction between private and public ownership (Sassen 2006). This is why monuments, urban and rural sites and even cultural sceneries together with rare and beautiful natural phenomena are presently nicely maintained in the core EU democracies regardless of the ownership type, while the same is not true of monuments, artefacts and even natural phenomena that are "protected" in South-East European (SEE) countries. The absence of protection and preservation is especially evident in Croatia, a country that belatedly accessed the EU, or in those SEE countries which are not yet eligible for accession to the EU. Therefore, it is of interest to elaborate, from the perspective of the social sciences, why heritage protection represents, in Croatia, a structural problem. It is hoped that the structural understanding of the barriers to coherent heritage protection policies in Croatia will help solve not only the situation in Croatia but also in the SEE countries aspiring to join the EU. Ultimately, such an understanding might increase the capacity of the EU to enhance security on the European continent by means of implementing heritage urbanism as a part of the accession arrangements.

5.2 The Structural Barriers to Informed and Effective Heritage Protection in Croatia

In order to understand the structural barriers to heritage protection in Croatia, comprehensive understanding of the political dynamics that shaped Eastern Europe (EE) at the end of the twentieth century is needed. Prior to the fall of the Berlin Wall in 1989, the political map of EE looked considerably different from how it appeared in 1993. The political map of EE had already changed in 1991 when Lithuania, Latvia, Estonia and Ukraine declared independence from the then Soviet Union. In 1991, at the southern end of EE, two ex-Yugoslav countries, Slovenia and Croatia, were internationally recognised and admitted to the United Nations. Bosnia and Herzegovina and the former Yugoslav Republic of Macedonia declared independence in 1992 and 1993 and were subsequently admitted to the United Nations. Czechoslovakia split in a "velvet" divorce in 1993. Therefore, after the fall of the Berlin Wall, EE consisted of 13 instead of 8 states and had Lithuania, Latvia, Estonia, Ukraine, Slovenia, Croatia, Bosnia and Herzegovina, Slovakia and the former Yugoslav Republic of Macedonia as new sovereign states even though they were (constituent) parts of the Soviet Union, Czechoslovakia and Yugoslavia only 4 years before.

Obviously, the destruction of a totalitarian regime that had been imposed on EE after World War II unleashed social forces that pushed for state-building and the shaping of national identities, at the same time reconfiguring political authority along putatively national lines (Brubaker 1996, p. 3). The process did not stop in 1993 but continued to produce new nations particularly in SEE throughout the last decade of the twentieth century and well into the first decade of the twenty-first century. Montenegro, a former Yugoslav Republic, declared independence in 2006 by divorcing from Serbia which also became independent in 2006. Kosovo, a former Yugoslav autonomous province, was the last to declare independence in 2008. The process of nation and state-building went in parallel with the process of triple transition (Offe 1991) which pushed for changing state identities, political regimes and economic systems in all EE countries.

The social and political dynamics briefly described here stand in hetero-chronic relation to the modernising and postmodernising temporal dynamics that occurred in Western Europe and was outlined in the first section of this chapter. It shows that a number of EE societies and almost all SEE societies (the exception is Albania) engaged in the ultimate modernisation project of state-building (Weber 1999) only at the end of the twentieth century while most of the nations on the western side of the European continent had completed the process of nation-building by the end of the nineteenth century. However, this was not the only political *differentia specifica* between the western and eastern side of the continent. The other specificity was also that the belated state-building dynamics occurred in parallel with accession to the EU. From the sociological perspective, this very fact suggests that the newly formed EE nations simultaneously opted for belated modernisation (i.e. state-building) as well as for timely postmodernisation (i.e. European integration), a fact symbolically corroborated by acceptance to include in newly formed legal systems the crucial elements of the acquis communautaire. These hybrid social dynamics of parallel modernisation and postmodernisation could be labelled as "neomodernisation" and they could be felt across EE in the newly established states. However, it was particularly felt in Croatia, a country that accessed the EU only in 2013, and in non-integrated SEE democracies. The evidence from Croatia shows that the pervasive feature of "neomodernisation" is a permanent hetero-chronic metanarration. In fact, the very hetero-chronic metanarration consists of two types of narratives that configure the political climate of Croatian democracy.

The first insists on a retro-nostalgic rhetoric which aims at the affirmation of a suppressed Roman Catholic tradition and on reinventing cultural memory through the commemoration of historic actors that opposed political oppression and publicly questioned political persecution and/or the functional brutality of the massive rural–urban transfer that occurred during the period of authoritarian rule in the name of the "permanent socialist revolution" and the monumental character of the working class (Rogić 2000). These dynamics were supported by the newly legitimised political elites that brought about democracy in newly established democracies/nations.

The second vector is visibly postmodern as it insists on the individual perspective and lifestyle freedoms and is co-generated by pressure from above, i.e. by the implementation of the "acquis communautaire". Therefore, the process of "neomodernisation" in Croatia differs substantially from the process of modernisation which dominated the western part of Europe two centuries previously, not only in its timing and transitional features but also in the collective political attitude towards the church. Contrary to the western mainstream where one of the principal vectors was secularisation, i.e. the separation of the powers of the state and the church, "neomodernisation" in Croatia is substantially co-created by the Catholic church. Such a situation is of course also present in other countries with dominant Roman Catholic majorities, such as Lithuania, Slovakia and Slovenia. However, in Croatia it is particularly strong, not only due to the fact that the Catholic church strongly opposed communist dictatorships and supported pro-democratic activities in the era of totalitarian rule in the world (Huntington 1991), but also because Croatia was the only EU country that was internationally recognised thanks to the leadership of the Catholic church[1] while waging war for independence.

The described specific internal "neomodern" social dynamics produce a situation which is not a zero-sum game in Croatian society. The evidence shows that with regard to public choices (Buchanan 2000) this society visibly opts for the re-traditional side. This is why Croatia is a country which holds a special position within this "neomodern" group of countries with a Roman Catholic sociocultural background and also why it is sometimes hard to connect Croatian internal social dynamics with those observed in similar "neomodern" EE societies such as Slovakia and Slovenia. Given the fact that there are structural as well as historical reasons that a "neomodern" climate is particularly felt in Croatia, it should not be surprising that the political emphasis on the epic dimensions of the creation of a nation-state that underpinned Croatian independence often colludes with the efforts to perform according to EU

[1]The Vatican was the first state to internationally recognise Croatia on 13 January 1991.

standards, all of which structurally arise from a postmodern condition and fragmented social value system.

Only this perspective helps explain why Croatian society, in particular, lives in an accentuated climate of simultaneous credulity of historic nation-building metanarratives and incredulity of history. The situation of the permanent mixing of historic and ahistoric metanarratives creates a local climate of immense value confusion intensified by the fact that the whole setting, due to the war, has been strongly influenced by the substantially belated "triple transition". By virtue of their origin and substance (Pal 1989), "triple transition" policies undermine public spending as they focus on the contraction of the public sector, deregulation and privatisation. In their "belated" version applied in Croatia, they are even more intense, because they have to deal with the social costs of the war which substantially postponed the timely transitional adjustments that occurred in the countries eligible for accession to the EU in 2004. In such conditions, efforts of the state to give incentives for long-term institution building are rare and weak. On the other hand, institution building, in order to be legitimate, depends vitally on the "rule of law". Given that the "rule of law" disturbs the forces of deregulation and the privatisation of tangible assets, local society is engaged in a vicious circle where the chances for structured policies to stabilise the conflicting historic, i.e. state-building, and ahistoric integration co-narratives are minimal. With the flow of time, such a chaotic climate becomes a pertinent feature of political life and a source of (political) instability. This is a deep reason why comprehensive heritage protection policies are virtually non-existent in Croatia, even though the country is paradoxically becoming a leading European tourist destination. An ever-contracting public sector, a hybrid system of heritage protection which by and large depends on the institution of "monument rent",[2] and a weak civil society living in a climate of chaotic value systems cannot cope with the cost of the sane and comprehensive protection of cultural heritage over the whole territory of the country. *Summa summarum*, the problem lies in the fact that the "neomodern" system of protection is (a) underfunded due to the belated triple transition effect and (b) socioculturally incoherent due to its leaning to the "re-traditional" side. Such a situation has consequences that undermine social cohesion and compromise the practices of cultural memory production. Therefore, it should be tackled as a political and institutional problem in Croatia.

On the other hand, path-dependent (historic) problems of belated state-building and substantially postponed integration prospects are pertinent features of all SEE democracies. This very fact implies the high possibility of even more intense problems with heritage protection and cultural memory production due to the "neomodern" vector in these societies. Such a situation could be especially delicate in Bosnia and Herzegovina and Macedonia because of their pronounced multiethnic social and religious traditions, and particularly in Kosovo which is ethnically compact but trea-

[2]Essentially, it is a tax on locational rent which is by default higher in the touristic (littoral) part of the country and is mostly collected at the local level (cf. Zakon o zaštiti i očuvanju kulturnih dobara [Act on the Protection and Preservation of Cultural Goods] NN 69/99, 151/03, 157/03, 100/04, 87/09, 88/10, 61/11, 25/12, 136/12, 157/13, 152/14, 98/15, 44/17).

sures some of the most important elements of the Serbian cultural heritage. Therefore, it could be hypothesised that the structural problems observed in Croatia and related to the undermining of social cohesion can be expected in all SEE countries aspiring to EU membership.

5.3 Attempts at Overcoming the Barriers to Sane Heritage Protection Policies in Croatia

It is not surprising that the lack of a structured social value system and ever-contracting public-sector incentives to protect heritage leads to the neglect of exceptional as well as "ordinary" cultural heritage, contempt for artefacts related to industrialisation per se and even iconoclastic behaviour towards specific memorabilia linked to actors and events celebrated for their socialist and/or totalitarian character and monuments commemorating epic dimensions of World War II (Pavlaković 2014). Even though all problems of neglect and contempt are urgent as they undermine the sane and truthful policies of cultural memory production, the neglect of some artefacts that are related to the epic dimensions of World War II is politically challenging because they sometimes question the commemoration of the defeat of fascism, the positive social dimensions of modernisation processes that occurred during the period of socialism/authoritarian rule and, ultimately, the commemoration of the Holocaust. Such an insight calls for an informed assessment of the problem and the revisiting of basic principles on which third-wave democratisation occurred in EE, thus also in Croatia. In short, it calls for a culture of informed heritage mapping and the swift inclusion of heritage urbanism into urban, rural and space development policies that regulate the social standards and economic life of Croatia and especially of non-EU SEE countries.

In order to socially and culturally accept the need for heritage urbanism to impartially treat exclusive, ordinary, industrial and aesthetically and historically relevant socialist artefacts as a tool for social and economic development and ultimately as a tool for the enhancement of national well-being, it would probably be wise to consider the fact that not all aspects of modernisation that occurred under socialism/totalitarian rule were repudiated by constituent voters (Columbus 1998). Among the positive aspects of modernisation, it should be acknowledged that "urbanism" as such was instrumental for the appreciation and/or creation of the "sense of place" (Shamai and Ilatov 2005) and especially for the sustainable use of space. Therefore, "urban planning" should be reintroduced as a policy tool to help prevent the production of negative externalities related to neglect and unsustainable use of space even though it sometimes disturbs market forces. The fact that contestation of the Holocaust was not "the credo shared by most of the citizens of Eastern Europe" when they aspired to independence and democracy (Beyme 1998, p. 11) should be culturally embedded in all newly formed democracies. This, in the era of deregulation, is the only way to maintain artefacts with a socialist character that commemorate the Holocaust. Further, an insight that some social consequences of international exchange that occurred during the socialist period benefited today's EE democracies should also be re-emphasised. This is especially true with regard to the dimensions of social capital (Putnam 1993) created by the early Yugoslav ratification of the Convention concerning the Protection of the World Cultural and Natural Heritage (UNESCO 1972) which was transmitted into the democracies that were formed after the dissolution of Yugoslavia. Through the Convention, UNESCO globalised the principles of the protection and conservation of heritage which was already integrated into the public domains of all proto-democratic countries and, thus, transmitted the fundamental responsibility of conservation and protection to the public sectors of all countries in the world that ratified the Convention. In this way, UNESCO spread the modern democratic credo of heritage protection throughout the global community and ensured the global reception of the positive aspects of heritage protection for local economies. Croatia is today one of the post-Yugoslav democracies that benefited from such a development through the early inscription of Dubrovnik, Split and the Plitvice Lakes on the UNESCO World Heritage List in 1979. The evidence shows that the mentioned inscription enhanced protection and preservation schemes and helped control negative externalities related to intense tourism pressure on these sites during the deregulation period immanent to "triple transition". In short, it helped maintain the latent influence of heritage urbanism principles in the times of deregulation in sites which are today vital for national identity and economic prosperity.

Therefore, the way out of the value chaos which represents a barrier to structured, meaningful and inclusive heritage protection in Croatia which has belatedly overcome the paradoxes of transition through accession to the EU is to challenge the current "neomodern" tide at levels of the truthful cultural memory policies and to adhere to progressive heritage urbanism policies as a vital part of public-sector activities. Inclusive and impartial heritage urbanism devoted to the concept of sense of place should be reintroduced as a policy tool of a welfare state in urban, rural and space development projects, at least in countries which belatedly accessed the EU such as Croatia, and particularly in non-integrated SEE countries.

Consequently, the future of heritage protection in Croatia and mutatis mutandis in newly formed non-integrated SEE democracies is highly dependent on the internal capacities of the respective societies to fully accomplish the modernising project of state-building by overcoming the transitional problems of the ever-increasing contraction of the public sector. It is also dependent on the capacity of the EU to acknowledge and properly evaluate the hetero-chronic dynamic immanent to newly formed EU nations and SEE democracies. The first sign of mutual understanding with regard to this issue between core EU countries and new EU nation states and SEE democracies should be the implementation of heritage protection policies based on informed and progressive heritage urbanism. This becomes urgent as the EU prepares to discuss issues related to the accession of other ex-Yugoslav countries such as Serbia, Bosnia and Herzegovina, the former Yugoslav Republic of Macedonia and ultimately Kosovo where questions of cultural memory production will be crucial for internal social stability and ultimately for European security.

References

Bauman Z (1992) Intimations of postmodernity. Routledge, London

Beyme K (1998) Parties and party systems in Eastern Europe. In: Columbus FH (ed) Central and Eastern Europe in transition. Nova Science Publishers, Commack, NY, pp 1–19

Brubaker R (1996) Nationalism reframed: nationhood and the national question in the New Europe. Cambridge University Press, New York

Buchanan MJ (2000) Politics as public choice. Liberty Fund Inc, Indianapolis

Columbus FH (ed) (1998) Central and Eastern Europe in transition. Nova Science Publishers, Commack, NY

Conger S (1973) Social inventions. Saskatchewan Newstart, Prince Albert

Hewison R (1987) The heritage industry: Britain in the climate of decline. Metheuen Publishing Ltd., London

Huntington PS (1991) The third wave: democratization in the late twentieth century. University of Oklahoma Press, Norman

Inglehart R (1997) Modernization and postmodernization: cultural, economic and political change in 43 societies. Princeton University Press, Princeton, NJ

Malraux A (1959) Décret n° 59-889 sur la mission et l'organisation du ministère chargé des Affaires culturelles. Journal officiel du 26 juillet 1959. Assemblée Nationale, Paris

Offe C (1991) Capitalism by democratic design—democratic theory facing the triple transition in East Central Europe. Soc Res 58(4):865–892

O'Shaughnessy J, O'Shaughnessy JN (2000) Treating the nation as a brand: some neglected issues. J Macromarketing 20(1):56–64

Pal LA (1989) Public policy analysis: an introduction. Nelson Canada, Calgary

Pavlaković V (2014) Fulfilling the thousand-year old dream: strategies of symbolic nation-building in Croatia. In: Kolstø P (ed) Strategies of symbolic nation-building in South Eastern Europe. Ashgate, Farnham, pp 19–50

Putnam R (1993) The prosperous community: social capital and public life. Am Prospect 4(13):35–42

Republic of Croatia (1999–2017) Zakon o zaštiti i očuvanju kulturnih dobara [Act on the protection and preservation of cultural goods]

Ritchie JRB, Crouch GI (2003) The competitive destination: a sustainable tourism perspective. CABI Publishing, Wellingford, UK

Rogić I (2000) Tehnika i samostalnost (Technics and self-determination). Hrvatska sveučilišna naklada, Zagreb

Sassen S (2006) Territory, authority, rights: from medieval to global assemblages. Princeton University Press, Princeton, NJ

Shamai SH, Ilatov Z (2005) Measuring sense of place: methodological aspects. Tijdschrift voor Economische en Sociale Geografie 96(5):467–476

Stiglitz EJ, Rosengard KJ (2015) Economics of the public sector, 4th edn. W.W. Norton and Company, Inc, New York

Titmuss MR (1958) Essays on the welfare state. Allen & Unwin, London

UNESCO (1972) Convention concerning the protection of the World Cultural and Natural Heritage. Adopted at the General Conference (17th session) in Paris on 16 November 1972

Weber M (1999) Politika kao poziv (Politik als Beruf). In: Katunarić V (ed) Vlast i politika [Politics and power]. Naklada Jesenski i Turk i Hrvatsko sociološko društvo, Zagreb, pp 161–214

Chapter 6
A Comprehensive Approach to Urban Heritage Regeneration

Iva Kostešić, Jana Vukić and Fedja Vukić

Abstract Urban regeneration has been the dominant practice in revitalising urban heritage over the last three decades. It represents a comprehensive, holistic and integrated approach, encompassing material and immaterial dimensions of space that make up the city's identity and image, considering the wider city area and systems relevant in the interaction between the city and the citizens. Therefore, it should be grounded on sociological insights into stakeholders' opinions and accomplished with experts from relevant fields offering a result focused on all elements, "from the spoon to the town", as EN Rogers defined designing in different scales. Urban and planning practices often focus on the bigger scale, but for the successful reintegration of urban fragments into a functional whole, it is necessary to establish a comprehensive strategy encompassing elements from the smaller scale as well. Simultaneously planning and designing in different scales and grounding the result on both material and symbolic elements of urban space and life would give way to a coherent and meaningful environment supporting the integrity and identity of a place. A comprehensive and parallel approach embodying sociological, psychological, economic and spatial research could provide a strategy for creating revitalised space through the synthesis of urbanism, architecture and design.

Keywords Urban regeneration · Urbanism · Architecture · Design · Interdisciplinary approach · Citizen participation

I. Kostešić (✉) · F. Vukić
Faculty of Architecture, University of Zagreb, Zagreb, Croatia
e-mail: ikostesic@arhitekt.hr

F. Vukić
e-mail: fvukic@arhitekt.hr

J. Vukić
Faculty of Humanities and Social Sciences, University of Zagreb, Zagreb, Croatia
e-mail: jana.vukic@ffzg.hr

M. Obad Šćitaroci et al. (eds.), *Cultural Urban Heritage*,
The Urban Book Series, https://doi.org/10.1007/978-3-030-10612-6_6

6.1 Introduction

The urban regeneration strategy as a concept and practice of revival in the historic urban environment appeared in the early 1990s. Since then, it has been a dominant practice in the domain of restoring historical cores. Some of the more famous examples of the application of this model represent the reconstruction of the cities of Genoa and Barcelona (El Menchawy 2008). Unlike earlier urban strategies, such as urban reconstruction (the 1950s); urban revitalisation (the 1960s); urban renewal (the 1970s); and urban transformation (the 1980s), urban regeneration assumes a comprehensive and integrated approach to the simultaneous rebuilding and revitalising of the urban environment. It takes into account all the dimensions of space and its material and immaterial data, which form the backbone of sustainable life, identity and image of the city.

Urban regeneration is defined as a long-lasting and sustainable, comprehensive and integrated action aimed at solving the economic, social, ecological and physical degradation of a given space (Roberts 2000). A potential disadvantage and the problem of implementing this strategy is the emergence of gentrification (Čaldarović and Šarinić 2008). It could be avoided if the strategy is in line with socio-economic and ethical norms, especially if the project in the development phase also includes citizens' participation.

Urban regeneration, as a comprehensive process, not only relates to the reconstruction of an individual building, but involves a wider urban area. Moreover, it understands the elements (systems) that are crucial to the process of interaction on the relation between the city and its citizens. The project solution is based primarily on the attitudes of the local population residing within the area foreseen for simultaneous rebuilding and revitalisation, as well as other stakeholders from the public and private domains that work together for the benefit of the community. The urban regeneration project in this context should be comprehensive: implemented and realised in collaboration with experts from relevant areas including sociology, urbanism, architecture, and design of products and communications, so as to achieve simultaneous rebuilding and revitalisation, with a focus on all key elements—"from the spoon to the town", as design at different scales was defined (Rogers 1946). The concept of comprehensive design, or simultaneous design in different scales, is one of the major innovations of modernism within the general industrial modernisation of society, with the intention of eliminating the boundaries between different design models (Kostešić 2017). Is it possible to link this historical idea and urban regeneration strategy today?

Planning practices related to the simultaneous rebuilding and revitalising of historic town cores mostly focus on design on a larger scale—the town level. For the successful reintegration of parts of the urban environment into a functional unit, however, it is necessary to establish a comprehensive method in the horizontal and vertical hierarchies, in order to include the elements of a smaller scale in the process of reconstruction (Vukić et al. 2017). Parallel designing in different scales, "from the town to the spoon", taking into account all material, non-material and symbolic components of the urban space, functional objects and communications, would achieve a coherent environment that supports the integrity and identity of a given space. In the process of regeneration in order to revive and preserve the historic urban space, it is essential to create an environment that meets the needs of the population at a functional and symbolic level. In this regard, a comprehensive and simultaneous approach that combines sociological, psychological, economic and spatial research provides the basis for the realisation of a revitalised urban environment through the synthesis of design methods of urbanism, architecture and design of objects and communications.

The possibilities and the need to link the urban regeneration strategy and the idea of comprehensive design are mostly clear today within the context of current problems in historical town centres, where residents have daily problems due to the strong pressure of tourism activities. Is a comprehensive approach really possible as a strategic method of regenerative urban activity in which various areas of expertise arrive at a synthetic solution? This will be shown in the example of a city in Croatia. But first, the basic features of the concept of comprehensive design need explanation.

6.2 Comprehensive Design—History of the Concept and Its Realisation

The concept of comprehensive design is derived from Buckminster Fuller's "comprehensive designer" whom he defined as a specialist, a "synthesis of artist, inventor, mechanic, objective economist and evolutionary strategist" (Fuller 2001), and combined with the idea of simultaneously planning on different scales, as summarised in the phrase "from the spoon to the town" (Rogers 1946), often ascribed to the mid-twentieth-century Italian architect Ernesto Nathan Roger. In other words, it is a strategy in planning and design practice that integrates methods from urbanism, architecture and design (from large- to small-scale planning, and vice versa), while being grounded upon insights from various scientific fields and sciences relevant to a specific project.

Historically, the idea of simultaneous planning on different scales can be connected to the idea of improving the quality of life by renewing and even reconstructing society through the physical improvement of the built environment by means of integrating art and industry, to which the members of the Deutscher Werkbund were dedicated. To some extent, this idea (and ideal) of designing "from the sofa cushion to town planning" was realised at the Werkbund exhibition in 1927 entitled "Die Wohnung" ("The Dwelling") organised in Weissenhof near Stuttgart. The Weissenhof Estate showcased around 60 housing designs from an international group of architects. Under the direction of Mies van der Rohe, who was in charge of the overall site plan as the appointed architect and coordinator, the exhibition was set to demonstrate the modern family flat equipped with complete furnishings that served as an integrated part of the display. Mies was thus engaged in the project as an urbanist, architect and (industrial) designer.

The exhibition, a testimonial of the emerging International Style, formally focused on social and economic housing that was in part achieved by means of standardisation, and prefabricated, serial architecture. The idea of mass-produced architecture had already been elaborated in Le Corbusier's Maison Citrohan developed in the early 1920s and exhibited at the Weissenhof Estate. The Citrohan house was the result of Le Corbusier's vision of a house as a machine for living, and the name "programmatically referred to that of 'Citroën', thus suggesting the idea of a house designed, produced and marketed like a bus, a ship's cabin, or a car" (Moos 2009), meaning that the house was conceived and treated as a mass-produced industrial product.

In the domain of urban heritage renewal, a comprehensive approach was advocated by Ernesto Nathan Rogers during the post-World War II reconstruction period. In his article from 1946, originally in Italian and translated as "Reconstruction: from objects of common use to towns", Rogers tries to reconcile opposite concepts of individualism and collectivism, utility and beauty, mass production, handicraft and art, and calls upon the responsibility of the architect and all of those who are to transform "any formal representation of existence, from a spoon to a town" (Rogers 1946), to bear in mind both individual freedom and social justice that equally need to be satisfied. Rogers contextualises and identifies key problems and relations regarding the private and the public, and the old and the new in the process of rebuilding towns, societies and civilisations. This kind of a humanistic approach in Rogers' writing is also evident in the way he refers to the home as the "house of man" implying continuously the task at hand—satisfying the material and spiritual need of man and society in general, and, in order to do so, it is inevitable to embrace a method of interdisciplinary, integrated, collaborative and coordinated actions. In this sense, urban regeneration serves as an appropriate strategy.

As a comprehensive strategy in terms of scope and methodology, urban regeneration improves physical, environmental, economic and social issues of spaces and places through actions that integrate multiple disciplines, including architecture, planning and design, as well as social and other relevant sciences. If grounded on sociological insight and through citizen participation, design, urban design, architecture and planning form urban environments that enhance social interaction and cohesion, convey meanings and act as a mean of communicating the non-material and symbolic through material, physical forms. In this way, the results of urban regeneration modify but do not degrade urban heritage and create towns and communities that respectfully fulfil the concepts of sustainability, liveability and identity (Butina Watson and Bentley 2007) that are regarded as essential issues of new urbanism and smart growth. It has been recognised that the success of urban regeneration projects relies on multiple factors, and the public participation of citizens that are day-to-day users of the regenerated space is often regarded as crucial (Colquhoun 1995; Galdini 2005; Elnokaly and Elseragy 2011; Bevilacqua et al. 2013; Clemente et al. 2015; Ertan and Eğercioğlub 2015, 2016). Historic urban areas are especially sensitive in this regard, because they carry values of history, tradition, memory, identity and belonging. As such, they tend to be focal points of the tourism industry, which in many cases causes urban transformations that are oriented towards tourists and to their mental models or presumed images of places, pursuing "authentic experience", often disregarding and/or neglecting the experiences, attitudes, opinions or daily routines of those who live and work in a particular place, simultaneously distorting the practical and symbolic dimensions of urban heritage sites. It is not intended to argue that towns should shut themselves off from tourists, but that the construction of the material and the symbolic should be grounded on the participation of the local community.

Urban regeneration that stands as good examples includes projects conducted in the Mediterranean cities of Barcelona, Genoa, Alexandria, Istanbul, Malaga and İzmir that, like the majority of historic cities, suffered from urban degradation caused by social, economic and technological transformations during the processes of modernisation, industrialisation, de-industrialisation and globalisation. As for Genoa, Barcelona and Alexandria, the problems the cities faced were summarised by El Menchawy (2008) as follows: high urban density, depopulation and isolation of the historic centre, poor housing, economic and environmental conditions, insufficient services and infrastructure, etc. Thus, the regeneration processes had to include a set of actions oriented towards multiple sectors and across several urban areas that resulted in economic and environmental (natural and man-made) improvement, enhanced social equality, accessibility and quality of life. This was especially the case with the urban regeneration project in Barcelona (Elnokaly and Elseragy 2011) which has been referred to as the "Barcelona Model", a concept formed as the Olympic City was being constructed (Montaner 2012).

The city of Barcelona went through several transformations, while the most extensive one started in the 1980s, during the preparations for the 1992 Olympic Games, and continued with urban development projects that began in the early 2000s (Forum 2004). The transformation included improving and building new infrastructure, renovation of the historic city centre and waterfront development (El Menchawy 2008), but an essential part of the "Barcelona Model" was large-scale citizen participation (Elnokaly and Elseragy 2011). Still, recent studies carefully examine the evolution of the "Barcelona Model" and point to some contradictions and difficulties that arose from the regeneration process (Bellini and Pasquinelli 2017) such as excessive tertiarisation in some parts that led to the migration of local citizens from the historic centre, gentrification, transforming public space to private land, and eliminating parts of architectural, social and urban heritage (Montaner 2012).

On the other hand, urban regeneration projects of historic city centres in Malaga (Spain) and İzmir (Turkey) serve as examples of good practice that maintained the balance between urban heritage preservation and urban development, and public and private interests. The projects were set on the principle of the active participation of the local community as an essential part of the process. In the case of both cities, the historic city centre had severe issues regarding physical, functional, economic, demographic and symbolic degradation. The regeneration process in Malaga started in the middle of the 1990s and was based on strong citizen participation. The aim was to increase the quality of urban life by physical, functional, environmental and economic upgrades. In particular, the project included the physical regeneration of buildings, public spaces and infrastructure accompanied by social and economic measures. Similarly, the regeneration process in İzmir (which began in 2008), as in Malaga, included the involvement of the local community in the decision-making and planning processes and encompassed the restoration and renewal of buildings and deteriorated facades, the construction of new infrastructure and urban equipment (Ertan and Eğercioğlub 2016).

As for the state of citizen participation in urban regeneration and planning in Croatia, as a country rich in historical towns, research has shown that the level of public participation is considerably lower (Mišetić 2016), both in comparison with other countries and in general. Another concern that the study revealed is the divergence of theory and practice when it comes to the realisation and implementation of interdisciplinary projects in urban planning. The problem is multifaceted and includes both global changes in type of planning (Madanipour 2010) as well as the local and regional specificities of (post)war and (post)transition experiences. During the socialist period (especially during the 1970s and 1980s) and state controlled urbanism in Croatia, there was a significant number of sociological studies conducted for the purpose of collecting inputs for urbanists about inhabitants' needs which could then be addressed and implemented in different kinds of plans. During and after the transition (democratic and economic change), there was a significant change in approach to planning as well as in addressing the needs of inhabitants. NGOs became the only force to oppose powerful political and economic actors led by their own interests in changing the urban tissue:

> Research findings show an obvious misbalance of power and influence among urban actors in Croatia. Some are very powerful (economic and political actors), some less powerful and sometimes completely useless (civil and professional actors). This is not problematic per se because it is a normal part of social reality, but when it endangers public interest and public space, it is a different matter (Svirčić Gotovac and Zlatar 2015).

Since contemporary post-industrial towns face numerous and complex problems ranging from environmental issues, social and economic challenges, to structural, functional and symbolic degradation, formulating strategies devised on collaborative, integrated and interdisciplinary activities in which citizen participation is a critical factor seems to be the means by which these challenges could be overcome. In the current context of intensive town transformations (social, economic and technological), historical practices based on the concept of comprehensive design might prove encouraging for elaborating new strategies that make citizen participation an imperative.

6.3 Citizen Participation as a Key Factor of Comprehensive Design

"Contemporary city design is a matter not only of iconic architecture, flagship projects or ambitious master-plans, but also of formal and informal practices that shape urban environments, produce and address urban problems, organise people as well as ordering space" (Tonkiss 2013).

There are two crucial points about participation and designing contemporary city and urban space: one is city making as a social process, and the other is social sustainability. Since the 1970s, when the idea of "the right to the city" was introduced by Lefebvre (and maintained by others like Harvey and Soja), the focus on everyday life and the need for something other than a "top-down" view on the city became prevalent. In the 1980s, when Giddens (1984) introduced the theory of structuration in the social sciences, bringing up individual agency and the process of constant structuration of social institutions, relationships and social system, breaking up with the dichotomies of the objective versus the subjective, this new insight into social dynamics influenced the way of thinking in many fields, including urbanism. It is now common to think of urban planning and design as a process concerned with people and their everyday use and (re)creation of urban space (Carmona et al. 2003; Madanipour 2007; Gehl 2010; Tonkiss 2013, etc.), but aside from theoretical approaches the results in creating real "cities for people" (Gehl 2010) are not so great, especially at the global level, and there are few cities with a constant high quality of life and high levels of satisfaction of their population.

However, cities are all (and only) about people after all—"the everyday inhabitants who make and re-make their cities, although not always, it might be said, under circumstances of their own choosing" (Tonkiss 2013). This sets the focus on participation and social sustainability which was recognised as an important part of the regeneration process (Colantonio and Dixon 2011; Manzi et al. 2010), but also a part of European tradition (Carpenter 2013). An integrated approach to regeneration could help defeat the "dark side of regeneration" (Porter and Shaw 2009), because all good activities and improvements come hand in hand with other consequences of regeneration: increased land value and displacement or exclusion of lower income users (the gentrification process in New York that was described back in 1989 by Zukin). As Porter and Shaw point out,

> the negative consequences of urban regeneration strategies (intended or unintended) are relatively unexplored by their advocates. This is not surprising: the beneficiaries of major projects are hardly likely to advertise their disadvantages. We hear few reflections from politicians, city boosters and property developers on the immediate and long-term effects of elimination of low-income people from city centers. There is little evidence in government decision-making of recognition that urban regeneration affects different people differently. Policy-makers rarely display understanding of the social, economic, cultural, environmental and political complexities of urban regeneration (Porter and Shaw 2009).

The problem is even greater in underdeveloped countries. Urban regeneration is a term that is used for all activities ranging from small-scale interventions that improve the quality of life to large-scale activities promoting economic growth. The term is so ambiguous that it can embrace practically every initiative that brings renewal of any kind—physical or economic—but participation of citizens in the process of regeneration can bring a balance and can aim the regeneration forces at the sustainability of the local community. We must note that the notion of social sustainability is a dynamic category that includes relationships between individual actions and the created environment (concerning environmental issues, public health, sustainable production, consumption and transport, as well as social inclusion—both at the local and global levels). According to Manzi et al. (2010), sustainable communities are active, inclusive and safe, well designed and built, well run, thriving, well connected, well served, fair for everyone and environmentally sensitive (Manzi et al. 2010).

Many of the protests and grassroots movements we have witnessed in the last few years in Barcelona, Venice and Dubrovnik could have been avoided if the citizens and their rights and needs were taken into account. Global processes of deterritorialisation and reterritorialisation make fragmented spaces (in combination with social fragmentation) across the globe, as many authors, such as Castells, Sassen, Urry, Soja, Baumann and others, pointed out at the beginning of the twenty-first century, and in this fragmented, complex and dynamic world, local—space and people, community—is becoming increasingly important.

In this context, we agree with Carmona et al. (2003) that urban design as inherently collaborative and interdisciplinary, involving an integrated approach and skills as well as expertise of a wide range of professions is an important tool for urban regeneration based on sustainability principles since "urban design operates at and across a variety of spatial scales rather than at any particular one" (Carmona et al. 2003).

The comprehensive approach as a model of urban regeneration rests on the understanding of regeneration as a continuous process, not individual "episodes" or time-framed "actions". This model of urban regeneration enables the incorporation of sustainability in every phase (or cycle of regeneration) through the integration of ecological, economic and social sustainability, through the continuous process of urban regeneration with the aim of enhancing the quality of life in a certain urban area or the city as a whole. It is important to avoid the problem, according to Madanipour (2007), that comprehensive rational planning faced in an attempt to take into account all relevant actors, activities and connections and "provide a set of connected ideas and actions by taking all relevant matters into account in planning", but "despite heavy reliance on sophisticated information and communication technologies, the result has often been creating rigid and simplistic orders, unable to take into account the complexity and spontaneity of social life" (Madanipour 2007). Another problem was the focus on the aesthetic aspects of the regeneration, but the idea of comprehensive urban design that we promote is based on social concern, in such a way that it can promote "a socio-spatial agenda in which both social and aesthetic concerns matter" and "it can provide arenas in which use value of urban space can be better understood and enhanced", to use Madanipour's (1996) words.

With participation in focus, we must point out another problem, the problem of trust and communication between the main actors in the regeneration process. In Croatia, as in many other post-transitional countries, the level of general trust is low; for example, in Zagreb, according to the Eurobarometer Quality of life in European Cities 2015 (European Commission 2016), people generally feel safe in their city and their neighbourhood and they generally trust their fellow citizens, but they do not trust the city administrative services to help them efficiently (Q2.6): 27% of respondents strongly disagree with the statement, 26% somewhat disagree, and only 7% strongly agree, while 37% somewhat agree, in sum less than 45% of respondents trust the administrative services of Zagreb to help people efficiently. On top of this, there is another potential problem—the problem of communication between citizens and professionals and between citizens and local government. Participation can take different forms, so it is important to ensure that the less powerful actors in the regeneration process are well understood and well represented. Research shows that citizens respond well when asked about local (neighbourhood) problems and their everyday needs, so in a comprehensive regeneration process this would be a starting point (repeatedly in every "micro cycle" of urban regeneration). In the following section, we will present some aspects of the potential of participation in some survey results on the quality of life in the historic core of Dubrovnik.

Social aspects of urban regeneration are closely connected to the process of revitalisation, especially in the historic parts of tourist cities like the Croatian cities of Dubrovnik, Split or Zadar (Vukić et al. 2017).

> Successful tourism-led regeneration policy hinges on the extent to which city governments and local urban elites can sell an authentic city identity, based on physical monuments, cultural heritage and place qualities. If only a city can establish its 'brand', the tourist industry offers a way out of poor economic performance (Porter and Shaw 2009).

However, without a fine balance, historic cities (especially those on UNESCO's World Heritage List) face serious problems with "turistification", depopulation and other problems, because city governments are mostly oriented towards the interests of short-term visitors, forgetting about their own citizens (Bellini and Pasquinelli 2017). This has led to a massive change in population rates and generated other problems like "museification" or "disneyfication" of historic cities. Terms such as the "carrying capacity" and "attraction capacity" of historic centres have been coined, and there are tendencies to finally start thinking about the inhabitants (if there are any left) and trying to find a balance between tourism and everyday life—or the social sustainability of the local community.

Based on a socio-demographic study and survey of the quality of life in the historic core of Dubrovnik[1] (Čaldarović et al. 2018), it is obvious that the local community, facing significant depopulation and deterioration of the quality of life, is on the one hand disappointed (which is in tune with the general lack of trust in institutions) but on the other hand is willing to do something and start the change. They do not see tourism as an obstacle, but ask to be listened to and to finally establish some kind of order and to balance their needs with others. The study was conducted using quantitative and qualitative methods (a survey of 240 inhabitants out of a total population of approximately 1500 people, focus groups and interviews with inhabitants and representatives of NGOs and other stakeholders in the historic core of Dubrovnik). The survey research was conducted in June and July 2017 with 240 participants ($N = 240$), inhabitants of the historic core, of whom 138 (57.5%) were women and 102 (42.5%) were men, 65 (27.1%) were 18–39 years old, 77 (32.1%) were 40–69 years old, and 57 (22.5%) were 70+ years old. The research was based on a questionnaire with 61 questions (most of them scaled questions and 10 open-ended

[1]The study was conducted for the Institute for the Restoration of Dubrovnik as a part of the Management Plan for the historic city core.

questions). The aim of the research was to investigate the quality of life and the level of satisfaction of the inhabitants, as well as to gain information on their attitudes towards tourism and tourists.

The level of satisfaction concerning life in the historic core of Dubrovnik in general is in significant decline if we compare the results with a similar survey conducted in 1986 (Rogić et al. 1986). In 1986, over 80% of inhabitants of the historic core of Dubrovnik were satisfied compared with only 30% today. Nevertheless, when asked to name the most attractive location for living in Dubrovnik, 50.9% of respondents in the 2017 survey answered—the historic core. Very importantly, the majority (60.78% of all respondents) are willing to participate in different kinds of actions aimed at improving quality of life.

Concerning attitudes towards tourism, back in the 1986 survey, the ideal type of tourism in Dubrovnik for inhabitants of the historic core was elite tourism, meaning that all activities, facilities and services in the historic part of the city should be related to culture and high-class tourism. In contrast, in the 2017 survey there is almost no mention of elite tourism; the focus is on the need to regulate cruising tourism (the need to determine the carrying capacity of the city core and to implement relevant measures) and to shift from one-day visitors to tourists who stay longer in the Old City of Dubrovnik.

When asked to name three main disadvantages of living in the historic core of Dubrovnik, respondents pointed out that the historic core is too crowded and too noisy, and there is a serious lack of shops and services for everyday needs. On the other hand, the advantages of living in the historic centre were more abstract and primarily rested on milieu characteristics and the symbols of the Old City of Dubrovnik. These answers are very similar to the answers of the 1986 sociological study of the historic core of Dubrovnik. Based on the survey results, we can conclude that the quality of life in the historic city centre of Dubrovnik is endangered and has deteriorated, but there is significant participatory potential among inhabitants who have concrete suggestions about what should be done and who are willing to participate in any improvement action.

6.4 Conclusion

The concept of comprehensive design is one of the most proactive strategies in creating new liveable places and the renewal of urban heritage. However, to enable a relevant positive change for citizens and other actors, especially in the context of dynamic social, technological and economic turmoil, it is necessary to include these actors and increase their participation in the planning and design process.

Given the complexity of the contemporary social world, the urgent need for creating more sustainable (urban) space and sustainable (local) communities, we argue that it is important to address the issue of urban regeneration in terms of a comprehensive design approach, meaning that the city is a social construct and should be treated as a constant process of (re)structuring (in which every small detail and every citizen is equally important).

Only by creating a high level of satisfaction among inhabitants of a certain urban space is it possible to create a sustainable local community. Choosing between small-scale transformations and large-scale regeneration depends on many factors, but it should always be accomplished through citizens' participation and should not be viewed as contrasting approaches (bottom-up vs. top-down) but as complementary methods with the same goal—urban regeneration based on social sustainability.

Urban heritage sites that stand as places of great historical and symbolic value often tend to succumb to physical, functional and symbolic degradation. As they represent points of constructing and maintaining the sense of identity and belonging of local communities, the regeneration process through which urban heritage sites regain both their material and symbolic functionalities needs to be comprehensive—interdisciplinary and participatory—if it is to fulfil this twofold aim.

References

Bellini N, Pasquinelli C (eds) (2017) Tourism in the city. Towards an integrative agenda on urban tourism. Springer, Heidelberg

Bevilacqua C, Calabrò J, Maione C (2013) The role of community in urban regeneration: mixed use areas approach in USA. In: Proceedings REAL CORP 2013. CORP—Competence Center of Urban and Regional Planning, Schwechat, pp 1361–1365

Butina Watson G, Bentley I (2007) Identity by design. Elsevier, Oxford

Carmona M, Heath T, Taner O et al (2003) Public places—urban spaces. The dimensions of urban design, 1st edn. Architectural Press, Boston, MA

Carpenter J (2013) Sustainable urban regeneration within the European Union. A case of "Europeanization"? In: Leary ME, McCarthy J (eds) The Routledge companion to urban regeneration. Routledge, London and New York, pp 138–147

Clemente M, Castagnaro A, Oppido S et al (2015) Cultural heritage for collaborative urban regeneration: community and stakeholders activations for the historical centre of Naples. BDC Bollettino Del Centro Calza Bini 15(1):93–112

Colantonio A, Dixon T (2011) Urban regeneration and social sustainability: best practice from European cities. Wiley-Blackwell, Chichester

Colquhoun I (1995) Urban regeneration: an international perspective. BT Batsford Limited, London

Čaldarović O, Šarinić J (2008) First signs of gentrification? Urban regeneration in the transitional society: the case of Croatia. Sociologija i prostor 46(3–4):369–381

Čaldarović O, Vukić J, Bogadi S (2018) Life in the historic core of Dubrovnik: a sociological and demographic study. Dubrovnik (in press), Institute for the Renovation of Dubrovnik

El Menchawy A (2008) Urban regeneration in Mediterranean cities: an integrated urban development of brownfield sites. WIT Trans Ecol Environ 117:115–127

Elnokaly A, Elseragy A (2011) Sustainable urban regeneration of historic city centres—lessons learnt. Finnish Association of Civil Engineers RIL. VTT Technical Research Centre of Finland, Helsinki

Ertan T, Eğercioğlub Y (2015) An integrated approach to urban preservation. A suggestion to combine two methods of urban regeneration in order to achieve further developed urban historic center preservation. GSTF J Eng Technol 3(3):57–63

Ertan T, Eğercioğlub Y (2016) Historic city center urban regeneration: the case of Malaga and Kemeraltı, Izmir. Procedia—Social Behav Sci 223:601–607

European Commission (2016) Quality of life in European cities 2015. Flash Eurobarometer 419, Brussels
Fuller RB (2001) The comprehensive designer (1949). In: Krausse J, Lichtenstein C (eds) Your private sky: R. Buckminster Fuller—discourse. Lars Müller Publishers, Zürich, pp 243–246
Galdini R (2005) Urban regeneration process—the case of Genoa, an example of the integrated urban development approach. Available via: ECONSTOR. https://www.econstor.eu/bitstream/10419/117623/1/ERSA2005_426.pdf. Accessed 26 Feb 2018
Gehl J (2010) Cities for people. Island Press, Washington
Giddens A (1984) The constitution of society. Outline of the theory of structuration. Polity Press, Cambridge
Kostešić I (2017) Sveobuhvatni dizajn: podrijetlo i razvoj ideje [Comprehensive design: its origin and evolution]. Prostor 25(1):128–139
Madanipour A (1996) Design of urban space. An inquiry into a socio-spatial process. Wiley, Chichester and New York
Madanipour A (2007) Designing the city of reasons. Foundations and frameworks, Routledge, London
Madanipour A (2010) Whose public space? International case studies in urban design and development. Routledge, London
Manzi T, Lucas K, Jones TL et al (eds) (2010) Social sustainability in urban areas: communities, connectivity and the urban fabric. Earthscan, London
Mišetić A (2016) Društveni aspekti urbane regeneracije: participacija i koncept socijalne održivosti. Hrvatski zavod za prostorni razvoj, Zagreb
Montaner JM (2012) The evolution of the Barcelona model (1973–2004). In: Montaner JM, Álvarez F, Muxí Z (eds) Critical files: the Barcelona model 1973–2004. Ajuntament de Barcelona, Department of Architectural Composition of the ETSAB-UPC, Barcelona, pp 11–26
Moos SV (2009) Le Corbusier: elements of a synthesis, revised and expanded edn. 010 Publishers, Rotterdam
Porter L, Shaw K (eds) (2009) Whose urban renaissance? An international comparison of urban regeneration strategies. Routledge, London
Roberts P (2000) The evolution, definition and purpose of urban regeneration. In: Roberts P, Sykes H (eds) Urban regeneration: a handbook. SAGE, London, pp 9–36
Rogers EN (1946) Ricostruzione: dall'oggetto d'uso alla città. Domus 215:2–5
Rogić I, Sarić B, Kovačević R et al (1986) Dubrovnik - sociološka studija stare gradske jezgre [Dubrovnik – old city core. Sociological study] Urbanistički institut SRH, Zavod za prostorno planiranje, Zagreb
Svirčić Gotovac A, Zlatar J (2015) Urban processes in Zagreb. Residential and commercial developments. Hrvatski geografski glasnik 77(1):29–45
Tonkiss F (2013) Cities by design: the social life of urban form. Polity, Cambridge
Vukić F, Jukić T, Podnar I et al (2017) Living historical city strategy: sustainable tourism as creative practice. In: Proceedings of the 6th international conference of Arte-Polis. Imagining experience: creative tourism and the making of place. Springer Singapore, Singapore, pp 279–288

Chapter 7
Virtual Restoration and Preservation of Anthropogenic Nineteenth-Century Landscapes Based on Historical Land-Use Data

Tomaž Berčič

Abstract Historical maps are a valuable source of information. The Franciscan cadastral maps comprehensively cover the entire territory of the Austro-Hungarian Empire. They are a trustworthy source of historical land-use data and information regarding past urban-planning concepts, especially due to their high geometrical precision. Studying and analysing historical land use makes it possible to reconstruct the historical cultural landscape, which is itself a significant element of cultural heritage. In studying historical maps, the most widely used analytical approach is the comparative method. Matching and overlaying historical and contemporary map data makes it possible to see and analyse changes in land use. This technique has been widely adopted and is now well established, but advanced technologies like high-resolution Lidar terrain data offer new opportunities. It is postulated that especially in rural areas, unlike land use, the terrain has not changed significantly, and the combination of detailed three-dimensional terrain data and historical land use can yield novel views on the man-made landscape of this era. The methodology presented can generate other interpretations of the available data. A combination of new technologies and techniques can provide an accurate virtual historical view and comparison with the same location in the present. The research presented offers fresh insights into what the cultural landscape looked like in the past and what actions can be taken to preserve it in the future.

Keywords Land use · Landscape · Franciscan Cadaster · Lidar · Urban simulation

7.1 Introduction

Vast amounts of historical data exist that are unquestionably useful for the study and interpretation of history but are usually too complex to be narratively interesting. Until recently, research on the comparison and visualisation of historical land-use maps combined with current land-use data also had this problem. Work in

T. Berčič (✉)
Faculty of Architecture, University of Ljubljana, Ljubljana, Slovenia
e-mail: tomaz.bercic@fa.uni-lj.si

M. Obad Šćitaroci et al. (eds.), *Cultural Urban Heritage*,
The Urban Book Series, https://doi.org/10.1007/978-3-030-10612-6_7

this field constantly seeks to pursue the ideal objective: to reconstruct an accurate three-dimensional view of the past.

Heritage denotes everything that has been handed down from the past. Although not all heritage is uniformly desirable, it is widely considered a precious and irreplaceable resource, essential to personal and collective identity (Lowenthal 2005). A valuable historical source frequently used for identifying and analysing cultural landscapes in Central Europe is the repository of maps of cadastral districts prepared under Emperor Francis I of Austria, known as the Franciscan Cadaster. Historical maps present a state in time and are therefore part of cultural heritage. To make sense of them, they are converted to digital format (Dobernik 2002). Land-use research can have a profound impact on the study of urban, agricultural, environmental, and local history, genealogy, traffic history, urban and landscape planning, regional planning, legal cases, archaeology, hydrology, environment and landscape protection, and registration and protection of cultural heritage, to name just a few areas (Berčič 2016). The consumption and use of digital data are an exponentially growing part of today's society; GIS technology is useful because it offers a powerful set of tools for collecting, storing, analysing, and visualising spatial data. Over the past decades, the concept of cultural heritage has expanded to include various materials and other objects that are significant to specific groups of society (Maiellaro and Lerario 2005).

The European Landscape Convention asks countries to recognise that all landscapes possess heritage values of one kind or another, at one level or another, and that these values all demand careful consideration in the long-term management and evolution of all landscapes (Council of Europe 2003). Terraced landscapes are a part of the vernacular landscape that evolved through use by the people whose activities or presence shaped that landscape. Terraces are part of cultural history that still exist and in many places are still used today. In the past, making terraces was labour intensive and communities depended on them for food production or water retention. Today, in some areas in Slovenia, terraces are still being built because of their advantages and aesthetic properties (Ažman Momirski 2018). The use of construction and agricultural machinery has changed the modes of terrace production and construction, and consequently the image of the terraced landscape. Some of the most beautiful terraced landscapes in Slovenia emerged only after the 1960s (Loeper et al. 2016).

In combining historical data sources with new technologies such as light detection and ranging (Lidar) and new geospatial software and techniques, it is possible to more accurately simulate past landscapes, and this new historical urban simulation can provide new interpretations and opportunities to restore and preserve man-made nineteenth-century landscapes based on historical land-use data. In 2011, Slovenian Government agencies commissioned the laser scanning of the country's territory with a resolution between two and ten points per square metre. This method resulted in a real revolution in topographical surveying (Mongus et al. 2013). Thanks to Lidar digital terrain models (DTMs), it is now possible to detect and define a terrace's range, size and boundary quickly and with great accuracy. The anthropological component of Lidar has extremely important research value in discovering abandoned overgrown landscape features, which was previously not possible without a great deal of fieldwork and enormous amounts of time. Lidar has surpassed other topographical data in its usefulness, and it has become the most important platform for landscape research, monitoring, and management, landscape archaeology, and landscape anthropology (Berčič 2016).

Humans have always wanted to visualise the problems they face. John Snow's (1813–1858) two-dimensional 1854 map of the cholera outbreak in London is generally considered the first example of the analogue use of geospatial data to solve a problem (Johnson 2006). Professionals in the past also tried to visualise changes in the landscape from the human point of view. The landscape designer Humphry Repton (1752–1818) was the first to present his designs to investors in three-dimensional form with watercolour pictures of before-and-after vistas of the projects (Vercelloni 1994). Accurate three-dimensional models have appeared relatively recently due to the availability of data and computer power. In 2004, a team from Japan produced a bird's-eye view of old Tokyo by relating historical maps to a DEM (Fuse and Shimizu 2004). A great achievement was made with a then-and-now view of a vanished landscape represented by the *Mannahatta Project*, which compares the landscape changes in the Manhattan Peninsula between 1609 and today (Sanderson and Brown 2007). Italian and German researchers that were part of the European CyberWalk project developed the largest virtual reality (VR) walking platform in the world. For this purpose, they produced a novel shape grammar for the procedural modelling of computer generated (CG) architecture. CGA can efficiently generate large urban models with an unprecedented level of detail, with the virtual rebuilding of the archaeological site of Pompeii (Müller et al. 2006). Cornell University's Department of City and Regional Planning was awarded a grant to explore the potential for four-dimensional GIS modelling to support historical preservation in the context of the important cultural landscape of Flushing Meadows Corona Park in New York's Queens Borough (Minner et al. 2017).

7.2 Hypothesis

Today we are beyond the initial stages of historical land-use comparisons and visualisations. Based on experience and research on land use, it is possible to recognise the developmental stages of this digital but still mainly manual process. Researchers are converging from a time-consuming process to a semi-automated procedure of converting digitised historical sources and preparing them for analysis and simulation. New technologies make it possible to convert troves of recently digitised archival data for computer analysis and visualisation, and to transition from two-dimensional to three-dimensional synthetic environments.

The claim is that, in areas outside urban centres in Slovenia, the terrain (disregarding the infrastructure) has not changed significantly in the last two hundred years. The combination of detailed three-dimensional terrain data and historical land use can yield novel views on the man-made landscape of a past era. An assortment of new GIS technologies such as Lidar data and new geospatial software can provide an accurate virtual window into the past and allow comparison with the present. It represents a significant improvement to a well-established analysis procedure for the virtual restoration and preservation of anthropogenic nineteenth-century landscapes. Through this process, changes in land use can be seen, interpreted and compared in a more familiar, understandable three-dimensional form. This offers new leverage for preserving, presenting, interpreting, managing and protecting cultural landscape heritage.

7.3 Methodology

The workflow to analyse the historical data is well established and widely used with some variations. The basic data needed for the process are the digital orthophotograph images available from the Slovenian Surveying and Mapping Authority (GURS1 2018; GURS2 2018), the Franciscan Cadaster available as scanned pages from the Archives of Slovenia and current land-use data available online from the Ministry of Agriculture, Forestry, and Food.

First the separate raster files of the Franciscan Cadaster sheets data need to be assembled in Adobe Photoshop. It is important not to digitally distort the original data when assembling the sheets, which have not all aged in the same way over the last two hundred years. Through the years, the physical pages have gained or lost a few millimetres depending on the amount of watercolour used, the storage conditions and the paper composition. The composed digital raster needs to be imported into GIS software; in the case at hand, this is ESRI's ArcGIS Desktop 10.5 and ArcGIS Pro 2.1. The files need to be georeferenced and rectified based on orthophotograph imagery and cadastral or parcel borders. The raster files need to be analysed according to the key (Table Fig. 7.2b), and land-use attributes are added. When this work is completed, the historical and current land use can be compared. To accomplish this, the land-use attributes need to be matched and the historical and contemporary land-use categories need to be harmonised. It is necessary to create a separate spreadsheet file that joins the relevant (*.shp) files or geodatabase with the new harmonised attributes.

Years of involvement in research on man-made cultural landscapes, especially terraced areas, have shown a gradual increase in complexity of visualisation. There are noticeable stages, or "generations", in visualising the landscape and development usually based on the hardware and software tools needed to accomplish it.

The "first generation" of the comparison of land-use data was a confluence of various developments in digital data and IT technologies. The first generation coincides with the digitisation of various data, the rise in computer power, and the first useful and feature-rich GIS software. The greatest achievement was to transfer the fragile historical data to a digitised environment and make it accessible to researchers in digital form while preserving the originals for posterity. The digitisation of the historical sources was the first step to popularising historical land-use research. The Archives of the Republic of Slovenia was the first institution on the territory of the former Austro-Hungarian Empire to completely digitise such holdings and make them available on the Internet. The project started in 1998 and lasted until 2008 (Kačičnik Gabrič 2008).

At about the same time as the historical data became available, the satellite data essential for georeferencing the historical sources also became accessible. Slovenia was fully covered for the first time in August 2001 with black-and-white orthophotographs (DOF5), and in 2006 with colour and colour infrared orthophotographs (GURS1 2018).

The first map of land use for all Slovenian territory was created in 1993 by the Statistical Office of the Republic of Slovenia (Petek 2004). Since 2002, the Slovenian Ministry of Agriculture has offered online access to vector GIS land-use data.

Essential for the comparison analysis is the harmonisation of land-use categories between the Franciscan Cadaster and contemporary data. After two hundred years, the categories are still similar enough for a direct comparison to be possible.

The general workflow of the first generation starts with georeferencing the historical source files (in this case, the Franciscan Cadaster) with the ESRI ArcGIS Desktop software. Then, the georeferenced cadaster is transformed manually to a vector form. Each polygon drawn is assigned an attribute value that corresponds to the land-use value on the historical source. At the end of the process, a digital copy of the raster file is generated. One of the most important criteria for interpreting land-use changes in the first generation is the unification table created by Franci Petek (Fig. 7.1b). It was primarily made to compare agricultural land use, but some of the basic categories were missing or were joined together. The first generation of land-use comparison methods did not focus on mapping symbols such as colour and visibility, or on other aspects such as detailed analysis of built-up areas or water bodies.

The "second generation" improved the visualisation results of the first generation with an expanded unification table, colour definition, and terrain analysis of slope, aspect, and elevation models and the first attempts at a three-dimensional visualisation form. In this phase, an expanded version of the harmonisation table was prepared. It expanded on some of the land-use attributes that were previously joined under the attributes. The attributes of water and built-up land gained in importance. The focus of the comparison table shifted from being completely agriculturally oriented to a broader spectrum of attributes offering new contextual information, especially for researchers not strictly interested in agricultural development but also in studying the urban form and socio-economic development of settlements. The colour values for the various attributes for two-dimensional visualisations were also agreed upon among the researchers; the CMYK values for the separate attributes can be seen in Fig. 7.2b.

For the first time, it was possible to simulate the three-dimensional morphology of the terrain with orthophotograph imagery or land-use data mapped straight onto the terrain, derived from a DEM in a 5 × 5 m grid resolution. The visualisation offered a general three-dimensional overview of the terrain morphology without much detail. The strengths of the additional three-dimensional terrain modelling were that it made it possible to conduct a new analysis of terrain slope, aspect and elevation. The maps also provided a new visualisation underlay, enabling new interpretations and research conclusions on land-use changes to be made.

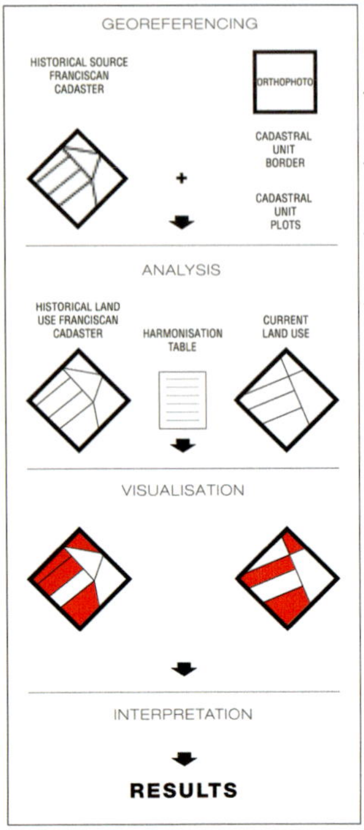

(A) THE FIRST-GENERATION WORKFLOW

BASIC CATEGORIES FRANCISCAN CADASTER	BASIC CATEGORIES OF CURRENT CADASTER	HARMONISED AND GROUPED CATEGORIES	CMYK VALUES (C, M, Y, K)%
ARABLE LAND, GARDEN	ARABLE LAND, GARDEN, TEMPORARY MEADOW	ARABLE LAND	RANDOM
ARABLE LAND WITH VINEYARD, VINEYARD WITH OLIVE TREES	VINEYARD, OLIVE GROVE	VINEYARD	RANDOM
MEADOWS WITH FRUIT TREES, FRUIT GARDEN	EXTENSIVE ORCHARD, INTENSIVE ORCHARD	ORCHARD	RANDOM
MEADOW, PASTURE, PASTURE WITH TREES	PERMANENT MEADOW AND PASTURE	GRASSLAND	RANDOM
FOREST, SHRUBS	AGRICULTURAL LAND OVERGROWN BY FOREST TREES, LAND OVERGROWN WITH TREES AND SHRUBS, FOREST	FOREST	RANDOM
BARREN LAND, BUILDING COURTYARD, TRAIL, ROAD, WATER BORDER	WATER, DRY OPEN LAND WITH SPECIAL PLANT COVER, OPEN MARSH LAND, BUILT-UP AND RELATED LAND	OTHER	RANDOM

(B) HARMONISED COMPARISON TABLE

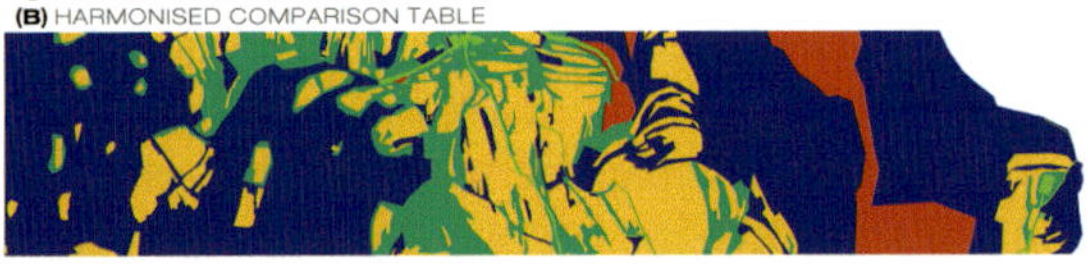

(C) THE FIRST-GENERATION VISUALISATION

Fig. 7.1 **a** Diagram of the first-generation land-use comparison workflow; **b** harmonised comparison table for converting historical land use and contemporary categories to match; **c** visualisation of the first-generation land-use analysis

The current "third generation" of cultural landscape visualisation is a significant improvement with Lidar terrain data and new software techniques in three-dimensional visualisations.

The land-use comparison process follows the same workflow for georeferencing the historical maps as the first two generations. A noteworthy advance from the second generation is the use of Lidar data to generate terrain data. A quantum leap in software development based on Lidar also makes it possible to visualise and interpret a nearly perfect representation of the character of the historical landscape (Fig. 7.3).

7.4 Results

The different visualisation workflows were tried on the pilot area of Golo. The cadastral district is twenty kilometres from the centre of the capital of Slovenia, Ljubljana. Although the purpose of the research is historical visualisation, a by-product of the land-use comparison can also be quickly summarised here.

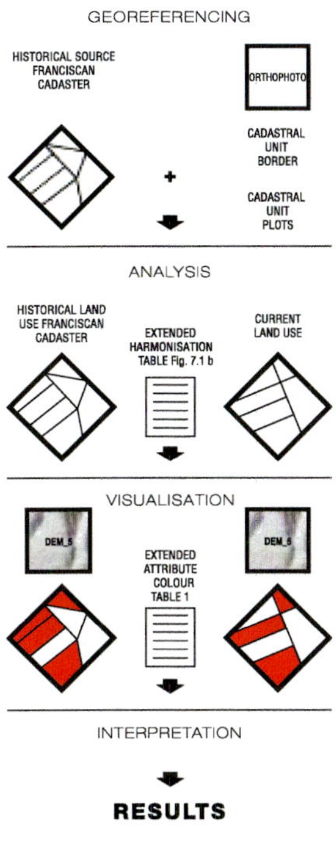

BASIC CATEGORIES FRANCISCAN CADASTER	BASIC CATEGORIES OF CURRENT CADASTER	HARMONISED AND GROUPED CATEGORIES	CMYK VALUES (C, M, Y, K)%
ARABLE LAND, GARDEN	ARABLE LAND, GARDEN, TEMPORARY MEADOW	ARABLE LAND	30%, 50%, 100%, 10%
ARABLE LAND WITH VINEYARD, VINEYARD WITH OLIVE TREES	VINEYARD, OLIVE GROVE	VINEYARD	80%, 100%, 20%, 10%
MEADOWS WITH FRUIT TREES, FRUIT GARDEN	EXTENSIVE ORCHARD, INTENSIVE ORCHARD	ORCHARD	0%, 40%, 100%, 0%
MEADOW, PASTURE, PASTURE WITH TREES	PERMANENT MEADOW AND PASTURE	GRASSLAND	40%, 0%, 100%, 0%
FOREST, SHRUBS	AGRICULTURAL LAND OVERGROWN BY FOREST TREES, LAND OVERGROWN WITH TREES AND SHRUBS, FOREST	FOREST	80%, 30%, 100%, 20%
BUILDING COURTYARD, TRAIL, ROAD	BUILT-UP AND RELATED LAND	BUILD-UP LAND	0%, 0%, 0%, 100%
WATER	WATER	WATER	80%, 0%, 0%, 20%
BARREN LAND, BORDER	DRY OPEN LAND WITH SPECIAL PLANT COVER, OPEN MARSH LAND	OTHER	0%, 0%, 0%, 20%

B. EXTENDED HARMONISED COMPARISON TABLE 1

C. THE SECOND-GENERATION VISUALISATION

Fig. 7.2 **a** Diagram of the second-generation land-use comparison workflow; **b** expanded harmonised comparison Table Fig. 7.1b for converting historical land use and contemporary categories to match; **c** visualisation of the second-generation land-use analysis

The cadastral unit of Golo consists of two small settlements, Golo and Škrilje, in the municipality of Ig and has an area of around 1609 ha. Golo today is a mainly rural settlement of a suburban commuter character. A significant difference in elevation creates a different microclimate. Golo is on average 350 m higher than the Ljubljana Basin. The name *Golo* refers to barren, cleared land with no forest. It also has some very good examples of terraced landscapes close to the centre of Slovenia.

One of the most obvious results of the visualisation and analysis of land use is the age of the terraced areas. Today the terraced areas are still farmed as large meadows for fodder. In the past, according to the Franciscan Cadaster created in 1825, they were used as fields. The parcel boundaries still match, which means the terraces were created by hand and not by mechanisation. They were not a post-Second World War phenomenon, as in some other parts of Slovenia (Ažman Momirski 2018). An interesting detail is that the terraces mostly have a northern exposure, probably because of the more favourable microclimate. This finding only confirms that the terraces are at least two hundred years old, but it does not indicate whether they are significantly older than that.

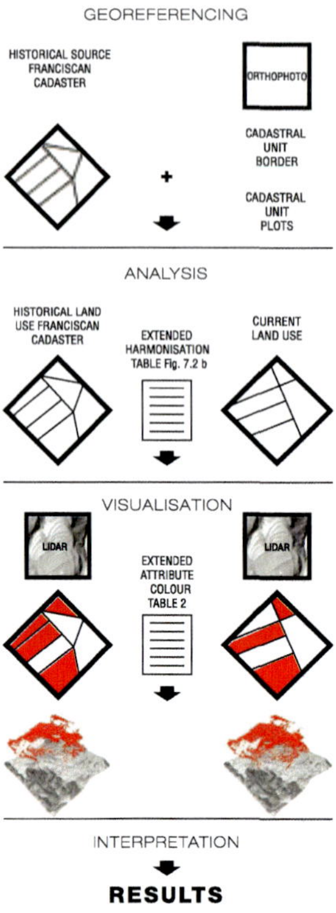

A. THE THIRD-GENERATION WORKFLOW

BASIC CATEGORIES FRANCISCAN CADASTER	BASIC CATEGORIES OF CURRENT CADASTER	HARMONISED AND GROUPED CATEGORIES	CMYK VALUES (C, M, Y, K)%
ARABLE LAND, GARDEN	ARABLE LAND, GARDEN, TEMPORARY MEADOW	ARABLE LAND	30%, 50%, 100%, 10%
ARABLE LAND WITH VINEYARD, VINEYARD WITH OLIVE TREES	VINEYARD, OLIVE GROVE	VINEYARD	80%, 100%, 20%, 10%
MEADOWS WITH FRUIT TREES, FRUIT GARDEN	EXTENSIVE ORCHARD, INTENSIVE ORCHARD	ORCHARD	0%, 40%, 100%, 0%
MEADOW, PASTURE, PASTURE WITH TREES	PERMANENT MEADOW AND PASTURE	GRASSLAND	40%, 0%, 100%, 0%
FOREST, SHRUBS	AGRICULTURAL LAND OVERGROWN BY FOREST TREES, LAND OVERGROWN WITH TREES AND SHRUBS, FOREST	FOREST	80%, 30%, 100%, 20%
BUILDING	BUILDINGS	BUILDINGS	0%, 0%, 0%, 100%
COURTYARD, TRAIL, ROAD	BUILT-UP AND RELATED LAND	ROAD NETWORK	0%, 0%, 0%, 100%
WATER	WATER	WATER	80%, 0%, 0%, 20%
BARREN LAND, BORDER	DRY OPEN LAND WITH SPECIAL PLANT COVER, OPEN MARSH LAND	OTHER	0%, 0%, 0%, 20%

B. EXTENDED HARMONISED COMPARISON TABLE 2

C. THE THIRD-GENERATION VISUALISATION

Fig. 7.3 **a** Diagram of the third-generation land-use comparison workflow; **b** further expanded harmonised comparison Table Fig. 7.2b for converting historical land use and contemporary categories to match; **c** visualisation of the third-generation land-use analysis

A quick historical land-use comparison shows that the extent of forest has increased from 857 ha (53% of the cadastral area) in 1825 to 1309 ha in 2017. This corresponds to 81% of the cadastral area and is an increase of 52% in absolute numbers, or an increase of 452 ha, corresponding to almost a third of the entire cadastral area. The most significant decrease was in the extent of the arable land (i.e. fields), from 92 ha in 1825 to only 4.6 ha in 2017, or a decline of 95% of the total area. Orchards decreased from 55 ha to only 8.6 ha, and grassland decreased from 37% of the area (597 ha) to 15% (237 ha) in 2017. The only land use that has increased in addition to forest is built-up land, from 1% in 1825 to 3% in 2017. In general, it can be observed that the arable land and orchard have been transformed into grassland, and the previous pastureland has become overgrown with forest (Fig. 7.4).

During the years covered by the study of terraced cultural landscapes, a rise in complexity of visualisation was observed. There are also noticeable "generations" in visualising the past landscape and in the development of hardware and software along with it.

Fig. 7.4 **a** View of the terrain morphology of the Golo settlement based on Lidar data; **b** historical three-dimensional view of the Golo settlement according to land-use analysis from 1825; **c** contemporary three-dimensional view of the Golo settlement according to land-use analysis from 2017

The "first-generation" landscape visualisation results follow the established land-use comparison method presented in the workflow diagram (Fig. 7.1). The visualisation results of this first generation show the difference between land-use changes. It was an effective tool to show the biggest impact of land-use change that could be shared with the general public and not just with experts with a trained eye. Generally, no effort was made to contextualise the information with colour or explain the abstract maps in any other way. The colours in use were usually the default values given by the software. The fact that an analysis was even possible, that the data were available in digital form and that the hardware was starting to be able to support this kind of research was exciting.

In the "second generation", much more detail was given to presenting the results. This is why the harmonisation table was expanded with further categories to better present and interpret values other than merely agricultural change. Researchers from other fields, not just geographers, started to adopt GIS technology. They started to look at other factors that also influenced land-use changes, such as the shape and size of settlements, in conjunction with the aspect and slope analysis of the terrain, the road network and accessibility, and how all of these criteria impacted on social and economic changes. The three-dimensional terrain model, which mainly served as a visual prop for presenting the two-dimensional images, was not useful as an interpretation tool because it was too abstract. The main reason for this was that it was too coarse. The resolution of 5×5 m terrain DEMs was too imprecise to be useful, and there was also always the computer power to consider.

The main improvements in the "third generation" lie in the three-dimensional realm. The key difference is a switch to ESRIs ArcGIS PRO, which replaces the ArcGIS Desktop and has new software for visualising geodata called CityEngine. CityEngine can directly access ArcGIS PRO's geodatabases, which ensures deep integration of two-dimensional vector and raster data. It can import a detailed terrain raster created in ArcGIS PRO, and all of the other data can be fitted to this terrain. The other significant upgrade to CityEngine is its parametric nature. The benefit can be seen in using or mapping real photographic images of the ground for ground cover, matching the land-use attributes. To upgrade the planar ground cover, the software allows the user to define in detail the vegetation cover in three dimensions. For different attributes, the CE can define a very specific density and mix of plants and trees. The third upgrade from the second-generation three-dimensional visualisation is the generation of buildings and roofs based on only outlines and parametric information.

7.5 Discussion

Visual representations are an essential part of how architects and landscape designers carry out their work. However, researchers, especially archaeologists and historians, are also interested in visual representations and visual models of things from the past that became lost or buried, or are uncertain (Rekittke and Paar 2010).

One of the major weaknesses is the amount of manual labour a researcher must still put into analysing the historical source. CityEngine has shown potential as a platform for storing, analysing and visualising vegetation data, especially if it is combined with the vegetation library that it contains and the ability to add custom elements to enhance the visualisations. However, CityEngine does not go as far as a GIS in supporting the ability to form queries based on attributes or spatial analysis. This is where the further development of a GIS that has built-in capabilities for query and analysis, such as two-dimensional GIS functionality, could be better supported and integrated with three-dimensional modelling capabilities. CityEngine does allow for interoperability with some commonly used architectural platforms. However, difficulties were experienced in importing textured architectural models from Rhino, as well as from Maya, into CityEngine. Textures were often stripped from the buildings upon import. Detailed information from building information models (BIM) cannot be imported into CityEngine. CityEngine was developed as an engine to easily generate cityscapes, and its focus is on accurate visualisation. Unlike two-dimensional GIS platforms, such as ArcGIS or QGIS, it does not include analytical tools (Minner et al. 2017).

Technology is coming increasingly closer to attaining almost perfect visualisation. To perfect it, contextual information is needed to recreate the character of the visualisation. Interesting research is taking place in this field. The proposed spatial identity presence index is one of the indicators through which stakeholders in cultural heritage management could monitor and even manage the public perception of the wider context of built heritage (Juvancic and Verovsek 2017).

7.6 Conclusion

In a 1998 speech, Al Gore said, "I believe we need a 'Digital Earth'. A multi-resolution, three-dimensional representation of the planet, into which we can embed vast quantities of geo-referenced data" (Gore 1998). His statement has partly already become a reality. Much work has been invested in the conversion from analogue to digital data. Large troves of these data are still fragmented and unconnected to a larger database and go completely unused or are even repeated multiple times because researchers are unaware of one another and the work that others do. Accurate scientific three-dimensional landscape visualisations make sense. They are an appropriate medium for translating data from past landscapes into understandable pictures for specialists and the public. There is also the potential for crowdsourcing elements of past and present landscapes, which was recognised by the research team as an important way to get people involved in stewardship of the landscape (Minner et al. 2017).

We as society are at the next step of evolution regarding the regulation, evaluation, management and preservation of cultural heritage data manipulation. The first industrial revolution used heat and steam to manufacture. The second used electricity for mass production. The third used electronics and IT to automate production.

The fourth industrial revolution is characterised by a blend of technologies that is blurring the lines between the physical, digital and biological domains. The speed of current breakthroughs has no historical precedent. The current industrial revolution is evolving at an exponential rather than a linear rate and is disrupting almost every field everywhere. The scope of these changes signals the transformation of entire systems of production, management and governance (Schwab 2017). With the developments in today's almost unrelated technological fields such as gaming, procedural modelling, and hardware, a phase is approaching in which automatic visualisation combined with added attribute data will become effortless and automatic.

However, we are still at the start of the new industrial era, with new developments in artificial intelligence and machine learning, new methods of gathering digital data, such as photogrammetry, Lidar and different monitoring methods like ground penetrating radar, or magnetic resonance. All these tools give researchers new insights into this opulent data trove of physical and digital artefacts. As physical and digital data continue to converge, new tools and platforms will enable people to engage with regulating bodies, voice their opinions and coordinate their efforts over cultural heritage matters.

At the same time, regulators will not only gain new technological powers to increase their control over the gathered data, based on surveillance systems and the ability to control the digital infrastructure, but they can, if they so choose, use the same newly gained powers to maintain, preserve and manage cultural heritage, by embracing "agile" governance. This means regulators must continuously adapt to a new, fast-changing environment, reinventing themselves, so they can truly understand what it is they are regulating (Schwab 2017).

The course of human life in the past was greatly different from its course today. Life went by without digital technologies, which modern times take for granted. If the large repositories of today's digital information are not lost or corrupted, images and video will preserve today's way of life in incredible detail, and these vast amounts of digital data will become the cultural heritage of tomorrow.

References

Ažman Momirski L (2018) Slovenian terraced landscapes. In: Varotto M, Bonardi L, Agnoletti M, Tarolli P (eds) World terraced landscapes: history, environment, quality of life. Springer (in print)

Berčič T (2016) Discovering terraced areas in Slovenia: reliable detection with Lidar. Annales, Series Historia et Sociologia 26(3):449–468

Council of Europe (2003) European landscape convention. https://www.coe.int/en/web/conventions/full-list/-/conventions/rms/0900001680080621. Accessed 2 June 2018

Dobernik M (2002) Reproduciranje franciscejskega katastra. Arhivi 25(2):65–76

Fuse T, Shimizu E (2004) Visualizing the landscape of old-time Tokyo (Edo City). In: Gruen A, Murai Sh, Fuse T, Remondino F (eds) ISPRS XXXVI-5/W1. Processing and visualization using high-resolution images. Pitsanulok, Thailand, November 2004. http://www.isprs.org/proceedings/XXXVI/5-W1/papers/21.pdf. Accessed 2 June 2018

Gore A (1998) The digital earth: understanding our planet in the 21st century. Aust Surveyor 43(2):89–91

GURS1 (2018) E-prostor – Ortofoto. http://www.e-prostor.gov.si/zbirke-prostorskih-podatkov/topografski-in-kartografski-podatki/ortofoto/. Accessed 17 Feb 2018

GURS2 (2018) E-prostor – Digitalni model višin z ločljivostjo (DMV 12,5, DMV 25, DMV 100). http://www.e-prostor.gov.si/zbirke-prostorskih-podatkov/topografski-in-kartografski-podatki/digitalni-model-visin/digitalni-model-visin-z-locljivostjo-dmv-125-dmv-25-dmv-100/. Accessed 24 Feb 2018

Johnson S (2006) The ghost map: the story of London's most terrifying epidemic—and how it changed science, cities, and the modern world. Penguin, London

Juvancic M, Verovsek S (2017) Spatial identity (re)constructed from web-sourced imagery: comparing expert opinion with quantitative query. J Cult Heritage Manage Sustain Dev 7(2):193–207

Kačičnik Gabrič A (2008) Dragoceni original ni več izpostavljen obrabi: digitalizacija gradiva franciscejskega katastra v Arhivu RS. Delo, January 10th

Loeper N, Ott M, Momirski LA (2016) Terraced landscapes: new design solutions within the transformation of artificial landscapes. Annales, Series Historia et Sociologia 26(3):523–536

Lowenthal D (2005) Natural and cultural heritage. Int J Heritage Stud 11(1):81–92

Maiellaro N, Lerario A (2005) Cultural heritage and web-mapping. Paper presented at the 8th AGILE Conference on GIScience, Estoril, Portugal, 26–28 May 2005

Minner J, Chusid J, Shi X, Feng Y, Sun F, Azel N, Taru T (2017) Visualizing the past, present, and future of New York City's 1964–5 world's fair site using 3D GIS and procedural modeling. Cornell University, Ithaca

Mongus D, Cekada MT, Zalik B (2013) The analysis of an automatic method for digital terrain model generation from Lidar data on Slovenian test cases. Geodetski vestnik 57(2):245–259

Müller P, Wonka P, Haegler S, Ulmer A, Van Gool L (2006) Procedural modeling of buildings. ACM Trans Graph 25(3):614–623

Petek F (2004) Land use in Slovenia. In: Orožen Adamič M (ed) Slovenia: a geographical overview. Association of the Geographical Societies of Slovenia, Ljubljana, pp 105–108

Rekittke J, Paar P (2010) Past pictures: landscape visualization with digital tools. In: Bloemers T, Kars H, Van der Valk A, Wijnen M (eds) The cultural landscape & heritage paradox: protection and development of the Dutch archaeological-historical landscape and its European dimension. Amsterdam University Press, Amsterdam, pp 309–320

Sanderson EW, Brown M (2007) Mannahatta: an ecological first look at the Manhattan landscape prior to Henry Hudson. Northeast Nat 14(4):545–570

Schwab K (2017) The fourth industrial revolution. Crown Business, New York

Vercelloni V (1994) Historischer Gartenatlas: Eine europäische Ideengeschichte. Deutsche Verlags-Anstalt, Stuttgart

Chapter 8
On the Edge of Protection: Archaeology and Territory, Culture and Landscape

Giovanni Azzena

Abstract This paper proposes an evaluation and re-reading of the common problems that characterise the difficulties in the practice of protecting and enhancing the territory, with particular focus on the archaeological field. Despite the widespread idea that archaeology should always be included in the territorial project, the link between protection and planning seems to be more evoked and required than programmed and realised. Current solutions define a concept of structured protection for the undifferentiated preservation of goods, where archaeological evidence with different characteristics (position, materiality, collective perception) seems to match univocal and indifferent solutions. This depends, of course, on what is meant to be protected: a single antique object may be subject to simple protection. However, everything is more complicated if the single monument turns out to be an area, an archaeological area, or where the object to be protected is the landscape. First of all, this is so because it is still extremely complicated to give it a unique and shared definition, secondly, because it is a theme that requires a review of the analytical approach, to subordinate it and make it operational on the observation scale. Therefore, this contribution attempts to provide a different approach to territorial analysis and to archaeological and landscape protection.

Keywords Archaeology · Landscape

Some time ago, I titled one of my works "*Le nozze di Cadmo e Armonia, ovvero: storia del paesaggio e progetto del territorio*" [The Wedding of Cadmus and Harmony: landscape history and the territorial project]. It was an attempt to discuss the protection and project of the territory (in other words, to discuss the relationship between the past and the future of places). Even if the title was changed in the meantime,[1] I thought it was meaningful in order to recall the first wedding in the myth, which was so complicated and also so rich in unexpected developments, although not all of

G. Azzena (✉)
Department of Architecture, Design and Planning, University of Sassari, Sassari, Italy
e-mail: azzena@uniss.it

[1] The title of that paper has become: "Fragilità dei forti. Conoscenza, tutela, progetto del paesaggio agrario 'storico'" (Azzena 2010a).

M. Obad Šćitaroci et al. (eds.), *Cultural Urban Heritage*,
The Urban Book Series, https://doi.org/10.1007/978-3-030-10612-6_8

them were positive, and not always were they so. Furthermore, I was thinking about a particular gift that Cadmus received from Apollo, namely: being able to convert sound into harmony and the *right music*. In brief, I had imagined an auspicious title, although one based on the acceptance of a fundamental problem.

Nowadays, the marriage of protection and planning seems to be more evoked—-more and more dangerously—than celebrated and consummated. What people usually say about two people who have never met: how nice they would be together! Such a perfect couple! Then, a marriage is not just contrasted, or interdicted, but totally asystematic, in spite of legislation that finally seems to sanctify these sacraments. A marriage that seems to be accidentally linked to a casual meeting between people, or extremely wise public offices, but that is never connected to a definitively operative decision-making policy. However, a path would seem to have been traced by law, although the use of the Italian word *co-pianificazione* (co-planning) sounds promising and it seems to evoke work to be carried forward together. So what? As always, it depends.

For example, it depends on what people intend to protect. An ancient object (it might be a fragment or the Roman Colosseum) can be easily protected. The topic becomes more complicated when moving from an individual monument to an area[2] or a zone. I believe an archaeological area (the "*letter m*", from the "Galasso Law"[3]) implies both theoretical and practical uncertainties from its definition. Indeed, it includes both meanings of the word: on one hand, it means concrete delimitation, and, on the other hand, a limited concept. Nevertheless, this idea moves within boundaries that are not exactly certain, but which are at least flexible. Starting from the idea that "*il diritto ha bisogno del dove*" [also rules need to be located] (Irti 2006, p 3), the law recognises the historical condition of a monument or an area due to the presence of objects no more in use. The idea that needs to be negotiated focuses on the geometry of the boundaries that define the space of respect (the buffer zones). Boundaries that should usually define a living space, although it always surrounds almost functionally dead things. It is the same space that often claims us, as individuals and social structure. A condition that fits in the clash between "*i diritti collettivi che prendono il nome di pubblico interesse e il diritto di proprietà e d'impresa, che agisce in nome del profitto individuale*" [collective rights that take the name of public interest and the right to property and business, which acts in the name of individual profit], as defined by Settis (2010, p 304). Furthermore, Settis adds: "*il pubblico interesse è lungimirante, il profitto privato di regola non lo è*" [public interest is farsighted, private profit as a rule is not]. Then, what happens when the conflict is between the public and the public? And what about a project not able to assimilate the ancient in a new territorial vision, but just able to border the heritage?

It depends also on the scale. So far, as said, the easy part has been discussed(!). Everything becomes more complicated when the object to be protected is landscape. The first question: What is landscape? Flipping through the first pages of any book related to this topic, it will be possible to find reference to a problem about defini-

[2]About archaeological parks, see De Caro (2008).

[3]Law 8 August 1985, n. 431 (Italian Republic 1985).

tions and the semantic ambiguity of the word. When the book by Zagari (2006) was published with its forty-eight definitions of landscape, I had already put together, modestly, about twenty of them. Just a few definitions had a similar meaning, while many of these were totally different. For this reason, it was easy to share what Zagari wrote: "*È difficile non urtare i sentimenti di qualcuno quando gli si parla di paesaggio, figurarsi quando è la sua disciplina a essere sfiorata. Ma la violazione diventa decisamente insopportabile quando addirittura si cerca di delimitare il paesaggio nella freddezza di una definizione*" [When it comes to landscape, it is hard not to hurt someone's feelings. It is even worse when some discipline is touched upon. These kinds of troubles certainly become unbearable in an attempt to delimit landscape in a clear and precise definition] (Zagari 2006, p 13). I believe that we must give into the suggestion that "*ciascuna disciplina consolida il proprio distinto concetto di paesaggio, basandosi però su un vocabolo che è rimasto uguale per tutti*" [each discipline usually reinforces its own concept of landscape, but it is based on a word that has remained the same for everyone] (Azzena 2011, p 199). However, I was struck by a brief, but dense, definition written by Eugenio Turri (2002, p 11): "*il paesaggio è il volto visibile del territorio*" [the landscape is the visible face of the territory]. Especially so, because I connected this approach with another composed, many years earlier by a famous jurist, Alberto Predieri: "*il paesaggio […] è il modo di essere del territorio nella sua percezione visibile*" [(the landscape […] is a way of being of the territory in its visible perception] (Predieri 1981, p 506; Carpentieri 2004). After overlapping them, I tried to draw the unified definition (that of the geographer and the jurist) closer to a statement related to archaeology, not far from the theme of landscape. Ian Hodder, after having defined the concept of *structure*, continues: "Style is very close to structure, since it refers to the way something is done. Style may be seen as the surface appearance of structure, the acting out of the deeper codes" (Hodder 1987, pp 1–10).

Since landscape is (also) regarded as the visible surface of the environment and of the territory and these items can be interpreted as structures, I asked myself if it was possible to state that landscape is the style of the territory.[4] This is at the same time beautiful and terribly disturbing. Because the famous Italian style and landscape (as proposed: the Italian style) are so ineffable, unique and indefinable, as elusive as concrete, they are also definitely *at risk*. If, hypothetically, this idea is right, it should admit that landscape exists only in the sight and feelings of those who look at the territory. Consequently, it would probably be easier to understand why it is so difficult to protect, and to understand that it is possible to protect the territory. In order to protect the landscape, more attention would be required in the thinking of the men who live, work, cross, watch and produce the landscape with activity and sensitivity. In other words, to understand that protection of the landscape is a cultural issue, before being a legislative, technical or scientific one. Once again, it is necessary to refer to Turri's point of view: "*bisogna riportare il paesaggio nell'alveo delle manifestazioni culturali e quindi dentro l'universo rappresentativo degli individui e*

[4]Indirectly, the starting point for this approach arises, from a helpful lecture by Luca Cerchiai. I obviously take responsibility for this correlation.

della società" [we need to bring the landscape back to a cultural issue and, therefore, within the representative universe of individuals and society] (Turri 2006, p 11).

So many things depend on the path that people dealing with territorial analysis and landscape protection want to pursue. For example, instead of continuing to denounce the problem, an attempt could be made to deal with the conflict between protection and landscape and, more importantly, to see the ambiguity of the term landscape as a positive characteristic. We could try to do something to bring our knowledge (so expert, so sharp, but also so *disciplinary* and ready to define everything) closer to the needs of the territory.

Perhaps what has never been clarified and certainly not well perceived[5] is that this is a fundamental path, but not a final result. Synthesis and interpretation of objective data are the keys for the transmission of the sense of context, according to a reading of history in the territory (and not about the history of the territory). This is a relevant question in archaeological methodological literature that appears, however, not much frequented by archaeologists themselves in an operational perspective. Communications about use, visual and emotional impact, the ineffable charm of ruin (Manieri Elia 2006; Carta 2006; Augè 2005; Choay 1995) or about spaces that seem ancient even if they do not contain *old objects* are the tools of a representation of the unrepresentable. This is the challenge that archaeologists and historians of the city and the territory (especially young researchers) will have to face in the coming years. Nevertheless, the challenge has already been grasped in other areas (Castelnovi 1998),[6] but it is obviously affected by the ambiguity of the term: "landscape is not the territory, nor the country, nor the site. Here, the problem and the paradox arise about its iconic and verbal representation (its description), including an individual and empirical one" (Jacob 2009, p 27).

Finally, it depends on when we will stop believing we are only acting because we declare ourselves indignant (some among us) for the agony of archaeology, the environment, the landscape and Italian culture. We could try—at least try—to make *our knowledge* available not only to the academic and scientific world, but also for the benefit of the famous *common welfare*. It is possible to do this through the democratisation and the sharing of specialist information and, for example, through networks, in order to suggest the equivalence of the great landscape and the normal one. The latter should really have to be improved (but as a recognition of its value) compared to the globalised advertising landscapes or to the glossy ones of tourist guides and, mostly, compared to those rather elitist ones, already ironically outlined by S.T. Coleridge: "ladies reading Gilpin's & c, while passing by the very places instead of looking at the places".[7]

[5]It is a subject that has always been interesting to me and which I have faced in Azzena (2010b).

[6]A further operational update can be found at http://www.landscapefor.eu/home.

[7]Coburn (1957), n. 760: cit. Translated in Jakob (2005, p. 149 e n. 192).

References

Augè M (2005) Rovine e macerie. Bollati Boringhieri, Torino
Azzena G (2010a) Fragilità dei forti. Conoscenza, tutela, progetto del paesaggio agrario "storico". In: Bonini G, Brusa A, Cervi R (eds) Il paesaggio agrario italiano protostorico e antico, Storia e didattica, Summer School Emilio Sereni, 1st edn, August 2009. Istituto Alcide Cervi, Reggio Emilia, pp 145–154
Azzena G (2010b) Archeologia no global. La Topografia Antica e i ripensamenti disciplinari. Journal of Ancient Topography (JAT) 19:7–20
Azzena G (2011) History for places/La storia per i luoghi. In: Maciocco G, Sanna G, Serreli S (eds) The urban potential of external territories. Franco Angeli, Milano, pp 194–225
Carpentieri P (2004) La nozione giuridica di paesaggio. Rivista trimestrale di diritto pubblico. http://www.giustizia-amministrativa.it/documentazione/studi_contributi/Carpentieri4.htm#_ftn1. Accessed 23 Mar 2018
Carta M (2006) L'armatura culturale del territorio. Franco Angeli, Milano
Castelnovi P (1998) Il senso del paesaggio. Relazione introduttiva. In: Il senso del paesaggio, Internatinal meeting, May 1998. Torino, pp 1–22
Choay F (1995) L'allegoria del patrimonio. Officina, Roma
Coburn K (ed) (1957) The notebooks of Samuel Taylor Coleridge. Routledge & Kegan, London
De Caro S (2008) *Ma esistono i parchi archeologici?* In: *Antichi sotto il cielo del mondo. La gestione dei parchi archeologici: problemi e tendenze.* Edizioni dell'Assemblea, Firenze, pp 159–166
Hodder I (1987) The contextual analysis of symbolic meanings. In: Hodder I (ed) The archaeology of contextual meanings. Cambridge University Press, Cambridge, pp 1–10
Irti N (2006) Norma e luoghi. Laterza, Roma-Bari
Jacob M (2009) Il paesaggio. Il Mulino, Bologna
Jakob M (2005) Paesaggio e letteratura. Leo S, Olschki, Firenze
Landscapefor (2014) Associazione Landscapefor. Piattaforma del paesaggio utile: per capire dall'esperienza diretta, per nutrire la curiosità e l'attenzione, per partecipare ad azioni sul bene comune. http://www.landscapefor.eu/home
Manieri Elia M (2006) Il polivalente senso del rudere. In: Billeci B, Gizzi S, Scudino D (eds) Il rudere tra conservazione e reintegrazione. Gangemi, Roma, pp 18–160
Predieri A (1981) Paesaggio. In: Enciclopedia del Diritto. Giuffrè, Milano, p 506
Repubblica Italiana (1985) legge 8 agosto 1985, n. 431
Settis S (2010) Paesaggio Costituzione cemento. Einaudi, Torino
Turri E (2002) La conoscenza del territorio. Marsilio, Venezia
Turri E (2006) Il paesaggio come teatro. Marsilio, Venezia
Zagari F (2006) Questo è paesaggio. 48 definizioni. Gruppo Mancosu, Roma

Chapter 9
Archaeological Heritage Enhancement in the City and in the Landscape

Marko Rukavina and Roberto Busonera

Abstract The chapter examines the possibilities of enhancement and planning of immovable archaeological heritage in situ from the perspective of urban and spatial planning. The aim of the research is to develop scientific starting points for the enhancement, preservation and sustainable use of archaeological heritage. The research is based on the deductive method whose results are confirmed by qualitative measurement obtained from field research conducted in Croatia and Italy. The survey included selected sites (case studies) in urban areas and in the landscape. The results of the research are defined spatial models of archaeological heritage planning: archaeological heritage in suburban recreational and leisure areas, archaeological heritage in protected natural areas, urban integration of archaeological heritage, archaeological heritage in tourism areas, archaeological heritage in the vicinity of important road infrastructure or traffic nodes, and the combined model. In addition to the basic planning models, the research includes specific archaeological heritage planning models (Council of Europe Cultural Routes, the European Heritage Label and World Heritage) and the cultural landscape model as a contemporary archaeological heritage planning model.

Keywords Archaeological heritage · Heritage enhancement · Spatial planning · Planning models · Planning criteria

9.1 Introduction

This chapter examines the possibilities of enhancing and planning immovable archaeological heritage in situ, i.e. archaeological sites in urban and rural areas from the perspective of urban and spatial planning. The starting points for the conducted research are: the fact that archaeological heritage is one of the most endangered

M. Rukavina (✉)
Faculty of Architecture, University of Zagreb, Zagreb, Croatia
e-mail: mrukavina@arhitekt.hr

R. Busonera
University of Sassary, Sassary, Italy
e-mail: rbusonera@uniss.it

M. Obad Šćitaroci et al. (eds.), *Cultural Urban Heritage*,
The Urban Book Series, https://doi.org/10.1007/978-3-030-10612-6_9

types of cultural heritage; the lack of cooperation between the conservation and planning sectors, both in the Republic of Croatia and Italy; and the global lack of research of archaeological heritage management (preservation in situ and enhancement possibilities) from the viewpoint of spatial and urban planning. Archaeological heritage is most often seen as a problem in planning—it is not recognised as a spatial development resource, and in planning documents, it is in most cases left as isolated islands excluded from contemporary life.

The aim of the paper is to explore planning models for the integration of archaeological heritage into contemporary life, as well as to determine the planning (spatial and functional) prerequisites for its sustainable use and management based on the hypothesis that the enhancement of archaeological heritage is based on the logic of contemporary land use and the integration of archaeological heritage into contemporary life through urban and spatial planning.

So far, research of archaeological heritage from the planning viewpoint has focused mainly on specific topics (urban integration, cultural landscapes, site design, etc.) without a general consideration of archaeological heritage planning issues (Fazzio 2004, 2005; Gücer 2004; Alpan 2005; Saruhan Mosler 2005; Rico 2010; Busonera 2014; Rukavina 2015, etc.). Such a type of research is generally rare, in spite of international conservation documents that recognise spatial and urban planning, from the 1960s onwards, as one of the most important tools in the conservation process—*integral conservation* (Rukavina et al. 2013). The research builds on: a paper exploring the prerequisites of archaeological heritage management in the territory of Sesvete (Zagreb, Croatia), in which some of the archaeological heritage planning models were explored (Rukavina et al. 2015); a paper exploring the planning of the archaeological site of the Roman town of *Andautonia* in the vicinity of Zagreb (Rukavina and Obad Šćitaroci 2013); a paper exploring the problems between juridical inertia and territorial management in Italy (Busonera 2016); and a paper exploring the necessity of a project-oriented view, no longer dominated by the cumbersome impediment that, often, archaeological heritage constitutes (Azzena 2011).

9.2 Research Methodology

Based on deductive research methodology, the general planning prerequisites, i.e. criteria for planning archaeological heritage, were defined—a *determined general planning model* that was tested on selected examples of the presented archaeological sites in Croatia, Italy and Germany. The defined planning criteria are: (1) accessibility of the area; (2) use/visiting of the area; and (3) visual quality of the area. In addition to the planning criteria, the general planning model includes archaeological and other sociocultural and economic criteria (Fig. 9.1).[1] Case studies, on which the theoretical general model was tested, were selected according to the criteria of their

[1]Cultural, social and economic criteria are not covered in this research—a multidisciplinary approach is required in further research of the topic.

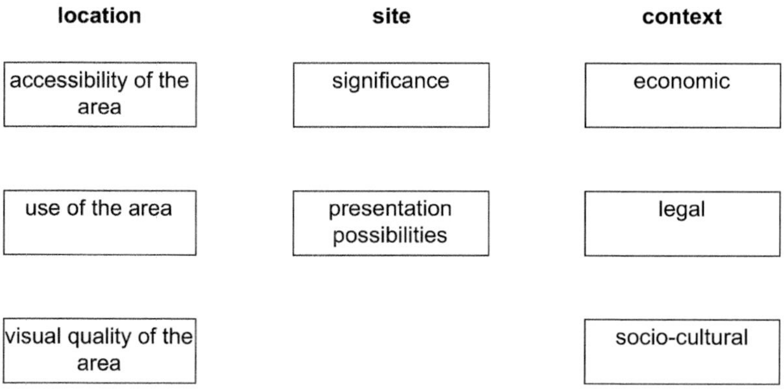

Fig. 9.1 Groups of criteria which define the general planning model

representativeness in terms of existing functional models of archaeological heritage planning (examples of good practice) and the criterion of non-overlapping multiple models in the same analysed case, although such overlapping is possible and preferable in practice/reality. One of the selection criteria is the number of selected cases in Sardinia and Croatia, as well as the possibility of field surveys of the sites.

In addition to the basic planning models, the research includes specific archaeological heritage planning models (Council of Europe Cultural Routes, the European Heritage Label and World Heritage) that are not based on determined planning criteria, and the cultural landscape model as a contemporary archaeological heritage management model that includes historical territorial system values.

9.3 Models of Archaeological Heritage Planning

The analysis of case studies aims to show how archaeological heritage depends on the contemporary context and how it should be differently planned in each situation. The need to identify planning models arises in order to trace some common elements of reflection, connected to different planning issues and to the process of protection and enhancement of the archaeological heritage. Obviously, these examples do not cover all the possible conditions, but they can be useful for a general survey of the topic and as a possible theoretical reference.

9.3.1 Xanten, Germany—Archaeological Heritage in Suburban Recreational and Leisure Areas

The planning of large recreational areas (*Grosserholungsgebiet*) and leisure centres (*Freizeitzentren*) in the Ruhr region began in the late 1960s (Mittelbach 1972; Baud-

Bovy and Lawson 1998). Within this concept, the city of Xanten was planned as one of several leisure centres in the Ruhr region. One of the planning starting points for determining leisure centres was the visual quality of the landscape—the attractiveness of the area (Land Nordhein-Westfalen 1976). The strategy for the town of Xanten included the revitalisation of the medieval town core, establishing an archaeological park and the construction of new sports and recreational facilities along the water surfaces (Bundesrepublik Deutschland 1975). The archaeological park[2] developed on the unbuilt area of the ancient Roman colonial town of *Ulpia Traiana* opened to the public in 1977, and the Xanten leisure centre immediately adjacent to the archaeological park in 1982 (Stadtverwaltung Xanten 2015). In the European Year of Cultural Heritage (1975), the city was selected as one of the five examples of good practice from Germany.

The recognised planning model is primarily aimed at residents of the city and the region,[3] and it is based on the use of recreational and leisure facilities, good accessibility and the visual quality of the area. Recreational and other facilities contribute to the number of visitors in the area, and thus to visits to the archaeological heritage sites.[4] Numerous examples of archaeological sites located in the vicinity of larger towns, where different recreational and leisure facilities should be planned, fall into this model.[5]

9.3.2 Žumberak, Croatia—Archaeological Heritage in Protected Natural Areas[6]

The Žumberak area gained the first form of protection in 1978 as the memorial area of *Žumberak-Gorjanci* in the cross-border area of Slovenia and Croatia, while the *Žumberak-Samoborsko gorje Nature Park* was proclaimed in 1999 on the territory of the Republic of Croatia occupying an area of 333 km^2 (Klemencic 2006). As part of the project of presentation and interpretation of natural and cultural heritage, in 2004

[2]The authors do not seek to evaluate the partial reconstructions done in the archaeological park, which are not a subject of this research.

[3]The total population of the region since the 1960s is about 5 million (Regionalverband Ruhr n.d.).

[4]The Roman route *Römer-Lippe-Route* was established in 2013 throughout the Ruhr region connecting numerous archaeological sites at a total length of 479 km (Lippeverband 2018).

[5]In the Republic of Croatia, this model can be applied to the archaeological parks of *Andautonia* near the capital city of Zagreb and *Salona* near the city of Split.

[6]According to the *Croatian Nature Protection Act* (NN 80/13, 15/18), protected natural areas include: national parks, nature parks, special reserves, strict reserves, regional parks, significant landscapes, nature monuments, monuments of park architecture and forest parks and internationally designated areas (MaB, Ramsar, WHS, Geopark). For national parks and nature parks in the Republic of Croatia, it is mandatory to develop a special spatial plan of the protected area (*PPPPO*), and for protected natural areas to establish a public management institution and develop a management plan (Republika Hrvatska 2013a, b).

the educational trail *The Trail of the Princes* was opened for visitors.[7] The hiking trail connects several archaeological sites in the settlements of Budinjak and Bratelji. It includes a prehistoric settlement and a necropolis with more than 140 tumuli from the early Iron Age (ninth–sixth centuries BC, Hallstat), a necropolis from Roman times (first to second century) and other spatial values of the area (Želle 2006). In addition to this archaeological trail, the *Žumberak Archaeological Route* which includes archaeological sites from prehistoric times to the Middle Ages throughout the park was also established (Park prirode Žumberak—Samoborsko gorje n.d.).

The model is primarily aimed at visitors of the protected natural areas and is based on the preserved landscape and other natural and cultural values of the area.[8] A special regime of management of the protected areas contributes to the enhancement of these areas and to the inclusion of archaeological heritage in the tourist offer, to increased visits of heritage sites, and to the preservation of the sites' setting.[9]

9.3.3 Pula, Croatia—Urban Integration of Archaeological Heritage[10]

Pula is a Roman town built on the foundations of a prehistoric proto-urban *Histri* settlement on a hillside at the bottom of the deep sea bay. From it, the town took over a circular-radial urban matrix that has been preserved until today (Matijašić and Ujčić 2005). The colony *Julia Pola* was founded in Caesar's time, but it was damaged during the Civil War, so Augustus restored it at the beginning of the first century because of its strategic importance (Matijašić and Ujčić 2005). The town has maintained the continuity of urban settlements since Classical times, although its significance declined significantly during the Middle Ages and in later periods, and at the end of the period of Venetian rule (end of the eighteenth century), the town had only a few hundred inhabitants. The town's recovery and the first archaeological excavations began at the time of the Austro-Hungarian administration (nineteenth century—early twentieth century) when the town became the seat of the imperial navy, which contributed to the economic recovery and growth of the town. Archaeological research continued during the Italian administration when the ancient town walls were excavated and numerous interventions on ancient monuments were performed (*Slavoluk Sergijevaca*, the *Arena*, a small Roman theatre, a Roman mausoleum). The urban integration of ancient and early Christian monuments and archaeological research has lasted continuously from the period after World War II to the present

[7] The *Žumberak Samoborsko gorje Nature Park* has about 50,000 visitors annually (Parkovi dinarida n.d.).

[8] Protected natural areas are usually characterised by lower population density and a relatively low number of users/visitors, and they can be located in remote areas (Baud-Bovy and Lawson 1998).

[9] The obligation to protect, maintain and promote all values of the protected area (Želle 2006).

[10] Research on the urban integration of archaeological heritage was the subject of unpublished doctoral research titled "The archaeological heritage integration method in urban planning" in which the town of Pula with 19 visible archaeological sites was analysed (Rukavina 2015).

day, prompted by the post-war reconstruction of the town.[11] According to statistics, the Roman Amphitheatre is today the most visited monument[12] in the Republic of Croatia (Eurostat 2016).

The model is aimed not only at town residents, but also at its visitors, based on the location of the archaeological remains and their integration into contemporary life. Access to archaeological heritage, other cultural and historical values and various contents contribute to visits to the archaeological heritage. Numerous examples of the urban integration of archaeological heritage in Europe and in the world, irrespective of whether the archaeological remains are located in the historical core of the town or outside it,[13] fall into this model.

9.3.4 *Santa Teresa/Arzachena, Sardinia—Archaeological Heritage in Tourism Areas (Coastal and Mountain Resorts)*

The exceptionality of many archaeological sites in the north-east of Sardinia is due not only to the material testimonies of the sites, but also to the environmental and landscape conditions in which they are inserted. One of these is the Lu Brandali complex (Antona 2005), close to the small town and tourist resort of Santa Teresa di Gallura. The archaeological area is very diverse and includes a nuraghe, surrounded by a wall with several towers, a village of huts and a tomb. The importance of the area is also testified to by numerous testimonies of the reuse of Nuragic structures in the following periods, which prove the presence of a checkpoint for connections with Corsica and the other villages and towns of the island. However, the site is just one representative of a much larger territorial organisation,[14] which includes other important areas such as Vigna Marina, Stirritoggju, La Colba, La Testa and many other sites.

The development of tourism in the coastal area connects the archaeological sites in a solid reality that already characterises the north-eastern part of the island.[15] Sustainable tourism development, in order to extend many benefits to individual areas as well as to the whole territory, should be oriented not just to a single place or activity but towards the whole territory. This would secure the protection and conservation of the archaeological heritage, and at the same time, it would ease tourism pressure on the coastal areas, redistributing it all over the territory. Therefore, the model aims at

---

[11] The process of reconstruction and revitalisation of the historic core has not yet been fully completed.

[12] 473,406 admissions in 2016.

[13] In some cases, the urban setting of archaeological sites can be inadequate and its enhancement is needed.

[14] In Sardinia, there are almost 9000 nuraghi, located with a density of about 3 per km^2 (Lilliu 2002, 428–429).

[15] Tourism development of the area started in the 1960s.

a balanced spatial planning project capable of including the archaeological heritage within a "territorial system" (Caravaggi 2002).

In brief, the identified model is primarily aimed at visitors from tourist resorts, based on the vicinity of archaeological sites, good accessibility and the visual values of the area.

9.3.5 Santa Cristina, Sardinia—Archaeological Heritage in the Vicinity of Important Road Infrastructure or Traffic Nodes

The Nuragic complex of Santa Cristina testifies to the presence of a site used as a sanctuary, extending over about 14 ha, situated 4 km from the modern village of Paulilatino (Oristano) and next to the most important road in Sardinia (E25). The archaeological site consists of two main sectors: a temple with its structures and a small nuraghe with the remains of a village. The area is managed by a local cooperative, which was created with the aim of conserving and enhancing the historical, archaeological and other cultural heritage of the territory.[16] Within this territory, tourist use is an important resource and an essential tool for local development.

This model is based on the good transport accessibility of archaeological sites and intensive use of the main road infrastructure that creates planning opportunities and conditions for visiting and for the contemporary use of the sites by tourists and travellers.[17] In this sense, it is essential to consider how the use and enhancement of the cultural asset can be closely connected to the possibility of accessing the site.

9.3.6 Porto Conte, Sardinia—The Combined Model

In 1999, the *Porto Conte* natural park was born (Regione Autonoma della Sardegna 1999). It is located in the territory of Alghero, in the north-west of Sardinia, and covers about 5000 ha. The main purpose of the park is to protect and manage ecosystems and many historical sites of great importance (53 archaeological sites and nine historical-architectural monuments), which cover a time span from the Neolithic

[16] *Cooperative Archeotour*.

[17] The extra-urban road can become a solid starting point for a large-scale project, which from the Santa Cristina complex can enable the whole territory to be enhanced. Starting from the main road, an important system of connection could be created between the numerous archaeological sites in the territory. However, in Sardinia, the weakness of the infrastructure and public transport network represents a serious weakness for the development of cultural sites. The supra-local road close to the area of Santa Cristina also has a negative effect on heritage because it leads visitors to cross the territory from one point to another (to "skip the space"), without paying attention to numerous "minor sites" situated in the area.

to the modern age (Moravetti 1996).[18] The model combines three identified basic models of archaeological heritage planning, thus creating planning preconditions for the better use and enhancement of the sites.[19] The most important archaeological sites in the park are the nuraghe of Palmavera and the area of Sant'Imbenia. While Palmavera is one of the most important Nuragic archaeological sites of the whole of Sardinia (Moravetti 1992), the area of Sant'Imbenia consists of two independent complexes: the remains of a Roman villa (with some traces of use up to the medieval age) (Maetzke 1959; Manconi 1999; Teatini 1993) and a nuraghe, with the remains of the village (Bafico 1993; Bafico et al. 1995, 1997; D'Oriano 2001). The main objective of the park is to protect the natural and historical heritage not only in its physical and structural expression, but mostly in order to connect it with contemporary and compatible uses (Parco di Porto Conte n.d.). However, even the main archaeological sites have problems with regard to access and use.[20] In brief, it is possible to state that, inside the park, the archaeological areas show an exceptional cultural complexity, which is still not adequately used and managed despite the existing spatial planning prerequisites.[21]

In general, the combined model combines at least the two identified basic models of archaeological heritage planning, thus creating the spatial planning preconditions for the better use and enhancement of the archaeological heritage.

9.4 Specific Models of Archaeological Heritage Planning

Despite the variety of cases studied, it seems necessary to focus on some specific cases that are significant or deserve strengthened protection that might be guaranteed through the action of international organisations. The analysis aims to study the degree of protection and enhancement of these sites, and tries to assess whether the action and intervention of bodies of international relevance have contributed to strengthening protection and enhancement processes from the planning viewpoint, and whether they have included spatial planning criteria in the process. The study proceeds by analysing certain sites with different strategies for protection: Roman

[18]Further information on the historical and cultural assets of the park is summarised in a report written by Giovanni Azzena (Azzena n.d.).

[19]At the same time, the park (protected natural area) represents the most important suburban recreation and leisure area for the residents of Alghero, which also includes a few coastal tourism resorts.

[20]At the Palmavera nuraghe, there is no space for parking, adequate access or visitor facilities (even no spaces for an expansion of the archaeological excavation). This kind of organisation is even worse in the Sant'Imbenia area, where the archaeological remains can only be visited "by appointment". This is an even more serious weakness considering that the main archaeological sites are located near the "gate of the park".

[21]Most of the sites within the park are not accessible and are often invisible, located on private land and, therefore, deprived of a minimum interpretative infrastructure and possibilities of being perceived.

Emperors and the Danube Wine Route (Croatia), the Krapina Neanderthal Site (Croatia) and Su Nuraxi in Barumini (Italy).

9.4.1 Roman Emperors and the Danube Wine Route, Croatia—Council of Europe Cultural Routes (CRs)[22]

In 2015, the European Cultural Route dedicated to archaeological heritage linked to Roman emperors was proclaimed in Croatia, Serbia, Bulgaria and Romania. Archaeological sites included on the territory of Croatia are: Varaždinske Toplice (*Aquae Iassae*), Ščitarjevo (*Andautonia*), Pula and Brijuni, Zadar and Nin, Split (Diocletian's palace) and Solin (*Salona*), Vid (*Narona*) and the Ilok and Baranja wine regions (Hina 2015). In the nomination of the route, no planning criteria were addressed. The chosen sites represent completely different management and planning issues ranging from urban integration, integration in protected areas, World Heritage, the proximity of larger cities to isolated sites with bad accessibility and the relatively low number of visitors.[23]

Although the model is aimed at preserving European natural and cultural heritage, at sustainable development and at the development of cultural tourism (COE 2016, 2017)[24] and that the nomination includes an assessment of the general spatial planning criterion—*sustainable land use* (Republik Österreich 2013)—this model does not include identified planning criteria and thus does not solve numerous management issues.[25]

[22]"... model for promoting the principles of the Council of Europe, stimulating trans-border cooperation, encouraging innovative approaches in the fields of cultural heritage, intercultural dialogue and sustainable local development" (COE 2017). The CRs model should not be confused with the ICOMOS concept of cultural routes.

[23]Brijuni 181,560 visitors in 2016; Split (basement of Diocletian's Palace) 277,598 admissions in 2016; Narona 17,057 in the same year (Eurostat 2016; Republika Hrvatska 2017; Arheološki muzej Narona n.d.). Of the included sites, Zadar and Split are listed on the United Nations Educational, Scientific and Cultural Organization (UNESCO) World Heritage List (WHL) and the Brijuni islands are a national park.

[24]Until 2018, a total of 33 European cultural routes were established, five of which are based on archaeological heritage: Viking Routes; The Phoenicians' Route; The Prehistoric Rock Art Trail; The European Route of Megalithic Culture; and Roman Emperors and the Danube Wine Route (COE 2018).

[25]There is still no methodology developed to measure the "network effect" (UNWTO 2015).

9.4.2 Krapina Neanderthal Site, Croatia—European Heritage Label (EHL)

By 2016, 29 sites in 16 European Union (EU) member states were awarded the European Heritage Label (Eurostat 2016), of which only three can be classified as archaeological sites. In 2016, the Krapina Neanderthal Site in the Republic of Croatia—a paleontological and archaeological site—received the label.

The model aims at the joint promotion of heritage sites of European significance and those that are of importance for European history and culture and/or for the creation of the European Union (EU 2011).[26] Nomination does not include spatial planning criteria, but assessment is made only of the accessibility of sites in reference to site design and the existence of accompanying infrastructure for visitors (EU 2011).

9.4.3 Su Nuraxi, Sardinia—UNESCO World Heritage Sites (WHS)[27]

The area of Su Nuraxi of Barumini represents the most important and representative Nuragic complex of Sardinia. The complexity of the structure, together with the extension of the area and its excellent state of conservation, allowed it to be declared a WHS by UNESCO in 1997.[28] The designation of Barumini refers to UNESCO's statement that in order to ensure the most effective and active protection and conservation some fundamental conditions must be respected, including "... a general policy which aims to give the cultural and natural heritage a function in the life of the community and to integrate the protection of that heritage into comprehensive planning programmes".[29] However, with regard to Su Nuraxi, data on tourist inflows recorded after the inclusion of Su Nuraxi in the World Heritage List (WHL) testify clearly to how the UNESCO nomination was not useful. Compared to a significant increase in visitor numbers recorded in the year of its inclusion in the List, site access fell significantly in the following years. Indeed, the trends show an average of 71,197 visitors between 1997 and 2006, lower than the numbers registered before

---

[26]Due to the only recent development of the model (2011), there are no relevant data on its effect.

[27]A more complete paper on the case is "The future (?) of effective protection" (Azzena et al. 2017).

[28]"The Committee decided to inscribe this property on a basis of cultural criteria (i) (iii) and (iv), considering that the nuraghe of Sardinia, of which Su Nuraxi is the pre-eminent example, represent an exceptional response to political and social conditions, making an imaginative and innovative use of the material and techniques available to a prehistoric island community" (UNESCO 1997).

[29]See article 5, paragraph A of the Convention Concerning the Protection of the World Cultural and Natural Heritage (UNESCO 1972). The most recent World Heritage guidelines address urban and regional planning instruments in effective management; sustainable use; and sustainable development principles that should be integrated into the management system (UNESCO 2017).

the UNESCO nomination (75,783 visitors in 1996).[30] Furthermore, there is also a problem with inadequate development in the setting of the site due to a small buffer zone.[31]

A further aspect, implicit in the key principles of UNESCO, may also be considered ambiguous. Inclusion in the WHL is exclusively for cultural and natural heritage that expresses an outstanding universal value. In order to protect the territory as much as possible, this aspect seems to reduce considerably the chances of enhancement, making the UNESCO WHS status insufficient for the survival of "minor heritage" (Tosco 2014, 109).

9.5 Archaeological Heritage as an Element of (Cultural) Landscape—A Contemporary Model of Archaeological Heritage Planning

Protection of cultural landscapes in the Republic of Croatia is still at its beginning (Dumbović Bilušić 2014). By 2018, only ten protected cultural landscapes were proclaimed,[32] half of which include archaeological heritage and one of which is on the World Heritage List—the Cultural Landscape *Starigradsko polje* on the island of Hvar with well-preserved traces of the Greek land division—*Chora* from the fourth century BC (Republika Hrvatska Ministarstvo Kulture 2018). Legally declared cultural landscapes in the Republic of Croatia are generally small scale (up to 1400 ha),[33] including uninhabited areas, uninhabited islands or areas under other protection regimes (protected natural areas, UNESCO WHS), thus avoiding complex management problems.[34]

In Italy, the regulatory capacity to manage and protect cultural landscape has been developed since 1985[35] through laws that have structured the protection system on the use of bounded areas to be associated with sites with particular physical characteristics and recognised by the central authority (Fazzio 2005, 57). Only recently has there been an attempt to extend the usual forms of protection to a real act of

[30] Data for this section refer to the surveys of the Ministry of Cultural Heritage, Activities and Tourism on the number of visitors to Italian museums and archaeological sites (Repubblica Italiana 2015).

[31] In the proximity of the site, there area privately owned hotel, amusement park and recently built heritage centre G. Lilliu, all disrupting the integrity of the site.

[32] There are also some other protected cultural landscapes that are classified as cultural heritage areas or natural heritage areas (significant landscapes) which are not included in the survey.

[33] Without a buffer zone (UNESCO 2014).

[34] The obligation to develop a management plan is obligatory for cultural heritage sites on the World Heritage List.

[35] Law n. 8 provided a decisive impulse towards protection and gave formal protection to each area belonging to a specific legal category including, for the first time, those of archaeological interest (Repubblica Italiana 1985).

planning.[36] However, the division between "cultural heritage" and "landscape heritage" has prevented (and is still preventing) the active involvement of archaeology within spatial planning policies. The Sardinia Region began to structure its tool for landscape protection in 2004 (the first step of the plan was approved in 2006), and it was the first Italian Region to have a landscape plan meeting the requirements of the Ministry of Heritage, Culture and Tourism. Today, with only one heritage site recognised as a WHS (Su Nuraxi, Barumini) and a few other important archaeological sites, the greatest challenge for Sardinia is to find a solution for the protection and enhancement of "minor heritage", widespread all over the territory which represents an integral part of cultural landscapes.

The (cultural) landscape "model" should be aimed at sustainable development, the preservation of natural and cultural heritage, and tourist visits to a particular area/territory. It should be based on a holistic approach and integral planning combining several sectors (nature protection, cultural heritage protection, spatial planning, etc.). It represents a contemporary field of multidisciplinary and interdisciplinary research prompted by the European Landscape Convention (2000). It can include not only cultural landscapes on the World Heritage List,[37] but also cultural landscapes of lesser importance. It is an approach to management and planning of archaeological heritage based on the management of cultural landscapes, which allows the planning of sites as elements of historical territorial systems including "minor heritage" and a step forward from the site-oriented approach. The possibilities of application and the potentials of the approach are yet to be fully explored.

9.6 Conclusion

The conducted research can offer some meaningful insights into the role of urban and spatial planning within the process of archaeological heritage enhancement and management. In order to achieve sustainability of the presentation and use of archaeological heritage, apart from an evaluation of the sites based only on archaeological significance, it is also necessary to meet spatial planning criteria as the basis for a decision on the presentation and use of archaeological heritage and its integration into contemporary life. A holistic approach to planning (an integrated approach) and the cooperation of all sectors which have consequences on archaeological heritage should be achieved to create the appropriate management conditions and to avoid numerous management problems. Obviously, none of the identified planning models can be considered and exported as the only correct way of intervention but just as a general reference point in the planning process. The identified models show how an enhancement strategy must necessarily be structured on the relationship between the archaeological heritage and the territory, i.e. its contemporary context. The analysis of the relationship between archaeology and territory has shown how the complex-

[36]Code of Cultural Heritage and Landscape, Art. 143–145 (Repubblica Italiana 2004).

[37]In 1992, UNESCO introduced a new category of cultural heritage—cultural landscapes.

ity of the territory, its organisation and planning might be able to convey not only the design choices on individual places, but also the fruition processes. Within the complex contexts such as those studied, the complexity of the territorial structure (accessibility to the sites, use of the territory, and the visual and perceptual quality of the area) requires the definition of smart planning strategies for the real integration of heritage protection, enhancement and planning. In brief, it seems important that a project interested in the protection of archaeological heritage can respond to the stresses indicated in this chapter, i.e. to the stimuli no longer considered as independent themes, but as an integral and necessary part of any project connected to the protection and enhancement of archaeological heritage.

Acknowledgements The research is part of the scientific project "Heritage Urbanism—Urban and Spatial Planning Models for Revival and Enhancement of Cultural Heritage". It is financed by the Croatian Science Foundation [HRZZ-2032] and is carried out at the University of Zagreb, Faculty of Architecture.

References

Alpan A (2005) Integration of urban archaeological resources to everyday life in the historic city centres Tarragona, Verona, Tarsus. Master of Science thesis, Middle East Technical University, Ankara

Antona A (2005) Il complesso nuragico di Lu Brandali e i monumenti archeologici di Santa Teresa Gallura. Delfino editore, Sassari

Azzena G (n.d.) Piano del Parco Naturale Regionale di Porto Conte: Relazione preliminare - Assetto storico-culturale. Available via Parco di Porto Conte. http://www.parcodiportoconte.com/public/docs/gli_aspetti_storici_e_culturali_del_territorio_del_parco_di_porto_conte.pdf. Accessed 26 Feb 2018

Azzena A (2011) History for places. In: Maciocco G, Sanna G, Serreli S (eds) The urban potential of external territories. Franco Angeli, Milan, pp 194–227

Azzena G, Busonera R, Perini C (2017) The future (?) of effective protection. Archeologia e calcolatori 28(2):549–560

Arheološki muzej Narona n.n. Izvješće o radu za 2016. godinu

Bafico S (1993) Il nuraghe S.Imbenia di Alghero. In: Sardegna: civiltà di un'isola mediterranea. Bologna, p 59

Bafico S, D'Oriano R, Lo Schiavo F (1995) Il villaggio nuragico di Sant'Imbenia ad Alghero (SS). Nota preliminare. Actes du IIIe Congrès international des Études Phéniciennes et Puniques, Tunis 11–16 Nov 1991, pp 87–98

Bafico S, Oggiano I, Ridgway D, Garbini G (1997) Fenici e indigeni a Sant'Imbenia (Alghero). In: Bernardini P, D'Oriano R, Spanu PG (eds), Phoinikes b Shrdn/I Fenici in Sardegna: nuove acquisizioni. Oristano, pp 45–53

Baud-Bovy M, Lawson F (1998) Tourism and recreation handbook of planning and design. Architectural Press, Oxford

Bundesrepublik Deutschland (1975) Xanten. Detschen Nationalkomitee für das Europäische Denkmalschutzjahr 1975, Berlin

Busonera R (2014) Leggere il territorio dell'archeologia: l'area archeologica di Neapolis (OR) tra politiche di tutela e processi di valorizzazione e fruizione, Dissertation, University of Sassari

Busonera R (2016) Landscape and archaeology in Sardinia. Between juridical inertia and land management. In: Lai L, Mastinu M, Saiu V, Schirru M (eds), Ricerca in vetrina. PhD in Sardinia: higher education, scientific research and social capital. Franco Angeli, Milano, pp 288–296
Caravaggi (2002) Paesaggi di paesaggi. Meltemi, Roma
COE (2016) Council of Europe cultural routes. Council of Europe/European Institute of Cultural Routes, Strasbourg/Luxembourg
COE (2017) Declaration by the Committee of Ministers on the 30th anniversary of the cultural routes of the Council of Europe (1987–2017)
COE (2018) Cultural routes. https://www.coe.int/en/web/cultural-routes/by-theme. Accessed 26 Feb 2018
D'Oriano R (2001) L'Emporion di Sant'Imbenia. In: Argyróphleps nesos. L'isola dalle vene d'argento, Esploratori, mercanti e coloni in Sardegna tra il XIV e il VI sec. a.C. Bondeno, pp 35–36
Dumbović Bilušić B (2014) Prepoznavanje i razvrstavanje krajolika kao kulturnog naslijeđa. Godišnjak zaštite spomenika kulture Hrvatske 36:47–66
EU (2011) Decision no 1194/2011/EU of the European Parliament and of the Council of 16 November 2011 establishing a European Union action for the European Heritage Label
Eurostat (2016) Culture statistics. Publications Office of the European Union, Luxembourg
Fazzio F (2004) Archaeology and urban planning. Urbanistica 124:2–5
Fazzio F (2005) Gli spazi dell'archeologia. Temi per il progetto urbanistico. Officina, Roma
Gücer E (2004) Archaeology and urban planning—a consensus between conservation and development: Aphrodisias and Geyre. Master of Arts thesis, Izmir institute of Technology
Hina (2015) Hrvatska na Ruti rimskih careva i dunavskoj vinskoj ruti. Available via Culturenet. http://www.culturenet.hr/default.aspx?id=65526. Accessed 26 Feb 2018
Klemenčić M (2006) Temeljna geografska obilježja Žumberka. Žumberak & Latobici. Arheološki muzej u Zagrebu, Zagreb, pp 11–17
Land Nordhein-Westfalen (1976) Landesentwicklungsplan III. Düsseldorf
Lilliu G (2002) Uomo e ambiente in Sardegna nel suo percorso storico. In: Mattone A (ed) G. Lilliu, La costante resistenziale sarda. Ilisso, Nuoro, pp 424–438
Lippeverband (2018) Unterwegs auf der Römer-Lippe-Route 2018. Available via Römer Lippe Route. http://www.roemerlipperoute.de/informieren/kostenfreie-downloads.html. Accessed 26 Feb 2018
Maetzke G (1959–61) Scavi e scoperte nelle province di Sassari e Nuoro. Porto Conte. Resti di necropoli in località S. Imbenia. StSardi, XVII, pp 656–657
Manconi (1999) Römische Villa von Sant'Imbenia. BetaGamma, Sassari
Matijašić R, Ujčić Ž (2005) Pula antički grad. Arheološki muzej Istre/Buvina/Difo/Laurana, Zagreb
Mittelbach, HA (1972) Die Planung von Erholungs- und Freizeitanlagen für das Gebiet des Siedlungsverbandes Ruhrkohlenbezirk. In: Olschowy G (ed) Landespflege im Ruhrgebiet. Deutschen Rates für Landespflege, Bonn, pp 38–40
Moravetti A (1992) Il complesso nuragico di Palmavera. Sassari
Moravetti A (1996) Il territorio dal Neolitico all'età romana. Alghero e il suo volto, Sassari
Parco di Porto Conte (n.d.) Il piano del Parco. http://www.parcodiportoconte.it/piano-del-parco.aspx?ver=it. Accessed 26 Feb 2018
Parkovi dinarida (n.d.) Park prirode Žumberak – Samoborsko gorje. https://parksdinarides.org/me/park/park_prirode_zumberak_samoborsko_gorje/. Accessed 26 Feb 2018
Park prirode Žumberak - Samoborsko gorje (n.d.) http://www.pp-zumberak-samoborsko-gorje.hr/aktivnosti/pjesacenje.html. Accessed 26 Feb 2018
Regionalverband Ruhr (n.d.) Metropole Ruhr - das neue Ruhrgebiet Differenzierte Bevölkerungsstruktur. http://www.metropoleruhr.de/land-leute/daten-fakten/bevoelkerung.html. Accessed 26 Feb 2018
Regione Autonoma della Sardegna (1999) Regional Law n. 4, February 26, 1999—establishment of the "Porto Conte" Regional Natural Park

Repubblica Italiana (1985) Law n. 8, August 1985—provisions for the protection of areas with particular environmental interest
Repubblica Italiana (2004) D.Lgs. 42/04—code of cultural heritage and landscape
Repubblica Italiana (2015) Visitatori e introiti di Musei, Monumenti e Aree Archeologiche Statali. Available via: Ministero dei Beni e delle Attività Culturali e del Turismo - Ufficio Statistica. http://www.statistica.beniculturali.it/Visitatori_e_introiti_musei_15.htm. Accessed 26 Feb 2018
Republika Hrvatska (2013a) Zakon o zaštiti prirode
Republika Hrvatska (2013b) Zakon o prostornom uređenju
Republika Hrvatska (2017) Turizam u brojkama 2016
Republika Hrvatska Ministarstvo Kulture (2018) Registar kulturnih dobara. http://www.min-kulture.hr/default.aspx?id=6212. Accessed 26 Feb 2018
Republik Österreich (2013) "European cultural routes": a practical guide. Federal Ministry for European and International Affairs/Federal Ministry of Economy, Family and Youth, Department for Tourism and Historic Objects, Vienna
Rico, JMF (2010) Formas de inserción de los yacimientos arqueólogicos en áreas fuertemente antropizadas de la Costa del sol: Una aproximación metodológica previa al aprovecamiento territorial da la ruina. Dissertation, Universidad Politécnica de Madrid
Rukavina M, Obad Šćitaroci M (2013) Arheološki parkovi kao pejsažni prostori grada - Andautonija u kontekstu Zagreba i okolice. In: Božičević J, Nikšić M, Mlinarić TJ, Missoni E (eds) Zelenilo grada Zagreba Zbornik radova s međunarodnoga znanstvenog skupa održanog 5. i 6. lipnja 2013. u Zagrebu. HAZU, Zagreb, pp 108–116
Rukavina M, Obad Šćitaroci M, Petrić K (2013) Prostorno-urbanistički aspekti zaštite nepokretnog arheološkog naslijeđa - Međunarodni i nacionalni dokumenti o zaštiti. Prostor 21(46):312–325
Rukavina M (2015) Metoda integracije arheološkog naslijeđa u urbanističkom planiranju (Archaeological heritage integration method in urban planning). Dissertation, University of Zagreb
Rukavina M, Petrić K, Obad Šćitaroci M (2015) Studija zaštite i prezentacijskog potencijala arheološkog nalazišta Kuzelin i bliskih arheoloških nalazišta (Izabrani djelovi teksta i grafičkih priloga). Muzej Prigorja, Zagreb
Saruhan Mosler, A (2005) Landscape architecture on archaeological sites—establishing landscape design principles for archaeological sites by means of examples from West Anatolia, Turkey. Dissertation, Technische Universität München
Stadtverwaltung Xanten (2015) Luftkurort Xanten Erleben. Xanten
Tosco C (2014) I beni culturali. Storia, tutela e valorizzazione. Il Mulino, Bologna
Teatini A (1993–94) Il clomplesso edilizio di Porto Conte. Almanaco Gallurese
UNESCO (1972) Convention concerning the protection of the world cultural and natural heritage. Paris
UNESCO (1997) World Heritage Committee Twenty—first session, committee report no. 833. Naples
UNESCO (2014) Periodic report—second cycle, section II Stari Grad Plain
UNESCO (2017) Operational guidelines for the implementation of the World Heritage Convention
UNWTO (2015) Global report on cultural routes and itineraries
Želle M (2006) Arheološki park u parku prirode. Hrvatska revija 6(3) Available via Matica hrvatska. http://www.matica.hr/hr/343/arheoloski-park-u-parku-prirode-20987/. Accessed 26 Feb 2018

Chapter 10
Urban Transformation and Sustainable Development of Small Historic Towns

Nikša Božić, Biserka Dumbović Bilušić and Jasenka Kranjčević

Abstract Small historic towns in Europe face many common challenges that remain relatively unknown and unrecognized. Small towns make an important element in a settlement network. Their urban functions provide a range of key services to rural areas surrounding them and are an important developmental stability and vitality factor of the wider area. In Europe, the majority of small towns are historic towns whose urban heritage significantly contributes to local identity, from local to national and wider levels. Small historic towns face common challenges, including the lack of administrative capacities and the lack of flexibility to adapt to complex demands of modern development. This leads to a reduction in economic activities and service functions, demographic problems and general economic, social and physical decline. Urban transformation of small historic towns is a specific topic requiring innovative approaches. Successful urban renewal models for small historic towns are models that are capable of achieving a balance between the preservation of their cultural heritage and the demands of development, strengthening their competitiveness while preserving their authenticity.

Keywords Small towns · Small historic towns · Urban heritage · Urban transformation · Sustainable development

N. Božić (✉)
Croatian Section of ECOVAST (European Council for the Village and Small Towns), Zagreb, Croatia
e-mail: niksa@ecovast.hr

B. Dumbović Bilušić
Ministry of Culture of Republic of Croatia, Directorate for the Protection of Cultural Heritage, Conservation Department in Rijeka for the area of the Primorje—Gorski Kotar County, Rijeka, Croatia
e-mail: biserka.dumbovic-bilusic@min-kulture.hr

J. Kranjčević
Institute for Tourism, Zagreb, Croatia
e-mail: jasenka.kranjcevic@iztzg.hr

M. Obad Šćitaroci et al. (eds.), *Cultural Urban Heritage*,
The Urban Book Series, https://doi.org/10.1007/978-3-030-10612-6_10

10.3 Small Historic Towns

In Europe, small historic towns play an important role in preserving local identities. Cultural heritage is at the very core of what constitutes the spirit of a space, contributing fundamentally to the creation of the identity of peoples and local communities. Despite their scale, small towns have a special relationship with their immediate environment, which is mostly composed of cultural landscapes, spaces that contain historically characteristic structures that testify of human presence in space, and represent the common work of man and nature, illustrating the development of the community and its territory throughout history.

In Croatia, the majority of small and medium-sized towns are historic towns with protected historic areas (Fig. 10.2) and individually protected buildings. Even the rare planned towns built in the twentieth century are now valued as cultural heritage, and therefore not even they make an exception to the general rule that a *small town* equals a *historic small town*.

This is the case in most European countries, where small towns are generally all historic and have grown up since (at least) medieval times (Carter 2014). Not overwhelmed by modern development, they remain compact and reflect architecture and history of their historical development. Protected historic areas and historic buildings in small towns pose specific development challenges, but can also be seen as the potential for development, especially for the development of heritage-based economic activities like tourism (Kranjčević 2009).

10.4 Common Problems of Small European Towns

There are no specific policies in Europe related to small towns. Urban policies are predominant, and there are also a significant number of rural policies (although many are devoted to agriculture and food production rather than rural settlements). The sheer number of people living in small towns should give them a much stronger voice in developing policies to reinforce their position.

Changes in the economy, modern way of living, labour and traffic, and changes in the demographic structure of the population result in numerous changes in small towns. In Europe, most small towns share the same problems; however, there are gaps in European development policies when it comes to small-town challenges. Large cities are relatively well represented in the European Regional Development Fund, and the same applies to rural areas covered by the European Agricultural Fund for Rural Development. Small towns have found themselves in the void between European policies devoted to regional development, urban areas and spatial development. These policies include measures for small towns, but often marginal ones. The mechanisms that enable the simultaneous use of different sources from several European funds for specific projects are still ineffective and burdened by numerous administrative barriers that do not guarantee the efficiency of use.

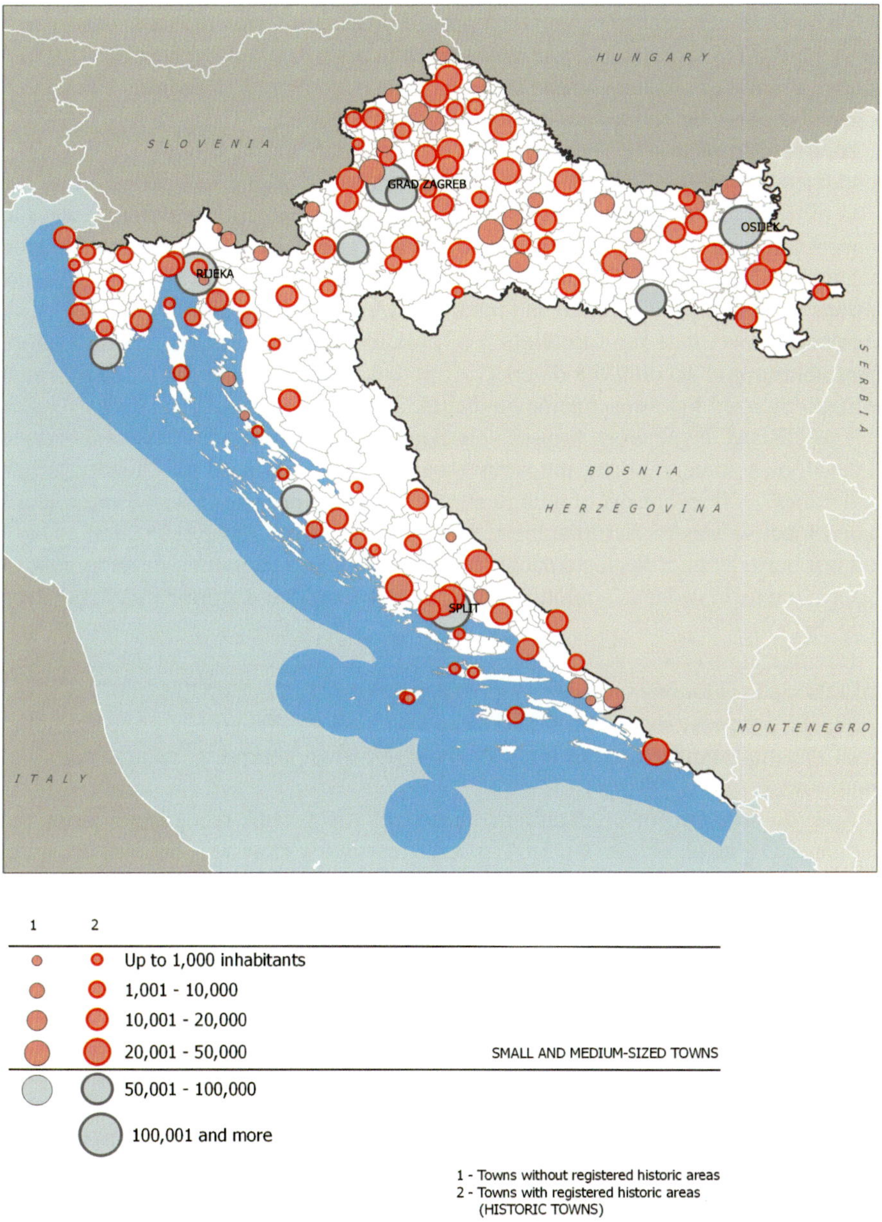

Fig. 10.2 Croatian towns by population size and existence of registered historic area

These challenges faced by small towns in Europe have been the topic of research for some time by experts and non-governmental organizations such as ECOVAST.[3] Lately, a number of pan-European research and cooperation projects have been devoted to common challenges of small European towns.

A large European scientific project "Roles of Small and Medium-sized Towns",[4] conducted within the ESPON 2000–2006[5] framework programme, focused on the roles of small and medium-sized European towns. The objective was to find ways to conceptualize small or medium-sized towns and to describe the main roles played by small and medium-sized towns in territorial development. The aim was to derive options for typologies for small and medium-sized towns and their regional context, which could facilitate the formulation of policies oriented towards urban development, taking into account the diversity of potentials and challenges faced by small or medium-sized towns (Schneidewind et al. 2006).

The URBACT[6] network project "Heritage as Opportunity" (HerO) was aimed at developing integrated and innovative management strategies for "historic urban landscapes", which consist of a large number of small historic towns. The project's main objective was to facilitate the right balance between the preservation of built cultural heritage and the sustainable, future-proof socio-economic development of historic towns in order to strengthen their attractiveness and competitiveness (Ripp et al. 2011).

Revitalization concepts of small European towns were the topic of the 2013 Central Europe Programme Qualist ("improving quality of life in small towns"), carried out by 11 partners from Central European countries (* 2013). The project emphasized the importance of developing lively small-town centres as central points of small towns.

Specific problems of small and medium-sized towns were recognized during the Latvian EU presidency in 2015. A special research report was devoted to small and medium-sized urban areas (SMUAs). The report found that such urban areas, while prominent in the EU territory, are largely unexplored in terms of their social, spatial and economic trends because of the lack of comparable data and relative political disregard. The report focused on main challenges, barriers and potentials for the economic development of small and medium-sized urban areas, including their contribution to territorial development (Valterbergs et al. 2015).

At the beginning of 2017, the European Economic and Social Committee (EESC), an advisory body of the European Union providing expert advice to the EU's main institutions, set up a working body to deal specifically with problems of small towns and their links with rural areas. An opinion paper was published stating specific

---

[3]ECOVAST—European Council for the Village and Small Towns—is a pan-European non-government organization established in 1984. Croatian section of ECOVAST was established in 1993. All three authors are members of the Croatian section of ECOVAST.

[4]The used acronym SMESTO stands for small and medium-sized towns and is now more widely used.

[5]ESPON—European Observation Network for Territorial Development and Cohesion.

[6]URBACT is a European exchange and learning programme promoting sustainable urban development.

challenges of small towns as well as opportunities for their development, stressing the importance of urban–rural linkages (Jones 2017).

These examples illustrate the importance of recognizing common problems faced by European (historic) small towns, as well as the need for a more systematic approach. They can form the starting point for taking the challenges of small European towns into account during the formulation of future European development policies and programmes.

ECOVAST has actively advocated specific needs of small European towns for the past three decades (Moore 2009; Carter 2014). In 2014, a study was published on the importance of small towns on the European level (Carter 2014). In Croatia, the Croatian section of ECOVAST has organized a number of conferences devoted to small towns, including the ones in Samobor in 2007,[7] Mošćenice in 2009[8] and Ivanić-Grad in 2011.[9] Conclusions from the Samobor conference, known as the *Samobor Declaration* (* 2007), and the *Position statement on small European towns* (Turner 2009) are still relevant today and constitute the starting point of the ECOVAST's position on the subject of small towns.

Through the analysis of the aforementioned projects, and based on an extensive work carried out by the Croatian section of ECOVAST, the following common problems of small European towns are recognized:

- *Changes in the economy,* in particular globalization leading to global production and transport of goods (including food), which leads to the weakening of the traditional role of small towns as markets for local products from the surrounding rural areas.
- *Changes in service organization and centralization of functions.* This is a consequence of efforts to reduce costs and results in the closing of administrative functions such as courts, state administration offices and health centres in small towns. In addition to losing traditional functions in this way, small towns also lose skilled professionals that move to larger urban centres.
- *Reduction of traditional economic activities* due to changes in the economy, leading to the closure of traditional businesses, changes in trade patterns and the inability to adapt to new demands. As a result, small towns are gradually losing their inhabitants and their traditional economic base (crafts and other traditional service and trade activities).
- *Loss of attractiveness for economic activities.* Built structures in small towns are often limited for adaptation to new spatial needs as they are located in sensitive buildings (often under formal protection) that are outdated and lacking adequate traffic or infrastructure systems. Demographic problems also lead to the shortage of qualified workforce. This is not attractive to new investors who are looking for

---

[7]The International Conference 'Small European Towns—the Role in Developing Rural Areas and Heritage Conservation', Samobor, 13–14. October 2007.

[8]International conference 'Revitalisation of Historical Small Towns and Their Hinterland, Mošćenice', 23–24 October 2009.

[9]International conference 'Models for Management of Historical Towns Revitalization And Development Processes', Ivanić-Grad, 11 November 2011.

more stimulating environments that are, generally, easier to find in larger urban centres.

- *Changes in the demographic structure of the population and ageing of the population*, a problem shared by many European countries but especially visible in rural areas and small towns entering the processes of natural ageing of the population that are often irreversible.
- *The imbalance between the supply and the demand of flats*—reduction of economic activities and the departure of young people cause disturbance in the housing market, falling housing prices and the appearance of many abandoned housing units.
- *Imbalances in social infrastructure*—there is a weakening and closure of basic social functions such as childcare, education, primary health care and the like. These systems apply strict rules that pursue exclusively economic efficiency without taking into consideration the impact of their closure on the perspectives of small communities.
- *The abandonment of historical areas and pressures on space and landscape.* Many of the small towns' historic centres are losing their appeal because they do not meet the today's requirements for residence or business. New areas in which settlements are spreading create pressure on the surrounding landscapes and lead to the loss of valuable agricultural or forest areas.
- *Disadvantages of inadequate planning systems*—from strategic planning to spatial planning, the systems are largely unmodified when applied to small towns. This leads to an inadequate approach, further contributed to by consultants and experts who are often not sensitive to the specific problems of small towns and propose general solutions difficult to apply in such fragile local environments.
- *Problems of the protection of historic sites and buildings that are viewed separately from development plans*—the protection of historic areas and individual buildings is often considered separately from future development strategies. As a result, cultural heritage is seen a burden instead of as a potential for development.
- *Inadequate accessibility and lack of public transport*—one of the most important causes of rural depopulation is the unavailability of basic services such as education, health care, trade, culture and entertainment, traditionally offered by small and medium-sized towns. The problem of traffic isolation of rural areas is related to the shortcomings of quality public transport. On the other hand, fast networking of large urban centres with high-speed services (high-speed trains, direct coach services via motorways) leads to the reduction or abolition of services in intermediate areas, and thus small towns lose not only their former traffic connections but also their former functions of transport centres.
- *Unsustainable tourism development and unfavourable impact on the area*—in many historic towns, especially in the coastal area, great importance is given to the development of tourism, often without any critical reflection on all the consequences. This leads to unfavourable pressures on space and infrastructure systems as well as economic imbalances, especially in towns where tourism has a seasonal character.

All of these challenges have led to the use of the concept of *shrinking areas* (Haase et al. 2012) related to the phenomenon of population collapse, economic activities, knowledge and potentials of small towns (Moore 2009).

10.5 Models for the Management of Small Historic Towns' Urban Transformation and Development Processes

Numerous projects and studies are seeking a new methodological approach in the research, planning and implementing of the sustainable urban development of existing urban structures with a complex aim at development as well as protection of specific values of cultural heritage in different historic urban structures (Fister 2007, 2011).

Sustainable urban development of small historic towns in this context is seen through sustainable urban economics, preservation of the urban environment, social context and good governance measures (Czischke et al. 2015).

Models for the management of historic towns' revitalization and development processes were one of the topics of the *Restoration and sustainable development of small historic towns in Croatia* project, conducted by the Croatian section of ECOVAST. Heritage of small historic towns has been recognized as an important part of Croatian spatial identity and as the potential for growth and development of the new quality of life within them (Božić and Dumbović Bilušić 2011).

The project stressed the importance of small historic towns as essential factors of spatial and cultural identity. They are significant because of their role in the preservation of local identities as well as their important role in relation to rural areas surrounding them.

The following are possible models for the management of small historic towns' urban transformation and development processes:

- *Polycentric development:* Subregional policies and planning have increasingly relied on partnerships between the network of towns and rural areas surrounding them. Power is based on cooperation and realized through the network of small towns and rural areas linked to them.
- *Encouraging awareness of the quality of life in small towns:* Small-town branding often targets small-town visitors, with little effort put into emphasizing the attractiveness of small-town living for local residents.
- *Responsive governance and dialogue:* The size of communities in small towns offers possibilities for open and tolerant dialogue between the town government and residents. This must include building awareness of the importance of greater participation of the local community in decision-making. Civil society organizations can have an important role in articulating the public voice.
- *Protecting local businesses:* A prominent peculiarity of small towns is the survival of small businesses, especially small private retailers and craftsmen. Measures

must be taken to ensure the efficiency and survival of local entrepreneurs who, besides economic purposes, also have a significant social and cultural function.

- *Entrepreneurial spirit development:* Individual entrepreneurial initiatives are important drivers of economic progress of an area. Therefore, it is necessary to constantly build trust in entrepreneurial initiatives, especially in business forms based on the principles of social, environmental and economic sustainability.
- *Local approach to economic development*: The business environment of today is characterized by complex interdependencies and complex systems of cooperation and connectivity. Knowledge transfers, learning from good practices, investing in collaboration and recognizing local development potentials are important elements for stimulating the economic development of small towns.
- *Care for the youth and the elderly:* Care for young people does not end with childcare and primary education. The ability to provide affordable accommodation and employment for young people and families can preserve the vitality of small towns. At the same time, as people nowadays live longer, elderly people seek education and healthcare services. Care for the elderly also includes policies to integrate elderly population into the society.
- *Integrated planning:* The existing system of spatial and urban plans is not well adapted to the historical environment of small towns. Often, the integration of measures for the protection of cultural heritage is merely formal. Obligatory public involvement in planning processes needs to be upgraded with active participation. It is important to strive to implement plans which would demand integrated approaches that combine models of planning, financing and sustainability of restoration programmes.
- *Public participation:* Professional studies of historic town structures and strategic decisions for the restoration of small historic towns must seek to include other stakeholders actively involved in these processes, namely local citizens, business owners, civil society associations, institutions, politicians, professional and administrative bodies, and professionals in town administration.
- *Preserving local heritage and identity:* Most European small towns are historical places. Permanent education and raising awareness of the value of historical heritage are of the utmost importance. Such awareness, combined with active public participation in all processes, may become the most efficient protection mechanism of the local built (urban) environment.
- *Local identity as a starting point for tourism development:* Small towns are associated with their landscape and have identities that contribute to the character of their wider regions. These identities across Europe are attractions on which cultural tourism is based. Quality models for the restoration of cultural property as the basis for tourism development are important. At the same time, tourism development must not ruin the harmony of social and economic conditions of local life and the identity of the place.
- *Creating new attractions:* The complex relationship between culture and the forming of local identities could also include the creation of new attractions in new or remodelled buildings which transfer new messages and meanings of local culture which inevitably intertwine with tourism.

- *Improving infrastructure:* Small towns are often faced with the constraints of the existing technical infrastructure. New environmental requirements often prove challenging for the limited local budgets, but should be also seen as the possibilities for economic development. Renewable energy infrastructure can stimulate economic development and improve living conditions, as well as strengthen the links with the surrounding rural areas.
- *Improving local transport:* The problem of traffic isolation, and in particular the lack of quality public transport in a rural area, reflects in the (un)availability of basic services for rural population, especially vulnerable groups such as the poor and the elderly. It is necessary to work on the development of public transport to ensure its availability and, in return, the use of such services.
- *Using the possibilities of cooperation and connectivity:* Opportunities for development through programmes and cooperation projects constantly open up on the national, European and wider international agenda. Small towns face many obstacles on their way to achieving development ideas so participation in projects that include the exchange of knowledge and networking is often of great importance in finding solutions.

Urban transformation of small historic towns has economic, social and environmental aspects.

Economic aspects of urban renewal of small historic towns include support to the local economy through works carried out, opportunities for new facilities in restored areas, improved image of the site (which contributes to the development of creative activities, culture and tourism) and attracting educated workforce and young people.

Social and cultural aspects of urban renewal include preserving cultural heritage and local identity, creating places for cultural and social events, improving the quality of life and encouraging residents to actively participate in the social life of the place.

Finally, environmental aspects of urban renewal include preventing the occupation of unbuilt areas by using the already occupied spaces, reducing the need for additional traffic loads and contributing to the reduction of the suburbanization process.

10.6 Conclusion

Small European historic towns face various common challenges that lead to the reduction in economic activities and service functions, demographic problems as well as general economic, social and physical decline. Renewal and sustainable development of small towns are specific topics that require innovative approaches, including integrated planning, local stakeholder participation, focus on clear goals and implementation, monitoring and management of measures.

In Europe, small towns play an important role in preserving local identities. The majority of small and medium-sized towns are also historic towns with protected historic areas and individual protected buildings. These protected historic areas and architectural buildings in small towns pose specific development challenges, but can also be seen as the potential for development, especially for the development of heritage-based economic activities like tourism.

For small towns, therefore, the goal is to find new active policies, strategies and renewal models which would provide balance between the preservation of cultural heritage and the sustainable socio-economic development in order to strengthen their attractiveness and competitiveness while preserving the authenticity of fragile local communities.

Basic strategic goals of urban transformation of small historic towns must respect the specifics of the place since the purpose of renewal is not always to reconstruct historical forms or elements in full, but to find solutions for reading historical contents, signs and symbols in the structure and cultural landscape of the settlement. The history of the place, with a symbolic, emotional charge, carries the potential that can and should be an inspiration for creating new values.

Acknowledgements The research is a part of the scientific project "Heritage Urbanism—Urban and Spatial Planning Models for Revival and Enhancement of Cultural Heritage". It is financed by the Croatian Science Foundation [HRZZ-2032] and carried out at the University of Zagreb, Faculty of Architecture.

References

Božić N (2017) Obnova i održivi razvitak malih gradova. In: Korlaet A (ed) Urbano-ruralne veze. Stručni skup (conference) Urbano-ruralne veze, Sveti Martin na Muri, p 66

Božić N, Dumbović Bilušić B (ed) (2011) Modeli upravljanja procesima obnove i razvoja povijesnih gradova. Zbornik radova međunarodnog znanstveno-stručnog skupa, Ivanić-Grad (Conference proceedings—International conference models for management of historical towns revitalization and development processes, Ivanić-Grad, November 2011). Available via Hrvatska sekcija ECOVAST-a. http://ecovast.hr/index.php?option=com_content&view=article&id=85%3Akonferencija-ivani-grad-zbornik-radova&catid=9%3Apub-download&Itemid=81. Accessed 1 Jun 2018

Carter V (2011) Encouraging Economic Success in Historic Small Towns. In: Božić N, Dumbović Bilušić B (ed) Conference proceedings—international conference models for management of historical towns revitalization and development processes, Ivanić-Grad, November 2011. Available via Hrvatska sekcija ECOVAST-a. http://ecovast.hr/index.php?option=com_content&view=article&id=85%3Akonferencija-ivani-grad-zbornik-radova&catid=9%3Apub-download&Itemid=72. Accessed 24 Jun 2018

Carter V (ed) (2014) The importance of small towns. The European Council for the Village and Small Town (ECOVAST). Kent, UK

Czischke D, Moloney C, Turcu C (2015) Setting the scene: raising the game in environmentally sustainable urban regeneration. In: Sustainable regeneration in urban areas. Available via URBACT. http://urbact.eu/sites/default/files/04_sustreg-web.pdf. Accessed 30 Jun 2018

ECOVAST (2007) Samoborska deklaracija (Samobor Declaration), Available via Hrvatska sekcija ECOVAST-a. http://ecovast.hr/index.php?option=com_content&view=article&id=14&Itemid=62. Accessed 30 Jun 2018

European Commission (2011) Territorial agenda of the European Union 2020. Towards an inclusive, smart and sustainable Europe of diverse regions. Available at EC. http://ec.europa.eu/regional_policy/en/information/publications/communications/2011/territorial-agenda-of-the-european-union-2020. Accessed 30 Jun 2018

European Urban Knowledge Network (2013) Urban vision Central Europe—small town 2020, project Qualist-improving quality of life in small town. Saxon State Ministry for Economic Affairs, Labour and Transport. Available via EUKN. http://www.eukn.eu/fileadmin/Files/News/2015/Handbook_Urban_Vision_Small_Town_Central_Europe.pdf. Accessed 15 Nov 2017

Fister P (2007) Reurbanizacija/prenova naselbin in arhitekture (metodologija načrtovanja)—Reurbanisation of architecture and urban structures (planning methodology). Univerza v Ljubljani, Fakulteta za arhitekturo, Ljubljana

Fister P (2011) Basic aims and new tasks of urban renovation. In: Božić N, Dumbović Bilušić B (ed) Conference proceedings—international conference models for management of historical towns revitalization and development processes, Ivanić-Grad, November 2011. Available via Hrvatska sekcija ECOVAST-a. http://ecovast.hr/index.php?option=com_content&view=article&id=85%3Akonferencija-ivani-grad-zbornik-radova&catid=9%3Apub-download&Itemid=72, Accessed 24 Jun 2018

Haase A, Horpers GJ, Pekelsma S et al (2012) Shrinking areas. European Urban Knowledge Network, Haag

Hrvatski zavod za prostorni razvoj (2017) Strategija prostornog razvoja Republike Hrvatske (Strategy of spatial development of Republic of Croatia). Available via Hrvatski zavod za prostorni razvoj. http://www.hzpr.hr/UserDocsImages/strategija/SPRRH_e-knjiga.pdf. Accessed 30 Jun 2018

Jones T (2017) Villages and small towns as catalysts for rural development—challenges and opportunities (own-initiative opinion, European Economic and Social Committee). Available via EESC. https://www.eesc.europa.eu/en/our-work/opinions-information-reports/opinions/villages-and-small-towns-catalysts-rural-development Accessed 30 June 2018

Kranjčević J (2009) Turizam – mogućnost za razvoj malih gradova. In: Mohović R, Dumbović Bilušić B, Lipovac N (ed) Mošćenički zbornik No.6. Katedra Čakavskog sabora Općine Mošćenička Draga, ECOVAST

Kranjčević J, Lukić A, Kušen E et al (2014), Regionalni razvoj, razvoj sustava naselja, urbani i ruralni razvoj i transformacija prostora. Stručna podloga za izradu Strategije prostornog razvoja Republike Hrvatske. Available via Hrvatski zavod za prostorni razvoj. www.hzpr.hr. Accessed 30 Jun 2018

Moore P (2009) The Small Historic Towns of Europe in the Economic Downturn; In: Mohović R, Dumbović Bilušić B, Lipovac N (ed) Mošćenički zbornik No.6. Katedra Čakavskog sabora Općine Mošćenička Draga, ECOVAST

Ripp M, Bühler B, Shakhmatova K (ed) (2011) The road to success—integrated management of historic towns, guidebook. URBACT HerO network, Stadt Regensburg. Available via URBACT. www.urbact.eu/hero. Accessed 30 Jun 2018

Schneidewind P et al (ed) (2006) The role of small and medium-sized towns (SMESTO)—Final Report, ESPON 1.4.1. Österreichisches Institut für Raumplanung (ÖIR)—Austrian Institute for Regional Studies and Spatial Planning

Turner P (2009) Izjava o stajalištu o malim europskim gradovima, ECOVAST - Europsko vijeće za sela i male gradove. Available via Hrvatska sekcija ECOVAST-a. http://ecovast.hr/index.php?option=com_content&view=article&id=15&Itemid=61. Accessed 30 Jun 2018

Valterbergs V et al (2015) Challenges of small and medium-sized urban areas (SMUAs), their economic growth potential and impact on territorial development in the European Union and Latvia, Research report to support the Latvian EU Presidency 2015. Available via EUKN. http://www.eukn.eu/fileadmin/Files/EU_Presidencies/Latvian_Presidency/Final_Report_26.05.2015_EXEC_SUMMARY.pdf. Accessed 30 June 2018

Chapter 11
Planning as a Function of Preserving the Identity of Place

Nenad Lipovac, Gojko Nikolić, Svetislav Popović and Nikolina Gradečki

Abstract People create *Places*, but can hardly distinguish *Places* from *Space* as the definition of *Place* is a complex integration of nature and culture manifested in physical terms, which has been developed and is still developing in particular locations. *Places* are linked by a flow of people and goods between them, and each of them has its attributes—altogether creating the *Identity of Place*. Proper respect of cultural and natural heritage values, along with adequate safeguarding and maintaining, will help in preserving the *Identity of Place*. The most effective tool a planer can use in managing and controlling the *Place Identity* and its valuable identity attributes is a clearly defined planning process that will result in setting clear and omni-understandable planning ordinances and proposed methodologies for conserving/preserving (protecting) the *Identity of Place*. This paper proposes possible planning steps that represent a planning model researched and developed within the HERU, scientific project, which could help achieve this goal. The planning process today is a multidisciplinary one, with professionals from different fields taking part. The planner's role is to coordinate and combine all of their efforts within a single document (physical or urban plan), a document that will create a unique searchlight for the *Place* development. To make it possible and reach that planning model, all participants must speak the same (professional) language. We need something we all shall agree upon—a relevant glossary within a field that will be prepared by professionals, not by lawmakers.

N. Lipovac (✉)
Faculty of Architecture, University of Zagreb, Zagreb, Croatia
e-mail: nlipovac@arhitekt.hr

G. Nikolić
Department of Geography, University of Montenegro, Nikšić, Montenegro
e-mail: gojkorn@t-com.me

S. Popović
Faculty of Architecture, University of Montenegro, Podgorica, Montenegro
e-mail: svetislav@ac.me

N. Gradečki
Faculty of Architecture, University of Zagreb, Zagreb, Croatia
e-mail: ngradecki@arhitekt.hr

M. Obad Šćitaroci et al. (eds.), *Cultural Urban Heritage*,
The Urban Book Series, https://doi.org/10.1007/978-3-030-10612-6_11

Keywords Identity of Place · Identity with Place · Place Appearance · Identity attributes · Planning ordinances · Planning steps

11.1 Introduction

One of the goals of the HERU scientific project was the search for applicable planning and implementation models that could be used in different levels of plan-making, the planning models or procedures that could help in preserving the *Identity of Place*. These models should assist in recognizing local and regional changes, as well as their impacts over the *Identity of Place*, *Identity with Place,* and *Place Appearance*, its natural and cultural values, social life, the economy, and the necessary steps in protection and preservation of these *Place* issues. Physical and urban planning should represent a process of finding out development needs, possibilities, and limits for a particular region or settlement. Proper identification, evaluation, and implementation of existing Place and curtilage values should be of crucial importance in the decision-making process for the planned development of a region or a settlement. And this is the point where we can talk about and discuss the *Identity of* or *with Place* and the *Place Preservation/Protection/Conservation* (whatever the meaning of these words is). Authors of this paper will define the elements that help in distinguishing space from a particular place and interpret the legal framework in which physical and urban planners have to prepare the plans nowadays. As the next step, the authors will, by reviewing research results acquired through their own professional experience in making plans for heritage areas, define the possible and necessary models and planning steps needed to ensure proper and sustainable planning, along with preparing planning ordinances, to preserve the *Identity of Place*.

11.2 Place Versus Space

Today's physical and urban planning should result from common sense and orderly approach to finding out the local needs, setting goals and priorities, and taking action to give meaning to the space with natural and cultural environment values. To do this through the process of planning any human settlement, a planner must recognize and understand the existing environment, and what needs to be done during the planning process and the forthcoming plan implementation. What are the most efficient planning steps and methodologies, regarding both the economic development of a particular region and the protection and preservation of the *Identity of Place* (specific settlement and its curtilage)? Natural and cultural heritage of the observed area and their preservation are mentioned in today's generation of legal documents and plans, but the planning ordinances are not written, and lately also not understood, as an obligatory legal tool. As a result, places lose their traditional sense of place.

It is difficult to define a unique and universal planning methodology that would preserve all of the values of a certain *Place*. From the very beginning of *Place* (settlement) planning, there have been recognizable steps in the planning process: *monitoring*, *reading*, *evaluating*, and *implementing*. Since then, centuries and millennia have passed, and Man continued on planning within the environment, changing settlements and their curtilage, trying to find an ideal way to stay in this living-in world. He learned that *planning is more than a tool; it is a philosophy for organizing actions that enable people to predict and visualize the future of any land area. ... It is up to us to plan with vision. Our responsibility is to retain what we treasure because we are merely guests on those spaces of the earth that we inhabit* (Steiner 1991). The afore-mentioned planning steps are still in use in contemporary planning models dealing with heritage that has to be preserved.

Within the planning process, the planner must understand the meanings of two important terms, essential for the future being of a *Place*: *Identity of Place* and *Identity with Place*. Human activity and creation within the natural and cultural environment play the most critical role in the planning process. But at the very beginning, one question has to be posed: what makes one *Place* so unique and different from other *Places*? To answer this question, a planner must be capable of distinguishing *Place* from *Space*, acknowledging its values, and reading its identity attributes. But to do that, a planner firstly has to reveal the elements which make that difference—elements of the *Place Identity*—and accept the most important rule: without people, there are no *Places* as human participation is essential in creating the *Identity of the Place*, regardless of the fact that different people differently perceive and experience the *Place* and its curtilage. All that, undoubtedly, leads us to the subjective feeling of being the *insider* or the *outsider* of a *Place*, a feeling that should not be forgotten in the planning process (Lipovac 1997). That feeling is known as the *Identity with Place*.[1]

At the very beginning of the planning process, a planner must be capable to clearly distinguish and then understand the meaning of these two words: *Space* and *Place*. The next important planner's issue is to be able to capture and understand the *Sense of Place*, and its attributes that make any area or region (within the Plan boundaries) so different and special from the others, protecting and keeping that sense throughout the entire planning process. A plan, as a step-by-step process, must protect the unique physical resources of each place through *time* by solving problems or at least by giving guidelines on how to avoid problems (Lipovac 2000) or diminish their impact. An urban or a physical plan must protect the *Place* by safeguarding the entire environment (natural, cultural, and/or built) and its identity attributes. To be able to conserve, preserve, and/or protect the *Place* throughout the planning process, a planner must clearly understand how to read the *Place Appearance* and how to distinguish it from the Place Identity.[2] And finally, it is crucial for a planner

[1] *Insiders versus Outsiders* is one of the *Ten Properties* of *Identity with Place* revealed by Prof. Violich (1996).

[2] *Place Appearance* can be described as the pure visual experience of the Place, while the Place Identity consists of visible and invisible source elements of identity.

to understand the meanings of the omnipresent and worldly accepted terms for the methodologies implemented in the maintenance of the Identity of Place, along with cultural or natural values within the Place and its curtilage.

The *Place Appearance* is tightly connected to all settings and situations in which we live; it carries all possible visual diversity and the intensity of human experience of any part of space. *Place Concept* is essentially a term from geography. An excellent discussion on this subject can be found in an article by Lukerman (1964) in which he wrote about six major of *Place* components concepts.[3] On the other hand, according to the geographer Yi-fu Tuan, *Place* possesses *perceptual uniqueness* that is given to it by an individual's experiences with places (Tuan 1975). In our living-in world, a *Place* cannot be experienced independently, clearly, as a defined entity, an entity that could be easily and simply described in terms of location and appearance. A *Place* must be understood in the context of other places, other features, attributes and appearances or settings within the observed environment, surrounded by different elements from the natural environment. All these features create a different *Identity of Place*, *Identity with Place,* and *Place Appearance*. They all simultaneously overlap, forming different *Place Identities* and different *Places*, creating endless varieties of their interpretations. Places are incorporated into intentional structures of all-human consciousness and experience, which is very likely to be a vital source of both individual and collective identities *of* and *with a Place*, a source point from which we place and orient ourselves within the world around us. Considering a *Place* as a multifaceted phenomenon of experience, a planner must deal with the following place elements: *location*, *landscape*, *time,* and *personal involvement* (Lipovac 2000) as these attributes are critical to the *Identity of Place* or the *Genius Loci*.

There are eight, very recognizable *Genius Loci* attributes that should be referred to during the implementation of certain planning process steps in order to preserve the identity attributes. The first seven belong to the so-called physical features of a place, while the last one has a subjective meaning. They are as follows:

1. Location;
2. Boundaries;
3. Geographic distinctiveness;
4. Frequency versus appearance uniqueness;
5. Integration within the environment;
6. Diversity;
7. Composition;
8. People.

[3](a) The idea of location, location as it relates to other things and places, is absolutely fundamental,

(b) Place involves an integration of elements of nature and culture; this undoubtedly implies that every place is a unique entity,

(c) Although the places are unique, they are interconnected by a system of spatial interactions and transfers, part of a frame of circulation,

(d) Places are part of larger areas and are the focuses in a system of localization,

(e) Places are emerging and becoming: with historical and cultural change, new elements are added, while some old disappear,

(f) Places have meaning: They are characterized by the beliefs of individuals.

Location is a physical attribute of every *Place* within *Space*. It helps to describe where a particular *Place* is (its connection with the curtilage), and nothing else. When defining the location of a *Place*, the definition of **boundaries** is necessary. Within the planning process, three kinds of boundaries can be recognized (*first, second,* and *third*). They represent linear delineation between the *Place* and its curtilage. The *first boundary* represents an administrative (political) boundary, the *second one* covers the area on the other side of the first boundary, while the *third one* represents the visual boundary[4] that can be observed while moving through the *Place* (within the first boundary). **Geographic distinctiveness** is a physical attribute of a *Place* that has to be recognized and observed regardless of the *first boundary,* and it defines the geographic position of a *Place*. **Frequency versus appearance uniqueness** describes the existence of the attributes that are either unique within a *Place* or are repeated within its curtilage (surrounding area). **Integration within the environment** represents the interconnection and interrelation between the *Place Identity attributes* and elements and their features from different environments. The term **diversity** consists of various features that can be described merely as: *scale*, *proportion*, *texture*, *color,* and *contrast* (of existing *identity attributes*), while **composition** represents the arrangement of all previously mentioned *identity attributes*. The final one, **people**, covers a very wide range of human behavior and the way of living and working within a certain *Place*.

As it is very difficult to determine a strong borderline between nature and the man-made world, many of the elements from nature that represent sources of place identity can be easily viewed through the glasses of human culture throughout the history of mankind. According to Ken Yeang, ecologist and environmentalist, the planner who wishes to protect and preserve nature (natural and ecological identity of a place) must not see the built environment as a separate "file" from the natural environment and the ecosystems that surround it. *Ecologically, people together with the built environment must be perceived to be part of the (biotic) components and functioning of the ecosystems within the biosphere, even though we may find that their presence may cause conflicts with the ecosystems* (Yeang 1995). The inventory of natural elements that create the identity of a place or a region represents an enormous range of different types of natural attributes. There are elements related to the earth activity (geomechanical and hydrothermal features) and another one, known as *geodiversity*, which combines *pedology*, *topography (physiography),* and *relief (landform).* As an undividable part of this inventory, an analysis of *water and related sources*, *plant,* and *wildlife*, as well as *climate*, must not be neglected. All of these elements create what is known and usually described with a straightforward word: *landscape*. To do the proper evaluation of our landscape, the landscape itself has to be perceived: to see what nature is offering us. This can be achieved almost from anywhere, but still, there are certain points from which we can get the best of it. These points are known as *scenic points* from which there are *scenic views*. *Existing land use analysis* should be a part of this inventory, too.

---

[4]In landscape architecture, this borderline is very often defined as a *borrowed landscape* (Lipovac 2014).

Next to these natural *identity attributes* are the attributes from the so-called *cultivated environment* (also known as man-made landscape or *cultural environment*) that are a result of the agricultural land use[5] (*Cultural Heritage Protection and Preservation Act—CHPPA*). The major groups of the *identity attributes* still connected with the ones from natural environment are: historic landscapes (farmland, cemeteries and parks and gardens) and greenbelts[6] (man-made greenways and buffers). On the other hand, slowly approaching the built environment range, there are the elements from vernacular architecture. *Vernacularism* is a term that originates from the cultural environment and plays a vital role in creating the *Identity of Place*. It can just be described as the traditional way of building and creating settlements that reflects the characteristics of local communities and represents their response to the surrounding environment. It encompasses something that was not planned but created in accordance with a sense of living in a specific region. It carries an effective response to functional, social, and environmental constraints. *Vernacularism* is, undoubtedly, a *sine qua non* in creating the *Identity of Place*. Neglecting *vernacularism* within a certain area would cause incurable damage to the *Identity of Place* that had sometimes been created for centuries. The most important issue regarding vernacular architecture is its unique connectedness with the surrounding environment, making a *Place* inseparable from its curtilage. Vernacular architecture (buildings or other built features with ethnographic or toponym importance) can be observed through three different types: rural, residential, and sacral architecture. Other elements from the cultural environment that should be evaluated are: settlements (town, village, hamlet, or parts thereof); buildings and building parts along with their curtilage; historic urban facilities; area, site, monument, or landmark connected with a historic event or a person; archeological areas and sites (including underwater sites); industrial architecture and equipment (*Cultural Heritage Protection and Preservation Act* 1999).

Finally, human activity and creation within the entire environment (natural and cultural) definitely plays one of the most important roles in the process of creating an *image*, the *spirit*, the *Identity* of a particular *Place*. A person is creative as he or she builds, composes, and reorganizes *Space* in order to create *Places*.

11.3 Legal Framework and the Identity of Place

For managing and maintaining *Places*, people develop, more or less successfully, policies, acts, and ordinances that will help in ruling, preserving, and protecting valuable *Place Identity attributes*. In many cases, legal documents contain glossaries

[5] According to the *Cultural Heritage Protection and Preservation Act* from Croatia, this term encompasses the landscape or part thereof that comprises historic structures as a witness of human interaction with nature.

[6] Greenbelt policies were discussed for the first time in England, for the London area in the sixteenth century where the settlement was sprawling out onto agricultural lands.

of terms used in them. But definitions of the two terms mentioned so many times in this paper, *Space* and *Place*, are not available in any of Croatian or Montenegrin legal documents dealing with physical and urban planning.[7] The *Spatial Planning Act* of Croatia does not use any recognizable terms regarding the protection or preservation of *Identity of Place* or *Place Appearance*, nor does it state any obligatory ordinances for preserving the *Identity attributes*.

Today's legislative documents are putting different demands before planners for preparing physical or urban planning documentation, along with planning ordinances (specifically, physical plans for local territorial units, or urban development plans for parts of a settlement with certain historic or cultural values). Three most important legislative acts in Croatia, closely connected with development within the environment, are the *Spatial Planning Act*, the *Cultural Heritage Protection and Preservation Act*, and the *Nature Protection Act*.

According to the *Spatial Planning Act*, the physical planning process is based upon 16 physical planning goals, among which two are concerned with the sensible usage and protection of natural assets, nature preservation, and environment protection, as well as the cultural assets and values protection. These goals support sustainable development based on monitoring and evaluation of activities within space and its sensibility in order to enable, among the others goals, the quality of the living-in environment and the protection of natural and cultural values. This should be part of the long-term preservation of the environment as an essential common benefit. The *Identity of Place*, as a term, is not mentioned in the Act. A similar legal act from Montenegro—the *Space Planning and Building Act*—SPBA (2017), through its 13 main planning goals, defines this topic more precisely than its Croatian counterpart. One of them very explicitly mentions the *Identity*, *Preservation,* and *Regional Recognition*, while the other explains the importance of natural, cultural, and built environment.[8]

For planning within and managing the existing cultural heritage, in both countries, a *Conservation Study* is a must for historic areas or places, but they seldom result in stating only cultural and natural values within the planning areas that have already been protected by the law. The only "new and usable" part may be the professional guidelines for their protection and maintenance. The Croatian *Cultural Heritage Protection and Preservation Act* states that the *Conservation Study* guidelines should be implemented in a physical or urban plan. If the *Study* is not mandatory for a certain area, protection measurements for any development within the plan boundary

[7] Authors of this paper are engaged in planning in the Republic of Croatia and the Republic of Montenegro, and that is why relevant legal documents discussed here originate from these two countries.

[8] SPBA, article 2:

3. Development of regional spatial attributes and preservation of the Identity and recognition of the region;

5. Protection and the improvement of cultural goods and protected environment along with preservation of cultural values integrity and authenticity, creating conditions for sustainable usage of cultural goods and highly valuable built environment along with respect and development of specific attributes, integrity and values of natural and urban areas and ambient.

- An effective instrument that would lead to maximum coordination of all existing legal documents and their implementation in practice (the physical planning process).

As we have learned from this paper, the recognition and correct reading of cultural heritage attributes are of utmost importance in physical and urban planning documents. According to their values, conditions, and significance, the planers and other specialists from the field will decide on the level of their *conservation*, *preservation*, *protection*, *safeguarding*, or *rehabilitation* (Lipovac et al. 2015). However, they should all constantly bear in mind that protection of cultural heritage and natural assets does not necessarily mean just the conservation of an abandoned old building or prohibition of any development within the natural environment. The first step toward the proper safeguarding of cultural and natural heritage (property) is the existence of a clear and unambiguous definition of the planning terms for any kind of intervention over the said property. According to the authors' experience, the planning ordinances are the most referred to and used by more people than any other parts of the planning documentation. The planning ordinance text, in addition to being the prime source of information on the most widely employed legal form of land use, development control, and heritage preservation and protection, is also an instrument which represents a link between the idea and the reality. A problem will arise if an ordinance is not worded clearly and understandably. The elements noticeably absent in most planning ordinances are illustrations. According to the research carried out by the authors, less than 2% of the observed planning ordinances from European countries contain any kind of illustrative material which would help in understanding the provision, proposed methodology, model for action, … (Lipovac 2018).

Until now, only a small number of terms were described in legislative documents, while the rest was up to the urban and physical planners. If we are ready to "solve the problem" of understanding the planning provisions and professional terms, and set up the basic models in physical and urban planning for conserving, preserving, and protecting cultural and natural heritage values, the very first step should be a *Glossary* of urban and physical planning terms, architecture and landscape architecture terms, along with cultural and natural heritage terms at the state level. This *Glossary* should not be part of any legal document, but to the contrary, it should be a separate bylaw document (issued by a professional association and approved by the government) independent of legislative documents or physical/urban plan ordinances (Lipovac et al. 2017). It should be a mandatory tool document for each professional in the respective field. This comprehensive *Glossary* on the state level will help in avoiding situations in which some professional terms are no longer in use just because a certain legislative act has ceased to exist, or even worse situation when the same term in physical plan provisions of two neighboring administrative units is described differently because two different planners defined and wrote it.

The purpose of planning should be to help decision-makers reach informed and thoughtful decisions. Good information, the basis of proper planning, allows decision-makers predict what might happen in certain situations. The planning process must be able to help, to provoke people to ask questions such as "what do you want your community to look like and how should it function?", and to get the right answers. This process should help people to accept and take responsibility for their community and help create the kind of *Place* they want to be in. To perform this, a precise definition of the planning model with distinctive planning steps and proposed methodologies should be set.

11.4 Conclusion

Human population creates and differs *Places* from *Space*. A *Place* is a result of the relationship between its physical attributes from the natural, cultural, and built environment and human behavior, while the definition of *Place* is a complex integration of nature and culture manifested in physical terms, which has developed and is still developing in particular locations. *Places* are linked by a flow of people and goods between them. *Place* is not only the answer to the question *where is something*, but it is rather a location plus everything that occupies that location in *Time* and *Space* as an integrated phenomenon.

The richness of cultural and historical, natural, and landscape heritage, along with the worldwide efforts in understanding, evaluating, and protecting all these values, is more than enough reason to pay maximum attention to proper inventorying and evaluation of all relevant elements taking part in creating the unique image and identity of a particular *Place*. The adequate respect for inherited cultural and natural heritage values through the planning process, along with adequate safeguarding and maintaining, will help, by all means, to preserve the identity of a *Place* or region. This will also result in evident economic progress, along with the progress of the most valuable type of tourism—cultural tourism.

The planning steps that serve as the *sine qua non* of nowadays planning process include the following: identification of problems and opportunities; establishment of goals; inventory and analysis of biophysical, historic, and cultural elements from the environment; human community inventory and analysis; suitability analyses, development of concepts and selection of options; community participation; adoption of a plan; and finally, its implementation and administration. This list should be supplemented with a few more planning steps that will be crucial in understanding the importance of all other elements, steps that will move us toward the understanding of the meaning and importance of the *Identity of Place*. One of these steps is undoubtedly needed: a *definition and establishment of borders* within the environment—not just administrative borderlines, but boundaries of influence, including capacity and limitations (in regard to natural and cultural heritage) and a definition of the possible development within the observed environment. After recognition, naming, and evaluation of numerous natural and cultural elements and their impact on the creation of *Place Identity*, as well as discussing the planning process itself, the following eight planning steps should be taken within the planning process in order to achieve the major goal: preservation of the *Identity of Place*:

1. Identification of planning problems and opportunities;
2. Identification of borderlines and conducting of an appropriate inventory of source identity attributes (landscape and cultural environment analysis at regional and local levels);
3. Evaluation of collected data through their uniqueness, appearance, and integration into the planning goals;
4. Establishment of planning goals in collaboration with *insiders* and *outsiders*;
5. Setting up planning concepts and options; choosing the proper planning methodology;
6. Preparation and carrying out a public presentation, discussion, and hearing;
7. Completing the plan, with continued studies and constant citizen involvement and community education;
8. Checking the adopted plan implementation and preparing further correction and its amendments.

These eight planning steps represent a planning model, researched and developed within the HERU scientific project, which will (if followed) help in preserving the *Identity of Place*. They reveal what must be done during the planning process (observing, evaluating, implementing, monitoring) to ensure the efficient preservation of the environment (natural and cultural) as a whole. These eight steps, if really adopted in the planning practice, undoubtedly need one more step, their implementation into the contemporary legislation. Some of them have already been included in some of the acts, bylaws, and other legal documents, but being mostly scattered, they are not readable as a whole. In the very near future, it is necessary to work more on the integration of this planning model into different legal documents dealing with physical and spatial planning and the preservation of the existing values.

The most effective tool that someone can use in controlling *Place Identity* and its valuable identity attributes is a **clearly defined planning process** that will result in establishing clear and omni-understandable planning ordinances and proposed methodologies in preserving (protecting) the *Identity of Place*. Today, the planning process is a multidiscipline, a battlefield on which many professionals meet and fight for their ideas from the fields they come from and for which they are responsible. Instead, all of us should meet in a "conference room," not only to confront our standpoints but to try to understand each other and reach a compromise for the common benefit. The planner's role is to be a sort of a moderator in this "conference room," To make this possible, all of the participants must speak the same (professional) language. Before implementing the proposed new planning model, we need something we all shall agree upon—we need a relevant glossary of terms prepared by professionals and not by lawmakers.

Acknowledgements The research is part of the scientific project "Heritage Urbanism—Urban and Spatial Planning Models for Revival and Enhancement of Cultural Heritage." It was financed by the Croatian Science Foundation [HRZZ-2032] and carried out at the University of Zagreb, Faculty of Architecture.

References

Lipovac N (1997) Space and place. Prostor 1(13):1–35

Lipovac N (2000) Planning as a function of preserving the identity of place. Dissertation, University of California, University of Zagreb

Lipovac N (2014) English-Croatian professional glossary for urban and physical planners, architects and landscape architects. Acta Architectonica, University of Zagreb Faculty of Architecture

Lipovac N (2018) English-Croatian professional glossary of cultural heritage. Acta Architectonica, University of Zagreb Faculty of Architecture

Lipovac N, Popović GS, Robina M (2015) The understanding of professional and legal terms in physical planning—better protection of cultural heritage. In: Obad Šćitaroci M (ed) Cultural heritage—possibilities for spatial and economic development. Proceedings of international scientific conference heritage urbanism, Zagreb. University of Zagreb Faculty of Architecture, pp 48–53, Oct 2015

Lipovac N, Popović GS, Gradečki N (2017) Pojmovnik kulturnog naslijeđa—model za primjenu u zakonodavstvu. In: Obad Šćitaroci M (ed) Modeli revitalizacije i unaprjeđenja kulturnog nasljeđa Economic development. Abstracts of international scientific conference heritage urbanism, Zagreb. University of Zagreb Faculty of Architecture, p 104, Oct 2017

Lukerman EF (1964) Geography as a formal intellectual discipline and the way in which it contributes to human knowledge. Can Geogr 8(4):167–172

Republic of Croatia (2017a) Cultural heritage protection and preservation act. National Gazette No. 69/99, 151/03, 157/03, 100/04, 87/09, 88/10, 61/11, 25/12, 136/12, 157/13, 152/14, 98/15, 44/17

Republic of Croatia (2017b) Spatial planning act. National Gazette No. 153/13, 65/17

Republic of Croatia (2018a) Environment protection act. National Gazette No. 80/13, 153/13, 78/15, 12/18

Republic of Croatia (2018b) Nature protection act. National Gazette No. 80/13, 15/18

Republic of Montenegro (2017) Space planning and building act. Official Gazette No. 64/17

Steiner RF (1991) The living landscape: an ecological approach to landscape planning. McGraw Hill Inc. Arizona State University

Tuan YF (1975) Space and place: humanistic perspective. In: Gale S, Olsson G (eds) Philosophy in geography. Theory and decision library (An international series in the philosophy and methodology of the social and behavioral sciences), vol 20. Springer, Dordrecht

Violich F (1996) Identity: key to meaningful place-making, the case for Berkeley. Prostor 2(10):201–216

Yeang K (1995) Designing with nature: the ecological basis for architectural design. McGraw Hill, New York, NY

Chapter 12
Physical Branding and Heritage

Marina Pavković and Jesenko Horvat

Abstract Physical branding is a method of managing architectural/urban systems which, in its significance, design and purpose, represents individual spatial identities and can affect the competitive identity of the site. The aim of the research presented in this chapter is to identify physical branding models in the field of cultural heritage and the designed natural environment in relation to the renewal of heritage and/or the application of urban modernity. The expected results are typologically defined physical branding models. The networking of specific spatial characteristics, through targeted architectural/urban planning, contributes to the establishment of individual spatial identities which, in their significance, affect competitive identity. The obtained results may be applied to strategic planning procedures, both in spatial planning and in the planning of economic development.

Keywords Physical branding · Spatial identity · Designed environment · Competitive identity

12.1 Introduction

This chapter studies in what way architectural and urbanistic interventions, as individual spatial identities, influence competitive place identity. The aspect of the study relates to heritage urbanism and the activation of heritage values in a physical, economic and social sense. The aim of the study is to introduce methods of physical branding to heritage urbanism by developing related models. Insufficiently related physical branding and strategic development planning result in a low utilisation level of heritage potential compared to what is optimally achievable. The consequences are dissociated spatial interventions which lack a synergistic effect. The hypothesis

M. Pavković (✉) · J. Horvat
Faculty of Architecture, University of Zagreb, Zagreb, Croatia
e-mail: marina.pavkovic1@zg.t-com.hr

J. Horvat
e-mail: jhorvat@arhitekt.hr

M. Obad Šćitaroci et al. (eds.), *Cultural Urban Heritage*,
The Urban Book Series, https://doi.org/10.1007/978-3-030-10612-6_12

set in this chapter proposes that there is a demand for systematic management of heritage potentials as individual spatial identities.

12.2 Starting Point

By anticipating the simultaneity of strategic-economic and physical-planning development platforms, the term ***physical branding*** means the method of managing architectural and urbanistic interventions which represent individual spatial identities with their meaning, shaping and function which can affect competitive place identity. The method results from elaborate doctoral research[1] used to introduce the principle of architectural and urbanistic intervention as an individual spatial identity in the integrated area of physical planning and strategic development programming. This individual spatial identity both independently and in a network of individual spatial identities has a positive influence on ***competitive place identity***.[2] The method of physical branding is equally applicable to inherited buildings and newly formed architectural and urbanistic interventions.

The approach of Anholt (2007), theoretician and practitioner, is relevant in the creation of this method. In his approach, the platform of place branding is upgraded in the general sense by introducing the term and defining the factors of competitive identity.[3] For this chapter, the works of other scholars (on city marketing and city branding) are also relevant. Such authors include Paliaga (2007), Kavaratzis (2004), Kavaratzis and Ashworth (2005, 2008), van den Berg et al. (2006) and Braun (2008). All deserve credit for shifting the perception of place branding from merely promotion and advertising to strategy and vision. Nowadays, city marketing and city branding are contemporary communication tools for cities to compete for investment and human capital, and, as such, deserve the attention of the scientific and professional public.

This paper, as did the doctoral research, took the following step: after reviewing the specific role of architecture and urbanism in this area, the term ***individual spatial identity***[4] was introduced into scientific terminology together with the method and physical branding models.

---

[1] Pavković (2018) Doctoral thesis: "Method of physical branding—the architecture and urban development impact on the competitive identity of the place".

[2] The term and concept of a single spatial identity, which implies an individual architectural/urbanistic approach, is introduced in scientific terminology in the aforementioned doctoral work.

[3] The concept of competitive identity has been developed by Simon Anholt in *Competitive identity: the new brand management for nations, cities and regions*, published in 2007, and based on the concept of national identity as the comparative advantage of a particular country in a globalised world. Anholt's model contains six factors of competitive identity that cover the key segments of social and economic activities, directly related to the creation of a nation's reputation in the perception of target groups: tourism, economy, politics, investment, culture and people.

[4] Individual spatial identity is an architectural and urbanistic intervention whose significance and characteristics differentiate it from others, and it holds a unique place in the perception of the general

Physical branding, which forms a programme package and prepares it for being embedded into the concept of physical planning, finds a basis for spatial development and protection also in the elements of heritage (Obad Šćitaroci 2018). While it is possible to analyse and then directly apply the identification procedures of economic potentials to material heritage, this is not possible in the case of intangible heritage since matter does not exist[5] or is present indirectly and is extremely dispersed in space.

The properties of intangible heritage can be preserved and maintained only by adding new innovative programmes that will add value to the heritage, becoming thus new spatial identities. Since the practice of physical planning and its operative framework is non-selective (some programmes support identity formation, and some simply follow directions without affecting the identity), the creators of the physical-planning concept are responsible for recognising economic generators, their dimensioning and placing them in an optimal position in space.

12.3 Research Aim

The aim of the research is to establish models of physical branding in the area of cultural heritage and a designed natural environment related to heritage regeneration or applied urban modernity. Individual architectural and urbanistic intervention does not necessarily imply creating an individual spatial identity which will influence competitive identity in a positive sense. The positioning of a unit intervention, as an individual spatial identity, depends on the originator's level of creativity, that is, the degree of agreement when making decisions, and it also depends on economic and social functionality.

Another goal of the research is to point to potential planning procedures when applying the elements of intangible heritage for constructing new material values and to develop an application method usable in everyday physical-planning practice. The method of physical branding in the area of cultural heritage is based on separating typical revitalisation case studies and activating heritage, as well as on a comparative function analysis, the results achieved and their effects on the social, economic and physical environment.

In addition to cultural heritage, the investigation also includes examples of the designed natural environment that present new cultural heritage through their creation.

---

and targeted public. Networked in a system, and individually dependent on the function, individual spatial identities affect place competitiveness.

[5]The instruments used for embedding the values of heritage in the methodology for making plans and codified by legal regulations are not sufficient to protect the values of intangible heritage, so it is only established as a value in name, although without concrete measures for implementing this protection.

12.4 Cultural Heritage in the European Context

The rich portfolio of European cultural heritage depends on many elements which affect its sustainability, including elementary survival in space. For centuries, civilisations have introduced changes in the creation and treatment of cultural heritage which have left a valuable legacy in Europe, a legacy which seeks equally appropriate treatment both in the physical and social/economic sense. The preservation of cultural heritage varies from country to country. European countries of stronger economic power (Germany, France, Scandinavian countries and other developed countries) set aside substantial funds. On the other hand, European countries with lower GDP find it difficult to preserve and maintain the quality of their cultural heritage. Since the EU has recognised this as a common problem, it allocates substantial means for preserving and maintaining cultural heritage within the Competitiveness and Cohesion Operational Programme 2014–2020, according to thematic priority TO 06 regarding environment and resource efficiency.[6] EU instruments for heritage revitalisation make programmes and functions which contribute to overall development a condition for its activation, evaluating the effects on the physical, economic and social environment. This is a principle for regeneration which should be respected by every environment, independently of the funding source.

12.5 The Method of Physical Branding in the Context of Heritage Urbanism

This method is used to introduce in the integrated area of physical planning the principle of creating an architectural and urbanistic intervention as an individual physical identity which both individually and in a network of individual spatial identities has a positive effect on competitive place identity. The method is equally applicable to inherited buildings as well as newly formed interventions. The assessment of heritage values for constructing such planned guidelines used for upgrading base programmes as well as other supplementary programmes includes the following variables:

- Does the basic value belong to a long-term and living traditional culture or is it perishing (perhaps it has already perished) and seeks regeneration?
- Is a new physical framework or a new physical intervention with a supplementary programme package required for its preservation?
- Has the value been registered (supported by protection measures regulated by law), recognised or only spotted in the planning bases?

[6]Since 2000, the European Commission has had several key funds to strengthen the competitiveness of the member states. The most used and important are the European Regional Development Fund (ERDF), the Cohesion Fund and the European Social Fund (ESF). The ERDF aims to strengthen economic and social cohesion in the European Commission by correcting imbalances between its regions (European Commission 2014).

- Is the value limited by space, dispersed or non-specific—which could limit the achievements and success of physical branding at a local level?
- Is the value inherent in the constructed space or does it belong to the anthropogenic environment?
- Are there positive (or negative) experiences in programming which rely on the heritage values (experience confirmed within a concrete physical environment or through analogy with other similar case studies)?

The basis of the method was established through mapping and programming studies implemented in an interdisciplinary procedure.[7] Using a selective approach, both in these and future studies, it will be possible to recognise material and intangible heritage resulting from the specificities of the geographical and cultural site, and which will be divided into groups and types in a further science-research investigation procedure through different types of heritage and different emanations of culture.

12.6 Physical Branding Overview and the Example of the City of Šibenik

Šibenik, a Dalmatian city on the Croatian coast of the Adriatic, was formally founded in the eleventh century. From a rich and tempestuous history, Šibenik inherited many material and non-material values. Unlike other Dalmatian cities established in the matrix of the antique period (Zadar, Split or Pula), Šibenik was founded on an authentic medieval matrix which was organically upgraded with structures along the main communication directions. Fortifications, the soil configuration or the terrain, and the bay marked its sights, and they have become symbols of its visual identity which make it different from other urban centres in its broader environment. For the past decade, with a population of 35,000, this city has been intensively renovating and revitalising its cultural and natural heritage. The revitalisation of its heritage for the purpose of tourism has enabled it to draw on European funds.

Heritage elements, as generators of development, have been extracted from this overview of examples of a single urban unity on a smaller scale. The reason is the procedure's applicability in making physical plans of a county or a city. The procedure is accessible in areas of smaller physical stratification where development programmes focus on the local level.

Since the subject matter of this chapter is related to the regeneration of heritage and its related models, three examples of heritage regeneration have been selected in the Croatian city of Šibenik. They are the revitalisation projects of St. Michael's Fortress, the City Beach of Banj and the promenade along St. Anthony's Channel.[8]

[7] For additional information on the method, see Horvat and Pavković (2013).

[8] The mentioned examples were studied in more detail within the scope of the doctoral research, partially conducted as part of the science-research project *Heritage Urbanism—Urban and Spatial Models for Revival and Enhancement of Cultural Heritage, 2032—HERU/2014–2018*.

The first is the revitalisation of St. Michael's Fortress, and it is related to the revitalisation of cultural heritage. The other two examples relate to intervention based on the natural heritage. Each of the examples presents an individual spatial identity. Their integration in a comprehensive system of individual identities has resulted in synergic effects on Šibenik's competitive identity.

12.6.1 Revitalisation of St. Michael's Fortress

The medieval Fortress of St. Michael was named after a church located inside its walls. The church was not preserved, but the fortress kept the name of the saint who is also the patron of Šibenik. The fortress is extremely important for the history and for the urban life of Šibenik. It was originally used as a watchtower with a view to the horizon over the bay. It also provided shelter for people, and, in time, a city was born, first mentioned in 1066.

The medieval fortress was destroyed and rebuilt several times through history. Its oldest part dates from the thirteenth century, although most of it was built in the fifteenth century when it fell under Venetian rule together with the city. After the conquerors' initial idea of demolishing the fortress, its double walls were repaired and rebuilt. In the sixteenth and the seventeenth centuries, the fortress was heavily damaged, but the Austrians, who ruled the city in the nineteenth century, repaired the damage and contributed to the revitalisation of the fortress and the city walls. The fortress went out of operation in the twentieth century both in the sense of providing defence and in the sense of culture. Visitors did not notice it as a historical landmark. For the people of Šibenik, it is a symbol of perseverance, but there were no signs of revitalisation inside its walls. The required changes took place at the beginning of the twenty-first century, when Croatia started its accession talks with the EU, gaining an opportunity to use EU pre-accession funds.

An awareness of the fortress's significance from the aspect of the architectural heritage of Dalmatia and Croatia, as well as the possibility of financing its revitalisation with EU funds, resulted in the project of the revitalisation of St. Michael's Fortress. This project started in 2012 and was completed in 2014.[9] The total value of the invested funds was slightly over €1.6 million, of which €1 million was provided by the EU. The main goal of the project was the economic development of Šibenik and its surroundings through the promotion of cultural tourism. The project's special goal was to revitalise the fortress and turn it into a new tourist attraction which would stimulate innovative cultural and artistic potential (Fig. 12.1). The ambitious project included infrastructural renovation over 2600 m^2 and a summer stage with 1077 seats. The fortress and the programme it offered were visited by over 110,000 visitors from the first year of its opening. That same year, during the Days of Croatian Tourism, St. Michael's Fortress was selected as the Cultural Attraction of the Year. It is important to mention that this revitalisation project found its place on the

[9]Project leader: Tomislav Krajina; Situs d.o.o.

Fig. 12.1 St. Michael's Fortress, Šibenik

list of projects of strategic importance for the city of Šibenik in 2011 as part of the City's Development Strategy. This example proves that development does not happen accidentally and that a successful project requires a transparent and functional idea, integral planning and synergistic realisation. St. Michael's Fortress has recorded a rising number of visitors in the following years. In 2015, it recorded 115,000 visitors; in 2016, 137,000; and the number climbed to 500,000 in July 2017.[10]

12.6.2 The City Beach "Banj"—Transformation of an Industrial Brownfield Site into a Bathing Area

From the late nineteenth century to the end of the twentieth century, the space, which now recalls Šibenik as a bathing area, actually served for industrial purposes. The area was once dominated by a factory producing electrodes and ferroalloys, which ceased working during the 1990s. The idea to restore its former function as a bathing area was born in 2010 when the erasing of the factory buildings intensified.

[10]Since 2016, St. Michael's Fortress and the Barone Fortress have been considered a single public institution under the name Fortress of Culture, which keeps records of the sum of visitors at both these locations.

Fig. 12.2 City Beach Banj, Šibenik, before intervention, *author* Željko Krnčević

The project was included in the City's Development Strategy 2011, and the beach opened in 2012.[11] The project revitalised the city's old centre and promoted the development of tourism. As a city on the sea, Šibenik finally had a city beach. A total of 12,000 m^2 of bathing area was brought back to life following an investment that totalled slightly over €1 million, of which 30% was financed by the EU and the remainder was covered by the city budget. Banj is now a favourite urban intersection. The view from the beach onto the historical centre has been recognised as a unique visual and ambiance experience (Figs. 12.2 and 12.3). The beach is open during the entire year, offering summer and winter bathing, recreation and various cultural and sporting events. Catering facilities on the beach operate throughout year, and to a smaller extent, they contribute to employment and revenues. At the same time, the revitalisation of the beach and the regeneration of the vacant area have had a multiple effect on the growing income of the locals who have ventured into tourism, primarily accommodation services. All this has resulted in the transformation of the urban area of Crnica, a past worker settlement beside the former factory, into a city block with better infrastructure and public functions. The city of Šibenik has become a city on the sea also in the sense of bathing.

[11]Project leaders: Damir Lasinović, Marko Paić, Ivona Ivčević and Ivana Lozić; ARX d.o.o.

Fig. 12.3 City Beach Banj, Šibenik, after intervention

12.6.3 Added Tourist Value of St. Anthony's Channel

The natural phenomenon of St. Anthony's Channel (Fig. 12.4), which separates Šibenik and its bay from the open sea, is one of the landmarks which characterise the area of Šibenik. It is protected under the category of Significant Landscape and belongs to the *Natura 2000* eco-network. It is 2000 m long and from 140 to 220 m wide. At the exit of the channel lies St. Nicholas' Fortress dating from the sixteenth century, built for the purpose of providing defence against the Turks. The fortress is inscribed in UNESCO's World Heritage List. Since Šibenik's Cathedral of St. James is already on UNESCO's list, this makes Šibenik one of the rare areas with two UNESCO monuments. With geomorphological features and Mediterranean vegetation, the channel also provides a beautiful view over Šibenik.

Although it is only 950 m away from the city centre, the channel was closed to the public for decades. The channel was managed by the military which left behind facilities in the shape of contemporary artefacts. When the army of the former state left the area, the city authorities decided to return the channel to the people and promote its public accessibility. A project worth €1.4 million was drawn up, with the EU financing €1 million of the total. The project was completed in 2013.[12] A

[12]Project leader: Nikola Bašić; co-leader: Zvonimir Bušić.

Fig. 12.4 Promenade along St. Anthony's Channel

total of 4.4 km of walking and cycling routes were designed with a viewpoint and a platform; urban equipment was installed; a pier below the church was renovated; and a small wooden bridge was constructed. In accordance with regulations for EU projects and visitor monitoring, 100,000 people have visited the promenade since its opening.[13] Since the project promotes the concept of sustainable development, eco-friendly economic activities are planned and opportunities will be presented to local companies and the community.

St. Michael's Fortress, the City Beach "Banj" and the promenade along St. Anthony's Channel represent attractions and landmarks of Šibenik, a beautiful city and tourist destination. The effect that the investigated individual identities have had on the growing number of visitors and overnight stays is measurable.[14] The events and potentials which follow the creation of spatial identities, as well as the growing accommodation infrastructure, all function as promoters of Šibenik's competitive identity. The final result is total urban product development.

[13] According to data provided by the Public Institution for Protected Area Management in the County of Šibenik and Knin (information published in the media).

[14] According to data provided by Šibenik Tourist Board, overnight stays increased by 9% in 2016 compared with 2015.

Table 12.1 Excerpt from table "classification of individual spatial identities based on typology, character and location"—case studies of Šibenik, Croatia

Individual spatial identity	Typology of architectural and urbanistic intervention		Character of architectural/urbanistic intervention		Location of architectural and urbanistic intervention		
	Heritage	Modernity	Cultural material heritage	Designed natural environment	Historical centre	Urban area	Wider city area
Fortresses							
St. Michael's	X		X		X		
Beaches							
Banj		X		X		X	
Promenades and corridors							
St. Anthony Channel		X		X			X

The original table was made as part of the doctoral thesis

12.7 Research Results

In compliance with the method of physical branding, the examples processed in this research are grouped based on the type of physical units and are divided according to typology, character and location of architectural and urbanistic intervention (Table 12.1).

Based on typology as a primary key, the models of physical branding are divided into models of physical branding of heritage on the one hand, and models of physical branding of modernity on the other hand.

The term heritage implies interventions in space which are implemented at the location of the existing cultural heritage or the existing artefact of the designed natural environment, where the language of architectural expression is not crucial. Whether it is a radical reconstruction with a modern approach or facsimile restitution, what defines a certain intervention in space as heritage is the existence of a physical entity prior to the new architectural or urbanistic intervention.

The term modernity implies architectural interventions which are a physical *novum*, with buildings emerging in places where they did not exist before. Just as with the model of physical heritage, the language of the architectural expression is not crucial. Therefore, independently of the type of architectural procedure, what defines a physical intervention as modern in this chapter is the lack of a physical entity before the architectural and urbanistic intervention.

Based on the character of the intervention, the models of physical branding of heritage and modernity are then divided into models of physical branding of cultural heritage and models of physical branding of designed natural environment.

Concerning the location of intervention, the models of physical branding of heritage and modernity are then divided into models of physical branding of the historical centre, models of physical branding of the urban area and models of physical branding of the city's wider area.

The examples, categorised according to the main models, and additionally divided according to the model type, show that the revitalisation and regeneration of heritage for development purposes is possible for each category. The benefit of the categorisation is the possibility of measuring the effects of an individual intervention on competitive identity, based on typology, character and location of the architectural and urbanistic intervention. This enables statistical recording of the effects, and their results will contribute to the quality of future investment decisions based on the results of past investments.

12.8 Conclusion

The networking of the specific characteristics of an individual space through targeted architectural and urbanistic interventions contributes to creating individual spatial identities. With their significance, they are able to influence competitive place identity. The method of physical branding deals with creating individual spatial identities and their networking into a competitiveness system.

The highlighted typical revitalisation and regeneration cases of cultural and natural heritage contribute to the strengthening of competitive place identity. The examples point to systemic revitalisation of cultural and natural resources of heritage with the goal of achieving sustainable development. The projects were designed, included in the physical plans and development strategies and then consistently realised. Considering their universal character, the results are applicable at a global level. In compliance with the physical branding method, the examples are grouped based on typology, character and location of the architectural and urbanistic intervention. The obtained results are models of physical branding and modernity.

Assessment of the methodology will lead to the setting up of models which, in addition to material and intangible heritage elements, might prove to be stimulating in the gradual construction of new material values. Since intangible value derives from and at the same time goes beyond the framework of artefacts, it is not necessarily related to a concrete site. It actually becomes a basis for new programme potentials and innovative interventions in space—the creation of new individual spatial identities.

The models can be applied to the heritage urbanism method for the purpose of activating cultural and natural heritage in the spatial, economic and social sense. At the same time, relating the discipline of physical planning to strategic development planning produces a synergistic effect and leads to the sustainable use of heritage potential with the goal of promoting the competitiveness and sustainable growth of the total urban product.

Acknowledgements The research is part of the scientific project "Heritage Urbanism—Urban and Spatial Planning Models for Revival and Enhancement of Cultural Heritage". It is financed by the Croatian Science Foundation [HRZZ-2032] and carried out at the Faculty of Architecture, University of Zagreb.

References

Anholt S (2007) Competitive identity: the new brand management for nations, cities and regions. Palgrave Macmillan, New York (ISBN 978-1-349-35243-2)

Braun E (2008) City marketing: towards an integrated approach. Doctoral thesis, Erasmus University Rotterdam (ISBN 978-90-5892-180-2)

European Commission (2014) Operational programme competitiveness and cohesion 2014–2020. http://ec.europa.eu/regional_policy/hr/atlas/programmes/2014-2020/Republic%20of%20Croatia/2014hr16m1op001. Accessed 5 June 2018

Horvat J, Pavković M (2013) Prostorno programska agenda/Physical planning and programming agenda: mapping and programming spatial potentials as a basis for integral development. Prostor: znanstveni časopis za arhitekturu i urbanizam 46(2/21):292–301 (ISSN 1330-0652)

Kavaratzis M (2004) From city marketing to city branding: towards a theoretical framework for developing city brands. Place Branding 1(1):58–73

Kavaratzis M, Ashworth GJ (2005) City branding: an affective assertion of identity or a transitory marketing trick? Tijdschrift voor Economische en Sociale Geografie 96(5):506–514

Kavaratzis M, Ashworth GJ (2008) Place marketing: how did we get here and where are we going? J Place Manage Dev 2(1):150–165

Obad Šćitaroci M (2018) Heritage as an active space and spatial resource. In: Grifoni RC, D'Onofrio R, Sargolini M (eds) Quality of life in urban landscapes—in search of a decision support system. Springer, Cham, pp 341–348

Paliaga M (2007) Branding i konkurentnost gradova. Independent edn. Rovinj (ISBN 978-953-99569-1-0)

Pavković M (2018) Method of physical branding—architecture and urban development impact on the competitive identity of the place. Doctoral thesis, University of Zagreb, Faculty of Architecture

van den Berg L, van de Meer J, Oligaar A (2006) The attractive city: catalyst of sustainable urban development. In: XVI Congreso de Estudios Vascos, Bilbao (2005), pp 485–491 (ISBN-10: 84-8419-022-6; ISBN 978-84-8419-022-6)

Chapter 13
Recycling Heritage Between Planning and Design Interventions

Vincenzo Paolo Bagnato and Nicola Martinelli

Abstract The topic of recycling and contemporary approaches to sustainable urban and landscape renewal are nowadays extended to those parts of our territory that, although characterised by important historical traces, are in a critical condition of *drosscapes*: this is the case of archaeological sites, abandoned places "in-between" city and landscape and past and present, whose relationship with the environment life cycle is interrupted and in which the concepts of "ruin" and "heritage" are easily confused with those of "residue" and "waste". The last important Italian National Research Programme (PRIN) called "Re-cycle Italy: New Lifecycles for Architecture and Infrastructure of Cities and Landscape" has connoted new complex issues in the triad Reuse/Reduce/Recycle and defined new interpretative matrices on the relationship between heritage protection, sustainable design and urban planning. Both under an epistemological and architectural perspective, a new interpretative approach emerges with a critical look entirely focusing on the archaeological sites' "life cycle" and on the modalities through which they "dialogue" with the social dimension of the environment. This chapter fits within this framework, through a critical analysis of the relationship between the concept of recycle and interdisciplinary actions in the archaeological landscapes between planning methods and design experiences.

Keywords Archaeological heritage · Architectonic heritage · Recycle · Urban planning · Design

13.1 General Framework

The context in which the idea of recycling plans and design projects have developed in the last few years is quite broad and crosscutting, regarding various disciplines and topics. The experience of the exhibition called "Re-Cycle: Strategies for Archi-

V. P. Bagnato (✉) · N. Martinelli
Department of Civil Engineering and Architecture Sciences, Polytechnic of Bari, Bari, Italy
e-mail: vincenzopaolo.bagnato@poliba.it

N. Martinelli
e-mail: nicola.martinelli@poliba.it

M. Obad Šćitaroci et al. (eds.), *Cultural Urban Heritage*,
The Urban Book Series, https://doi.org/10.1007/978-3-030-10612-6_13

tecture, City and Planet" held at MAXXI, Rome, in 2011 (Ciorra and Marini 2011), and that of the German Pavilion at the 13th Venice Biennale in 2012 whose name was "Reduce Reuse Recycle: Architecture as Resource" (Petzet and Heilmeyer 2012), as well as the Italian National Research Programme "PRIN Re-Cycle Italy: New Lifecycles for Architecture and Infrastructures of City and Landscape" (2013–2016), have agreed that recycling is something exquisitely contemporary and does not necessarily coincide with the traditional actions of rehabilitation and/or renovation (Fabian and Munarin 2017).

Normally but improperly used to express strategies of reusing ordinary materials, elements or building components, thanks to the above-mentioned experiences, nowadays within the meaning of recycling there is a natural relationship with pre-existence but not with its original configuration, nor with any kind of retrospective behaviour. Recycling approaches are more like "creative transformations" that work with different times in specific spaces, treating them as "places" in which the past, present and future achieve a new ethical configuration with new linguistic codes and aesthetic symbols. On the other hand, recycling strategies include cognitive and methodological tools aimed at controlling planning and design actions, with implications on the physical and social urban and environmental context (Losasso 2016).

13.2 Heritage Recycling

In this framework, the association of recycling with the concept of heritage has a double effect. On the one hand, heritage is no longer treated as "uniqueness" and, losing its status of "monument", it is no longer considered for its exceptional nature but becomes itself, in a certain sense, "ordinary". On the other hand, actions on historical evidence are aimed not only at preserving it but also at putting it into a new relationship with society and the environment and, after all, at giving it a new life cycle. Hence, intervention strategies have enriched places with new items and new priorities such as, for instance, ecological factors, energy efficiency and economic sustainability that go towards new paradigms whose aim is to reclaim the lost and forgotten dignity of the territory, citizens and social context (Germanà 2014). What happens is that places of heritage are no longer seen as isolated islands and "historical emergencies", but they fully belong to the urban environment with its structure and its contemporary landscape and they also actively participate in its dynamic and continuous transformations (Bagnato 2015).

A specific condition of heritage is given by archaeological heritage, a complex system of ancient historical remains produced at various moments in time through mechanisms of layering and/or continuity which now appear as a "unity" with a strong relationship with landscape and nature and as inerasable territorial landmarks.

When a ruin, an archaeological settlement or a heritage complex is abandoned, the whole territory suffers a sort of cultural gap that affects its identity and the quality of its social life because what is produced are nowhere places broken from the physical and social contexts (Clementi 1990). This is why ruins, relics, historical traces and entire archaeological settlements, if abandoned or separated (spatially, functionally and socially) from everyday urban or suburban life, lose their meaning for the collective memory and become residues, remains and wastes, sometimes even in conflict with the city and landscape themselves. Indeed, it is not easy to find compatibility between archaeological sites and actions of "reuse", but it is certainly possible to identify new attitudes, new interpretations and new behaviours which transform the condition of *drosscape* (Berger 2007) into a state of "dialogic landscape" for heritage. In order to return the archaeological heritage to the urban and landscape life cycle, to its spaces and its times, it is important that the intervention strategies integrate at least three fundamental elements, adapted to each specific cultural context: multidisciplinary knowledge, participative and integrated planning actions, and ethical design interventions (Bagnato 2017).

13.3 Planning and Urban Archaeology

Archaeology intended as awareness of historical value and a tool of knowledge of the past was born in recent times. Until the 1970s, urban archaeology was almost exclusively a passive practice depending on random findings or occasional controls of archaeological digs with no conscious scientific purpose. At the end of the twentieth century, excavations in Venice promoted by the architect and archaeologist Giacomo Boni and big operations at the *Fori Imperiali* in Rome were a hugely significant innovation for historical knowledge and for the definition of a new relationship between archaeological ruins and urban context.

In the course of the following decades, archaeological interventions became more autonomous and excavations more and more superficial, not always properly documented and, furthermore, careless of the existing urban structure. The problem was that the archaeological actions tended to have the aim of "releasing" the ruins from successive overlapping layers so as to "rescue" their original formal configuration: interest was focused on objects, elements, monuments but never on the entire urban processes.

In Germany and France, although archaeological settlements have never been as large or as important as in Italy or Greece, management approaches demonstrated the possibility also to respect the most recent stratifications, recovering and adequately describing them and avoiding their deletion, with a view to all the urban context's historical life cycle stages (which started to be carefully documented) and to the "material culture" aspects of the territory, particularly relevant to understand the complexity of the processes of urbanisation.

In Great Britain, the work of the archaeologist Martin Biddle on Winchester's medieval remains during the 1960s (1961–1971) represents an important step in the definition of the modern methodological approaches because for the first time the archaeological excavations were planned considering large urban areas and covering a sufficiently long period of time. From this moment on, in the UK integration between planning and archaeology developed a new model called "Implications

Survey" through which the archaeological potentialities of a city start to be taken into account by planning approaches, in order to rationalise the excavations and to avoid damage to the ruins: this new method, then supported by some important publications such as "The Erosion of History: Archaeology and Planning" by the Council for British Archaeology and the Urban Research Committee (1972) on the one hand, identifies areas where ancient layers are expected to be found, linking them to urban regeneration projects, and on the other hand transforms stratigraphic excavation from the only possible solution into part of an integrated system of actions included into a more general planning strategy.

Nowadays, an analysis of urban development processes, preliminary to the actions of every environmental transformation, finds in the contemporary city, more than in any other old configuration of it, its best place to be because an understanding and the study of the structure discover in the actual condition of the city the physical and cultural continuity between the past and present necessary to better comprehend its morphological and social nature. This aspect of the contemporary city is distinctly "archaeological" in the sense that it contains all the tracks and all the "times" of its memory, identity and cultural history; furthermore, archaeology itself can be seen by the contemporary city as a conscious tool of knowledge and a concrete benchmark integrated into urban and landscape planning: in other words, the presence of archaeology lies not only in ruins or ancient remains but also (and often prevalently) in the urban forms, in the pathways, in the relationship with the site, and in the individual and collective perceptions of the public spaces. In addition, the planning approaches have developed in the last decades a behaviour that now starts to share a new vision based on "integrated actions" where institutional public and private subjects operating at different territorial levels dialogue between each other in order to define policies and synergistic actions for physical, economic and social interventions on the territory (Secchi 2005).

The contemporary city and environment seen as a complex archaeological structure combined with integrated planning approaches lead us to consider our territory as a result of a complex system of "life cycles" the duration of which depends on general and specific social, economic and cultural factors and enables us (as planners, architects, designers and citizens) to be actively involved in its dynamic transformations. In this sense, the role of heritage becomes particularly important because of its intrinsic capacity always to be the backbone of society's ethical character, firm and steady beyond the rapid changes of its cultural shared values. As an example, we can look at Tarragona (Spain) where the archaeological heritage has been networked through a model that does not focus on the contemporary city itself nor on archaeological emergencies, but on the different routes of knowledge that come from the relationships between the various historical layers: on the one side, there are museums, exhibition spaces and research centres (primary and specialised information); on the other side, there are the ruins (intended as fragments) standing in the environmental context of the excavations, not separated from the functional rhythms of the contemporary city but forming with it a unique, integrated, open and collaborative "knowledge path": a model that consists of an urban ecosystem in which the archaeological settlement is not something that needs to be passively rehabilitated but

Fig. 13.1 Miracle Park in the archaeological site of Tarragona, Spain. Photograph by Federica Bonerba

can become a strong cultural impulse and a social stimulus (Fig. 13.1). Hence, this demonstrates that the only playground in which it is possible to find a strong ethical relationship between planning and heritage can be no other than the contemporary city. But "contemporaneity", in conclusion, is not only part of historical time, and still less can it be considered as a "style". Contemporaneity is actually a "condition", made of problems and difficulties but also full of challenges and opportunities, whose aesthetic power is measured by the cultural dimension of all the elements and actions developed over centuries through processes of production, conservation, accumulation and transformation, nowadays seen and experienced as "essential systems of urban facts".

13.4 Design Interventions

We have seen how multidisciplinary knowledge becomes essential to redefine the relationship between historical traces and the contemporary city and how appropriate planning strategies can build new spaces of sociability for local communities and new places for ruins through a new economic development model in which "doing better" is replaced by "doing less". Secondly, we have noted that planning activity with a vision on the processes can more easily reduce the likelihood that intentions remain mere theoretical principles instead of generating practical objectives. Thirdly, the epistemological interpretation of landscape has changed, now being intrinsically archaeological as a container of traces of heritage, requiring planning and design to orient eyes towards the large scale instead of being limited only to emergency interventions. Now, at the design scale, an approach of multiple dialogue is needed (Marini and Santangelo 2013).

The design process and all the choices linked to it have to take into account the need to establish the best condition for the management of archaeological sites, balancing between the conservation requirements and the handing down of heritage (intended as *pater + hereditas*) to future generations, as well as between the study, investigation and diffusion of historical significances and values of the archaeological goods (Oteri 2009). Furthermore, the problem of their function leads us to consider archaeological sites no more as borderline cases in architectonic heritage, and "recycling" no longer as a concept but as a concrete system of actions that can find a solution at least to the following dualities:

- Cancellation/densification
- Re-naturalisation/re-urbanisation
- Re-build upon/around/inside/into
- New additions/re-construction
- Traces, constructive elements, materials, structures/gaps, intervals, empty spaces
- Presence/absence
- Cultural belonging/tabula rasa
- Continuity/contrast.

The new life cycles, then, do not just concern materials and objects (ruins intended as physical elements) but mostly "places", spaces made of historical and social structures naturally or artificially designed by all the different rhythms of the past, present and future time (Norberg-Schultz 1979).

Otherwise, the relationship between concepts and strategies has to take into account the problem of "use", because the ethical approach to heritage needs to pass through the central role of human activity. Therefore, integrated and different functions, multiple uses (always compatible with conservation and protection purposes), co-adaptabilities as well as the correct utilisation of measures, proportions, orientation, typology, fences and original accesses become important in taking the ruins into a contemporary dimension with a new character and a new social identity (Ghersi and Mazzino 2009). Indeed, on the design scale, the concept of "recycle" associated to that of "heritage" can create risks for its protection because of the possible dangers resulting from excesses of individualism in the creative design processes. In order to prevent this, it may be helpful to go back to the reflections of Aldo Rossi who wrote that ancient buildings have a formal image that constitutes their permanent structure, independent of the functions that have cyclically over time been hosted in their constructions; this means that heritage is always a combination of permanence (structure, construction, space, form) and changes (uses, functions, needs, requirements, meanings) generated by cyclical processes (Rossi 1966), and it also means that a creative design approach cannot be dangerous if it demonstrates its ethical ability to act on each aspect and on the relationships between them, always with the aim of giving heritage new life cycles.

Moreover, the actions of recycling should not be "utilitarian" but should be linked to "recognition" of a meaning for ancient constructions among society that needs to be translated into a creative process able to define a new "temporality" for the ruins. In this sense, there are at least two main directions (which can overlap and be complementary) that can be identified by observing the contemporary scenario of architectural design experiences: on the one hand, there is a recycle model that goes through the addition of new buildings that dialogue with the site using compatible constructive systems, analogue forms, natural and renewable resources and recycled materials. On the other hand, there is a recycle model that passes though analysis, recovering and adaptation of the cultural heritage to contemporary social needs and requirements, as an active part of the everyday life of a territory (Capanna and Nencini 2016).

Therefore, the word "recycle" with reference to heritage may contain many declensions like recover, transformation, regeneration of cultural elements, etc., but all of them are always linked to a social dimension because their final challenge is to allow heritage's historical value to be comprehended by all generations and to be included in today's social dynamics as a key element of the contemporary physical context.

Now, the contact between the old fragment and new architecture can achieve unexpected relationships and new formal orders. This is the case of the Ningbo History Museum (China) by Amateur Architecture Studio (2008), where existing constructive elements and old tiles and bricks have been recycled and used in the new construction. The aim of the project is to recover lost urban identities and memories through a new "stratigraphic" architecture whose aesthetic rules are directly generated by the different used materials, resulting from the demolition of some ancient buildings (the most ancient is 1500 years old and comes from the Tang Dynasty). In this interpretation, the sign of the past is not something to venerate, but becomes an active element, reassembled with a traditional technique called "*wa pan*", recycled to keep on participating in the cultural life of a landscape in which, now, the present learns to accept its formal vulnerability and becomes ethically "archaeological".

Recycling also means a remedy for abandonment and rejection: the project of the Turó de la Rovira in Barcelona (Spain, 2011) by Imma Jansana and Jordi Romero works by inserting routes into a setting of ruins representing the remains of an anti-aircraft battery that helped defend Barcelona from air attacks during the Civil War and traces of the Canons district shantytown. A paved route through the ruins and another with a small path that goes along farmland areas have the dual purpose of protecting the ancient traces and designing a new landscape standing 262 m above sea level and boasting 360° views of the city. The intervention, based on the concept of "historical continuity", reinforces and conserves the existing structures with a minimum design system of actions, displaying a captivating and integrated landscape (Fig. 13.2).

The recycling of an archaeological site can also pass through temporary intervention in one of the most important places of European heritage: this is the case of the exhibition design in the *Palestra Grande* in the Archaeological Site of Pompeii by Francesco Venezia (2016), a project that demonstrates how necessary it is to guar-

Fig. 13.2 Imma Jansana and Jordi Romero, Turó de la Rovira, Barcelona, Spain 2011 Photograph by Lourdes Jansana

antee the conservation of historical sites, placing new cultural activities into them and giving social importance to contemporary architectural action. The temporary construction, made of corrugated metal sheets and developed along the whole north wing of the *Palestra Grande*, defines a sincere dialogue with the ruins that can be seen as an interpretation of the relationship between the continuous cycle of history and the rhythmic cycle of contemporary life, whose aim is to keep always socially alive such an important ancient context, avoiding the aberrations of mass tourism rituals and rules.

13.5 Conclusions

Is it possible to establish a methodology of analysis and evaluation of the relationship between archaeological settlement, urban or environmental context and architectural design, considering these elements together and not as independent subjects, bearing in mind that all of them belong to the same territory constantly under modification at the hands of human actions?

As argued by Marc Augé, far away from the evidence of the past (with erudition and restoration) as well as from the illusory evidence of the present (with spectacle and updating), the links with history of the places signed by the presence of archaeological ruins are something that does not necessarily depend on human interventions, because the relationship between the different space–time components of these peculiar places of our territories are too intertwined and strongly interdependent to be easily wiped out from the landscape systems. Otherwise, Augé underlines that ruins are not a subtraction or a lack but instead something added to nature and to the landscape; thereby, they cannot belong to history anymore and, on the contrary, they have to be considered parts of a "pure time", a time without history (Augé 2004). This means that heritage recycling should not simply aim at avoiding the loss of the physical presence of ruins, but primarily at protecting, maintaining or re-constructing the social, cultural and functional relationship of these places with individuals, people and society, so as to be able to believe in history through a new perception of time. This relationship is always influenced by dynamic and always changing paradigms, and this is why recycling actions must look carefully at the role that the ancient traces have for the present time and for the contemporary city and landscape (Bartolone 2013).

A critical approach to planning and architectural design, as well as ethical human behaviour in inserting new activities, is ultimately the most important ingredients of a dialogical approach in heritage recycling processes because the contemporary landscape is complex but only apparently contradictory, having intrinsic logics and ontological reasons. Therefore, the real main danger of a wrong attitude to the ruins comes from what Paul Ricoeur defines as the different forms of oblivion (Ricoeur 1983). From this perspective, it is possible to identify a system of planning strategies that can start making a small contribution to avoid oblivion. Even if not exhaustive, these strategies can pass through the following short abacus of models of heritage recycling at the planning scale:

- Critical review of the policy instruments
- Inclusion of ancient settlements in new regeneration or transformation planning guidelines
- Actions of territorial "microsurgery" for the purpose of resolving immediate problems of the serious condition of the separation of the ruins and the immediate surroundings
- Economic, educative and scientific development of the archaeological places through creative and collective activities
- Analysis and protection of the ancient remains and consequent recycling of the ruins.

On the other hand, the relationship between social groups of citizens and their environment, deeply related to collective memories, defines, through design intervention, its resilience and adaptability with always new significances and shared values, thus solving dualities and conflicts. In this perspective, recycling cannot exist without collective participation, without multidisciplinary and multiscale actions and where there is no unity between ethical and aesthetic demands. This means that a system of recycling models is needed also in relationship with the design actions, strictly connected to the dualities mentioned above and to the planning strategies. As a temporary conclusion and as an input for further reflections, this system of models can refer to the following points:

- Reassembly of existing ancient materials
- Re-connection of different single historical "episodes"
- Re-composition of the dialogue between the archaeological site and the local context
- Recognition and re-figuration of the existing urban and landscape "texts"
- Redefinition of the terms of balance between temporary and permanent structures
- Overwriting the historical frameworks
- Definition of new interpretative approaches for the places, giving new meanings to the archaeological sites
- Explanation for the historical, typological and constructive reasons of the settlements
- Definition of a new recognisable form for the settlement.

References

Augé M (2004) Rovine e macerie. Bollati Boringhieri, Turin

Bagnato VP (2015) Interventions on the archaeological heritage. An interpretative approach for a new landscapes ethics. In: Obad Šćitaroci M (ed) Possibilities for spatial and economic development—proceedings. Faculty of Architecture University of Zagreb, Zagreb, pp 20–23 (ISBN 978-953-8042-11-9 e-book)

Bagnato VP (2017) Architettura e rovina archeologica: etica, estetica e semantica del paesaggio culturale. Aracne Editrice, Roma

Bartolone R (2013) Dai siti archeologici al paesaggio attraverso l'architettura. Engramma 110:58–90

Berger A (2007) Drosscape: wasting land in urban America. Princeton Architectural Press, New York

Capanna A, Nencini D (2016) Progetti di riciclo. Cinque aree strategiche nella Coda della Cometa di Roma. Aracne, Rome

Ciorra P, Marini S (2011) Re-cycle: Strategie per la casa, la città e il pianeta. Mondadori Electa, Milan

Clementi A (1990) Il senso delle memorie in architettura e urbanistica. Laterza, Rome-Bari

Fabian L, Munarin S (2017) Re-cycle Italy, Atlante. LetteraVentidue, Siracuse

Germanà ML (2014) Technology and architectural heritage. Research experiences in archaeological sites. Techne 07:41–51

Ghersi A, Mazzino F (2009) Landscape & ruins: planning and design for the regeneration of derelict places. Alinea, Florence

Heighway CM (1972) The erosion of history: archaeology and planning in towns: a study of historic towns affected by modern development in England, Wales and Scotland. Council for British Archaeology, Urban Research Committee, London

Losasso M (2016) Historical context and contemporary design: technological innovation between memory and change. Techne 12:6–10

Marini S, Santangelo V (2013) Recycle Italy: new life cycles for architecture and infrastructures of city and landscape. Aracne, Rome

Norberg-Schulz C (1979) Genius Loci: Paesaggio, ambiente, architettura. Electa, Milan

Oteri AM (2009) Rovine: Visioni, teorie, restauri del rudere in architettura. Argos, Rome

Petzet M, Heilmeyer F (2012) Reduce reuse recycle: architecture as resource. Hatje Cantz, Ostfildern-Berlin

Ricoeur P (1983) Temps et récit. Seuil, Paris

Rossi A (1966) L'architettura della città. Marsilio, Padova

Secchi B (2005) La Città del XX secolo. Laterza, Bari

Chapter 14
Models of Heritage Tourism Sustainable Planning

Ana Mrđa and Hrvoje Carić

Abstract The study is based on the theory models of heritage tourism sustainable planning that act as a catalyst for the destinations' market positioning. From the spatial-planning point of view, the competitiveness of the destination is based on the specific cultural experiences, environmental quality and scenic landscape, that positions destination as a quality place for living, working and investing as result of (positive) tourism–heritage interaction. Models of heritage tourism sustainable planning imply the harmony and balance between the global tourism standards and the preservation of the cultural identity of the destination. They are comprised of two main components: an autochthonous place with its cultural–social–economic characteristics and a viable tourism scenario with its sustainable cultural product. The research, based on the visual perception questionnaire, reinforces the paradigm that the heritage tourism sustainable planning models should aim to support the development of tourism without jeopardizing the spatial and socio-economic characteristics of both natural and anthropogenic features of the area and without creating social or economic difficulties for the local community. At the same time, they should be empowered to regulate the tourism/visitor issues consistent with the destinations' lifescape image and cultural tourism experience.

Keywords Heritage tourism · Cultural landscapes · Sustainable planning · Spatial planning · Lifescape

14.1 Introduction

Cultural tourism is one of the largest and fastest growing global tourism markets (Farid 2015; McCain and Ray 2003). Culture and heritage are progressively used

A. Mrđa (✉)
Faculty of Architecture, University of Zagreb, Zagreb, Croatia
e-mail: amrdja@arhitekt.hr

H. Carić
Institute of Tourism, Zagreb, Croatia
e-mail: hrvoje.caric@iztzg.hr

M. Obad Šćitaroci et al. (eds.), *Cultural Urban Heritage*,
The Urban Book Series, https://doi.org/10.1007/978-3-030-10612-6_14

to endorse destinations and enhance their competitiveness and attractiveness. Many locations and destinations are now vigorously emerging their tangible and intangible cultural resources as a means of developing comparative advantages in a progressively competitive tourism marketplace, and to create local distinctiveness in the appearance of globalization. Cultural tourism can be one of theme-based tourist attractions. The development of tourism has been traditionally related to the development of culture. The unique cultural resources of a territory serve as a basis for the cultural supply required by the local community, as well as contributing to increased tourism attractiveness (OECD 2009).

Cultural tourism is partially rooted in the need/motivation to experience the ways of life in visiting communities,[1] thus gaining first-hand understanding of the physical environment, customs, traditions, the intellectual capital, ideas in the real context of architectural, historic, archaeological and other cultural values. Cultural tourism differs from mainstream tourism in that it seeks to gain an understanding or appreciation of the nature of the place being visited (ICOMOS 1997). This interest is profound[2] and requires a certain level of skill, knowledge, conditioning and/or experience (Stebbins 1996). As Silberberg (1995) states, cultural tourism can be defined as visits by persons from outside the host community motivated wholly or in part by interest in the historical, artistic, scientific or lifestyle/heritage offerings of a community, region, group or institution. Therefore, cultural tourism has implemented an educational value—a desire or an ability to perceive and learn about a place and its features.

The transformation of cultural tourism from the elite Grand Tourists to a large-scale tourism segment in the twentieth century has its consequences. Direct revenues from tourism and foreign investment in tourism sector are dominant discourse in economic development (Farid 2015). However, growing and no-limits visitation of major sites, especially those that are home to small and fragile communities, have raised bottom-up questions about all aspects of the sustainability (social, environmental, economic). As evidenced by both media and science research, the ‘over tourism’ of UNESCO WHS and other heritage sites still are in desperate search to deliver the balance between visitation and conservation (Li et al. 2008).

It has been challenging for destinations and tourism providers to profile culture among the welter of products on offer striving to position themselves as unique. There is a growing number of places searching to articulate synergy between culture and tourism (Farid 2015) that would help reinforce, rather than water down, local culture, raising the value accruing to local communities and improving the links between locally induced creativity and tourism (Richards 2013).

---

[1]Cultural tourists are motivated to travel for different reasons than other tourists and, therefore, feel that motivation must be considered an important element when defining cultural tourism (Richards 1996).

[2]In the case of cultural tourism, the issue is that of the ‘wise exploitation’ of the heritage for tourist use (Russo and van der Borg 2002).

14.2 Cultural Heritage as a Basic Resource for Cultural Tourism in the Days of Globalization

To survive, tourist destinations must be globally competitive as well as locally recognized. Special or unique features, overall attractiveness and spatial recognizably are the building blocks of tourist offer and demand. Hence, uncontrolled and unplanned tourism activities lead to the reduction or disappearance of specific *genius loci*, attractiveness and spatial recognition (Mrđa and Bojanić Obad Šćitaroci 2016).

Glocalization in cultural tourism is a mixture of two directions: one is standardization and traditionalism, and the other is uniqueness and diversity—balance between these two features, could be a starting point of cultural approach to tourism. Glocalization of cultural tourism entails providing harmony and balance between the global standards and the cultural identity in delivering sustainable destination management. The assumptions that cultural tourism strategies worldwide face major growth presents an opportunity that it can be directed to boost local-related culture needs, tackle seasonality and oversaturation/crowding via geographic distribution (Richards 1996).

The governance of cultural heritage as one of the more significant and fastest growing components of tourism (Li et al. 2008; Poria et al. 2003; Richards 1996) can be fraught with tensions between the requirements of protection and conservation and the opportunities for economic development (McKercher et al. 2005). On the one hand, the use of historic assets can promote local socio-economic development (Wang and Bramwell 2012). Subsequently, heritage resources and their conservation sometimes have acted as, or been part of, urban or rural regeneration and revitalization schemes to improve the competitiveness of places. More so, culture as a tourist attraction can be a powerful force in arguing that a region's historic, cultural, religious and industrial past should be conserved (McKercher et al. 2005).

Ho and McKercher warn that economic activities may damage historic resources, through physical damage caused by tourist use or through commoditization, trivialization and standardization (2004). Richards also points out that culture as process is transformed through tourism into culture as product (1996).

Yet, attempts to preserve local historic resources that deny socio-economic development can condemn places to economic impoverishment (Yang et al. 2010), as the heritage economic use may also be considered the only funding source for preservation (Chhabra 2009). Thus, the unsolvable dilemma, the use of heritage for tourism can involve mutual benefits, but it may also entail trade-offs and the loss of irreplaceable features of our past.

The symbiosis of tourism and cultural heritage has become the major objective in the tourism management and planning (Garrod and Fyall 2000) with overarching idea that they must respect, protect and upgrade heritage, culture and identity. Cultural tourism, presented as a solution to these issues, rises more than tourism planning and management issues for developing destinations, they are fundamentally the problems of spatial development (Mrđa and Bojanić Obad Šćitaroci 2014).

On the same note, there is no doubt that the commercialization of heritage can generate revenue to help conserve heritage resources, but for this to happen there must be measures to facilitate this transfer of funds (Abu Hassan et al. 2014). Too often, income from publicly owned heritage simply disappears into general budgets, rather than being reinvested in the heritage resources themselves (Swarbrooke 2002).

In that way, the concern for cultural heritage should be upgraded from basic conservation and protection to the higher level of heritage planning and management. From the urban planning point of view, it is highly important to integrate the possible scenarios of identity protection of heritage places into the planning process that will trigger a new method of specific planning approach (Mrđa and Bojanić Obad Šćitaroci 2016).

This research presents an analysis of the paradigms that have been followed in assessing perceived heritage values and identifies the theoretical or conceptual bases which underlie these approaches on the case of the Old City of Dubrovnik, Croatia—UNESCO World Heritage Site (WHS). The aim of the paper is to: (1) identify key criteria for opportunities and obstacles for a stronger cooperation between tourism and culture and (2) to discuss new framework (heritage tourism sustainable planning)—that strengthen the socio-economic links, reinforce mutual engagement in the promotion, protection/safeguarding and revitalization of culture and heritage, and build up cultural tourism for sustainable development.

14.3 Tourism in World Heritage Sites

Number of studies have identified factors that have an impact on the demand for tourism—they have concluded that the cultural and natural heritage sites are the one of tourist's attractions (Lim 2006). In the process of economic development, many cultural and natural heritage sites were increasingly threatened by the traditional causes of decay and industrialization. Yet, the protection of these heritage sites often remains incomplete at the national level, especially in developing and less developed countries.

In noting this formidable phenomenon of damage as well as inadequate conservation, the World Heritage Convention of UNESCO launched an initiative in 1972 to preserve heritage sites considered to be of great value to humanity. The Convention enacted an international treaty entitled Protection of the World Cultural and Natural Heritage and began to ratify sites on the World heritage list in 1978. Since the adoption of the Convention concerning the Protection of World Natural and Cultural Heritage in 1972, many sites throughout the world have been designated and then protected by strict regulations as World Heritage Sites (World Heritage Centre—WHC 2005).

Nowadays, UNESCO WHSs are recognized by tourism demand as the visitation priority attractions,[3] raising the need to allocate tourism sector's responsibility in safeguarding of the outstanding universal value. In order to approach tourism at WHSs in comprehensive and integrated manner, more inputs from culture to tourism development is required.

The WHSs are exploited for national tourism marketing campaigns (Li et al. 2008) since they effectively increase the international visibility and in result earn foreign exchange, thus serving as a major driving force for growth in many developing countries (Herbert 2001; McIntosh and Prentice 1999; Wager 1995). Specifically, once sites are inscribed on the list of WHSs, UNESCO provides support and aid.

The success of the Convention has, however, turned to a potential curse—increased tourist visitation is now the primary threat to WHSs. Visitation increases the threat of damage or destruction of the environmental and cultural integrity focusing the discussion on how to achieve a balance between protection and tourism development (Garrod and Fyall 2000; Leask and Fyall 2006; Li et al. 2008).

Cultural tourism should responsibly, inclusively and sustainably contribute to the host community. At the same time, it should contribute to the cultural identity by promoting cultural exchanges as well as preservation of cultural heritage. This emphasizes the necessity for implementing new methods for affecting the viability and competitiveness of a tourist destination's physical attractiveness. Heritage urbanism here is introduced as a model for integrating sustainable tourism and urban planning principles, strengthening the natural and cultural active protection by advocating broad stakeholder engagement and promoting quality tourism products and services.

The research is confined to a case study heritage site in urban settings: The Old City of Dubrovnik—an UNESCO WHS. Data were gathered and analysed by using urban planning tools to establish the connection between commercialization and conservation towards sustainable planning of heritage tourism and to answer the questions: Do the tourists choose to visit Dubrovnik more because of the inscribed status as the UNESCO World Heritage city? What would be the understanding of the tourists towards UNESCO World Heritage status and their interest in Dubrovnik? What would be the improvement in the development of the city and its inhabitants' life quality?

14.4 Materials and Methods

From a spatial-planning perspective, tourism consists of two major components (Mrđa 2015):

[3]In 1998, Hall and Piggin conducted a survey of 44 World Heritage Sites, in which over two-thirds of the site managers reported there had been an increase in visitor numbers after their site gained World Heritage status.

1. Authentic place and its physical and socio-economic characteristics—must be able to support tourism development without compromising the physical characteristics and peculiarities of the area (natural and anthropogenic, either physical or socio-economic) and without creating a sociocultural and economic problems to the local community.
2. The tourism scenario and the tourism product—the visitor capacity that are consistent with the image of the tourism product and related experience.

The importance of local specificities/attractions is a fundamental factor in the method of determining the tourist capability in spatial planning. The unprecedented changes in landscapes raise concerns over the environmental and cultural integrity of WHSs and have also led to a re-examination of tourism growth in the light of sustainable development (Drost 1996).

In this context, sustainable development and sustainable cities are mainly considered the theoretical basis of sustainable WHSs. In addition, theories of sustainable tourism, including carrying capacity, life cycle and community participation also provide the basis for research on resource-, activity- and community-based traditions of sustainability (Saarinen 2006). The topic of sustainable WHSs encompasses a variety of aspects, such as sustainability assessment (Ko 2005), legal study (Boer and Wiffen 2006), policy research (Maikhuri et al. 2001) and conflict analysis (Maharjan 2013). A conflict between authentic conservation and commodification is usually involved, particularly for WHSs in cities (Lee 1996; Pendlebury et al. 2009).

Conflicts over sustainable development (which aims to provide new development without jeopardizing the legacy of WHSs) according to Zhang et al. (2015) can be classified into those between economic benefit and protection, social benefit and protection, and social and economic benefit.

In summary, the application of the game theory to WHS conflicts mainly refers to communities, governments, tourists and managers (Zhang et al. 2015). However, there has been little research to date considering the conflicts from the spatial perspective. This study seeks to rectify this by using geographic information systems (GIS) as the primary analytical method with the software of ArcGIS. This tool examines the spatial distribution patterns of the heritage site and its spatial relationship comparing how to optimize city-space allocation within WHSs by analysing the conflicts involved and how they can be resolved.

The aim is to compare the crowding effects and the city capability with the state of the core attractions protection. For achieving such, the data from a key stakeholder—visitors—should be taken into consideration, particularly, the attitude towards the WHS inscription. Visitors' perceptions will be understood via visitor surveys based on the evaluation of the onsite (in situ) locations.

14.4.1 The Case of Dubrovnik

Dubrovnik, a historic city and a UNESCO World Heritage Site since 1979, at the south end of the Croatian Adriatic coast, is a flagship Croatian tourism attraction. The old mediaeval core, surrounded by almost 2 km of intact city walls which encircle the city is one of the most beautiful and strongest fortification in Europe and the main attraction of the city making it a 'must visit' destination on many itineraries. Within the walls, there is an array of museums, galleries, churches and monasteries and a concentration of shops and restaurants. The Old City also functions as the city centre for 42,615 of its residents Croatian Bureau of Statistics (2017). The Old City has its own harbour with many touristic functions including a boat transfer to the island of Lokrum (distance app 500 m) that is a nature reserve—a unique situation in the Mediterranean. Around the Old City is a tightly built area on a steep mountain slopes, combining residential, retail and tourism functions. Most of the city's five-star hotels are in this area in the immediate vicinity of the Old City and from most of them guests can admire the sea and the Old City views. The main harbour is located a couple of kilometres from the Old City where cruisers dock.

In terms of tourism demand, Dubrovnik recorded over 1 million arrivals and 3.4 million overnights in 2016 (Central Bureau of Statistics 2017). Additional pressure comes from day-visitors whose number is unknown but could be gauged from the tourism statistics for the County of Dubrovnik–Neretva, where there were 5.5 million arrivals realizing 13.5 million overnights (Central Bureau of Statistics 2017) and likelihood that these visitors are also like to make an excursion to Dubrovnik. On top of it, there is a pressure by cruise tourism as Dubrovnik is 'a must-see' destinations for most of the companies operating Mediterranean cruise lines. Over the last year, Croatia recorded 813 cruise calls with over 1 million passengers. Of these, 67% of cruise calls were recorded in Dubrovnik (Central Bureau of Statistics 2017), while most these ships were large cruisers carrying 1000 or more passengers. On average, Dubrovnik has cruisers in its two ports for about 243 days in a year and, during the peak demand, it records up to three cruise calls with eight thousand passengers per day.

14.4.2 Conceptual Models and Questionnaire Design

This study has a threefold purpose: (1) to understand the popularity of WHS; (2) to understand visitors' perspectives on WHS designation and their attitudes towards the inscription; and (3) to find the spatial factors with the most influence on visitors' attitudes towards protected WHSs attractions.

To achieve that mapping approach is introduced that uses unsupervised classification to automatically cluster area sharing similar inquiry answers to land characteristics, as described by a set of spatial indicators. Heritage crowding maps of the

Fig. 14.1 UNESCO WHS Model of the Old City of Dubrovnik—Tourist use of the public space

sites are generated by considering five indicators: remoteness, area size, area shape, visitor perception and attraction quality (Fig. 14.1).

To shed light on the question of whether it is better for a site to be a WHS, we designed a questionnaire that could aid discussion on the opinion of tourists. It contained three parts: basic information (gender, age, education, residence, occupation, number of visits); cognitions on the impacts of WHS designation (heritage protection, social development and visitor experience); and attitudes towards the inscription. To address the question on the fever of inscription on the WHS in a broader context, we also interviewed on site some residents as well as experts. The survey was undertaken in places where tourists are concentrated. Primary data collection was mainly undertaken at location 1 (29), location 2 (31) and location 3 (28) (Fig. 14.1, Table 14.1). Data were collected over seven days from 1 to 7 August 2018. This is a peak tourist season in Croatia and Dubrovnik. Three research assistants provided respondents with a brief introduction and explanation about the purpose of the study before distributing the questionnaires. The average length of an interview was 10 min, and 100 visitors were sampled (88% response rate with 12 refusals). All items were measured on a five-point Likert scale, where 1 = strongly disagree; 2 = disagree; 3 = neutral; 4 = agree; and 5 = strongly agree.

The concept of sustainable development is widely interpreted as a need to achieve sustainability concurrently within environmental, economic, social and cultural spheres (The Mediterranean Strategy for Sustainable Development 2005). According

Table 14.1 Selected locations for interviews

Location 1: Square in front of the Onofrio Fountain

Location 2: Square in front of the Church of St. Vlaho

Location 3: Square in front of the Church of St. Ignacio

Table 14.2 Criteria for the evaluation of tourism resources

Criteria for the evaluation of tourism resources					
Topic	Score				
	1	2	3	4	5
Authentic heritage scenery (cultural vitality)	Authentic heritage scenery modified in general	>	>	>	Authentic heritage scenery preserved at large
Sense of a sociocultural setting (social equity)	Sociocultural value does not exist	>	>	>	Sociocultural value is very high
Functionality of a site for a tourist use (economic health)	Tourist use is not possible	>	>	>	Tourist use is very likely to happen

to that the complexity of the cultural heritage is expressed in four main components: natural, cultural, social and economic. Natural complexity is largely represented by an unbuilt area. Cultural and social complexity is intimately linked to the diverse human use of resources and to a wide spectrum of land use. Economic complexity is linked to the diversified use of local resources (Farina 2000, 313). Therefore, evaluation criteria (Table 14.2) are based on the state of:

- Authentic heritage scenery (cultural vitality and environmental responsibility),
- Sense of a sociocultural setting (social equity), and
- Functionality of a site for tourist use (economic health).

14.4.3 Results

Profiles of respondents

Respondents comprised more females (62%) than males (38%). Visitors were aged between 19 and 51. Most of the respondents (72%) had obtained a bachelor's, whereas a significant part (24%) earned higher degrees. Only 4% reported they were high school graduates and below. Of the respondents, foreign tourists (73%) outnumbered local tourists (27%). Most respondents (85%) reported that it was their first-time visiting Dubrovnik.

Cognition on the inscription of Dubrovnik on the World Heritage List

Respondents were asked to complete questions about WHS and illustrate their knowledge. The first question was to focus on awareness of WHSs based on four indicators, namely, WHSs, the WHS convention, criteria of WHS designation, and WHS that they had visited or know of. The results showed that 48% of respondents were aware of UNESCO's WHSs, although only 6% of respondents reported having a higher familiarity with UNESCO's WHSs. Thus, awareness was quite low. A little over

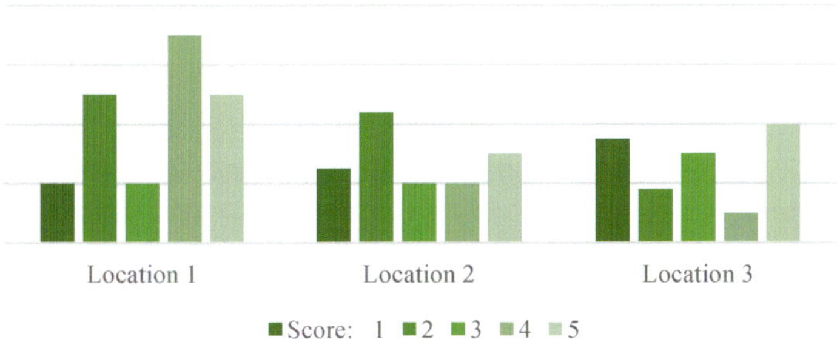

Diagram 14.1 Authentic heritage scenery—respondents' answers

half of the respondents (57%) knew that Dubrovnik is on WHS list. This is perhaps surprising given that visitors had selected to go to Dubrovnik because of its cultural attractions. However, most of the respondents (96%) agreed that Dubrovnik meets all the requirements to be enlisted as a WHS. Likewise, almost all the respondents (92%) agreed that the public space of the Dubrovnik Old City should be more protected from tourist congestion and crowding effect. Only 2% expressed neutrality and none who disagreed. Additionally, more than half of the respondents (52%) indicated that the Old City has been taken care of and is in a good condition.

Visitors' perspectives on impacts on visitor experience

First task was to evaluate the authenticity of the heritage scenery on site location using numeric values from 1 (Authentic heritage scenery modified in general) to 5 (Authentic heritage scenery preserved at large).

The concern with authenticity in tourist destinations like these is also a concern with place identity (Jamal and Hill 2004, 362). Respondents confirmed that the Location 1 is the most authentic heritage scenery (Diagram 14.1). Thus, these groups of respondents valued the traditional urban setting as authentic cultural heritage from an aesthetic point of view, but also as having the dominant heritage value.

Heritage sustainability can be characterized by ensuring the continuing contribution of heritage to the present through the thoughtful management of change responsive to the historic environment and the social and cultural processes that have created it (Tunbridge and Ashworth 1996). Put like that, for the cultural vitality of the WHS, it is important that new tourist sites must be in accordance with and respecting the existing peculiarities of the cultural heritage. The more authentic they are, the more tourists value them.

The second task was to select the preferred ratio of the sociocultural setting using numeric values from 1 (Sociocultural value does not exist) to 5 (Sociocultural value is very high).

Respondents confirmed that the Location 3 is rated with the highest sociocultural value (Diagram 14.2). Thus, these groups of respondents valued only the traditional historic setting as their most valued cultural landscape from primarily cultural point of view but also having social components in their minds. It seems that this group

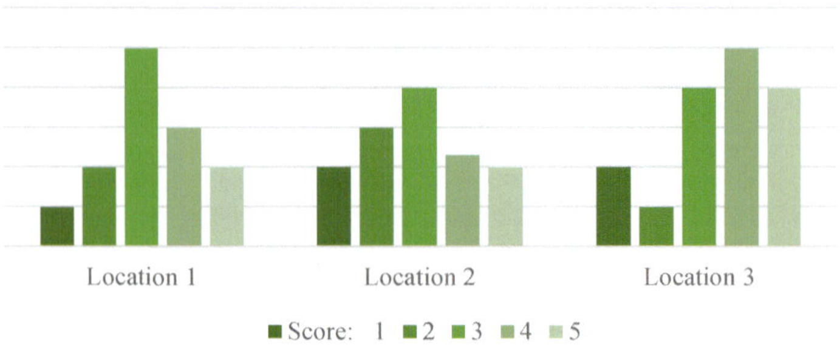

Diagram 14.2 Sociocultural setting—respondents' answers

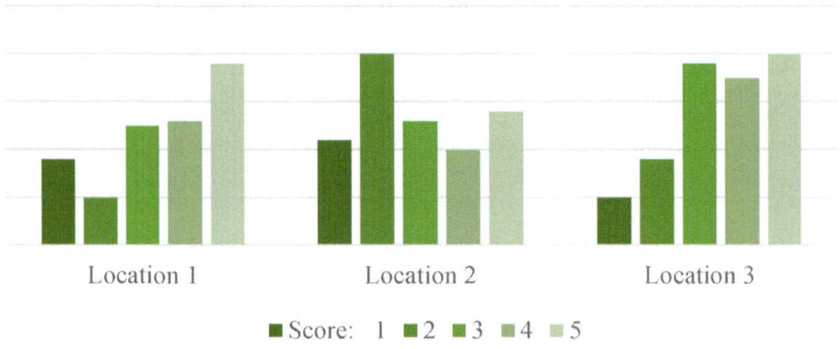

Diagram 14.3 Tourist use—respondents' answers

strongly valued the community significance of historic places and not the isolation of the solely tourist environment without locals–tourists interaction like other two locations, where tourists do not have insight into the real social component of the everyday life and community or, better to say, lifescape.

The unique expression of culture within each landscape provides a backdrop against which people—mostly unwittingly—structure their own identity. We develop together with our landscape. It gives us a sense of place and reveals our relationship with the land over time (Maessen et al. 2008, 551). Lifescape can be defined as both a place and a process or as a branding of the landscapes in a means of the sense of the place. It is a question how social structures and social context affect the livelihoods of communities and, in turn, how these factors shape the use of natural resources and the potential to manage them well within a landscape. Protecting the identity of lifescapes is a powerful way for capacity building and enhancing the social capital leading towards upgrading the social equity of a tourist region.

The last task was to evaluate the possibility for tourist use of the space using numeric values from 1 (Tourist use is excessive) to 5 (Tourist use is underdone).

This theme showed strong preference for the cultural heritage possessing visual quality. Respondents were thinking about the level of activation of a site for tourist use and the level of touristic congestion on the site. Consequently, they evaluated Location 3 as the optimal (Diagram 14.3).

This questionnaire shows that the protection of cultural heritage is of utmost importance and that the tourist services should not be planned without good evaluation of the area's identity and tourist potential. It is also interesting that all respondents who are not familiar with the area mainly perceive cultural heritage as a place rich in social and cultural values and dislike the appearance of excessive resource exploitation and tourist congestion. This kind of perception is consistent with the studies of community scape values such as the ones reported by Kaufman (1997), Davenport and Anderson (2005) and Jacobsen and Steen (2007). The concept of 'place attachment' (Kaltenborn and Bjerke 2002; Brown 2005) relates perception to actual places where people have interacted with and give meanings to such places. What the respondents not familiar with the area perceived as valuable are locations that constitute the identity of the cultural heritage when viewed but also having the authentically life going on and not being mainly occupied with tourists.

14.5 Conclusion

Although the authentication of the World Heritage List aims to protect historical and natural assets of humanity, many countries have adopted the UNESCO recognition as a marketing tool or assurance label to attract foreign tourists. The idea that having World Heritage Sites can promote tourism seems to be a common belief. Work presented proposes a theoretical framework to inform how human-heritage-outcome actions interrelate. Culture and heritage are of immeasurable value to host communities and are also important assets for tourism. Tourism can be a considerable force for the promotion and safeguarding of tangible and intangible heritage while encouraging the development of arts, crafts and other creative activities. However, if poorly managed, tourism can also cause negative externalities primarily on culture and heritage.

Results indicated that Dubrovnik has been a preferred tourist location in heritage tourism sector resulted from the political stability and long-time protection. Having UNESCO World Heritage status gives Dubrovnik extra boost to its award-winning strengths. Cultural diversity, multiculturalism and ample accommodation make Dubrovnik a good choice to visit. More development opportunities have been offered which may bring more economic stability in the state that creates more opportunities for local people. It can be said that from the findings, the local people in Dubrovnik have gain advantage result of the growth of tourism. The authenticity and uniqueness of the heritage buildings and monuments that have significant historical background intrigues tourists. More advance promotion has been done by Dubrovnik Tourist office by using electronic media approach and still implementing the traditional way of promotion.

As the Moy and Phongpanichanan (2014) state, opportunities provided for the place being on the UNESCO World Heritage List to attract the tourists are plentiful; on the other hand, the empirical results from Huang et al. (2012) indicate that there is no significant effect of World Heritage List on promoting tourism other than possibly a short-run tourism-enhancing impact.

Four technical sessions, addressing the following issues:

1. Promoting and protecting cultural heritage

If properly managed, tourism can be a key driver in the preservation and promotion of cultural heritage, while nurturing a sense of pride within communities and destinations around the world.

2. Living cultures and creative industries

Tourism is about experiencing the world and its living cultures. Creative industries such as music, cinema, crafts, architecture or design play an increasing role in creating a unique visitor experience, while tourism can foster innovation and the exchange of experiences that grow the creative industries.

3. Linking people and fostering development through cultural routes

Cultural routes offer important opportunities for tourism development, regional integration and economic development, while promoting the revitalization and promotion of cultural and economic links at national and international levels.

4. Cultural tourism and urban regeneration

With more than half of the world's population living in cities, the scope for urban tourism is growing rapidly. Urban tourism is critical in the preservation of cultural identity, economic revitalization and for enriching city life.

References

Abu Hassan Z, Bin Jailani M, Rahim AF (2014) Assessing the situational analysis of heritage tourism industry in Melaka. Procedia—Soc Behav Sci 130:28–36

Boer B, Wiffen G (2006) Heritage law in Australia. Oxford University Press, Melbourne, p 43

Brown G (2005) Mapping spatial attributes in survey research for natural resource management: methods and applications. Soc Nat Resour 18(1):17–39

Chhabra D (2009) Proposing a sustainable marketing framework for heritage tourism. J Sustain Tourism 17(3):301–320

Croatian Bureau of Statistics (2017) Statistical yearbook of the Republic of Croatia. Retrieved 21 March 2017. Web site: http://www.icomos.org/tourism/, https://www.dzs.hr/Hrv_Eng/ljetopis/2017/sljh2017.pdf

Davenport MA, Anderson DH (2005) Getting from sense of place to place based management: an interpretive investigation of place meanings and perceptions of landscape change. Soc Nat Resour 18:625–641

Drost A (1996) Developing sustainable tourism for World Heritage Sites. Ann Tourism Res 23(2):479–492

Farid SM (2015) Tourism management in world heritage sites and its impact on economic development in Mali and Ethiopia. Procedia—Soc Behav Sci 211:595–604

Farina A (2000) The cultural landscape as a model for the integration of ecology and economics. Bioscience 50(4):313–320

Garrod B, Fyall A (2000) Managing heritage tourism. Ann Tourism Res 27(3):682–708
Herbert D (2001) Literary places, tourism and the heritage experience. Ann Tourism Res 28(2):312–333
Ho P, McKercher B (2004) Managing heritage resources as tourism products. Asia Pac J Tourism Res 9(3):255–266
Huang C, Tsaur J, Yang C (2012) Does world heritage list really induce more tourists? Evidence from Macau. Tour Manag 33:1450–1457
ICOMOS (1997) Charter for cultural tourism. Retrieved 9 March 2017. Web site: http://www.icomos.org/tourism/
Jacobsen J, Steen K (2007) Use of landscape perception methods in tourism studies: a review of photo-based research approaches. Tourism Geographies 9(3):234–253
Jamal T, Hil S (2004) Developing a framework for indicators of authenticity: the place and space of cultural and heritage tourism. Asia Pac J Tourism Res 9(4):353–372
Kaltenborn BP, Bjerke T (2002) Associations between landscape preferences and place attachment: a study in Roros, Southern Norway. Landscape Res 27(4):381–396
Kaufman P (1997) Community values in cultural landscape decision making: developing recommendations for ensuring planning processes include differing expectations of communities of interest. Historic Environ 13(3 and 4):57–62
Ko TG (2005) Development of a tourism sustainability assessment procedure: a conceptual approach. Tour Manag 26(3):431–445
Leask A, Fyall A (2006) Managing world heritage sites. Butterworth-Heinemann, MA, USA
Lee SL (1996) Urban conservation policy and the preservation of historical and cultural heritage: the case of Singapore. Cities 13(6):399–409
Li M, Wub B, Cai L (2008) Tourism development of World Heritage Sites in China: a geographic perspective. Tour Manag 29:308–319
Lim C (2006) A survey of tourism demand modeling practice: issues and implications. In International handbook on the economics of tourism. Edward Elgar, Cheltenham
Maessen R, Wilms G, Jones-Walters L (2008) Branding our landscapes: some practical experiences from the LIFESCAPE project. In: 8th European IFSA symposium, Clermont-Ferrand (France), pp 551–561
Maharjan M (2013) Conflict in World Heritage Sites of Kathmandu Valley: a case study on the conservation of private houses in three durbar squares. Nepal Tourism Dev Rev 2(1):87–104
Maikhuri RK, Nautiyal S, Rao KS, Saxena KG (2001) Conservation policy-people conflicts: a case study from Nanda Devi Biosphere Reserve (a world heritage site), India. Forest Policy Econ 2(3):355–365
McCain G, Ray NM (2003) Legacy tourism: the search for personal meaning in heritage travel. Tour Manag 24:713–717
McIntosh A, Prentice R (1999) Affirming authenticity: consuming cultural heritage. Ann Tourism Res 26:589–612
McKercher B, Hoa PSY, du Crosb H (2005) Relationship between tourism and cultural heritage management: evidence from Hong Kong. Tourism Manag 26:539–548
Moy LYY, Phongpanichanan C (2014) Does the status of a UNESCO World Heritage City make a destination more attractive to Mainland Chinese tourists? A preliminary study of Melaka. Procedia—Soc Behav Sci 144:280–289
Mrđa A (2015) Method for determining tourism carrying capacity in spatial planning. Doctoral thesis, Faculty of Architecture, University of Zagreb, Zagreb
Mrđa A, Bojanić Obad Šćitaroci B (2014) The importance of the concept of tourism carrying capacity for spatial planning—previous research, development and methodological approaches. Prostor 22(2):212–227
Mrđa A, Bojanić Obad Šćitaroci B (2016) Heritage touristscapes: a case study of the Island of Hvar. Annales-Anali za Istrske in Mediteranske Studije-Series Historia et Sociologia 26(3):553–572
OECD (2009) The impact of culture on tourism. Retrieved 4 October 2016. Web site: http://www.oecd.org/cfe/tourism/theimpactofcultureontourism.htm

Pendlebury J, Short M, While A (2009) Urban World heritage sites and the problem of authenticity. Cities 26(6):349–358
Poria Y, Butler R, Airey D (2003) The core of heritage tourism. Ann Tourism Res 30(1):238–254
Richards G (1996) Production and consumption of European cultural tourism. Ann Tourism Res 23(2):261–283
Richards G (2013) Tourism development trajectories: from culture to creativity. In: Smith M, Richards G (eds) The Routledge handbook of cultural tourism. Routledge, London, pp 297–303
Russo AP, van der Borg J (2002) Planning considerations for cultural tourism: a case study of four European cities. Tourism Manag 23:631–637
Saarinen J (2006) Traditions of sustainability in tourism studies. Ann Tourism Res 33(4):1121–1140
Silberberg T (1995) Cultural tourism and business opportunities for museums and heritage sites. Tour Manag 16(5):361–365
Stebbins RA (1996) Cultural tourism as serious leisure. Ann Tourism Res 23(4):948–950
Swarbrooke J (2002) Heritage tourism in UK—a glance from things to come. From http://www.insights.org.uk/article.aspx?title=Heritage+Tourism+in+the+UK+a+Glance+at+Things+to+Comeandtype=7andlang=en_usandoutput=json
The Mediterranean Strategy for Sustainable Development (2005) UN Environmental program. UNEP/MAP, Atena
Tunbridge J, Ashworth G (1996) Dissonant heritage: the management of the past as a resource in conflict. Wiley, London
Wager J (1995) Developing a strategy for the Angkor world heritage site. Tour Manag 16(7):515–523
Wang Y, Bramwell B (2012) Heritage protection and tourism development priorities in Hangzhou, China: a political economy and governance perspective. Tour Manag 33:988–998
WHC (2005) Retrieved 5 October 2016. Web site: http://whc.unesco.org/
Yang CH, Lin HL, Han CC (2010) Analysis of international tourist arrivals in China: the role of World Heritage Sites. Tour Manag 31(6):827–837
Zhang X, Zhou L, Wu Y, Skitmore M, Deng Z (2015) Resolving the conflicts of sustainable world heritage landscapes in cities: fully open or limited access for visitors? Habitat Int 46(91):91–100

Chapter 15
Tourism Valorisation of Cultural Heritage

Vuk Tvrtko Opačić

Abstract Until recently, cultural heritage was treated as something of high value, primarily to be protected. Recently, however, this conservational approach has lessened and cultural heritage has been treated as a valuable resource that can be managed in a sustainable manner and as a function of economy, mostly via tourism. By incorporating cultural heritage into the tourism supply, the economic significance of cultural heritage becomes equally important to its social, scientific, and political significance. Economic significance is not solely limited to financial profit, developing tourism destination supply, and economic development of the community in question, but also for securing funds for financing of cultural heritage, which are necessary for its maintenance, revitalisation, and enhancement. Tourism valorisation of cultural heritage includes several steps: identification of cultural heritage suitable for conversion into tourism attractions; evaluation of the tourism attractiveness thereof; determination of its spatial distribution and; finally, application of an appropriate model for tourism valorisation. In evaluating tourism attractiveness of cultural heritage, it is necessary to define clear criteria and indicators, as well as an assessment scale. Four spatial models of tourism valorisation of cultural heritage are proposed based on identification of cultural heritage suitable for conversion into tourism attractions, the evaluation of its tourism attractiveness and determination of its spatial distribution: concentrated or point model; dispersed urban model; linear thematic model; and regional networked model. Each mentioned model is based on a specific principle and is suitable for attracting certain groups of cultural heritage tourists, depending on their motivation.

Keywords Cultural heritage · Tourism valorisation · Cultural tourism · Tourism attractiveness · Cultural tourists

---

V. T. Opačić (✉)
Department of Geography, Faculty of Science, University of Zagreb, Zagreb, Croatia
e-mail: vtopacic@geog.pmf.hr

M. Obad Šćitaroci et al. (eds.), *Cultural Urban Heritage*,
The Urban Book Series, https://doi.org/10.1007/978-3-030-10612-6_15

15.1 Introduction

Until around thirty years ago, cultural heritage was treated as something of high value, primarily as something to be protected. This conservational approach mainly encompassed material heritage, while intangible heritage, manifested in arts, languages, folkways, music and performing arts, religion, sport and festival and pageants (Timothy 2011), was part of a different plan. Lack of the necessary immense funding from the state for conservation and maintenance of cultural heritage has often led to the neglect and decay of important cultural-historic monuments, causing in turn the irreversible loss of all four of the fundamental areas of significance in cultural heritage: social; scientific; political; and economic (Hall and Mc Arthur 1993; Ashworth and Graham 1997; Timothy and Boyd 2003).

Neglect and decay of cultural heritage have opened the question of the necessity of its adequate repurposing, keeping in mind the cultural identity of a given cultural good. Jelinčić (2010) stated the three most common forms of repurposing of cultural heritage: repurposing for social-administrative aims; repurposing for cultural aims; and repurposing for economic aims.

The increasing need to promote local, regional, and national identity in the era of globalisation has motivated the research community to debate developmental (economic, touristic) use of primarily tangible, but also intangible cultural heritage (Nuryanti 1996; Graham et al. 2004; Jelinčić 2008). More developed European states, which started to apply a conservational approach to cultural heritage earlier, were also advanced in the implementation of the economic valorisation thereof (Poljanec-Borić 2017). Over time, the conservational approach in cultural heritage management in nearly all developed European states has abated somewhat, and it started to be treated as something of value that could be managed in a sustainable way and in the function of economy, i.e. tourism (Moulin and Boniface 2001). UNESCO has also given incentive to treat cultural heritage in a new way since the beginning of the 1990s; it has encouraged greater involvement of tourism into cultural heritage management with the reasoning that sustainable tourism valorisation contributes to revitalisation of cultural heritage and its integration into the economic development of local communities (Richards 1996; du Cros 2001; McKercher and du Cros 2002; Hampton 2005; Smith 2009).

The key question which emerges from this is how to developmentally manage cultural heritage in a sustainable manner while simultaneously preserving all four of its fundamental significances, with the goal of preventing cultural goods from becoming dead, forgotten capital excluded from contemporary local identity. Tangible and intangible cultural heritage should not be forgotten and left to decay; however, they should also not be treated as an inexhaustible resource in the creation of an area's tourism supply. Thereby, the examples of cultural heritage being haphazardly transformed into the tourist products serve as a warning of the danger of losing the identity dimension of cultural heritage, of physical degradation, as well as negative physiognomic, economic, and sociocultural implications in a certain area (van der Borg et al. 1996; Ashworth 2009; Timothy and Nyaupane 2009).

The economic significance of cultural heritage is reflected, above all, in its touristic value and integration within the overall tourism supply of tourism destination, whereby the goal is not only to create income and develop tourism, but also to maintain, revitalise, and further enhance of cultural heritage (Zeppel and Hall 1991; Throsby 2001).

15.2 Contemporary Context of Cultural Heritage Tourism

In the postmodern era, cultural heritage tourism has developed in the context of several key social processes and circumstances: the search for local identity; the dynamic relationship between global and local processes (glocalisation); pluralism of lifestyles in relation to the ability of free time usage and an ever more pronounced tendency of mass tourism mobilities.

Under the current conditions of globalisation and unification of social norms, cultural paradigms, and lifestyles, the expression of local identity is often an element crucial in the creating of a unique tourism supply (Millar 1989; Antolović 1997; Kušen 2002; Park 2014; Marković and Fuerst Bjeliš 2015). Cultural heritage accordingly, as a "protector" of spatial identity, has a first-rate meaning and can also be perceived as a sort of "local" answer to "global" expansion (Graham et al. 2004). Furthermore, tourism valorisation of cultural heritage promotes regional and local identity throughout the entire area of a given tourism destination, which should also be considered in the context of glocalisation. Glocalisation is understood to include the simultaneous presence of universal (global) values, which, in this case, are represented by tourism, and local specificity and identity, as presented via cultural heritage (Salazar 2005, 2010).

The deregulation of working hours and workplaces opened new possibilities for use of free time, which can be observed in the increased popularity of "city-break" trips, within which cultural heritage supply play an expressed role. Viewed from the aspect of tourism demand, cultural tourism, as well as cultural heritage tourism, is becoming ever stronger (Wall and Mathieson 2006; Jelinčić 2008). The main recognised reasons for this are: increasing interest of tourists in the local and regional spatial identity of a given tourism destination, along with a simultaneous decrease in the popularity of mass tourism; growth in the education level of tourists; increased mobility of tourists, which is directly affected by improvements in transit infrastructure; the emergence of an increasing number of low-cost air carriers; competition of an increasing number of tourism destinations with similar tourism supply; more retired persons engaging in tourism; etc.

Postmodern tourism supply increasingly rely on tourist products of alternative forms of tourism—within the structure thereof, in many tourism destinations, cultural tourism assumes a key role (Vrtiprah 2006), i.e. its component part: cultural heritage tourism (Kaminski et al. 2014), based mainly on tourism valorisation of tangible cultural heritage within its regional spatial framework (Park 2014). Cultural heritage often becomes a comparative advantage of a tourism destination in relation to similar competitor destinations (Tomljenović 2006). One of the fundamental characteristics of cultural heritage tourism supply is the vast number of different tourist products based on cultural heritage (Hughes 1996), not only in the leading tourist places but throughout their entire associated tourism regions, i.e. tourism destinations. Namely less known cultural heritage of local significance might not, according to value, significance, and overall attractiveness to tourists, measure up to cultural heritage of global significance, but they are indeed important for local identity, tourism and cultural policy, as well as the identity of a given tourism destination (Jelinčić 2005). Therefore, the tourism supply of cultural heritage tourism of a destination in a wider area is shaped on the basis of comparative attractive advantages of cultural heritage. The advantage of defining a regional concept of development of cultural heritage tourism this way lies in the more balanced distribution of tourism throughout a given destination, whereby smaller settlements are also touristically and culturally activated, as opposed to only the leading tourist and cultural centres.

The tourist product of cultural heritage tourism is based on both tangible and intangible cultural goods, as well as visitors' impressions (Jelinčić and Senkić 2017), which are the result of carefully composed presentation and valorisation of cultural goods of a given tourism destination. During tourism valorisation of cultural heritage, conservation and protection of cultural goods are of exceptional importance (Fyall and Garrod 1998), as well as an exact interpretation of resources (Moscardo 1996; Timothy 2011; Ennen and van Maanen 2014), and an authentic tourism experience (Teo and Yeoh 1997; Slunjski 2017).

One of the key challenges in further development of cultural heritage tourism is adequate tourism valorisation of cultural goods via structuring sustainable tourist products based on applicable models (du Cros 2001; Landriani and Pozzoli 2014), so that cultural heritage would be persevered for the future in its entirety and at the same time contribute to quality development of tourism in tourism destinations.

Appreciating the aforementioned processes and the significance of cultural heritage in modern tourism supply, emphasis is hereafter placed on tourism valorisation of cultural heritage along with four suggested models.

15.3 Integration of Cultural Heritage into the Tourism Supply

Tourism valorisation of cultural heritage is understood to include the act of its own conversion into a tourist product that includes the following steps:

(a) identification of cultural heritage suitable for conversion into tourism attractions;
(b) evaluation of tourism attractiveness of cultural heritage;
(c) determination of spatial distribution of cultural heritage;
(d) application of an appropriate model for tourism valorisation of cultural heritage (Fig. 15.1).

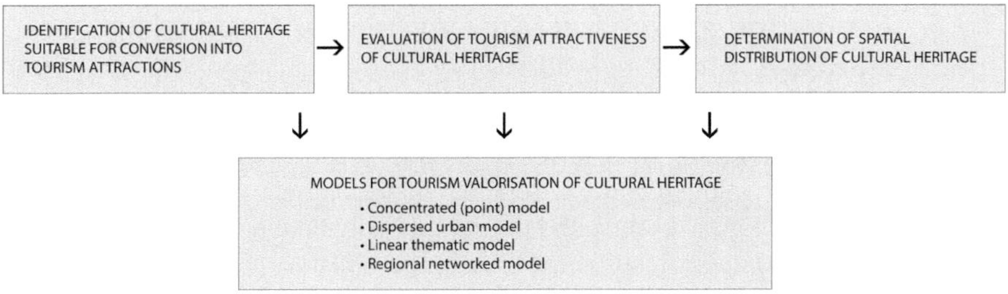

Fig. 15.1 Main steps in the scope of tourism valorisation of cultural heritage

15.3.1 Identification of Cultural Heritage Suitable for Conversion into Tourism Attractions

The starting point for identification of cultural heritage suitable for conversion into tourism supply is often official documentation of the competent ministry of a given state.

For example, in the Republic of Croatia, the Registry of cultural goods of the Ministry of Culture is composed of three lists: (a) the list of protected cultural goods; (b) the list of cultural goods of national significance; and (c) the list of preventively protected cultural goods (cultural goods whose protection is on hold due to various reasons) (Ministry of Culture of the Republic of Croatia 2018). For each cultural good mentioned in the Registry, the following data are included: its name or title; a designation stating which list it can be found on; registry order number; a designation of its type (immobile, mobile, intangible); its classification; its location (address, settlement, town/municipality, county); its creator; date of creation; the competent conservatorial department; a summary description; and photographs. Immobile cultural heritage in the Registry is classified into three groups of cultural goods: (a) individual cultural goods; (b) cultural-historical complexes; and (c) cultural landscapes.

Aside from official documentation, a basis for recognising appropriate cultural goods for integration into the tourism supply can also be lists of cultural goods based on the assessment of experts (Kušen 2017), as well as the perceptions of local population. For sustainable management of cultural heritage and sustainable development of cultural heritage tourism, data sources should be combined as much as possible during the identification of cultural goods (Šmid Hribar and Ledinek Lozej 2013; Šmid Hribar et al. 2015).

15.3.2 Evaluation of Tourism Attractiveness of Cultural Heritage

Cultural goods differ among themselves regarding their own value. It is important to emphasise that the value of cultural goods is a characteristic that changes in time (Dumbović Bilušić 2015). As the main reasons for the fluctuating value of heritage in time Jelinčić (2010) states an insufficient knowledge of history, the difficulty for local population to relate to the cultural good and intentional rejection of the connotation with certain historical eras relating to social and political circumstances.

In relation to the aforementioned fact that heritage has many significances, evaluation of its tourism attractiveness is, in most cases, a part of the evaluation procedure of the significance of a given piece of heritage. Numerous examples of evaluation of heritage indicate that there is no simple set of criteria and indicators and that criteria and indicators vary depending on the type of cultural heritage, the goal of the evaluation, as well as research area.

For example, during evaluation of cultural landscapes in the mountainous area of Žumberak in Croatia Dumbović Bilušić (2015) assessed their evident, historical, aesthetic, cultural, natural-environmental, and socio-economic values, and took the following criteria into account: pristine state/wholeness; authenticity and state of preservation; representativeness and recognisability; rarity/uniqueness; diversity and complexity; harmoniousness; aesthetics and visual characteristics; linkages with events, people, and movements; and status, ability (assessment of the entire area), and representation of the cultural goods (assessment of the entire area). Mrđa and Bojanić Obad Šćitaroci (2016) used the following as criteria for cultural landscape evaluation on the island of Hvar, Croatia: authentic heritage scenery (cultural vitality); authentic natural landscape (environmental responsibility); sense of sociocultural setting (social equity); and activation of the site for tourism use (economic health).

In many cases of evaluation of geoheritage via inventory cards method the characteristics most often evaluated are: scientific; ecological; cultural; aesthetic; economic; educational; and geoconservational value, using different criteria and quantitative indicators (Coratza and Giusti 2005; Reynard et al. 2007; Comănescu and Dobre 2009; Cocean 2011).

For assessment of the tourism potential of smaller cultural and heritage attractions, McKercher and Ho (2006) evaluate cultural, physical, product and experiential values on a sample of 16 secondary museums, historic sites and temples in Hong Kong.

Slunjski (2017) states four groups of cultural heritage values which should be considered during evaluation: scientific; cultural; educational; and tourism. The same author points out the following as criteria for evaluating tourism attractiveness: accessibility and ingress; recognisability outside of science and relevant professional field; representation in tourism brochures; level of services and equipment; length of usable season; and danger of degradation.

Opačić (2017) put forward the following indicators for assessment of tourism attractiveness and tourism accessibility of cultural heritage: level of cultural heritage; aesthetic value of the cultural heritage; recognisability of the cultural heritage in the

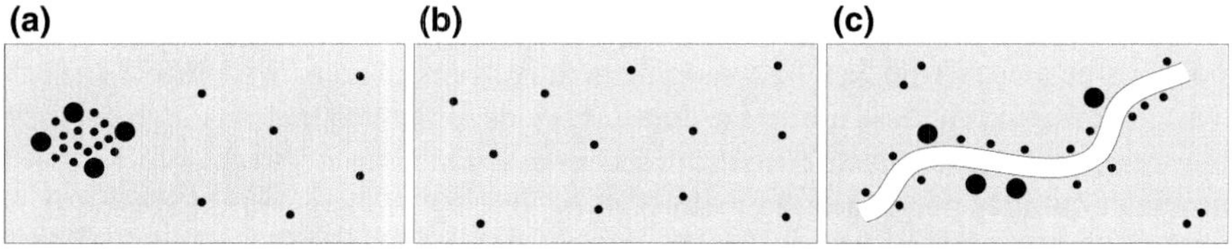

Fig. 15.2 Types of spatial distribution of cultural heritage in a tourism destination

space; access to cultural heritage; walking distance of the cultural heritage from parking/public transport stop; distance of the cultural heritage from available lodgings; distance of the cultural heritage from tourism/hospitality objects (e.g. specialised visitors' centre, tourist information centre, restaurants, fast food, cafes, and souvenir shops); possibility for tourists to visit the cultural heritage during the day; possibility for tourists to visit the cultural heritage during the year; representation of the cultural heritage in tourism communication channels/mediums (postcards, brochures, internet sites, tourist guidebooks); and the risk of degradation of the cultural heritage due to tourist visits.

From the aforementioned, it follows that tourism attractiveness of a cultural heritage is a broader idea than simply its aesthetic value, and during evaluation the following should be assessed: impressiveness of the cultural heritage in the space; its physical accessibility; its level of tourism and hospitality infrastructure; representation in tourism communication channels/mediums (postcards, brochures, internet sites, tourist guidebooks); etc. Furthermore, clear definition of objective and measurable indicators for each of the mentioned criteria, as well as a well-considered point system for assessment, is needed.

15.3.3 Determination of Spatial Distribution of Cultural Heritage

Following identification of potential cultural heritage suitable for tourism valorisation and evaluation of its potential tourism attractiveness, the spatial distribution of cultural heritage in a specific tourism destination should also be determined, whether it is a city (or single settlement) or a broader area (region) in question. The accessibility of cultural goods largely depends on their spatial distribution characteristics, as well as possible models for creating tourist products for different groups of cultural tourists in relation to motivation. According to typology of tourism attractions in relation to their spatial distribution, in a given tourism destination the following three types of spatial distribution of cultural heritage are most often recognised: (a) concentrated or point; (b) dispersed; and (c) linear (Wall 1997) (Fig. 15.2).

The concentrated or point type of spatial distribution is characterised by a concentration of the majority of cultural goods in one place in the destination or in a relatively small area; or by a concentration of the most valuable and attractive cultural goods in one place in the destination or in a relatively small area (Wall 1997). In such a case, the remaining, less attractive cultural goods in the destination can have their value relegated to secondary tourism attractiveness, while the most valuable, i.e. most attractive, cultural goods are the main goal of visiting tourists and the reason they come to the destination, which can overburden the area(s) in question. The classic examples of this kind of spatial distribution of cultural heritage are historical cities, such as Ghent in Belgium (Jansen-Verbeke and Lievois 2008), Venice in Italy (Russo 2008), or Dubrovnik in Croatia. In the urban area of Dubrovnik, the most valuable cultural goods are found in a relatively small area of the medieval city core that is surrounded by walls (a cultural-historical complex on UNESCO's World Heritage List since 1979), while the cultural goods outside of the core are less attractive. A similar spatial distribution can also be found in the wider area (the Dubrovnik region), due to the fact that the most valuable and attractive cultural goods are situated in the city of Dubrovnik itself, which is the explicit target of many tourists, while the other cultural goods in the area serve only to supplement the cultural tourism supply of the tourism destination Dubrovnik.

The dispersed type of spatial distribution of cultural heritage is characterised by diffusion of cultural goods of similar value and tourism attractiveness in a given tourism destination—in a city or region (Wall 1997). As there are no expressed differences in value or initial attractiveness between cultural goods, the number of visitors depends more on the quality of the tourist product, transit accessibility, and representation in the tourism supply. This type of spatial distribution of cultural heritage in Croatia can be observed in the case of roughly fifty forts, castles, villas, and manor houses in the hilly region of Hrvatsko Zagorje (Obad Šćitaroci 2005).

The linear type of spatial distribution of cultural heritage is characterised by a concentration of the majority or the most important and attractive cultural goods in a given tourism destination (city/region) along a linear feature present in the space—a road, a railway, sea coast, a river, etc. (Wall 1997). In this instance, it is expected that tourists will primarily have the goal of visiting cultural goods along the aforementioned linear feature within the designed heritage trails (Timothy and Boyd 2003), while the remaining cultural goods that are further away can be recognised as secondary tourism attractions. As an example of the linear type of spatial distribution of cultural heritage in Croatia, the old Austro-Hungarian roads which run from the Pannonian-Peripannonian region towards Kvarner region can be singled out. Along these roads, many of the largest settlements as well as cultural goods of the mountainous Gorski Kotar region are located.

15.3.4 Proposal of Models for Tourism Valorisation of Cultural Heritage

Four spatial models of tourism valorisation of cultural heritage can be proposed based on identification of cultural heritage suitable for conversion into tourism attractions, the evaluation of its tourism attractiveness and determination of its spatial distribution (Magdić et al. 2015; Opačić 2017):

(a) concentrated or point model;
(b) dispersed urban model;
(c) linear thematic model;
(d) regional networked model.

It is worth noting that, due to the increasing segmentation of demand in cultural heritage tourism, the same cultural good can be valorised simultaneously using one or more of the proposed models.

Tourist products for different groups of tourists, with respect to their motivations, are based on the proposed models. The motivation of cultural tourists, i.e. of cultural heritage tourists, is heterogeneous. Referring Gail Lord (1999 in Timothy 2011) states that around 85% of all tourists fall into the category of cultural tourists, while 15% of tourists in no way participate in the cultural supply of a given tourism destination. Timothy (2011) identifies four categories of cultural tourists: greatly motivated (15% of all tourists in a given destination); partly motivated (roughly 30% of tourists); adjunct (roughly 20% of tourists); and accidental (around 20% of tourists). For the "greatly motivated" group of cultural tourists, visiting cultural goods and/or cultural manifestations is a primary motive for tourist travel. "Partly motivated" cultural tourists are equally motivated by the cultural supply, as well as the remaining segments of the tourism supply when deciding which destination they wish to visit. For the "adjunct" tourists, cultural heritage and cultural manifestations do not represent the main motive for tourist travel, but they typically want to consume the most important parts of the cultural supply. "Accidental" tourists do not plan to visit cultural heritage sites or to take part in cultural manifestations during their stay in a given destination, but they might visit any of the aforementioned if something piques their interest.

Jelinčić (2008) puts forth a similar typology of cultural tourists, which served in the creation of the Strategy for the Development of Cultural Tourism in Croatia (Ministry of Tourism of the Republic of Croatia 2003). The aforementioned document differentiated three groups of cultural tourists in relation to motivation: tourists of accidental cultural motivation ("attracted by culture"), whose motivation for coming to the destination was not based on culture, rather most often on typical mass tourism appeal; "must see" cultural tourists ("inspired by culture"), whose motivation is expressly based on culture and primarily oriented towards large, trendy cultural events; and true cultural tourists (motivated by culture), whose fundamental motivation for travel is culture.

Therefore, the proposed models of tourism valorisation of cultural heritage are mostly based on two criteria: characteristics of spatial distribution of cultural heritage; and the motivation of cultural tourists.

Concentrated or Point Model

The concentrated or point model includes tourism valorisation of individual cultural goods and/or localities that, due to their exceptional tourism attractiveness or remote, isolated location, present a unique tourism attraction. It is possible to apply this to all types of tourism areas: urban; rural; coastal; mountain; spa; as well as protected areas. This model is primarily appropriate for tourism valorisation of cultural goods that are of great tourism attractiveness, most often of the global or international level, which can be transformed into stand-alone tourism attractions; but this can also be applied to cultural goods on lower levels (e.g. national) in the case of isolated, very attractive locations.

This model is based on the idea of uniqueness, being that the goal, during the application of this model of tourism valorisation, is to create a strong tourism impression, e.g. a sensation or the so-called wow-effect. By this, it is understood that a tourist product conceived using this model is stand-alone and is typically presented as the peak of tourist travel.

With respect to authenticity, merit, and significance, cultural goods which can be valued touristically according to this model are often found on the UNESCO World Cultural Heritage List. These kinds of cultural goods can have the role of symbol or metaphor in the collective consciousness and above all, they attract tourists who are inspired by culture, i.e. the so-called must see tourists and tourists attracted by culture (tourists of accidental cultural motivation).

Within this model historical monuments can change their function, particularly if their location or meaning is outstanding. As an example of tourism valorisation of individual historical monuments in isolated locations repurposed into accommodation facilities, we can point to the "Stone Lights" project for which some Croatian lighthouses (currently there are 10 included), mainly built in the nineteenth century during Austro-Hungarian times, have been partially or totally converted into accommodation capacities (Opačić et al. 2010).

Dispersed Urban Model

Within dispersed urban model dispersedly located cultural goods are connected in a given city into a linked and functional whole. It is appropriate for tourism valorisation of individual cultural goods on the national and regional levels, which are spread out in a larger number of locations within a given city. Therefore, for tourism valorisation using this model, those goods which contribute the most to "sense of place" should be chosen.

The model is based on the principle of "place identity" since the goal of tourism valorisation is to induce a reconstructed impression of a place in its historical context, whereby cultural heritage is only one segment of the overall attraction base of a given city. This principle is understood to include the choice of typical locations in a given city that are important for shaping the lifestyle of the local population, i.e. locations with which citizens identify themselves and that create urban identity in the context of their own past, present, and future.

The dispersed urban model is appropriate for tourism valorisation of cultural goods that are not sufficiently attractive to most tourists that they would be standalone attractions in the overall tourism supply, which means that they are primarily applied to cultural goods of the national and regional hierarchical levels.

This model is appropriate for attracting tourists who are attracted by culture, i.e. tourists of general accidental cultural motivation, who, for example, want to get to know the main facts of the urban identity (of a given place) on a "city-break" trip, so the tourist programme offers a combination of different cultural attractions—from sightseeing typical historical monuments, to visiting to authentic pubs and/or restaurants, getting to know legends about the city, enjoying a panoramic view of the city skyline, giving the tourist a total "sense of place" experience of a given tourism destination.

Linear Thematic Model

The linear thematic model includes tracing the cultural heritage trails which integrate thematically linked cultural goods in the creation of the tourist product. Apart from cultural goods, other attractions along the line can also be added. This model is appropriate for tourism valorisation of cultural heritage on the national, regional, and local hierarchical levels that are laid out in a linear pattern in a region or city (e.g. along a road, bicycle or walking path, railway, and river), and thematically linked into regional, interregional, and/or international cultural heritage trails.

This model is based on the principle of thematic homogeneity (e.g. functional, aesthetic, stylistic, and historical) and representativeness; and, during its application, the goal of tourism valorisation is to satisfy a specific cultural motivation.

The model is adequate for attracting primarily true cultural tourists who are motivated by specific culture motivation. The important ability of its application can also be found in tourism attraction of interested communities gathered around a specific common hobby or interest (e.g. groups interested in visiting historical monuments from the socialist era in former socialist states of Eastern Europe). Contemporary online communication enables fast and easy connection between individuals with specific common interests, which gives this model great development potential.

Regional Networked Model

The regional networked model is based on networking cultural goods in a wider area (region) with the remaining natural and anthropogenic attractions in the creation of the tourism supply of a given tourism destination. This model is appropriate for tourism valorisation of cultural goods which best present the individual makeup of regional cultural identity, and most often includes cultural goods and localities on the regional and local hierarchical levels.

Being that during the application of this model, the goal of tourism valorisation of cultural heritage is to enable tourists to get different tourism experiences in the region by getting to know heterogeneous tourism attractions; the model is based on the principle of regional cultural diversity. Namely the tourist, who visits a region in this manner, typically wants to experience as wide an array of things as possible in the time available to them; being economic—getting the most value for their money—is a priority during their tourist stay. This approach can be applied for tourists who visit a given tourism destination for a single, short time, i.e. those who do not plan to return in the near future.

This model is suitable for attracting primarily tourists who are attracted by culture, i.e. tourists of general accidental cultural motivation.

15.4 Conclusion

Precise identification of cultural heritage, specific trends and processes in contemporary tourism thereof, and the rise in the number of tourists with diverse motivations for visiting cultural heritage enables a given cultural good to be integrated into different tourist products via the application of the four proposed models for tourism valorisation of cultural heritage. The aforementioned models are largely based on two criteria: characteristics of spatial distribution of cultural heritage and the motivation of cultural tourists.

The following steps precede the structuring of the aforementioned models:

1. identification of cultural heritage suitable for conversion into tourism attractions;
2. evaluation of tourism attractiveness of cultural heritage;
3. determination of spatial distribution of cultural heritage;
4. after which models for tourism valorisation should be applied (Table 15.1).

The advantage of implementation of the aforementioned models of tourism valorisation of cultural heritage can be seen in the strong tourism pressure on the most visited cultural goods along with simultaneous inclusion of less visited cultural goods in the tourism supply. In this way, the local economy in a given tourism destination is supported and tourism demand is diversified, whereby the possibility to develop certain specific services, such as support for the development of the tourism supply, is opened.

It should be kept in mind that the goal of tourism valorisation of cultural heritage should not only be generation of income and development of tourism, rather sustainable management. Thereby, securing financing for sustaining, revitalising, and further improving cultural heritage is of key importance, so that it can retain all four of its fundamental significances: social; scientific; political; and economic.

Table 15.1 Fundamental characteristics of models for tourism valorisation of cultural heritage (adaption according to Opačić 2017)

Model	Hierarchical level of cultural heritage	Principle	Additional aspect for evaluation	Type of cultural tourists in relation to motivation
Concentrated or point model	World (UNESCO World Heritage List), international, and more rarely national	Uniqueness	Ability to provoke a strong tourism impression, i.e. the so-called wow-effect	Tourists inspired by culture ("must see" tourists) and tourists attracted by culture, i.e. tourists of general accidental cultural motivation
Dispersed urban model	National, regional	Place identity	Contribution to sense of place	Tourists attracted by culture, i.e. tourists of general accidental cultural motivation
Linear thematic model	National, regional, local	Homogeneity, representative-ness	Contribution to the theme	True cultural tourists motivated by culture, specific cultural motivation
Regional networked model	Regional, local	Regional cultural diversity	Presentation of regional cultural diversity	Tourists attracted by culture, i.e. tourists of general accidental cultural motivation

References

Antolović J (1997) Nepokretni spomenici kulture i turizam. Acta Turistica 9(2):136–154
Ashworth GJ (2009) Do tourists destroy the heritage they have come to experience? Tourism Recreation Res 34(1):79–83. https://doi.org/10.1080/02508281.2009.11081577
Ashworth GJ, Graham B (1997) Heritage, identity and Europe. Tijdschrift voor Economische en Sociale Geografie 88(4):381–388. https://doi.org/10.1111/j.1467-9663.1997.tb01632.x
Cocean G (2011) Inventory cards for regionally relevant geomorphosites. Rom Rev Reg Stud 7(1):131–136
Comănescu L, Dobre R (2009) Inventorying, evaluating and tourism valuating the geomorphosites from the central sector of the Ceahlau National Park. GeoJournal Tourism Geosites 2(1):86–96
Coratza P, Giusti C (2005) Methodological proposal for the assessment of the scientific quality of geomorphosites. Il Quaternario 18(1):307–313
du Cros H (2001) A new model to assist in planning for sustainable cultural heritage tourism. Int J Tourism Res 3(2):165–170. https://doi.org/10.1002/jtr.297
Dumbović Bilušić B (2015) Krajolik kao kulturno naslijeđe: metode prepoznavanja, vrjednovanja i zaštite kulturnih krajolika Hrvatske. Ministarstvo kulture Republike Hrvatske – Uprava za zaštitu kulturne baštine, Zagreb
Ennen E, van Maanen E (2014) Telling the truth or selling an image? Communicating heritage as an instrument in place marketing. In: Kaminski J, Benson AM, Arnold D (eds) Contemporary issues in cultural heritage tourism, 1st edn. Routledge, London, pp 45–55
Fyall A, Garrod B (1998) Heritage tourism: at what price? Manag Leisure 3(4):213–228. https://doi.org/10.1080/136067198375996
Graham B, Ashworth GJ, Tunbridge JE (2004) A geography of heritage: power, culture and economy. Arnold, London
Hall CM, McArthur S (1993) Heritage management: an introductionary framework. In: Hall CM, McArthur S (eds) Heritage management in New Zealand and Australia: visitor management, interpretation and marketing, 1st edn. Oxford University Press, Auckland, pp 1–17
Hampton MP (2005) Heritage, local communities and economic development. Ann Tourism Res 32(3):735–759. https://doi.org/10.1016/j.annals.2004.10.010
Hughes HL (1996) Redefining cultural tourism. Ann Tourism Res 23(3):707–709. https://doi.org/10.1016/0160-7383(95)00099-2
Jansen-Verbeke M, Lievois E (2008) Visiting patterns in historic cityscapes: a case study in Ghent, Belgium. In: Jansen-Verbeke M, Priestley GK, Russo AP (eds) Cultural resources for tourism: patterns, processes and policies, 1st edn. Nova Science Publishers Inc., New York, pp 17–29
Jelinčić DA (2005) Kultura kao pokretačka snaga redefiniranja imidža destinacije. Acta Turistica 17(1):56–78
Jelinčić DA (2008) Abeceda kulturnog turizma. Meandarmedia, Zagreb
Jelinčić DA (2010) Kultura u izlogu. Meandarmedia, Zagreb
Jelinčić DA, Senkić M (2017) Creating a heritage tourism experience: the power of the senses. Etnološka Tribina 47(40):109–126. https://doi.org/10.15378/1848-9540.2017.40.03
Kaminski J, Benson AM, Arnold D (2014) Introduction. In: Kaminski J, Benson AM, Arnold D (eds) Contemporary issues in cultural heritage tourism, 1st edn. Routledge, London, pp 3–18
Kušen E (2002) Turistička atrakcijska osnova. Institut za turizam, Zagreb
Kušen E (2017) Tourism attraction system. In: Dwyer L, Tomljenović R, Čorak S (eds) Evolution of destination planning and strategy: the rise of tourism in Croatia, 1st edn. Palgrave Macmillan, Cham, pp 119–148
Landriani L, Pozzoli M (2014) Management and valuation of heritage assets: a comparative analysis between Italy and USA. Springer, Cham
Magdić H, Obad Šćitaroci M, Bojanić Obad Šćitaroci B (2015) Metoda tablično-dijagramskog programiranja povijesnog naslijeđa: primjer dvorca u Nuštru. In: Obad Šćitaroci M (ed) Proceedings of the international scientific conference "Cultural heritage: possibilities for spatial and economic development". Faculty of Architecture, University of Zagreb, Zagreb, pp 144–149

Marković I, Fuerst-Bjeliš B (2015) Prostorni identitet kao pokretačka snaga razvoja turizma: komparativna analiza regija Bjelovara i Čakovca. Hrvatski Geografski Glasnik 77(1):71–88. https://doi.org/10.21861/HGG.2015.77.01.04
McKercher B, du Cros H (2002) Cultural tourism: the partnership between tourism and cultural heritage management. Haworth Hospitality Press, New York
McKercher B, Ho PSY (2006) Assessing the tourism potential of smaller cultural and heritage attractions. Journal of Sustainable Tourism 14(5):473–488. https://doi.org/10.2167/jost620.0
Millar S (1989) Heritage management for heritage tourism. Tour Manag 10(1):9–14. https://doi.org/10.1016/0261-5177(89)90030-7
Ministry of Culture of the Republic of Croatia (2018) Registar kulturnih dobara. http://www.minkulture.hr/default.aspx?id=31. Accessed on 20 Mar 2018
Moscardo G (1996) Mindful visitors: heritage and tourism. Ann Tourism Res 23(2):376–397. https://doi.org/10.1016/0160-7383(95)00068-2
Moulin C, Boniface P (2001) Routeing heritage for tourism: making heritage and cultural tourism networks for socio-economic development. Int J Heritage Stud 7(3):237–248. https://doi.org/10.1080/13527250120079411
Mrđa A, Bojanić Obad Šćitaroci B (2016) Heritage touristscapes: a case study of the island of Hvar. Annales-Series Historia et Sociologia 26(3):553–572. https://doi.org/10.19233/ASHS.41
Nuryanti W (1996) Heritage and postmodern tourism. Ann Tourism Res 23(2):249–260. https://doi.org/10.1016/0160-7383(95)00062-3
Obad Šćitaroci M (2005) Dvorci i perivoji Hrvatskoga zagorja. Školska knjiga, Zagreb
Opačić VT (2017) Modeli turističkog vrednovanja kulturnog naslijeđa. In: Obad Šćitaroci M (ed) Zbornik radova Znanstvenog kolokvija "Modeli revitalizacije i unaprjeđenja kulturnog naslijeđa – multidisciplinarni dijalog". Arhitektonski fakultet Sveučilišta u Zagrebu, Zagreb, pp 28–32
Opačić VT, Favro S, Perišić M (2010) Tourism valorisation of lighthouses on Croatian islands and along the coast. WIT Transactions on Ecology and Environment 130:37–48. https://doi.org/10.2495/ISLANDS100041
Park HY (2014) Heritage tourism. Routledge, London
Poljanec-Borić S (2017) Prikladni model razvojnog korištenja kulturne baštine. In: Obad Šćitaroci M (ed) Zbornik radova Znanstvenog kolokvija "Modeli revitalizacije i unaprjeđenja kulturnog naslijeđa – multidisciplinarni dijalog". Arhitektonski fakultet Sveučilišta u Zagrebu, Zagreb, pp 18–21
Reynard E, Fontana G, Kozlik L, Scapozza C (2007) A method for assessing "scientific" and "additional values" of geomorphosites. Geographica Helvetica 62(3):148–158. https://doi.org/10.5194/gh-62-148-2007
Republic of Croatia (2003) Strategija razvoja kulturnog turizma: "Od turizma i kulture do kulturnog turizma". Ministry of Tourism, Zagreb
Richards G (1996) Production and consumption of European cultural tourism. Ann Tourism Res 23(2):261–283. https://doi.org/10.1016/0160-7383(95)00063-1
Russo AP (2008) Cultural attractions and destination structure: a case study of the Venice Region, Italy. In: Jansen-Verbeke M, Priestley GK, Russo AP (eds) Cultural resources for tourism: patterns, processes and policies, 1st edn. Nova Science Publishers Inc., New York, pp 43–58
Salazar NB (2005) Tourism and glocalization: "local" tour guiding. Ann Tourism Res 32(3):628–646. https://doi.org/10.1016/j.annals.2004.10.012
Salazar NB (2010) The glocalization of heritage through tourism: balancing standardisation and differentiation. In: Labadi S, Long C (eds) Heritage and globalisation, 1st edn. Routledge, London, pp 130–146
Smith MK (2009) Issues in cultural tourism studies. Routledge, London
Slunjski R (2017) Turističkogeografski pristup u valorizaciji kulturne baštine. Podravina – časopis za multidisciplinarna istraživanja 16(31):163–172
Šmid Hribar M, Ledinek Lozej Š (2013) The role of identifying and managing cultural values in rural development. Acta Geogr Slov 53(2):371–378. https://doi.org/10.3986/AGS53402

Šmid Hribar M, Bole D, Pipan P (2015) Sustainable heritage management: social, economic and other potentials of culture in local development. Proc-Soc Behav Sci 188:103–110. https://doi.org/10.1016/j.sbspro.2015.03.344

Teo P, Yeoh BSA (1997) Remaking local heritage for tourism. Ann Tourism Res 24(1):192–213. https://doi.org/10.1016/S0160-7383(96)00054-0

Throsby D (2001) Economics and culture. Cambridge University Press, Cambridge

Timothy D (2011) Cultural heritage and tourism: an introduction. Channel View Publications, Bristol

Timothy D, Boyd SW (2003) Heritage tourism. Pearson Education Limited, Harlow

Timothy D, Nyaupane GP (2009) Cultural heritage and tourism in the developing world: a regional perspective. Routledge, London

Tomljenović R (2006) Kulturni turizam. In: Čorak S, Mikačić V (eds) Hrvatski turizam: plavo, bijelo, zeleno, 1st edn. Institut za turizam, Zagreb, pp 119–147

van der Borg J, Costa P, Gotti G (1996) Tourism in European heritage cities. Ann Tourism Res 23(2):306–321. https://doi.org/10.1016/0160-7383(95)00065-8

Vrtiprah V (2006) Kulturni resursi kao činitelj turističke ponude u 21. Stoljeću. Ekonomska Misao i Praksa 15(2):279–296

Wall G (1997) Tourism attractions: points, lines and areas. Ann Tourism Res 24(1):240–243. https://doi.org/10.1016/S0160-7383(96)00039-4

Wall G, Mathieson A (2006) Tourism: change, impacts and opportunities. Pearson Education Limited, Harlow

Zeppel H, Hall CM (1991) Selling art and history: cultural heritage and tourism. J Tourism Stud 2(1):29–45

Chapter 16
Recognition and Preservation of Associative Landscape Features

Ana Sopina and Bojana Bojanić Obad Šćitaroci

Abstract This study is based on exploring the relationship between examples of coastal settlements of the Eastern Adriatic coast in Croatia and their natural context—the sea and the mountain hinterland. These are landscapes with different intensities and qualities of associative features. Associative landscape provides mental connections of physical elements with intangible heritage through the experience of landscape meaning. Associative experience is possible when landscape is perceived as a whole, with recognised contextual values and a unity of intellectual and physical content. In this research, landscape identity is observed as the most prominent and comprehensive associative landscape feature. The research is based on the *Heritage Urbanism* approach, with the aim of defining models for the revival and enhancement of landscape identity from the landscape and spatial and urban planning points of view. Selected case studies, Starigrad Paklenica with South Velebit, and Makarska with Biokovo, are landscapes where strong relations exist between urbanscape and the natural context, through which means of the recognition and preservation of the landscape associative features and landscape identity are established. The research results define the present state of perceiving associative landscape features (the landscape identity model), identifying the means of perceiving the landscape identity which requires preservation (the landscape concern model) and setting a spatial development strategy from the aspect of the relationship between the settlement and its natural landscape (the landscape resilience model). Landscape perceived as a whole is presented as a new heritage dimension and as a process of the development of the perception of knowing a landscape. By establishing the perception, concern and resilience models of associative landscape features, the associative dimension is affirmed in landscapes as fundamental to their being retained, restored and redefined.

Keywords Landscape perception · Associative experience · Landscape identity · Landscape concern · Landscape resilience

---

A. Sopina (✉) · B. Bojanić Obad Šćitaroci
Faculty of Architecture, University of Zagreb, Zagreb, Croatia
e-mail: asopina@arhitekt.hr

B. Bojanić Obad Šćitaroci
e-mail: bbojanic@arhitekt.hr

M. Obad Šćitaroci et al. (eds.), *Cultural Urban Heritage*,
The Urban Book Series, https://doi.org/10.1007/978-3-030-10612-6_16

16.1 Introduction

Throughout history, people have recognised places that are suitable for life and specific use, places of unique aesthetics, and have associated certain places with meanings, events, artistic expression, eminent people, the interpretation of natural phenomena, (supernatural) beings and other intellectual constructs which have over time been incorporated into culture, forming intangible heritage. Recognition of landscape values (natural, cultural, functional, aesthetic and associative values) have resulted in: marking the place (art); establishing settlements, towns and sacred places; organising and constructing (in) the landscape. Cultural heritage, both tangible and intangible, first emerged from the landscape. Simultaneously, human culture transformed, in different intensities, the natural landscape into cultural landscape and urbanscape.

The history of the recognition and protection of heritage neglected the intangible and common dimension of heritage in favour of tangible heritage of outstanding universal value—cultural and natural heritage as defined by the World Heritage Convention of 1972[1] (UNESCO 1972). Contemporary rethinking and the re-evaluation of heritage and landscape recognised common heritage and everyday landscapes, experience, the intangible dimension and associative features of heritage and landscape as integral to the ability to understand them.[2]

This research focuses on defining the methodological approach for analysing associative landscape features. It is also part of the scientific project *Heritage Urbanism* (HERU).[3] In the framework of this research, landscape is considered both as heritage and as a heritage context, accepting all landscape dimensions: natural and cultural, tangible and intangible, quantitative (documentation and analysis) and qualitative (perception and concern) as equally valuable. In the context of the *Heritage Urbanism* project, the focus of research is on everyday landscapes: natural, cultural and urban.[4]

Associative landscape features are rooted in landscape identity and landscape perception. Landscape identity is an intertwining of natural, cultural and urban dimen-

[1]The Convention concerning the protection of the world cultural and natural heritage—World Heritage Convention in Paris, 1972—defined cultural heritage as monuments, groups of buildings and sites of outstanding universal value, and natural heritage as natural features, geological and physiographical formations, natural sites of outstanding universal value.

[2]Based on Operational Guidelines for the implementation of the World Heritage Convention, Paris, 2017 (UNESCO 2017); Recommendations on the Historic Urban Landscape, Paris, 2011 (UNESCO 2011); Convention for the Safeguarding of the Intangible Cultural Heritage, Paris, 2003 (UNESCO 2003), and the European Landscape Convention, Florence, 2000 (Council of Europe 2000).

[3]Scientific project Heritage Urbanism (HERU)—Urban and Spatial Planning Models for the Revival and Enhancement of Cultural Heritage (HRZZ 2032) financed by the Croatian Science Foundation, which is carried out at the Faculty of Architecture, University of Zagreb, under the project leadership of Prof. Mladen Obad Šćitaroci, Ph.D.

[4]The focus of the *Heritage Urbanism* project is placed on everyday cultural heritage present in all towns, settlements and landscapes and which is important at the regional and local level with no necessarily universal global or national value (Obad Šćitaroci 2017).

sions, where "landscape gives urban nature a context", with "urban nature emerging from the landscape in a context where elements of the urban landscape create a unique meaning" (Bojanić Obad Šćitaroci 2018). Associative experience is possible when landscape is perceived as a whole, with recognised contextual values and a unity of intellectual and physical content (Sopina and Bojanić Obad Šćitaroci 2015). The research is based on the theoretical framework of landscape identity, landscape concern and landscape resilience (Relph 1976; Lucas 2009; Turner 2001; Fein 1972; Andersson 2006). The aim is to determine models for the revival and enhancement of landscape identity.

Associative landscape features

The research identifies the associative dimension of cultural heritage, analysed through associative landscape features. According to Sopina and Bojanić Obad Šćitaroci (2015), associative landscape features refer to associations with the physical components of landscape—a connection with intangible elements through experience, perception and interpretation of landscape meaning. Associative landscape provides mental connections between the physical elements of landscape and intangible heritage. In the context where associative experience presents landscape perceived as a whole, landscape identity is researched as the most prominent and comprehensive associative landscape feature.

Associative experience and the perception of landscape as a whole are processes developed with the perception of knowing a landscape. Berleant (2013) defines knowing a landscape as knowing in the sense of appreciation, where "concern with the environment shows what is most significant is not the object of appreciation, but the process of appreciation". Marić et al. (2017) suggest "a process-driven appreciation of heritage, rather than a static, object-driven approach". In this context, the perception of knowing a landscape is presented as a process more than just as a first impression of a landscape or apparent landscape perception which is often superficial and one-dimensional (depending on individual interests). By gaining appreciation and by knowing a landscape, heritage and context, history and the present, the tangible and intangible, the natural and cultural, and the collective and personal are all perceived, and do not necessarily just lie in the sight of the eye. Thus, perception of landscape as a whole is presented as a heritage dimension of landscape.

The research focuses on everyday landscapes of coastal settlements and towns coexisting with two strong natural landscapes: the Eastern Adriatic and the Dinarides mountain hinterland. Landscape is defined as heritage in the context of contemporary UNESCO declarations and conventions (UNESCO 2003, 2011, 2017) and the European Landscape Convention (Council of Europe 2000).

The European Landscape Convention (ELC) defines **landscape** as an area, as perceived by people, whose character is the result of the action and interaction of natural and/or human factors (Council of Europe 2000). While the ELC covers natural, rural, urban, periurban, land, inland water and marine (land)scapes, UNESCO in its declarations and conventions distinguishes natural, cultural and urban landscapes. **Cultural landscapes** are defined by UNESCO (2017) as cultural properties which represent the combined work of nature and of man, and are categorised as follows: landscape designed and created intentionally by man, organically evolved landscape and associative cultural landscape. Cultural landscapes are illustrative of the evolution of human society and settlement over time, under the influence of the physical constraints and/or opportunities presented by their natural environment and of successive social, economic and cultural forces, both external and internal. In a context where there is no unmediated perception of nature (Ignatieff 1995) and where human influence is present in almost all natural landscape, all landscape is considered cultural landscape (Taylor 2008), subject to cultural influence and a source of cultural knowledge (Bridgewater and Bridgewater 2004).

The concept of **urbanscape** is based on the UNESCO (2011) definition of historic urban landscape. Historic urban landscape is an urban area understood as the result of an historic layering of cultural and natural values and attributes, extending beyond the notion of historic centre or ensemble to include the broader urban context and its geographical setting. Urbanscape encompasses contemporary urbanised landscape (both built and unbuilt areas, transformed and designed nature) with the historical and contemporary cultural space[5] of its inhabitants.

Landscape identity and associative landscape features can be comprehended by perception. In the framework of this research, **landscape perception** is defined as the act of apprehending landscape by means of the senses and of the mind, as an effort to understand landscape qualities. Perceived landscape is observed as a whole, including the historical and contemporary, natural, cultural, functional, aesthetic and associative context. In the urban environment, people observe cultural heritage by interpreting and experiencing the urban fabric through the urban morphology of historical sites, whereas in landscape we observe it in preserved landscape patterns as testimonies to human presence in the area (Bojanić Obad Šćitaroci 2015). Recognition as perception and experience defines how landscape attributes form and represent **landscape identity**.

Preservation defines the means of landscape perception which need to be protected from decay and maintained in order to retain, restore or redefine the perception of landscape identity. Preservation as a **landscape concern** presents a matter that engages society's attention, interest and care for the landscape.

Resilience is the ability of a system to return to its original state after being disturbed. In the context of this research, **landscape resilience** is defined as the ability of landscape to adapt to change and still retain the characteristics of own identity.

[5]Cultural space (UNESCO 2003) is associated with intangible cultural heritage which communities, groups and, in some cases, individuals recognise as part of their cultural heritage. Cultural space was established by Sopina and Bojanić Obad Šćitaroci (2015) as the origin of intangible cultural heritage, the location of its occurrence, the area of community presence and an area to which the community associates meaning.

16.2 Research Framework

Landscape Identity

Landscape identity is a complex concept, closely related to the sense of place, place identity and place attachment. Just as place is distinguished from space, landscape is distinguished from the environment. Places are spaces incorporated with meaning, individuals, groups or societies (Relph 1976). They are essentially centres of meaning constructed out of lived experience that would be perceived through time as significant to the lives of people (Tuan 1977; Ujang and Zakariya 2015). Defined as an area perceived by people, every landscape can be defined as place, or as an ensemble of places. Perception is essential to the definition of both place and landscape.

The concept of place, just like the concept of landscape, is physical as well as psychological (Ujang and Zakariya 2015). Associative landscape features connect the physical (material) to the psychological (intellectual, intangible) aspects of landscape.

Relph (1976) defines identity of place as a persistent sameness and unity which allows one place to be differentiated from others. This persistent identity is described through three interrelated components: physical features and appearance, activities and functions, meaning and symbols.

This threefold division, similar to the components of place identity of Relph (1976), is also present in other studies. Montgomery (1998) defines that physical form, activity and meaning together form the sense of place. According to Parris (2002), the visual perception of landscape can be grouped by how it relates to structure, function and value. Di Lucas (2009) proposes criteria for the assessment of heritage landscape, where heritage landscape is considered from within, from the vicinity of landscape, and far beyond, covering visual, existential and spiritual experience. The proposed criteria encompass all values of landscape experience.[6]

The perception of landscape identity is analysed further through the threefold categorisation of the character of landscape identity:

- Visual character (physical features and appearance)
- Functional character (activities and functions)
- Associative character (meanings and symbols).

Landscape Concern

The guiding principles, aims and role of design and planning are the continuous concern of architecture, landscape architecture, and urban and landscape planning theory. Design and planning are represented in the context of preserving the continuity of heritage, and creating excellence in contemporary interventions so that these interventions can someday become new heritage (Obad Šćitaroci 2015).

---

[6]Heritage landscape criteria (Lucas 2009) are: heritage fabric, natural science value, time depth, cultural and spiritual value, cultural diversity, legibility and evidential value, shared and recognised value, aesthetic value and significance.

Vitruvius Pollio (1999), in *Ten books of architectures* (De architectura libri decem), summarised the aims of the design process as firmness (*firmitas*), commodity (*utilitas*) and delight (*venustas*). Thompson (2000) identified three key areas of landscape architecture knowledge and reformulated Vitruvius's aims as ecology, community and delight. This re-classification was continued by Turner (2001) as the natural, social and visual aims of landscape architecture. With the development of the science of landscape ecology, Bell (1999) presented the role of landscape ecological patterns, the patterns of human use and the aesthetics of nature. Fein (1972) defined the concerns of landscape architecture as: aesthetics; ecological needs; public welfare and enjoyment; and comfort and pleasure for the individual.

The significance and role of urban landscape have numerous perspectives: the predominantly visual perception as defined in Lynch's *Image of the city* (1960), the natural and ecological approach to landscape planning of McHarg (1969) and the functional perception of landscape as urbanism of Waldheim (2006).

Landscape concern is seen through the three-fold categorisation of the aims of landscape design and planning:

- the ecological (natural) aim;
- the social (use, comfort and public welfare) aim;
- the aesthetic (pleasure) aim.

Landscape Resilience

Resilient and adaptable landscapes and towns refer to ecological, economic and social change. The term landscape resilience applies to ecology related to climate change and the need to restore landscape characteristics.

In the middle of the twentieth century, the ecologist Leopold (1949) defined the concept of ecological resilience thus: "Health is the capacity of the land for self-renewal. Conservation is our effort to understand and preserve this capacity". More than 50 years later, Gunderson (2000) defined resilience as nature's "capacity for renewal in a dynamic environment". Gunderson linked his definition to Leopold's, while updating it to incorporate the reality of change—a recognition fundamental to today's practice of conservation and sustainability. Andersson (2006) applies resilience theory to urban landscapes and sustainable cities, describing the resilience of urban landscape as the integration of ecosystem functions with social dynamics.

The concept of resilience complements the concept of sustainable development which is formulated thus: "development that meets the needs of the present without compromising the ability of future generations to meet their own needs" (World Commission on Environment and Development 1987), including social inclusion, environmental balance, economic growth and cultural diversity (United Nations 2015; United cities and local Governments 2008). Target 11.4 of the 2030 Agenda for Sustainable Development (United Nations 2015) is "to strengthen efforts to protect and safeguard the world's cultural and natural heritage" and "to make cities and human settlements inclusive, safe, resilient and sustainable". In a context where cities have always been dependent on their hinterlands for food and other ecosystem goods and services (Andersson 2006; Folke et al. 1997), the sustainable development of cities and settlements must be included within their surrounding landscape. Thus, sustainable and resilient cities encompass respecting the relationship between settlement and its natural context, concerning not only physical territory with its components but also history, local tradition and intangible heritage.

Landscape resilience, as the capacity of landscape to adapt to change and still retain the characteristics of its own identity, respects the principles of sustainable development and relationships between the settlement and its natural context.

Landscape Identity Factors

The factors of landscape identity determine the landscape character which makes it recognisable, authentic and original. The factors of landscape identity are differentiated in different ways in various sources. The division of landscape identity factors is based on an overview of UNESCO conventions and declarations and on professional documents related to the physical development strategy of the Republic of Croatia (Table 16.1).

In the World Heritage Convention, UNESCO (1972) differentiated cultural heritage from natural heritage. With the development of the operational guidelines for the implementation of the World Heritage Convention (UNESCO 2017), new categories of heritage were introduced: mixed cultural and natural heritage, cultural landscapes, historic towns, cultural canals and cultural routes. The inseparability of the intellectual, sensory and spiritual dimensions from the natural, cultural and urban dimension of landscape was confirmed by UNESCO in the Convention for the Safeguarding of the Intangible Cultural Heritage (UNESCO 2003) and the Declaration on the Conservation of Historic Urban Landscape (UNESCO 2005, 2011).

The professional documents for the Physical Development Strategy of the Republic of Croatia (Furlan Zimmermann 1999; Obad Šćitaroci et al. 2014) recognise natural factors, anthropogenic factors and factors of perception.

Criteria for the Perception of Landscape Identity

Perception defines how landscape factors form and represent landscape identity. The perception of landscape as a whole encompasses all landscape identity factors and determines the character of landscape identity categorised as follows: visual character, functional character and associative character.

The visual character of the perception of landscape identity, including the perception of physical features and appearance, is covered in the Framework for Analysing the Visual Landscape Character (Tveit et al. 2006, Ode et al. 2008; Tveit and Ode Sang 2014). The framework identifies ten key visual concepts: stewardship, coherence, disturbance, historicity, visual scale, imageability, complexity, naturalness, ephemera and safety.[7] Each of the concepts focuses on a different aspect of landscape important for visual quality, where a holistic experience of them all is essential (Tveit et al. 2006).

[7]The first nine visual concepts are defined by Tveit et al. (2006, 2008), and the tenth visual concept (safety) is defined by Tveit and Ode Sang (2014) in the application of the Framework for Analysing the Landscape Visual Character on Metropolitan Areas.

Table 16.1 Landscape identity factors in UNESCO conventions and declarations and in professional documents for the Physical Development Strategy of the Republic of Croatia

Source	Natural factors	Anthropogenic factors		Factors of perception	
World Heritage Convention (UNESCO 1972)	World natural heritage	World cultural heritage	/	/	/
Intangible cultural heritage convention (UNESCO 2003)	Tangible natural heritage	Tangible cultural heritage	/	/	Intangible cultural heritage
Historic urban landscape declaration (UNESCO 2005)	Natural values and attributes	Cultural values and attributes	Functional use	Visual relationships	City tradition
Operational guidelines (UNESCO 2017)	Physical natural environment	Cultural material heritage	Use and activities	Aesthetic qualities	Cultural tradition and sense of place
European Landscape Convention (Council Of Europe 2000)	Natural factors	Human factors	Human activity	Perception	Identity
Landscape factors (Furlan Zimmermann 1999)	Natural factors	Anthropogenic cultural and historical factors	Anthropogenic socio-economic factors	Factors of perception—aesthetic factors	Factors of perception—associations
Landscape components (Obad Šćitaroci Et Al. 2014)	Natural components	Cultural/anthropogenic components	/	Physiognomic—morphological components	Intangible and sensory components
	Natural factors	Anthropogenic factors—characteristics of space settlement	Anthropogenic factors—functions and activity	Factors of perception—aesthetic factors	Factors of perception—associative factors
	Material factors		**Intangible factors**		
	Landscape identity factors				

The functional character of the perception of landscape identity includes concepts of usage, connectivity and organisation. In research of the heritage landscape criteria (Lucas 2009) and landscape features (Sopina and Bojanić Obad Šćitaroci 2015; Sopina et al. 2015), concepts of usage and connectivity are identified. The concept of organisation is presented in the research of urban ambience (Osmond 2007) and in *The image of the city* (Lynch 1960).

The associative character of the perception of landscape identity includes concepts of meaning, associativity, ambience and legibility. These concepts are presented in research titled "Landscape and memory" (Taylor 2008), "Landscape features" (Sopina and Bojanić Obad Šćitaroci 2015; Sopina et al. 2015), "Heritage landscape criteria" (Lucas 2009), "Urban ambience" (Osmond 2007), *The image of the city* (Lynch 1960) and *Place and placelessness* (Relph 1976).

16.3 Case Studies

The research is based on the *Heritage Urbanism* approach with three main components: factors, criteria and models for heritage rehabilitation and enhancement. The research aim is to define models for the revival and enhancement of landscape identity from the points of view of landscape and spatial and urban planning.

Associative experience is possible when the landscape is perceived as a whole. Thus, the case study research encompasses all landscape identity factors:

- Natural factors: physical natural factors; living natural factors and protected natural areas
- Anthropogenic factors—the characteristics of space settlement: history of space settlement; settlement organisation, historical and contemporary constructs; the character of public space
- Anthropogenic factors—functions and activity: functions; use and activity; connections
- Factors of perception—aesthetic factors: dominant elements and views; visual relations
- Factors of perception—associative factors: meanings and genius loci; historical events and famous people; tradition of cultural places; migrations.

In the case studies, the perception of landscape identity factors is analysed through the dimensions of the character of the landscape identity: stewardship, coherence, disturbance, historicity, visual scale, imageability, complexity, naturalness, ephemera, safety, usage, connectivity, organisation, meaning, associativity, ambience and legibility.

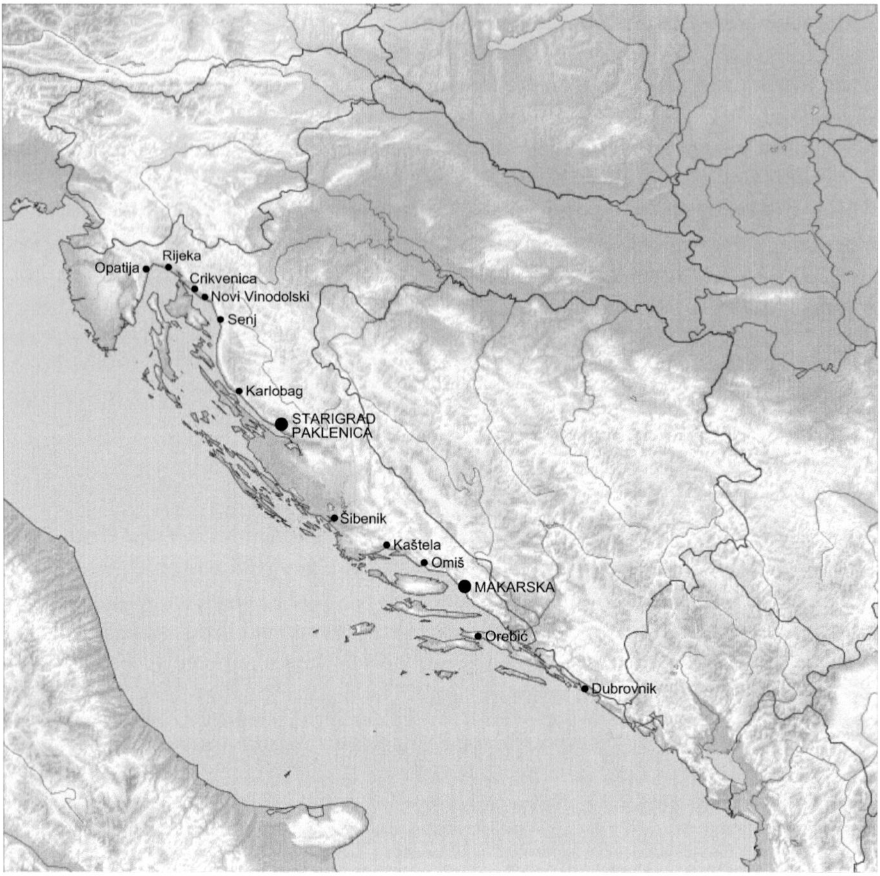

Fig. 16.1 Most prominent locations on the Eastern Adriatic coast in Croatia representing continental coastal settlements and towns established in the foothills of the Dinarides

16.3.1 Criteria for Case Study Selection

Along the continental coast of the Croatia Eastern Adriatic, there are numerous settlements and towns established in the foothills of the Dinarides. The most prominent locations were part of the field research under the *Heritage Urbanism* project (Fig. 16.1): Opatija below Učka; Rijeka below the Rijeka Hinterland; Crikvenica and Novi Vinodolski below the Vinodol highlands; Senj, Karlobag and Starigrad Paklenica below Central and South Velebit; Šibenik below the Zagora highlands; Kaštela below Kozjak; Omiš below Omiška Dinara; Makarska below Biokovo; Orebić below the Pelješac ridge and Dubrovnik below Srđ.

The analysed locations are landscapes where a strong relationship exists between the urbanscape and its prominent natural context, in other words towns and settlements coexist between a seascape and a mountainscape. (Land)Scapes of the sea and mountains are spatially (physically) and visually dominant and intertwined with a linear urbanscape.

In the research of associative landscape features, the first impression does not give a full associative experience and neither does it result in the perception of the landscape as a whole. The criterion for case study selection is knowing a landscape (Berleant 2013). Thus, landscapes explored multiple times are selected as case studies: Starigrad Paklenica with South Velebit (field research conducted since 2013) and Makarska with Biokovo (field research conducted since 2016).

16.4 Case Study Comparison: Starigrad Paklenica with South Velebit Versus Makarska with Biokovo

The perception of landscape identity is explored through the comparison of spatial frames on photographs.[8] Sixteen photographs were selected and analysed for the chosen case studies of Starigrad Paklenica with South Velebit on the one hand, and Makarska with Biokovo on the other. The photographs represent the visual, functional and associative character in the perception of landscape identity, and the characteristics and relations of and between seascape, urbanscape and mountainscape. The case study photographs are grouped according to common attributes: different position vistas, communication corridors, the spatial character and spatial contrasts.

The different position vistas give rise to diverse perceptions of natural and urban landscapes. Different standpoints and altitudes affect this perception. Views from the outside allow for the perception of a landscape in totality, and a view from the inside provides a perception of elements and places in the landscape. From the sea level view, urbanscape is often perceived as a line or a thin strip in contrast to a mountainscape, while the hilltop view reveals different relations between the seascape and urbanscape.

Communication corridors refer to the visual and/or physical connections in and between landscapes. This analysis observes the differences in corridors which are parallel (longitudinal) and those which are vertical (transversal) to the contours and the coast, finding specific and recognisable corridors along certain streets, channels, water flows and canyons in urbanscapes and in the mountainscape.

Spatial characteristics focus on the perception of nature[9] and the essence of landscape. The nature of landscape, in its essence, is either natural or built (man-made and urban). Types of characters distinguished in urbanscapes and mountainscapes are wilderness, a transformed natural character as cultural landscape, a designed natural character as gardens and that of a built character.

[8]Photographs are used only as a representation of the means of perception. Impressions through photographs cannot replace the perception of landscape in situ.

[9]The nature of landscape is based on the concept of three natures (Hunt 2000; Cicero 2008): the first nature (primeval), the second (cultivated) and the third (horticultural) and extended to one of a built character as the most evident influence of man and the essence of urbanscape.

Spatial contrasts refer to both differences among seascape, urbanscape and mountainscape in each case study, and to changes of characteristics in each landscape. Duality covers the weather, season and daily changes, differences in usage and users, differences in the level of maintenance and number of visitors, and others.

16.4.1 Starigrad Paklenica and South Velebit Mountain

Starigrad Paklenica is a settlement in Zadar County,[10] situated between the coast of the sub-Velebit Channel and the slopes of South Velebit Mountain, and in the proximity of the Velika Paklenica Canyon and a small cape. The Velika Paklenica Canyon and the South Velebit highlands are protected as a Nature Park, National Park and UNESCO World Biosphere Reserve, with two beech forest locations inscribed on the UNESCO World Heritage List. Numerous hiking trails on traditional trade and shepherds routes, mostly abandoned traditional housing and mountain dwellings, the ruins of hill fortresses and grave mounds, *mirilo* as sacred monuments—the cemetery of souls (Trošelj 1992, 2013; Valentić 2013), churches, and stone drywalls—are evidence of man's centuries-long presence and relationship with the mythical mountain.[11] The place of living descended from the mountain onto the coastal settlement half way through the twentieth century, and the traditional semi-nomadic way of life was abandoned[12] with the development of tourism on the Adriatic coast. In 50 years, Starigrad Paklenica has developed from a coastal village to the Starigrad Paklenica municipality. In 2011, Starigrad Paklenica had 1140 inhabitants.

The Perception of Starigrad Paklenica and South Velebit through Different Position Vistas

The natural landscapes between which the Starigrad Paklenica urbanscape is situated are perceived in outside views (from the mountain and from the sea) and in inside views. The totality of the South Velebit mountainscape, due to its indented topography and diversity of landscape forms and habitats, can be perceived through the integration of multiple and different outside and inside views. In the outside view from the sea, from the opposite side of the sub-Velebit Channel (Fig. 16.2, photograph 4), the perception of the complexity of the mountainscape is low, since Velebit is perceived as a single massif. The inside view of South Velebit (Fig. 16.2, photograph 3) reveals higher complexity, diversity and richness of the landscape elements.

[10]In Croatian: Zadarska Županija.

[11]For Croatians, Velebit Mountain is one of their national symbols, the subject of many works of poetry, literature, legends and stories.

[12]The semi-nomadic way of life was traditional for Velebit Mountain and its littoral foothills. This traditional way of life was called *prljanje* or *izdizi* (Bušljeta 2010; Belaj 2004; Faber 1995), meaning the ascent of the mountain by people and cattle (mostly sheep and goats). This upward movement was done yearly in search of summer pastures. Permanent settlements were located close to the seacoast where winters are mild, while seasonal settlements and dwellings were located high in the mountains, in the proximity of the grasslands.

Human interventions are of a small scale in comparison to the natural landscape, and they are merged in nature, so the coherence of the mountainscape is perceived in both views from the outside and from within. The mountainscape is present in all vistas of Starigrad Paklenica, giving the settlement a context and recognisability. The indented mountain topography, the versatile landscapes of deep canyons, rocky peaks, the forests of the mountain slopes, the pastures of the mountain highlands and distinctive karst forms establish both the mountainscape and settlement identity. From the inside view of Starigrad Paklenica (Fig. 16.2, photograph 2), complexity, coherence and disturbance are perceived in small and mixed patterns with a difference in the contextual fit between the traditional and contemporary built structure. From the Veliki Vitrenik hilltop of South Velebit (Fig. 16.2, photograph 1), the complexity, coherence and disturbance of the urbanscape are perceived in random interspersion and in the repeating pattern of contemporary built structures with little correspondence to the natural conditions. In the outside view from the sea, from the opposite side of the sub-Velebit Channel (Fig. 16.2, photograph 4), Starigrad Paklenica is perceived to be low in complexity, having an almost linear urbanscape.

The Perception of Starigrad Paklenica and South Velebit through Communication Corridors

Communication corridors refer to physical and visual corridors, including those that are longitudinal and transversal, and corridors specific and recognisable for Starigrad Paklenica and for South Velebit. The longitudinal corridors, which are parallel to the contours in the mountainscape (Fig. 16.2, photograph 5), offer a wide view on the South Velebit hinterland. In the transversal corridor vertical to the contours (Fig. 16.2, photograph 6), a distant view is blocked by the mountain slopes and focus is directed to the Velika Paklenica Canyon. Although physical connectivity in the mountainscape is in principle easier along the contours, most of the connections in South Velebit are set across the mountain range. One of the transverse corridors and a specific corridor for South Velebit is along the Velika Paklenica Canyon (Fig. 16.2, photograph 6), which is one of the most important local historic trade and shepherd routes connecting the coast of the sub-Velebit Channel with South Velebit and the Lika hinterland. The visual corridor on the cliffs of the Velika Paklenica Canyon and the main entrance to Paklenica National Park is one of the local symbols and an emblem of the municipality. The corridors in the Starigrad Paklenica urbanscape are mostly visually and physically more accessible when lying parallel to the coast. The longitudinal corridors in Starigrad Paklenica, especially along the coastline (Fig. 16.2, photograph 8), offer wide views and good physical communication, while the transversal corridors (Fig. 16.2, photograph 7) are often short in reach.

The Perception of Starigrad Paklenica and South Velebit through Spatial Characteristics

The spatial characteristics which represent the nature and essence of Starigrad Paklenica and South Velebit are: wilderness and the transformed natural character of the mountainscape; and the traditional and contemporary built character of the urbanscape. Human interventions in South Velebit, for example the *mirilo* (Fig. 16.2, photograph 10) and the traditional settlement of Marasovići (Fig. 16.2, photograph 11), are small in scale and correspond to the natural conditions. Traditional structures and settlements are abandoned and lost in vegetation succession, with only a few examples[13] used for tourism and secondary housing. The perception of South Velebit (Fig. 16.2, photograph 9), in totality and in most of its elements, is close to a wilderness. The South Velebit wilderness is a natural refuge, protected at both the national and international levels, and is present in the collective memory as a national symbol. With cultural spaces of transformed nature and built structures, the South Velebit mountainscape creates a strong visual image and forms a particular ambience and reflects historical continuity, the composition of space, and legibility. On the other hand, the spatial character of Starigrad Paklenica is presented in its built structure, differentiating the traditional from the contemporary. The transformed natural character and naturalness of Starigrad Paklenica are perceived in the segmented and transformed natural elements enclosed in its built structure. With the spread of contemporary settlements (Fig. 16.2, photograph 11), the traditional settlement of Marasovići, built in the foothills of the South Velebit slopes (Fig. 16.2, photograph 11), became part of the Starigrad Paklenica urbanscape. The traditional built structures of Starigrad Paklenica and South Velebit show historical layers, composed of an urbanscape and a mountainscape, of significance to the local community, and connected to an intangible heritage. They form a distinct ambience and provide for the legibility of tangible and intangible landscape attributes.

The Perception of Starigrad Paklenica and South Velebit through Spatial Contrasts

At first sight, the primal contrast between Starigrad Paklenica and South Velebit is a place of evident life versus a place of hidden life (Fig. 16.1, photographs 13 and 15). Starigrad Paklenica is a small settlement of almost 1200 (permanent) residents which is mostly built of secondary housing. It is a tourist-oriented settlement which, during the summer season, increases several times in population. Although oscillations in settlement population are distinct, life during the whole year in Starigrad Paklenica is evident in built and maintained structures, in the infrastructure and on the cultivated land, and in the activities of both residents and visitors. Another contrast lies in spatial size, where Starigrad Paklenica is a small settlement in comparison to the Velebit mountain range. South Velebit is a vast karst mountainscape of harsh weather and tough life conditions, where life is not evident in itself but only with deeper exploratory insights. As a border between the Mediterranean and continental climate, with distinctive geomorphologic structures, South Velebit creates unique life conditions and diverse habitats which provide a home for numerous flora and fauna, including endemic and rare species. A contrast between the mountainscape and urbanscape is the winters and summers, perceived primarily in the change of climate and weather conditions and in the change in the number of visitors. In Starigrad Paklenica, a prominent contrast is seen between the solitary winter (Fig. 16.2,

[13]Examples of traditional settlements which are kept and transformed with contemporary usage: Marasovići on Veliko Rujno; Ramići and Parići in Velika Paklenica Canyon; Marasovići inside Starigrad Paklenica.

photograph 16) and the overcrowded summer (Fig. 16.2, photograph 15). The same contrast can be perceived in South Velebit, between the cold, wet, windy and empty winter (Fig. 16.2, photograph 13) and the hot and dry summer with numerous visitors (Fig. 16.2, photograph 14).

The first impression of the contrast between Starigrad Paklenica and South Velebit is deepened and extended through knowledge of the landscape composed of the following contrasts: contemporary life versus a natural refuge; a place of contemporary life versus a place of life in the past; maintained versus abandoned places; a place of mild versus harsh weather changes.

16.4.2 Makarska and Biokovo Mountain

The Town of Makarska sits in the region of Makarska in Split-Dalmatia County[14] with a population of 13,834 according to the 2011 census. It is a tourist centre and, with the adjacent settlements of Brela, Baška Voda, Promajna, Krvavica, Tučepi and Podgora, forms the agglomeration of the Makarska Riviera. The Makarska Riviera is situated on a littoral slope between the Brač-Hvar[15] Channel and the cliffs of Biokovo Mountain. The microlocation of Makarska is a natural port of a horseshoe-shaped bay enclosed between two peninsulas: Osejava and St. Peter. The benefits of a natural harbour, the littoral sloping fields, the mountain hinterland suitable for livestock breeding, and the inaccessibility of the Biokovo cliffs acting as a natural defence were recognised from prehistoric and ancient times (Tomasović 2002, 2009, 2017; Vidović 2012). The historic urban centre of Makarska, the traditional settlements and fortresses along the Biokovo foothills, the traditionally grubbed and cultivated fields, and stone drywall structures are evidence of historical continuity. The contemporary town and tourist structures connect traditional settlements and have transformed mostly abandoned cultivated land structures.

The Perception of Makarska and Biokovo through Different Position Vistas

The totality of the Biokovo mountainscape, due to the difference between the littoral cliffs and the continental hinterland, the indented topography and the diversity of the karst forms, can be perceived through the integration of outside and inside views. In the outside view from the sea, from the Peninsula of St. Peter (Fig. 16.3, photograph 4), the perception of complexity, coherence and disturbance in the mountainscape is low, as Biokovo is perceived only as a vertical surface of cliffs. The inside view of Biokovo (Fig. 16.3, photograph 3) reveals the complexity and diversity of the indented topography, karst forms and the unity of the scene. Human interventions are small in scale compared to the natural landscape, and they are mostly adapted and merged in nature, so the coherence of the mountainscape is perceived in both views from the outside and from within. The strong karst mountainscape and the Biokovo cliffs

---

[14]In Croatian: Splitsko-dalmatinska Županija.

[15]Brač and Hvar are Croatian islands.

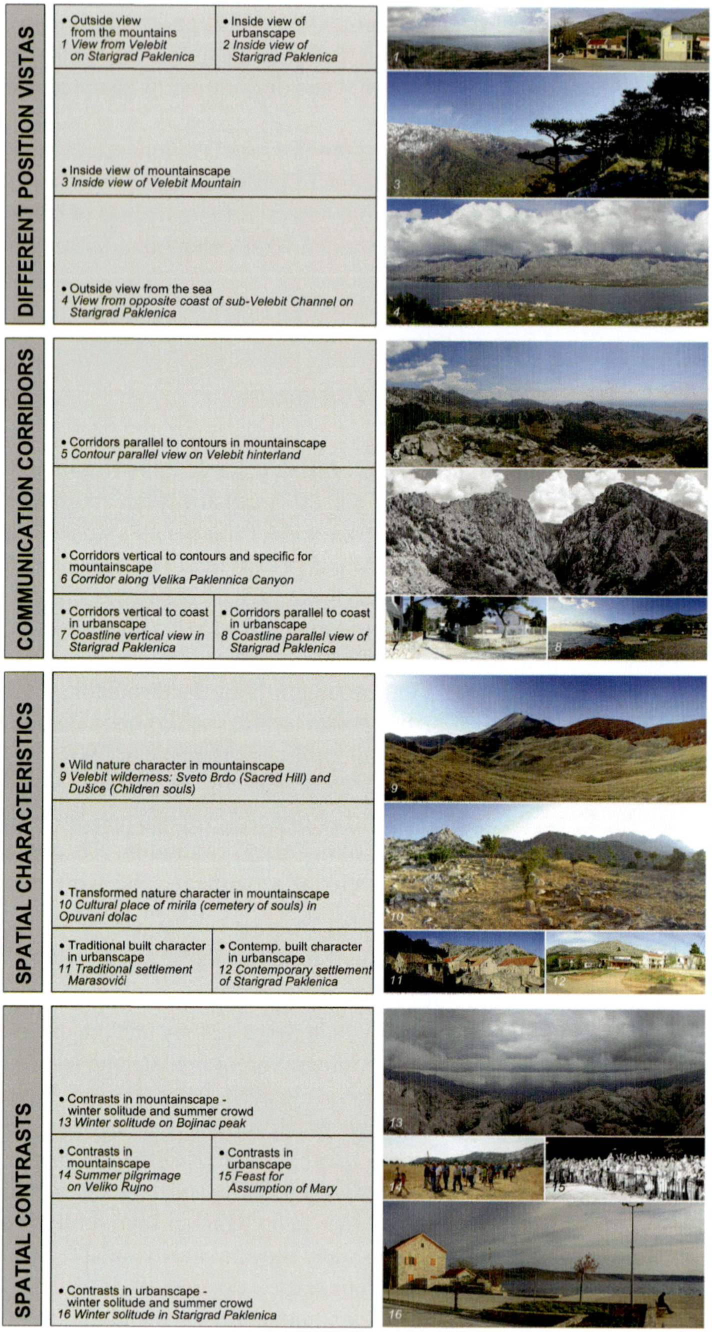
DIFFERENT POSITION VISTAS
• Outside view from the mountains
1 View from Velebit on Starigrad Paklenica
• Inside view of urbanscape
2 Inside view of Starigrad Paklenica
• Inside view of mountainscape
3 Inside view of Velebit Mountain
• Outside view from the sea
4 View from opposite coast of sub-Velebit Channel on Starigrad Paklenica
COMMUNICATION CORRIDORS
• Corridors parallel to contours in mountainscape
5 Contour parallel view on Velebit hinterland
• Corridors vertical to contours and specific for mountainscape
6 Corridor along Velika Paklennica Canyon
• Corridors vertical to coast in urbanscape
7 Coastline vertical view in Starigrad Paklenica
• Corridors parallel to coast in urbanscape
8 Coastline parallel view of Starigrad Paklenica
SPATIAL CHARACTERISTICS
• Wild nature character in mountainscape
9 Velebit wilderness: Sveto Brdo (Sacred Hill) and Dušice (Children souls)
• Transformed nature character in mountainscape
10 Cultural place of mirila (cemetery of souls) in Opuvani dolac
• Traditional built character in urbanscape
11 Traditional settlement Marasovići
• Contemp. built character in urbanscape
12 Contemporary settlement of Starigrad Paklenica
SPATIAL CONTRASTS
• Contrasts in mountainscape - winter solitude and summer crowd
13 Winter solitude on Bojinac peak
• Contrasts in mountainscape
14 Summer pilgrimage on Veliko Rujno
• Contrasts in urbanscape
15 Feast for Assumption of Mary
• Contrasts in urbanscape - winter solitude and summer crowd
16 Winter solitude in Starigrad Paklenica

◄**Fig. 16.2** Perception of the character of the landscape identity of Starigrad Paklenica and South Velebit Mountain through representative photographs. Guidelines: the size of the photograph indicates the intensity of perception, high (larger photograph) and low (smaller photograph). Bullets present the means of perception through representative photographs. *Numbers in italics* present a description of individual photographs

overlooking the town are present in all the vistas of Makarska, giving the settlement context and recognisability. From the inside view of Makarska, the urbanscape is perceived as recognisable in parts: the historic town centre and historic settlements integrated in the contemporary town structure, the urban seafront promenade and the forest-parks of the peninsulas of Osejava and St. Peter. From the Vošac hilltop of the Biokovo cliffs (Fig. 16.3, photograph 1), complexity, coherence and disturbance in the urbanscape are perceived through repeating patterns of residential, tourist and commercial built structures with little correspondence to the natural conditions, and as a part of the Makarska Riviera agglomeration. In the view from Biokovo, the indented seacoast of Makarska Bay and the peninsulas of St. Peter and Osejava give recognisability to the urbanscape. In the outside view from the sea, from the Peninsula of St. Peter (Fig. 16.3, photograph 4), the perception of complexity is low, where Makarska is perceived as an almost linear urbanscape.

The Perception of Makarska and Biokovo Character through Communication Corridors

Communication corridors refer to physical and visual corridors, including those that are longitudinal and transversal, and corridors specific and recognisable for Makarska and for Biokovo. Corridors parallel to the contours of Biokovo cliffs (Fig. 16.3, photograph 5) are perceived in wide views of the settled coast and Biokovo foothills, littoral cliffs and continental hinterland, while physical connections are mostly set across the vertical cliffs. The wavy topography in the highlands of the Biokovo hinterland (Fig. 16.3, photograph 6) does not provide longitudinal and transverse stretching of the contours; thus, the visual and physical corridors are adapted to the indented landscape. The corridors in the Makarska urbanscape are visually and physically more accessible when parallel to the coast. The longitudinal corridors in Makarska, especially along the coastline (Fig. 16.3, photograph 8), offer wide views and good physical communication, while the transversal corridors (Fig. 16.3, photograph 7) are often interrupted by built structures. All urbanscape views of Makarska have the specific and distinctive background of the Biokovo cliffs.

The Perception of Makarska and Biokovo through Spatial Characteristics

Spatial characteristics present the nature and essence of an individual scape: the wild, transformed and designed natural character and built character of Biokovo and Makarska. Human interventions in Biokovo are small scale and adapted to the natural conditions, such as the Kotišina Fortress (Fig. 16.3, photograph 10) and the settlement of Mali Vrv (Fig. 16.3, photograph 13). They are mostly abandoned and lost in vegetation succession; thus, the perception of Biokovo (Fig. 16.3, photograph 9), in totality and in most of its elements, is close to a wilderness. With cultural places of transformed nature and traditional built structures, the Biokovo mountainscape creates a strong visual image: it is connected to tradition and forms a unique ambience. The built and cultural elements of Biokovo reflect historical continuity, the composition of space and legibility. The transformed and designed natural character and naturalness of Makarska are most recognisable in public parks, the park-forests of the peninsulas of St. Peter and Osejava (Fig. 16.3, photograph 11) and the urban seafront promenade (Fig. 16.3, photograph 8). The most recognisable built character of Makarska is perceived in the historic town centre (Fig. 16.3, photograph 12) and in the historic settlements built on the foothills of Biokovo (Fig. 16.3, photograph 15) integrated in an urbanscape with the development of a contemporary urban structure. The traditional built structures of Makarska and Biokovo present historical layers, composed of an urbanscape and a mountainscape, of significance to the local community, and connected to an intangible heritage. They form a distinct ambience and provide for the legibility of tangible and intangible landscape attributes.

The Perception of Makarska and Biokovo through Spatial Contrasts

At first sight, the primal contrast between Makarska and Biokovo is the clash between the coastal town and its background of the almost vertical Biokovo cliffs (Fig. 16.3 photographs 4, 7, 8 and 16). The contrast in spatial size is also apparent between the almost linear coastal agglomeration of the Makarska Riviera and the extensive mountainscape of the Biokovo hinterland. Makarska is a small town and tourist centre of the Makarska Riviera which during the summer season increases several times in population. Makarska has expanded mostly in the last 50 years with the development of tourism on the Eastern Adriatic Coast and during the Croatian War of Independence with the arrival of refugees from neighbouring Bosnia and Herzegovina. Thus, the distinction between the historic town centre (Fig. 16.3, photograph 12) and the traditional foothill settlements (Fig. 16.3, photograph 15) on the one hand and the contemporary residential, touristic and commercial built structures (Fig. 16.3, photograph 16) on the other is noticeable. Biokovo is a Nature Park with a karst landscape, with numerous plant and animal spices, some of which are endemic. Biokovo was populated from prehistoric times with dwellings and settlements, most of which have been deserted in the process of depopulation. Abandoned traditional settlements on the cliffs and in the hinterland of Biokovo (Fig. 16.3, photograph 13) are in contrast to the traditional routes and dwellings transformed for the needs of recreation and hiking (Fig. 16.3, photograph 14). A contrast evident for both Makarska and Biokovo is between winter and summer, perceived primarily in the change of the climate and weather conditions and in a change in the number of visitors. In Makarska, the prominent contrast is between the solitary winters and the overcrowded summers, and the same contrast is perceived in Biokovo, between the cold, snowy, windy and empty winter and the hot and dry summer with the arrival of visitors.

The first impression of the contrast between Makarska and Biokovo is deepened and extended through knowledge of the landscape composed of the following contrasts: a concentrated contemporary life versus a vast natural refuge; a place of present life versus a place of life in the past; a maintained versus an abandoned landscape; a place of mild versus harsh weather changes.

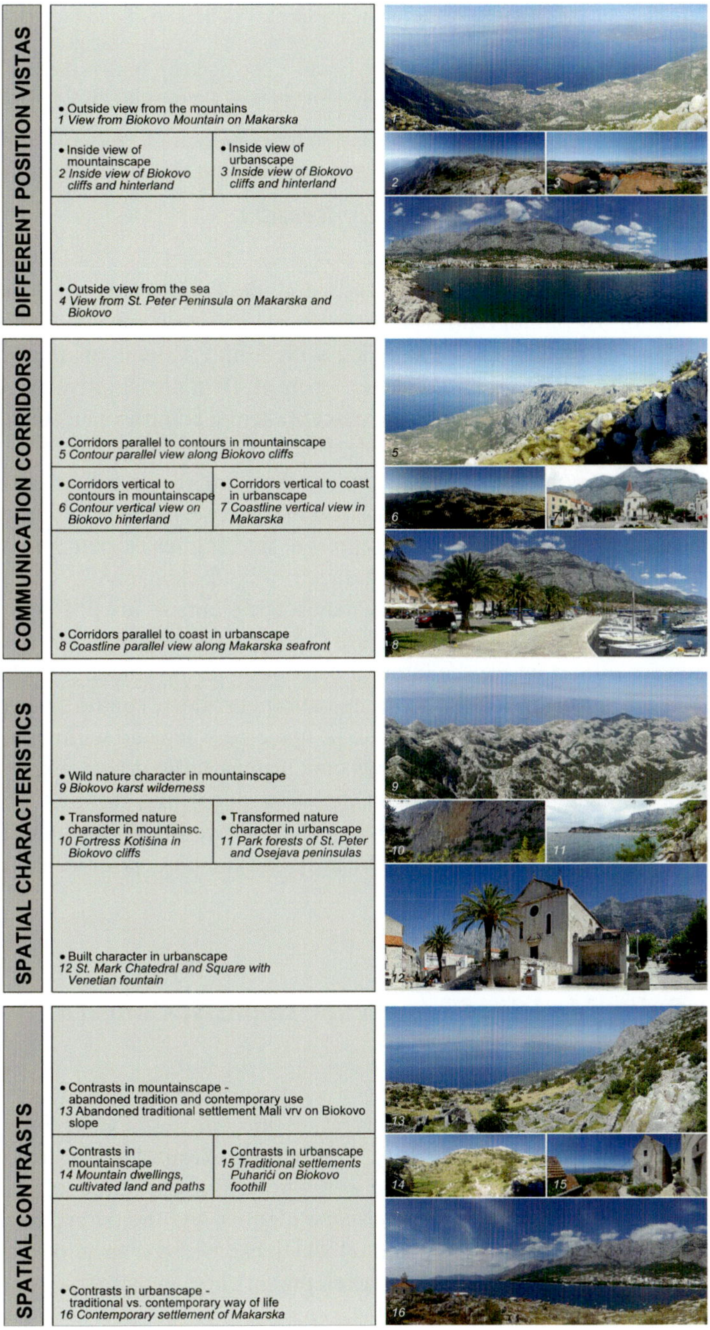
DIFFERENT POSITION VISTAS
• Outside view from the mountains
1 View from Biokovo Mountain on Makarska
• Inside view of mountainscape
2 Inside view of Biokovo cliffs and hinterland
• Inside view of urbanscape
3 Inside view of Biokovo cliffs and hinterland
• Outside view from the sea
4 View from St. Peter Peninsula on Makarska and Biokovo
COMMUNICATION CORRIDORS
• Corridors parallel to contours in mountainscape
5 Contour parallel view along Biokovo cliffs
• Corridors vertical to contours in mountainscape
6 Contour vertical view on Biokovo hinterland
• Corridors vertical to coast in urbanscape
7 Coastline vertical view in Makarska
• Corridors parallel to coast in urbanscape
8 Coastline parallel view along Makarska seafront
SPATIAL CHARACTERISTICS
• Wild nature character in mountainscape
9 Biokovo karst wilderness
• Transformed nature character in mountainsc.
10 Fortress Kotišina in Biokovo cliffs
• Transformed nature character in urbanscape
11 Park forests of St. Peter and Osejava peninsulas
• Built character in urbanscape
12 St. Mark Chatedral and Square with Venetian fountain
SPATIAL CONTRASTS
• Contrasts in mountainscape - abandoned tradition and contemporary use
13 Abandoned traditional settlement Mali vrv on Biokovo slope
• Contrasts in mountainscape
14 Mountain dwellings, cultivated land and paths
• Contrasts in urbanscape
15 Traditional settlements Puharići on Biokovo foothill
• Contrasts in urbanscape - traditional vs. contemporary way of life
16 Contemporary settlement of Makarska

◀**Fig. 16.3** Perception of the character of the landscape identity of Makarska and Biokovo Mountain through representative photographs. Remarks: the size of the photograph indicates the intensity of the perception, high (larger photograph) and low (smaller photograph). Bullets present the means of perception through representative photographs. *Numbers in italics* present the description of individual photographs

16.4.3 Comparison of Case Study Results

The perception of the character of the landscape identity is analysed through four framed groups of 16 photographs which represent different means of perception. In the case studies of Starigrad Paklenica with South Velebit on the one hand, and Makarska with Biokovo on the other, each of 16 representative photographs is described with their contribution to the perception of a complex landscape identity and the perception of landscape as a whole. Four framed groups and 16 representative photographs are connected to concepts of perceiving the character of landscape identity (Table 16.2), presenting a comparison of the case study research results.

From the same photographs, which represent live frames of perception in each case study, the same dimensions of the character of landscape identity are perceived but their intensities and connections to the landscape elements are different. Each of the live frames focuses on different aspects of seascape, urbanscape and/or mountainscape, which are important for the integral perception of the character of landscape identity. Different position vistas, communication corridors, spatial characteristics and spatial contrasts enable comparisons to be made between and within landscapes, and between the case studies. This allows conclusions to be drawn, expanding the perception of knowing a landscape and therefore deepening the first impression towards associative experience. The experience of all live frames is essential for the perception of comprehensive landscape identity and the perception of landscape as a whole.

16.5 Defining Models of Landscape Identity, Concern and Resilience

The research of associative landscape features, in the context of the *Heritage Urbanism* project, presents landscape identity as a heritage feature and a new heritage dimension—the perception of landscape as a whole. By setting models for revival and enhancement, the perception of landscape identity and the perception of landscape as a whole are identified as a potential which can become an active participant in social, cultural, economic and tourist development and can therefore contribute to landscape resilience.

A basis for setting models for the revival and enhancement of landscape identity is found in the theoretical framework: terminological definitions and research starting

Table 16.2 Perception of the character of landscape identity in the case studies Starigrad Paklenica with South Velebit, and Makarska with Biokovo as a result of a comparative analysis of framed groups and representative photographs

Framed groups	Perception of the character of landscape identity of Starigrad Paklenica with South Velebit through representative photographs	Perception of the character of landscape identity of Makarska with Biokovo through representative photographs	Dimensions of the character of landscape identity perceived in frames
Different position vistas	• The totality of South Velebit complexity, coherence and disturbance is perceived in the integration of multiple outside and inside views • From the sea, the perception of South Velebit complexity is lower (mountainscape as the surface of a massif) • From the sea, the perception of Starigrad Paklenica complexity is lower (urbanscape is almost a line) • From the hilltops and from within Starigrad Paklenica, the complexity, coherence and disturbance are perceived in totality and in elements	• The totality of Biokovo complexity, coherence and disturbance is perceived in the integration of multiple outside and inside views • From the sea, the perception of Biokovo complexity, coherence and disturbance is lower (only cliffs are perceived) • From the sea, the perception of Makarska complexity is lower (urbanscape is almost a line) • From the hilltops and from within Makarska, the complexity, coherence and disturbance are perceived in totality and in elements	Complexity Coherence Disturbance

(continued)

points. Models of landscape identity, concern and coherence are established in the analysed case studies (Table 16.3).

Landscape Identity Model

The landscape identity model defines the present state of perceiving associative landscape features. In the context of associative experience as described in the theoretical framework, identity perceived as a whole landscape is a new heritage dimension that could become a landscape, spatial and urban planning criterion.

Recognition as perception defines how attributes (factors) form and represent landscape identity. The question is: What is landscape identity perceived by? The answer is found in the theoretical framework regarding landscape identity which differentiates the character of landscape identity in three ways: the visual, functional and associative characters. A further theoretical definition determines the dimensions of perceiving the character of landscape identity: complexity, coherence, disturbance, stewardship, imageability, the visual scale, naturalness, historicity, ephemera, safety, usage, connectivity, organisation, meaning, associativity, ambience and legibility.

The present state of perceiving landscape identity and the outcome of the case study analysis of Starigrad Paklenica with South Velebit, and Makarska with Biokovo is the distinction between the coastal urbanscape and the hinterland mountainscape. The synergistic characteristics of seascape, urbanscape and mountainscape—a concentrated coastal settlement with a hinterland of a mountain massif and a linear coastal agglomeration with a cliff background and mountain hinterland—need to be perceived integrally in order to retain a comprehensive landscape identity. The landscape identity model is complete when all landscape identity characteristics are perceived as a whole and when synergy is achieved in perceiving the seascape, urbanscape and mountainscape.

Landscape Concern Model

The landscape concern model identifies the means of perceiving landscape identity which needs to be protected from decay and needs to be maintained in order to retain, restore or redefine the perception of landscape identity. One of the *Heritage Urbanism* project starting points is that the survival of cultural heritage is possible only if we integrate it into modern life (Obad Šćitaroci 2015). Applying the research of associative landscape features indicates that integrating tangible and intangible landscape heritage with the means of perception of landscape as a whole into the contemporary way of life can protect it from decay. This landscape heritage framework needs to represent the perception of landscape identity at the same time as being an object of protection and an integral part of the contemporary way of life.

Preservation as a concern presents a matter that engages society's attention, interest and care of the landscape. The theoretical framework regarding landscape concern distinguishes a set of three aims of landscape design and planning as follows: ecological, social and aesthetics. These aims are related to the preservation of landscape identity factors: natural and anthropogenic factors and factors of perception. The question regarding the concern of the core phenomenon—the perception of landscape as a whole—is: How to retain, restore and/or redefine the perception of

Table 16.3 Relation between terminological definitions, the research starting points, questions regarding the perception of landscape identity and models for the revival and enhancement of landscape identity established in the case studies of Starigrad Paklenica with South Velebit, and Makarska with Biokovo

Definitions	Research starting points	Question regarding perception of landscape identity	Models established in Starigrad Paklenica with South Velebit	Models established in Makarska with Biokovo
Recognition Perception	Perception of landscape as a whole	What is landscape identity perceived by? By the character of landscape identity	**Landscape identity model** Concentrated urbanscape of a small coastal settlement with a hinterland of a mountainscape massif	**Landscape identity model** Linear urbanscape of a coastal agglomeration with a cliff background and a mountain hinterland
Preservation Concern	Preservation through integration into a contemporary way of life	How to retain, restore and/or redefine the perception of landscape identity? By preserving the means of landscape perception which concern landscape identity	**Landscape concern model** Constrain development of the settlement from the seacoast, from the foothills and from the mountainscape	**Landscape concern model** Constrain development of the agglomeration from the seacoast, from the foothills and from the mountainscape
Resilience Adaptability Sustainability	Inseparability of landscape and town	What are the result of the perception of landscape identity and concern? Contributing resilient landscape and towns	**Landscape resilience model** Whole landscape resilience	**Landscape resilience model** Whole landscape resilience

landscape identity? The answer is by preserving the means of landscape perception which concerns landscape identity.

Different means of perception of landscape identity are analysed in the case studies of Starigrad Paklenica with South Velebit, and Makarska with Biokovo, through frames of perception grouped in different position vistas, communication corridors, spatial characteristics and spatial contrasts. The findings of the case study research suggest that experience of all frames is essential for the perception of comprehensive landscape identity and the perception of landscape as a whole. The preservation of all frames of perception, from the landscape, spatial and urban planning standpoint, can be summarised in one concern: to constrain the development of settlement from the seacoast, from the foothills and from the mountainscape. The summarised concern encompasses different aspects of preserving integral landscape identity: the preservation of the synergistic characteristics of seascape, urbanscape and mountainscape, the integration of associative landscape features into the contemporary way of life with the preservation of all frames necessary for the perception of landscape identity and the perception of landscape as a whole.

From the landscape, spatial and urban planning standpoints, the frames necessary for the perception of landscape identity and landscape as a whole can be preserved by:

- Sustainable construction (and not just construction) in the seascape, urbanscape and mountainscape, regarding different position vistas, communication corridors and spatial characteristics of an individual landscape;
- Maintaining existing vistas and communication corridors, their accessibility, and enabling new ones regarding inherited and contemporary landscape values;
- Sustaining existing spatial contrasts which increase landscape diversity, and restraining negative contrasts which distort the character of landscape identity (concern of sustainable landscape capacity and others).

The landscape concern model defines frames of perception as a dimension which needs to be preserved in order to retain, restore and/or redefine the perception of landscape identity as the most prominent associative landscape feature.

Landscape Resilience Model

Landscape resilience is the capacity of landscape to adapt to change, while respecting what has been inherited, adapting to the contemporary, anticipating the need for sustainable development and still retaining the characteristics of landscape identity. The resilience of landscape and towns is expanded to associative features and the whole landscape experience. The question regarding landscape resilience is: What are the results of the perception of landscape identity and landscape concern? The recognition and preservation of the perception of landscape identity can contribute to the sustainable development of resilient landscape and towns.

As set in the theoretical framework, the concept of landscape resilience is complemented by concepts of landscape adaptability and sustainability, which include respecting the situation or site of the town in relation to the surrounding landscape. The inseparability of town and landscape is also one of the research starting points.

"Cities are integral parts of their surroundings and are deeply connected with their landscape. This relationship acts like a driving force that creates the identity of an area. It suggests possible lines of development and points to certain problems that might arise in this context" (Gašparović and Sopina 2018). Landscape is of great significance for the cultural development of towns, and the future of towns depends on their landscape history (Timmermans et al. 2015). The relationship between the city and its landscape is symbiotic (Mumford 1961). In history, urbanity emerged from the landscape; in the present, landscape gives urbanity a context; and in the future, both landscape and urbanity need to be the basis of sustainable development.

The perception of the inseparability of the landscape and the town is presented in the case study research of Starigrad Paklenica with South Velebit and that of Makarska with Biokovo. Seascape, urbanscape and mountainscape give each other a context, concerning not only physical territory with its components but also history, local tradition and intangible heritage. The landscape resilience model sets the spatial development strategy from the aspect of the inseparability of the settlement and its natural landscape. Associative experience and the perception of the whole landscape, through the perception of the inseparability of the town/settlement and its landscape, contribute to the resilience of the whole landscape.

16.6 Conclusion

The research identifies the connection between associative landscape features, landscape identity and the perception of landscape as a whole, through the notion of perceiving the relationship between examples of coastal settlement of the Eastern Adriatic coast in Croatia and its natural context of sea and mountain hinterland.

Models for the revival and enhancement of landscape identity are defined as the research results. The landscape identity model defines the present state of recognising landscape identity; the landscape concern model identifies the means of perceiving the characteristics of landscape identity which require preservation; and the landscape resilience model sets the spatial development strategy from the aspect of the relationship between the settlement and its natural landscape. These models derive from a comparison of the perception of landscape identity through the case study of Starigrad Paklenica with South Velebit and that of Makarska with Biokovo, as landscapes with strong relationships among seascape, urbanscape and mountainscape. The case studies were analysed and compared through live frames and presented through photographs of perceptions of the character of landscape identity. The dimensions of the character of landscape identity are the criteria for the perception of landscape identity, which define how factors form and present landscape identity. The research is based on the *Heritage Urbanism* approach, with three main components applied to the research of associative landscape features: the factors of landscape identity; the criteria for the perception of landscape identity; and the models for the revival and enhancement of landscape identity.

Fig. 16.4 View of the Durmitor region from the peak of Bobotov Kuk

In the absence of man we do not have landscape,
we have a geographical area (Berleant 2013).

The inseparability of towns and their natural landscape relates to the inseparability of the perception of landscape and man. Berleant (2013) presents the condition of wholeness, which "recognises that the appreciator is in the landscape, an integral part of the landscape, and not an external spectator. It is an experience of being present and, by one's presence contributing to the formation of landscape and to its unique tonality". Since the birth of mankind, man's presence and culture have transformed the natural landscape in different intensities into cultural landscape, urbanscape and towns. Towns are important for the recognition of landscape, just as landscape is important for the recognition of towns. In the absence of culture and in the absence of nature, natural landscape and built urbanscape would be just geographical areas. In the example of almost indefinite but still accessible natural landscape, such as the Durmitor region in Montenegro (Fig. 16.4), all factors, criteria and frames of perceiving the character of landscape identity can be recognised. Still, Durmitor, due to the absence of any urbanscape and the immeasurability of human intervention, cannot be compared to a vast mountainscape.

Resilient landscape is one which retains its identity when human interventions, settlements and towns settle in. The wholeness, appreciation, knowing and resilience of landscape through the relationship between urbanscape and its natural landscape are thus themes for future research.

The recognition and preservation of the perception of landscape identity lead to appreciation and raising awareness of everyday landscapes and their cultural heritage, enhancing people's environment, raising the quality of life and contributing to the resilience of towns, settlements and landscapes. According to Berleant (2013), landscape appreciation is "entering into the experience as direct knowing, knowing that is engaged and replete. Its aesthetic is what makes a place come alive as a presence to those who live, work or visit it". The contribution of associative landscape features (Bojanić Obad Šćitaroci 2018) sets sustainability parameters which "are perceived through critical knowledge and interventions. Interpretation, analysis, clarification and reading of the meaning of the landscape must fall within a framework that will not compromise the needs of future generations".

Landscape perceived as a whole is presented as a new heritage dimension, developed with the perception of knowing a landscape. By establishing perception, concern and resilience models of landscape associative features, the associative dimension is affirmed in all landscapes as fundamental to their being retained, restored and/or redefined.

Acknowledgements The research is part of the scientific project "Heritage Urbanism—Urban and Spatial Planning Models for the Revival and Enhancement of Cultural Heritage". It is financed by the Croatian Science Foundation [HRZZ-2032] and carried out at the University of Zagreb, Faculty of Architecture.

References

Andersson E (2006) Urban landscapes and sustainable cities. Ecol Soc. https://doi.org/10.5751/ES-01639-110134

Belaj V (2004) Traditional mountain cattle breeding on the mountain of Velebit and ethnogenesis of Bunjevci. Stud Ethnologica Croat 16:5–31

Bell S (1999) Landscape: pattern, perception and process. E&FN SPON, London

Berleant A (2013) The art in knowing a landscape. Diogenes 59(1–2):52–62. https://doi.org/10.1177/0392192112469320

Bojanić Obad Šćitaroci B (2018) Urbanscape emanation versus types of landscape. In: Cocci Grifoni R, D'Onofrio R, Sargolini M (eds) Quality of life in urban landscapes. Springer, Cham, pp 337–343

Bojanić Obad Šćitaroci B (2015) Perceiving heritage versus awareness of heritage. In: Obad Šćitaroci M (ed) International scientific conference cultural heritage—possibilities for spatial and economic development—proceedings. Faculty of Architecture, University of Zagreb, Zagreb, pp 212–215

Bridgewater PB, Bridgewater C (2004) Is there a future for cultural landscapes? In: Jongman RHG (ed) The new dimensions of the European landscape. Springer, Berlin, pp 245–248

Bušljeta A (2010) Deruralisation of Southern Velebit—aspects of life of Velebit Piedmont dwellers in the first half of the 20th century. Senjski zbornik 37:397–428

Cicero MT (2008) The nature of gods (trans: Walsh PS). Oxford University Press, Oxford (Original work published 45 BC)

Council of Europe (2000) European Landscape Convention. Council of Europe. Available via COE. Retrieved from https://rm.coe.int/1680080621. Accessed on 22 Aug 2018

Faber A (1995) The life of the Velebit cattle shepherd and his relationship to death (Discussion with the *mirilo*). Senjski zbornik 22:157–170

Fein A (1972) A study of the profession of landscape architecture. The Gallup Organization, Princeton

Folke C, Jansson A, Larsson J, Costanza R (1997) Ecosystem appropriation by cities. Ambio 26:167–172

Furlan Zimmermann N (1999) Uvod. In: Krajolik – Sadržajna i metodska podloga Krajobrazne osnove Hrvatske. Ministarstvo prostornog uređenja, graditeljstva i stanovanja – Zavod za prostorno planiranje, Agronomski fakultet Sveučilišta u Zagrebu – Zavod za ukrasno bilje i krajobraznu arhitekturu, Zagreb

Gašparović S, Sopina A (2018) The role of landscape in planning the city of Zagreb from the early 20th to the early 21st century. Prostor A Sch J Archit Urban Plan 26(1):133–145

Gunderson LH (2000) Ecological resilience—in theory and application. Annu Rev Ecol Evol Syst 31:425–439
Hunt JD (2000) Garden perfections: the practice of garden theory. University of Pennsylvania Press, Philadelphia
Ignatieff M (1995) Walk on the wild side—review of Simon Schama's "Landscape and memory". The Independent on Sunday, 9 April 1995, pp 36–37
Leopold A (1949) A sand county almanac. Oxford University Press, Oxford
Lucas D (2009) Heritage landscape criteria. Available via Lucas Associates. Retrieved from http://www.lucas-associates.co.nz/assets/Heritage/HERITAGE-LANDSCAPE-CRITERIA-Lucas-March-2009.pdf. Accessed on 22 Aug 2018
Lynch K (1960) The image of the city. The MIT Press, Cambridge
Marić T, Palaiologou G, Griffits S, Bojanić Obad Šćitaroci B (2017) Gateway-pathway heritage and urban growth—Zagreb case study. In: Heitor T, Serra M, Pinelo Silva J, Bacharel M, Cannas da Silva L (eds) Proceedings of the 11th international space syntax symposium. Instituto Superior Técnico, Departamento de Engenharia Civil, Arquitetura e Georrecursos, Lisabon
McHarg IL (1969) Design with nature. Garden City, New York
Montgomery J (1998) Making a city: urbanity, vitality and urban design. J Urban Des 3(1):93–116
Mumford L (1961) The city in history—its origins, its transformations, and its prospects. Harcourt, San Diego
Obad Šćitaroci M (2017) Uvod u Znanstveni kolokvij Modeli revitalizacije i unaprjeđenja kulturnog naslijeđa. In: Obad Šćitaroci M, Bojanić Obad Šćitaroci B (eds) Znanstveni kolokvij Modeli revitalizacije i unaprjeđenja kulturnog naslijeđa – Zbornik radova – Multidisciplinarni dijalog. Faculty of Architecture, University of Zagreb, Zagreb, pp 6–11
Obad Šćitaroci M (2015) Introduction. In: Obad Šćitaroci M (ed) International scientific conference cultural heritage—possibilities for spatial and economic development—proceedings. Faculty of Architecture, University of Zagreb, Zagreb, pp 6–7
Obad Šćitaroci M, Dumbović Bilušić B, Bojanić Obad Šćitaroci B, Božić N (2014) Krajolik – čimbenik strategije prostornog razvoja. Faculty of Architecture, University of Zagreb, Zagreb
Ode A, Tveit MS, Fry G (2008) Capturing landscape visual character using indicators: touching base with landscape aesthetic theory. Landscape Res 33(1):89–117
Osmond P (2007) Quantifying the qualitative: an evaluation of urban ambience. In: Semat Kubat A et al (eds) Proceedings: 6th international space syntax symposium. ITU Faculty of Architecture, Istanbul
Parris K (2002) Agricultural landscape indicators in the context of the OECD work on agrienvironmental indicators. In: Dramstad W, Sogge C (eds) Agricultural impacts on landscapes: developing indicators for policy analysis. NIJOS/OECD Expert Meeting, Oslo, pp 10–18
Relph E (1976) Place and placelessness. Pion, London
Sopina A, Bojanić Obad Šćitaroci B (2015) Associative features of landscapes. Prostor A Sch J Archit Urban Plan 23(2):304–313
Sopina A, Bojanić Obad Šćitaroci B, Gašparović S (2015) Natural, anthropogenic, morphological and intangible elements of South Velebit landscape. In: Obad Šćitaroci M (ed) International scientific conference cultural heritage—possibilities for spatial and economic development—proceedings. Faculty of Architecture, University of Zagreb, Zagreb, pp 304–309
Taylor K (2008) Landscape and memory: cultural landscapes, intangible values and some thoughts on Asia. In: 16th ICOMOS general assembly and international symposium: finding the spirit of place—between the tangible and the intangible, Quebec, 29 Sept–4 Oct 2008
Thompson IH (2000) Ecology, community and delight—sources of values in landscape architecture. E&FN SPON, London and New York
Timmermans W, Woestenburg M, Jonkhof J, Annema H, Shllaku M, Yano S (2015) The rooted city—European capitals and their connection with the landscape. Blauwdruk Publishers, Wageningen
Tomasović M (2002) The Hvar culture of the late Neolithic and the neighbouring coast. Prilozi povijesti otoka Hvara 9(1):33–44

Tomasović M (2009) Archeological suggestions for ubication of cities from chapter 36 of De administrando imperio by Porfirogenet. Starohrvatska prosvjeta 3:293–313

Tomasović M (2017) Gothic art in Primorje, Gorska Župa and Radobilja in the time of the Herzog Stjepan Vukčić Kosača—between reality and folk tradition. Hercegovina 3:273–329

Trošelj M (1992) A review of mirila specifics on Velebit. Senjski zbornik 19:73–80

Trošelj M (2013) Southern Velebit *mirilo* from the research of Professor Ante Glavičić to today. Senjski zbornik 40:611–630

Tuan YF (1977) Space and place: the perspectives of experience. University of Minnesota Press, Minneapolis

Turner T (2001) HyperLandscapes. Landscape Design 304:28–32

Tveit MS, Ode A, Fry G (2006) Key concepts in a framework for analysing visual landscape character. Landscape Res 31(3):229–255

Tveit MS, Ode Sang A (2014) Landscape assessment in metropolitan areas—developing a visual indicator-based approach. SPOOL 1(1):301–316. https://doi.org/10.7480/spool.2013.1.641

Ujang N, Zakariya K (2015) The notion of place, place meaning and identity in urban regeneration. Proc Soc Behav Sci 170:709–717

UNESCO (1972) Convention concerning the protection of the world cultural and natural heritage. United Nations Educational, Scientific and Cultural Organisation. Available via UNESCO. Retrieved from https://whc.unesco.org/archive/convention-en.pdf. Accessed on 22 Aug 2018

UNESCO (2003) Convention for the safeguarding of the intangible cultural heritage. United Nations Educational, Scientific and Cultural Organisation. Available via UNESCO. Retrieved from https://ich.unesco.org/en/convention. Accessed on 22 Aug 2018

UNESCO (2005) Declaration on the Conservation of Historic Urban Landscapes. United Nations Educational, Scientific and Cultural Organisation. Available via UNESCO. Retreived from https://whc.unesco.org/document/6812. Accessed on 24 Dec 2018

UNESCO (2011) Recommendation on the historic urban landscape. United Nations Educational, Scientific and Cultural Organisation. Available via UNESCO. Retrieved from https://whc.unesco.org/uploads/activities/documents/activity-638-98.pdf. Accessed on 22 Aug 2018

UNESCO (2017) Operational guidelines for the implementation of the world heritage convention. United Nations Educational, Scientific and Cultural Organisation. Available via UNESCO. Retrieved from https://whc.unesco.org/en/guidelines/. Accessed on 22 Aug 2018

United Cities and Local Governments (2008) Agenda 21 for culture. Available via Agenda 21 for culture. Retrieved from http://www.agenda21culture.net/sites/default/files/files/documents/multi/ag21_en.pdf. Accessed on 22 Aug 2018

United Nations (2015) Transforming our world: the 2030 Agenda for Sustainable Development. Available via UN. Retrieved from http://www.un.org/ga/search/view_doc.asp?symbol=A/RES/70/1&Lang=E. Accessed on 22 Aug 2018

Valentić M (2013) The Velebit *mirilo* in the research of Ante Glavičić and others and several themes about the people which the Velebit *mirilo* belong to together with their dual faith. Senjski zbornik 40:583–610

Vidović D (2012) An overview of the toponymy of the Makarska region. Folia Onomastica Croatica 21:207–232

Vitruvius (1999) In Rowland I (ed). Ten books on architecture. Cambridge University Press, Cambridge

Waldheim C (2006) The landscape urbanism reader. Princeton Architectural Press, New York

World Commission on Environment and Development (1987) Report of the World Commission on Environment and Development: Our common future. Available via the UN. Retrieved from http://www.un-documents.net/our-common-future.pdf. Accessed on 22 Aug 2018

Chapter 17
Landscape Models of Enhancing the Inherited City Identity

Sanja Gašparović

Abstract The role of landscape planning in preserving the city's identity is researched. Heritage landscape is considered a concept of integrally planned landscape, recognized as an important city's identity factor. The landscape has always played a significant role in forming the image of Zagreb. Its strong landscape strategy determined by twentieth-century urban planning gradually loses its clarity over the past decades. The research starts from the thesis that the concept of an integral landscape image should be considered as urban planning heritage that contributes to the preservation of the city's recognizability. It is necessary to determine the landscape models of its revitalization, adaptation and improvement. Comparative analysis of contemporary landscape strategies of European cities resulted by the identification of landscape model reflected in:

- recognizing landscape as an important formative element of the city;
- overcoming the boundaries between the city and the region and their territorial intertwining;
- determining the starting point of the landscape strategy for the whole city level;
- networking of unbuilt spaces into a recognizable system of green infrastructure;
- 'conquering' brownfield areas for new public open spaces;
- including the dynamic features of natural systems in planning process.

By applying a landscape model on the example of Zagreb and other cities, it is possible to protect and enhance their recognizable landscape image, or inherited identity.

Keywords Landscape heritage · Landscape planning · City image · Landscape urbanism · City strategies

S. Gašparović (✉)
Faculty of Architecture, University of Zagreb, Zagreb, Croatia
e-mail: sanja.gasparovic@arhitekt.hr

M. Obad Šćitaroci et al. (eds.), *Cultural Urban Heritage*,
The Urban Book Series, https://doi.org/10.1007/978-3-030-10612-6_17

17.1 Introduction

In a certain way, cities are rooted into their *sit*—they are connected to their landscapes and the identity of space, possibilities and developmental problems stem from their mutual relationship (Timmermans et al. 2015). Mc Harg (1969) considers the basic identity of a city to stem from the location itself—the natural givens of the *sit (genius of the site)*, while the city's excellency is the result of preserving, using and emphasizing those features. As stated by Lynch (1960), landscape is one of the factors that form the image of a city, which affects its readability and recognizability. According to a research by Lendholt (1970), landscape areas represent specific places which contribute to the 'irreplaceability' of a city, the identification with it and the sense of belonging.

In the last few decades, due to the increasing pressure of urban development, there has been an increase in the need for revalorization of the role of landscape in city planning. Contemporary processes of densification in many urban areas have limited both the development and the preservation of open landscape areas within the city boundaries, and the rise of population density has affected the increasing awareness of the need for availability and proximity of public landscape areas.

Landscape planning has a key role in the concept of city development so as to reflect the natural features of the *sit* in the best possible way, networked through active public systems of open spaces and connected to landscapes of a wider city area. European cities are undertaking initiatives to become '*greener cities*' and to realize as sustainable as possible models of urban development. They are trying to achieve the concept of integrally planned landscape which has been recognized as an important factor of the city identity—the heritage landscape. One of the important prerequisites to achieve such a concept is the continuity of (landscape) planning.

Landscape has always had a significant role in the formation of the image of Zagreb. In the first half of the twentieth century, the recognizable landscape concept of emphasizing the landscape features of the natural *sit*—the river in the south and the mountain in the north—connected by landscape strips throughout the city was established.

In the recent thorough and systematic research of the urban planning of the city of Zagreb from the aspect of the development of the city landscape concept, it has been confirmed that the landscape component of the planning of Zagreb is gradually losing its importance (Gašparović and Sopina 2018). This has been determined based on quantitative and, in particular, qualitative changes in the features of the landscape planning city image. Certain master plans from the middle of the previous century partly moved closer towards the comprehensive, active approach to city landscape planning indicating the creation of the landscape conception, but it has never been entirely planned or realized in its true sense. Dynamic changes in the development of the city in the last three decades have been accompanied by the trend of reducing the landscape are neglecting the significance and role of landscape in the city planning. The clarity of the recognizable landscape image of the city is disappearing, which unquestionably contributes to the fading of the inherited city identity.

The research starts from the thesis that the concept of an integral landscape image should be considered as urban planning heritage that contributes to the preservation of the city's recognizability. One of the important prerequisites for achieving such a (heritage landscape) concept is the continuity of (landscape) planning. The aim of the research is to determine the landscape models of revitalization, adaptation or improvement of the integral landscape city image.

17.2 Theoretical Framework

The basis for the change of the city landscape paradigm should be sought in contemporary landscape planning theories and good urban planning practices which show that contemporary urbanism assigns greater and multifunctional roles to landscape. Landscape is seen as a system/infrastructure which has equally important roles of the environment quality preservation and the formation of the recognizable city image (Gašparović and Sopina 2018).

Czechowski et al. (2015), as well as Timmermans et al. (2015), point out that the concepts of 'green urbanism', incited by a stronger perception of the ecological crisis, draw attention to the role of landscape in the tasks of adaptation to climate changes, food provision, energy transport, etc. Benedict and McMahon (2006), as the most illustrative in that sense, provide examples of forming green infrastructure by articulating the complete and coherent multifunctional networks of open urban spaces and connecting them with much wider regional boundaries in the primary function of achieving ecological sustainability.

At the same time, landscape replaces architecture as a formative element of contemporary urbanism and bases its development primarily on rediscovering derelict and neglected city spaces (Bojanić Obad Šćitaroci and Matuhina 2012; Waldheim 2006). According to Corner (2006), apart from the role of reading and experiencing urban space, its 'green' scenography, landscape also becomes a formative medium of a city.

17.3 The Conducted Research Method

The research is based on a comparative analysis of Zagreb landscape planning and contemporary examples of good landscape planning practice in European cities.

Three comparative examples, the winners of the prestigious title of the European Green Capital, have been chosen: Essen, Oslo and Stockholm. They are comparable to Zagreb when it comes to the population and the city area. They also have common and emphasized natural-geographical features of the *sit*: water (blue) infrastructure and topography.

Green Capital City programme was created by the European Commission to recognize cities that have a 'consistent record of achieving high environmental standards' and are 'committed to ongoing and ambitious goals for further environmental improvement and sustainable development' (Beatley 2012). The criteria for the choice of the city to receive *the green capital* award every year are made of twelve environmental indicators, among which, for the purpose of this research it is important to emphasize the following: green urban areas—sustainable use of land, nature and biodiversity, and mitigation of and adaptation to climate changes.

The examples of the cities have been analysed and compared based on the following criteria:

- natural features of the *sit*—in order to determine in which way and to what extent landscape is ingrained into the city image and which landscape factors affect the creation of the city identity;
- historical development of landscape planning—in order to determine the most significant stages of the city planning in which landscape occupies an important place, the period and the starting points of the landscape conception formation, and in order to affirm the continuity of the city landscape planning;
- contemporary trends in the city development through the landscape aspect—in order to determine pressures on landscape and specificities of urban development compared to the landscape image;
- characteristics of the contemporary landscape strategies of the city—in order to determine the landscape planning starting points and to recognize the factors of the landscape image establishment.

17.4 Comparative Analysis of City Landscape Planning Examples

17.4.1 The Role of Landscape in Zagreb City Planning

The most important landscape features of the city of Zagreb's natural *sit*, which significantly determine its formation and development and define its image, are the river Sava in the south and the hills of Medvednica Mountain in the north.

One of the important characteristics of Zagreb urban planning from the beginning of the twentieth century was the establishment of the landscape concept which was based on the preservation of those natural givens.

The recognizable landscape concept is already present in the city plans from the 1930s, while in the master plans from 1949 and 1971 it is the closest to the recognizable landscape conception. At that time, the landscape planning image of the city was based on the establishment of landscape strips/breaches into the city tissue which connect the river Sava and Medvednica and have an important ecological-biological and aesthetic-experiential function. The characteristic of those plans is the acknowledgement of the multiple role of landscape in the city planning: the formation of the city, the creation of communications and vistas, the separation of functional areas and ecological sustainability.

In their research, Gašparović and Sopina (2018) find that in the last few decades, due to the increased pressures of taking over free spaces by urban spread, the planning attitude towards the role of landscape has been changing. The 1986 master plan brought changes in the theoretical foundation of the city planning—from the formally theoretical model of modernism to the recognition and preservation of the existing values of postmodernism—and it considered the influence of landscape strictly within the administrative city borders. The new approach to the city planning neglects the vision of the whole and focuses on the detailed standardization of fragments. Many of the roles of landscape in planning are being neglected. The priority is given to the quantification of the proportion of 'greenery' in the city, while the clarity and totality of the landscape concept of green strips between the river and the mountain is lost. The landscape system of the city is becoming increasingly fragmented, and the landscape concept is unrecognizable. The planning approach to landscape is becoming passive—planning protects the unbuilt areas from building and landscape does not get enough of the other roles which are a prerequisite for its perception as an active factor of urbanity (Fig. 17.1).

Today, Zagreb has no landscape strategy and landscape, as a formative element of the city, is neglected in the planning documentation.

17.4.2 The Role of Landscape in Essen City Planning

Essen is located in the Ruhr area between the rivers Emscher and Ruhr. The development of the city is mostly marked by its industrial past. Intensive development of underground coal mining in the second half of the nineteenth century was accompanied by largely unplanned growth of the city without the infrastructure or green areas keeping place.

Due to the obvious environmental problems, regional green spaces planning began in 1923 to protect open spaces and give structure to the Ruhr area conurbation. The landscape planning concept of Essen consisted of two recognizable parts. Northern half of the city was characterized dominantly by business and industry, which consisted of individual, isolated green and woodland areas, and city and district parks. Ruhr Valley—southern part of the city—was less industrialized and developed as a leisure area.

Comprehensive spread of greenery through the northern half of the city started in 1975. The aim of the green area development policy was to create a functional green area system, in order to promote the sustainable recolonization of the collieries and industrial premises that have fallen vacant, and to improve the environmental and living conditions in the city. The last colliery was closed in 1986, and since then, Essen has been making admirable efforts to establish itself as a 'City in transformation' that is overcoming a challenging industrial history to reinvent itself as a 'Green City' and a leading example for others (Fig. 17.2).

Green–blue infrastructure is the driver of Essen's sustainable urban development. Kleinebrahm and Lipsius (2015) emphasize that a comprehensive landscape restora-

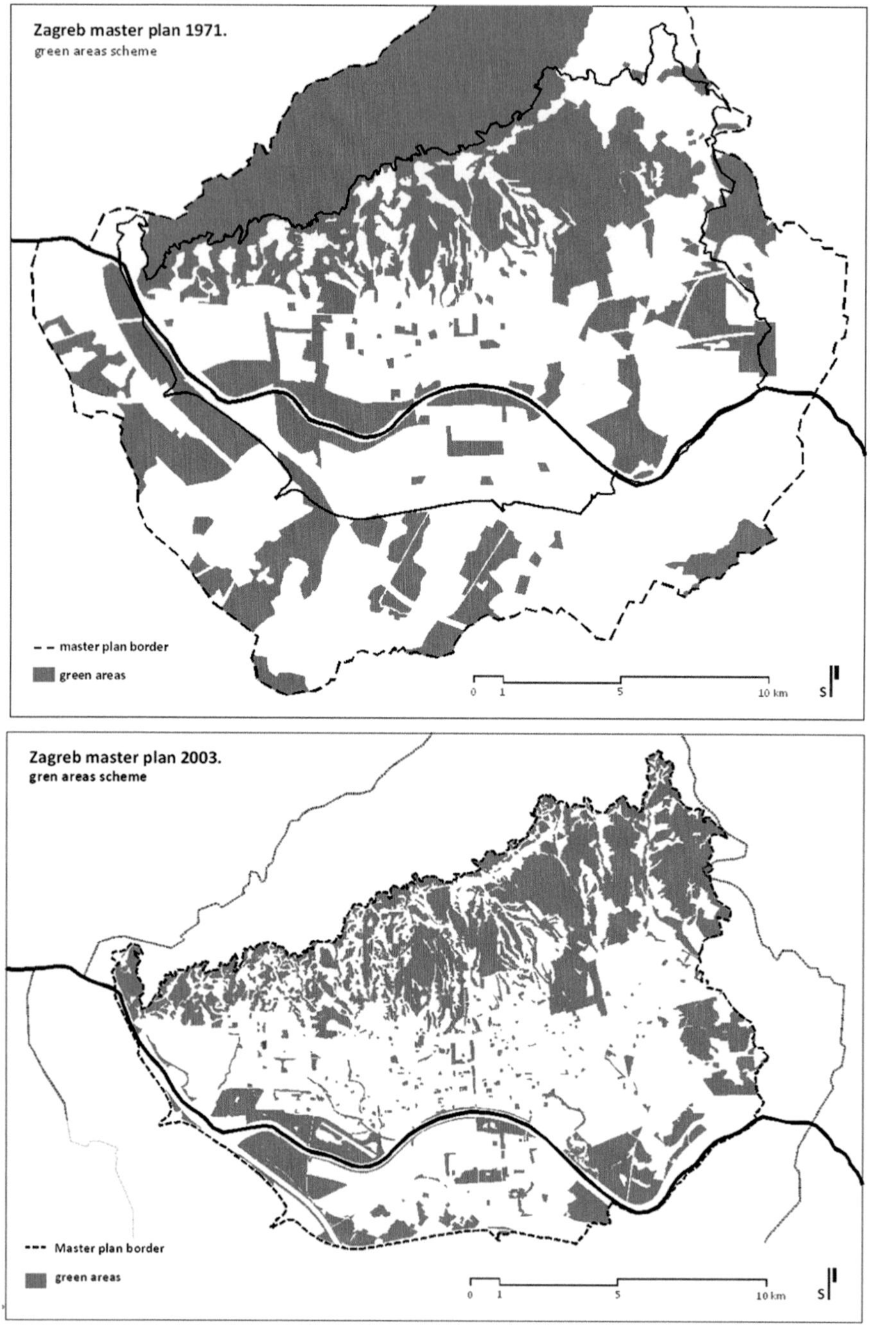

Fig. 17.1 Comparison of the landscape proportion in Zagreb master plans

Fig. 17.2 Landscape concept sketches of analysed cities Oslo, Essen and Stockholm

tion programme of the Ruhr region, as a part of IBA Emscher park, incited the creation of the city of Essen master plan in 2007, under the name of 'Open space creates city space' through which the city is planning a network of green areas and open spaces. This network follows the water bodies along the region's three north–south axes, which connect the more natural Ruhr Valley in the south with the industrial Emscher Valley and hence also countryside spaces and the regional open space system of the Emscher Landscape Park. The water line connects the valleys of the two rivers—Emscher and Ruhr—through a system of small waterways, streams and other types of water landscapes. The cultural line is connected with a system of cultural monuments, historical sites and other important city areas. The natural line connects the biggest number of parks, forests and other green areas in the eastern most part of the city.

17.4.3 The Role of Landscape in Oslo City Planning

Oslo is located on an arc of land at the northern most end of the Oslo Fjord and is surrounded by forested hills Marka, a nationally protected area. Ten waterways flow from Marka into the fjord, transecting the city.

The awareness of the landscape city image recognizability factors has been developing throughout the history of planning in Oslo. Large forest reservations and the Oslo Fjord have been protected, and importance has been given to the development of the culture of life in open spaces among the citizens (Luccarelli and Gunnar Røe 2016). 'Green fingers' and 'urban forests' have been the principles of planning since the first modern city plan in 1929. Green fingers were planned as corridors within urban built-up areas.

Current green infrastructure planning is based on preserving the landscape corridors, the fjord and the forest as important parts of the city as it strengthens the blue–green city character. (Falleth and Saglie 2012).

Since the 1990s, the development strategy of Oslo has been densification of the existing built-up area and seafront and brownfield areas transformation. In order to protect the green infrastructure from urban densification threats, the strategy involves separate land use plans of the green structure within the building zone (Jorgensen et al. 2016).

The city's waterways have been subject to a new revolutionary strategy with the vision of Oslo as 'a capital city in sustainable development, characterized by economic, social and cultural growth according to nature's ability to sustain that growth ecologically'. The strategy has completely reversed the previous approach of enclosing these channels. They are now being actively reopened in order to make them accessible for people, to efficiently manage storm water and facilitate development and restoration of habitat.

In 2009, a revision of the Green Plan for Oslo from 1993 was carried out. The *Grøntplan for Oslo* included parks, walking paths, waterways, green corridors, less well-maintained green areas and other green areas in the city and provided an important argument in withstanding the pressure of development in green areas (Oslo Municipality 2009) (Falleth and Saglie 2012).

This plan of the city area has determined aims, strategy, cartographic overview (of four recognizable city areas) and the binding regulations and guidelines for the blue–green structures in the building zone. The aims are based on maintaining and strengthening the recognizable character of Oslo as a 'blue–green city between the hill and the fjord', enhancing the fulfilment of the citizens' needs for green recreational areas and establishing urban development in accordance with the ecological principles. The plan has also determined the measures considered essential for establishing an integral landscape image of the city. Oslo's spatial plans, and the new draft of the Green Plan, place nature even more front and centre in the city. The forest of Marka holds special cultural and social importance (Beatley 2012).

17.4.4 The Role of Landscape in Stockholm City Planning

Stockholm is an archipelago of fourteen islands on the coast of the Baltic Sea. Water is part of the uniqueness of the Stockholm brand. The major waters are Lake Mälaren, Saltsjön Bay and 12 smaller lakes. Wedges of woods and greenery extend from the outskirts of the landscape towards the city nucleus.

The city of Stockholm began to develop a plan for urban growth in the 1930s. The development concept was established by long 'fingers' of built areas along the public transportation system and the structure of unbuilt 'green wedges' left in between. This radial development network pattern of ten green long corridors continued throughout the region, providing easily accessible and ecologically beneficial open spaces (Nelson and Malarstrand n.d.).

In order to preserve green areas, the city of Stockholm supports infill development in both its inner city and outer suburban communities. The main goals of City Plan 99 ('Build the City Inwards') are to preserve the most valuable landscape areas within and outside the city and to connect them, both for human recreation and to promote biodiversity. The Park Programme, an action plan for the development of Stockholm's parks and green areas, has been developed in order to maintain the amount of open space and to achieve three main goals: good supply of parks, sustainable park environment and rich park culture (Ståhle 2002). A purpose of The Green Map is to define the green structure element of the city better and to make it easier to coordinate and plan for green space preservation and urban development. At the wider regional level, the preservation of the ten green wedges structure, important urban ecological and recreational corridors, containing urban and peri-urban forests, wetland, agricultural land and parks are the goals of The Regional Plan (2001). Each of the green wedges has its own name and main function. These wedges are not only rich in natural resources, but also in biodiversity, cultural history and environment, which makes Stockholm's green wedges unique worldwide in several aspects (Kong 2012).

Urban growth and the pressure on land to facilitate new development constitute central topics in the comprehensive plan as a strategic backbone for green urban planning—The Green Walkable City (GWC 2013). Its purpose is to conceptualize future strategies for the planning of Stockholm's green space. The document states that green and blue infrastructure is very highly valued; it addresses accessibility and quality of green space for recreation, as well as the view that green space should facilitate a sense of belonging and social interaction necessary for urban development.

The basic objectives of the GWC plan are: to protect and develop the city's green character; to support accessibility and recreation; to support the ecological infrastructure; and to develop tools and processes for the government to work with green space.

Landscape management is focused on development, quality, accessibility and connectivity (both spatial and functional). One of the features of the plan is the potential for green areas to physically connect districts via the strategic placement of parks and public functions along natural green corridors. The GWC plan acknowledges many roles of urban green space: functions to provide public space; social cohesion and inclusion, safety, health, restoration, accessibility, management, seasonal changes, biodiversity, local climate mitigation and microclimate, and storm water management (Littke 2015).

17.5 Discussion

The beginnings of landscape planning in all the analysed cities, including Zagreb (Table 17.1), are determined by similar characteristics: the historical period of landscape planning formation around the 1920s and the 1930s, the concept of the establishment of the recognizable landscape image (radial, parallel, linear, etc) whose aim is to emphasize and connect landscape features of the *sit* (natural waterways or bays with the surrounding hills and forest areas), structuring and separation of functional areas of the city and its connection to the landscape of the wider area/region.

Thus established concepts gradually and continually developed in a planned way until the end of the twentieth century, the period characterized by various urban

Table 17.1 Basic comparative cities' features

		Zagreb	Essen	Oslo	Stockholm
Population: 500,000–1,000,000		790,000	589,000	590,041	925,934
City area: 200–450 km^2		394	210	426	381,6
European Green Capital award winning year		–	2017	2019	2010
Main natural features of the city	Waterscape	The river Sava	The rivers Emscher and Ruhr	The Oslo fjord, 10 water ways through the city	An archipelago of 14 islands, 2 big and 12 smaller lakes
	Topography woodlands	Forested mountain Medvednica	Densely built-up valley	forested hills Marka	5 'wedges' of woods and greenery

development processes which presented a threat to the landscape concept. In Essen, the collapse of coal mining was followed by a great economic, ecological and social crisis; in Oslo, urban densification of the existing built area directly threatened landscape areas; in Stockholm, infill development presented a danger to the preservation of open spaces, just like in Zagreb where spatial reserves for development were sought within the city itself. The tendency was to restore and fill in the already built parts of the city and to neaten the neglected periphery.

Essen, Stockholm and Oslo were aware of the importance and the role of landscape in the city identity at that moment, so they developed special measures for its preservation/restoration and enhancement. The national programme of the comprehensive restoration of the Ruhr area (IBA Emscher Park) focuses on landscape as a symbol of ecological, economic and urban restoration. The idea of the creation of a new regional park was also the driver of the creation of the Essen Master Plan, in which it was planned to set up thematic landscape strips as a part of the system of networked open city spaces ('Open space creates city space', 2007). The aim of the Stockholm Regional Plan (2001) is to preserve the ten green wedges structure, each with its own main function, as one of the crucial elements of the city's cultural history and uniqueness. Nelson and Malarstrand explain the advantages of new tools which the city develops for a better understanding of the value of open spaces and their planning (biotope and sociotope mapping). Oslo adopted a clear strategy of blue–green structure in the city zone (2009), with binding regulations for maintenance and strengthening of the recognizable character of Oslo as a 'blue–green city between the hill and the fjord' and measures essential for the establishment of the integral landscape image of the city.

Unlike them, the Zagreb plans (Master Plan 1986, 2003) criticize the previously too big proportion of landscape; they suggest the reduction in its area and abandon the concept of the complete consideration of the city, whose integral and formative part is landscape. The landscape image is fragmented, and the composition disappears.

Essen, Stockholm and Oslo in their strategic landscape documents point out the multiple roles of landscape which are important for the city. They emphasize the physiognomic/formative role which contributes to emphasizing the recognizable features of the *sit* and strengthening of the green character, and which affects the creation of the sense of belonging. In all three cities, there is a pronounced communicational role of landscape through the research of the possibilities to physically connect open spaces by strategical positioning of new parks or by establishing new systems of pedestrian and cycling paths. Public open spaces of the city take over significant roles of socialization, recreation, but also of connecting cultural-historical sights. The ecological role of landscape achieved through, for example, new systems of water landscapes and the enhancement of biodiversity is not negligible either.

Gašparović and Sopina (2018) conclude that, at the same time, Zagreb has neither a landscape strategy nor a landscape plan. The new approach to the city planning (Master Plan 2003 and onwards) focuses on detailed standardization of fragments while neglecting the totality of the city, so many of the roles of landscape in the city planning are neglected. The planning of public spaces and amenities within the landscape system is being neglected, while mostly ecological aspects are being more significantly emphasized. The role of landscape in achieving free movement and vistas and separating certain functional units is replaced by the conditions of achieving a satisfying relationship between the built and unbuilt areas. The planning approach to landscape is becoming passive—planning protects the unbuilt areas from building, and landscape does not get enough of the other roles which are a prerequisite for its perception as an active factor of urbanity.

17.6 Conclusion

Landscape is one of the key factors in establishing the recognizable image of a city—its identity. City planning implies an integrated perception of space, its built and unbuilt parts. Landscape cannot be perceived and planned separate from its urban context. The relationship and influences of the city and landscape are mutual, and integral planning results in benefits for both aspects of space.

Konijnendijk (2010) points out that, nowadays, branding of the green city is becoming an increasingly important condition for achieving strategic planning ideas by which contemporary cities, due to great pressures of globalization, increasingly compete for money and attention, recognizability, innovation and creativity. This branding is achieved, among other things, through systems of city parks and other green spaces, areas which contribute to healthy and attractive urban surroundings.

Landscape strategies are becoming an essential part of the development of cities; they ensure preservation and enhancement of green capitals. Planning and implementation of clearly conceptualized landscape strategies of cities and their wider regional area are becoming an imperative in sustainable planning of the contemporary age (Gašparović and Sopina 2018).

Successful landscape planning of a city is based on establishing a clear landscape conception—a comprehensive idea of the entire landscape system of a city which implies clear goals (multiple roles of landscape) and measures (establishing identity factors) for its implementation. It is based on a landscape strategy—spatially and temporally determined sequence of planned measures whose aim is to achieve the goal (the landscape conception) based on continuing long-term planning.

The landscape strategy and the landscape concept determined in a planned way represent a landscape model. Its aim is to activate (to contribute to the active use/multiple roles of landscape for the city) and to be sustainable.

The conducted comparative analysis of the role model cities (Essen, Oslo and Stockholm) (Table 17.2) has resulted in the identification of the landscape model reflected in:

- recognizing landscape as an important formative element of the city;
- overcoming the boundaries between the city and the region and their territorial intertwining;
- determining the starting points of the landscape strategy for the whole city level;
- networking of unbuilt spaces into a recognizable system of green infrastructure;
- 'conquering' brownfield areas for new public open spaces;
- including the dynamic features of natural systems in the planning process.

By applying the landscape model on the example of Zagreb and other cities, it is possible to protect and enhance their recognizable landscape image or inherited identity.

Table 17.2 Comparison of the landscape planning characteristics of the analysed cities

		Essen	Oslo	Stockholm
Strategic documents of city landscape planning		Master plan 'Open space creates city space' (new ways to the water) 2007	Grøntplan for Oslo (the plan for blue–green structure in the city zone) 2009	The Green Walkable City (future strategies plan for the landscape planning of the city) 2013
Landscape concept recognizability factors		– Networking of green areas and open spaces along three axes – Connection of countryside spaces in the Ruhr Valley with the Emscher Valley and the regional open space system of the Emscher Landscape Park	– Radial networking of 'green fingers' between the Oslo fjord and Marka urban forests – Establishing blue infrastructure and a system of public parks – Establishing a system of communications which networks the blue–green infrastructure	– Preservation of radial green wedges network continuing throughout the region
The role of landscape In city planning/the aims of landscape planning	Formative	– Networking of recognizable landscape features (of the river)	– Strengthening of the recognizable character of the city as a 'blue–green city between the hill and the fjord' – Determining measures for the establishment of the integral landscape image of the city	– Protection and development of the city's green character

(continued)

Table 17.2 (continued)

		Essen	Oslo	Stockholm
	Communicational	– Networking of landscape areas via pedestrian and cycling paths	– The footpaths connecting the various landscape areas	– Potential for green areas to physically connect districts via the strategic placement of parks and public functions along natural green corridors – Accessibility and quality of green space for recreation
	Functional	– Networking of cultural-historical sites	– Enhancing the fulfilment of the citizens' needs for green recreational areas	– Providing public space – Green space facilitating a sense of belonging and social interaction (cohesion and inclusion)
	Ecological	– Establishing a system of water landscapes	– Establishing urban development in accordance with ecological principles	– Supporting the ecological infrastructure, pertaining to biodiversity and ecosystem services – Providing local climate mitigation and microclimate, and storm water management

References

Beatley T (2012) Green cities of Europe: global lessons on green urbanism. Island Press, Washington DC

Benedict MA, McMahon ET (2006) Green infrastructure linking landscapes and communities. Island Press, Washington DC

Bojanić Obad Šćitaroci B, Matuhina N (2012) Landscape urbanism—new spatial paradigm. Prostor 20(43):106–117

Corner J (2006) Terra Fluxus. In: Waldheim C (ed) The landscape urbanism reader. Princeton Architectural Press, New York, pp 21–33

Czechowski D, Hauck T, Hausladen G (eds) (2015) Revising green infrastructure: concepts between nature and design. CRC Press Taylor & Francis, Boca Raton

Falleth E, Saglie IL (2012) Planning a Green Oslo. In: Luccarelli M, Gunnar Røe P (eds) Green Oslo, visions, planning and discourse. University of Oslo, Norway

Gašparović S, Sopina A (2018) The Role of landscape in planning the City of Zagreb from the early 20th to the early 21st century. Prostor 28(55):132–145

Jorgensen K et al (eds) (2016) Mainstreaming landscape through the European landscape convention. Routledge, New York

Kleinebrahm T, Lipsius K (2015) Essen: a green city of North-Rhine Westphalia. In: Paper presented at 6th global forum on urban resilience and adaptation. Bon

Kong L (2012) Break the green belt? The differences between green belt and its alternative green wedge—a comparative study of London and Stockholm. Master of science programme in urban planning, Karlskrona. Available via Diva-portal. Retrieved from http://www.diva-portal.org/smash/get/diva2:832039/FULLTEXT01.pdf. Accessed on 15 Jun 2018

Konijnendij CC (2010) Green cities, competitive cities—promoting the role of green space. Paper presented at IFPRA world congress—city Branding, Hong Kong

Littke H (2015) Planning the green walkable city: conceptualizing values and conflicts for urban green space strategies in Stockholm. Sustainability 7:11306–11320

Lendholt W (1970) Funkcije mestnega zelenja. In: Ogrin D (ed) Zbornik Zelenje v urbanem okolju. Ljubljana

Luccarelli M, Gunnar Røe P (eds) (2016) Green Oslo: visions, planning and discourse. Routledge, NY

Lynch K (1960) The image of the city. The MIT Press, London

Mc Harg I (1969) Design with nature. Natural History Press, New York

Nelson A, Malarstrand N (n.d.) Stockholm—city of water. Available via University of Washington. Retrieved from https://depts.washington.edu/open2100/Resources/1_OpenSpaceSystems/Open_Space_Systems/Stockholm_Case_Study.pdf. Accessed on 15 Jun 2018

Ståhle A (2002) Urban planning for a quality dense green structure. Paper presented at Stockholm sociotop map and park programme, green structures and urban planning 6th management committee meeting and working group meetings, Milan, Italy, 6–8 Oct 2002

Timmermans W et al (eds) (2015) The rooted city: European capitals and their connection with the landscape. Blauwdruk Publishers, Wageningen

Waldheim C (2006) Landscape urbanism. Princeton Architectural Press, New York

Chapter 18
Reuse and Revitalisation of Contemporary City Areas: Structural and Functional Transformation of Brownfield Sites

Examples from Zagreb and London

Tihomir Jukić and Georgia Butina Watson

Abstract Urban regeneration and revitalisation of brownfield sites and urban areas are the result of functional urban transformations and an approach to creating new methods for spatial change. This chapter presents various models of urban transformations with particular focus on ownership and the functional use of buildings and urban areas; strategic planning mechanisms for shaping and transforming brownfield areas; the role of citizens' participation and engagement in the decision-making processes; and potential funding models and the value of public/private partnerships. First, the paper presents key theories and practice-based examples of brownfield regeneration and revitalisation schemes, using cases from London and Zagreb that have been implemented or are waiting to be realised. Discussions are also framed by a critical analysis of the two planning systems that underpin these interventions. Special attention is given to the role of "urban projects" as potential models for the functional and structural revitalisation of brownfield areas. Such projects, it is argued, are of special interest to cities today where the "city" participates in the co-creation and realisation of such projects through land or building ownership or as a public infrastructure investor with the overall purpose of improving the quality of life of its citizens. The analysis also discusses the value of area regeneration models and presents examples from London and Zagreb as contrasting socio-political and planning systems.

T. Jukić (✉)
Faculty of Architecture, University of Zagreb, Zagreb, Croatia
e-mail: tjukic@arhitekt.hr

G. Butina Watson
Faculty of Technology, Design and Environment, Oxford Brookes University, Oxford, UK
e-mail: gbutina-watson@brookes.ac.uk

M. Obad Šćitaroci et al. (eds.), *Cultural Urban Heritage*,
The Urban Book Series, https://doi.org/10.1007/978-3-030-10612-6_18

Keywords Industrial heritage · Brownfield sites · Urban revitalisation · Urban projects

18.1 Introduction/Motivation for Research

Vacant industrial land, warehouses, and military barracks, popularly called *brownfields* in the professional literature (CABERNET 2006), present enormous potential for further urban development and raising the quality of life. The process and model of their transformation, as well as their functional and structural tailoring to the contemporary urban tissue, is questionable. Urban regeneration of brownfield sites represents a process of revitalising the consequences of spatial transformations, as well as ensuring the desired direction of the sites' changes (Jukić 2016). This chapter suggests that quite often such initiatives are conducted on a project-by-project basis, but there are opportunities, as evidenced in the case of London, that a more strategic view can be developed which also links to the broader strategic planning of cities. Equally, the case of London will show the link between urban planning and the political commitment to deliver housing, employment, and other services.

On the other hand, disorganisation, the lack of a systematic approach, a non-existent unique information system with all the required information on brownfield sites (Spirić 2015) (location, ownership, utilisation and function of the site, the site's value and significance at the city/regional level) leads to the present use of completely different brownfield transformation models in the same or similar situations, and Zagreb is a good example of this.

18.2 Research Aim and Hypothesis

The aim of the implemented research was to evaluate various brownfield transformation models (Đokić and Sumpor 2010) and to determine urban guidelines for the related process based on comparisons with positively evaluated urban practices of European cities, with particular reference to London and Great Britain as one of the countries with the longest tradition of urban planning and spatial management, as well as a stable legal system.

The starting point is the hypothesis that economic, political, cultural and traditional differences, as well as relations to historical heritage, result in different transformation approaches and models of former brownfield sites in Zagreb and London. The assumption is that considering the influence of the capital and politics, it is impossible to define for them a unique brownfield transformation model, especially concerning their significance in space, size, the degree of preservation and ownership. However, given the complexity of urban planning and the length of time needed to complete developments, it is useful to briefly discuss different periods within which, for example, London brownfield transformations have been achieved and are achieving London's broader socio-economic and civic goals.

18.3 Experience and Knowledge of Brownfield Site Revitalisation

The starting point of the research consists of theoretical texts as well as existing experience and knowledge (Svirčić Gotovac 2010; Smith 2002) on the revitalisation of brownfield sites in London and Zagreb relating to examples and locations on the basis of which the transformation was completed, is on-going or is in the planning stage.

In addition to desk study, research carried out in London also utilised primary fieldwork and original research carried out on behalf of the UK Government and discussions with key experts in the field (Butina Watson et al. 2006).

In the UK, a brownfield site is defined as "previously developed land that has a potential of being redeveloped" (Dixon 2006). It is the UK's priority and commitment to use such land for housing and other social, cultural and employment uses. As some of the land is contaminated, the UK has developed innovative methods for decontamination and is a world leader in using engineering methods to facilitate brownfield developments as well as in forming public/private partnerships to implement large regeneration projects with physical, social and economic benefits.

18.4 Applied Research Methods

Desk research and the comparative method were utilised in the research. Brownfield transformation models were investigated in relation to the ownership and utilisation of the site, the planned transformation strategy of these areas, and citizens' participation in the decision-making process. In addition, potential financing models (Kojaković 2016) and the use of public/private partnership were explored.

Where possible, the research also used post-occupation surveys using interviews and case-study evaluations regarding broader societal benefits.

Potential transformation models of vacant brownfield sites were examined using examples from Zagreb considering the already implemented transformation processes in relation to those which were in the planning stage. This was performed from the aspect of tracking changes in the sites' function and building structure.

With the aim of forming a transformation model, three types of structural transformation and four types of functional changes of the brownfield sites were recognised in the process of their transformation (Table 18.1). Therefore, it is possible to form a matrix of combinations of the mentioned types and obtain the most frequently used and/or planned models.

Table 18.1 Transformation models of the constructed structure and site function—as a basis for shaping the used or planned matrix of the brownfield transformation model

Structural and functional transformation models			
Structural transformation		Functional transformation	
1	Structure completely preserved	A	Industrial function—to industrial function
2	Structure partially preserved + new	B	Industrial function—to mixed function (various other functions)
3	Existing structure is removed—new structure	C	Industrial function—to social (public) function
		D	Industrial function to residential function

The aim is to show which transformation models of brownfield sites were used in already completed transformations and which models are planned during the future realisation of strategic urban projects. Based on the example of London, the investigation will focus on the influence of political decisions and the accompanying legal regulations on the transformation processes of the city. Focus will be placed on the characteristics of the implemented brownfield transformations for time periods dating from the 1970s to contemporary examples.

18.5 Zagreb Versus London and Vice Versa—Selected Research Examples

Despite the fact that the two cities discussed in this paper have different political, historical and developmental profiles, and that the scale of interventions is much larger in London, we have been able to derive some key principles that could lead to the transfer of theoretical knowledge as well as to practical interventions and solutions. We shall first provide research on Zagreb, as a smaller city, and then focus on London, outlining some of the key values of brownfield regeneration interventions achieved so far and compare future directions as well as the potential for the transfer of approaches.

18.5.1 Zagreb Brownfield Transformation—Realised/Planned in the Future

In the majority of European cities, well-organised and planned urban regeneration procedures are implemented with clear actors in the process, and, in Zagreb in particular, the procedures are implemented from case to case in the form of minor inter-

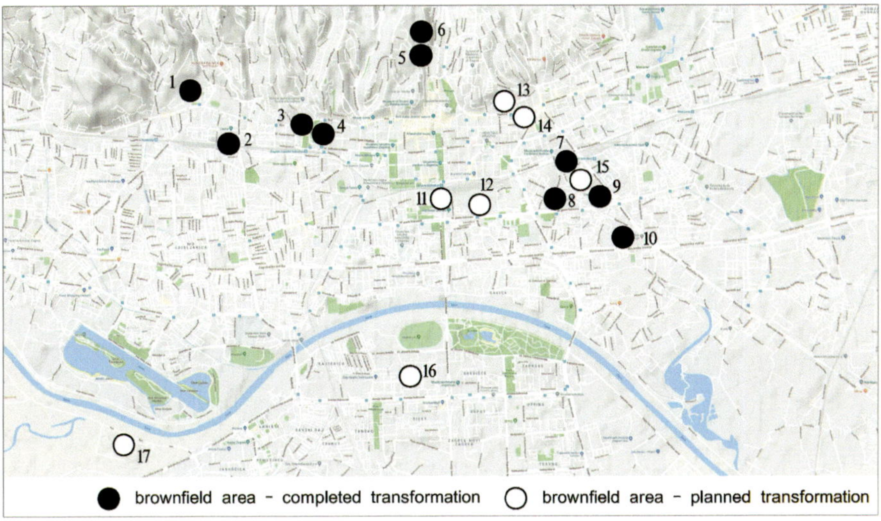

Fig. 18.1 Analysed brownfield sites in Zagreb: 1. Ciglana Črnomerec (brick factory); 2. Lauba culture centre; 3. Kamensko; 4. Rudolf barracks; 5. Kaptol centar/Astra; 6. Glyptotheque of the Croatian Academy of Sciences and Arts; 7. Zavrtnica—paper factory; 8. NIK; 9. Chromos; 10. TEŽ—lighting fixture factory; 11. Paromlin—steam mill; 12. Gredelj; 13. Military hospital; 14. Badel distillery; 15. Zagrepčanka; 16. Zagreb International Fair; 17. Blato hospital (*Source* Google maps)

ventions required for normal living in the city. For this same reason, the research is performed using the examples of two completely different cities—with different cultures, history, legal system, and economic and political organisation in order to use the acquired experience for finding new solutions for regenerating brownfield sites.

Brownfield transformations were investigated at the level of Zagreb (Fig. 18.1), including separately those that have been implemented already (10 sites) and those waiting to be realised at the level of strategic urban projects (Jukić and Smode Cvitanović 2011) of the Zagreb Development Strategy 2020 (City of Zagreb 2017).

18.5.1.1 Realised Transformation (Zagreb)—10 Projects

Sites on which brownfield transformations had been implemented previously were analysed. Bigger production complexes were analysed as well as vacant industrial buildings. The investigation was conducted on 10 sites with the goal of obtaining mutual characteristics which are the result of the implemented transformation (Table 18.2):

- it is interesting that the majority of the realised sites are under private ownership (8/10);

Table 18.2 Zagreb—Implemented brownfield regeneration—realised by 2017

Site	Ownership	Present function	Heritage protection	Preserved structure	Function		Model (Table 18.1)
					Original	Planned by GUP	
Rudolf Barracks	State/City of Zagreb	Ministry administration	Yes	Yes	Military	Business function K1	1C
Glyptotheque of the Croatian Academy of Sciences and Arts	State	Glyptotheque gallery	Yes	Yes	Production	Social function D	1C
Lauba	Private	Gallery, culture	Yes	Yes	Production	Social function D	1C
Ciglana Črnomerec (Brick factory)	Private	Trade	Yes	Yes	Production	Business function K1	1A
TEŽ—Lighting fixture factory	Private	Temporary business warehouses	Yes/No Partially	Yes	Production	Business function K1	1A
Zavrtnica—Paper factory	Private	Temporary business warehouses	No	Yes	Production	Business function K1	1A
Kaptol-centar/Astra	Private	Trade	No	No	Production	Business function K1	3A
Chromos	Private	Trade—business	No	No	Production	Business function K1	3A
NIK	Private	Bank	No	No	Production	Mixed business M2	3A
Kamensko	Private	Residential/business	No	No	Production	Mixed M	3D

- most of the sites are not protected industrial heritage (7/10);
- even though they are not protected under industrial heritage, the structure of the buildings was in most cases (6/10) partially or completely preserved;
- for all the sites, the original function was industrial, production or military (10/10), and, according to the Zagreb City Master Plan (GUP) (City of Zagreb 2016) the plan was to preserve the business function (K1) of the majority of the sites (6/10).

18.5.1.2 Planned Transformation (Zagreb)—7 Strategic City Projects

Seven out of 15 locations (7/15), listed in the Development Strategy of the City of Zagreb 2020 (City of Zagreb 2017), are included under the category of brownfield sites, that is, transformed areas of former production floors or vacant premises (Table 18.3). Using this procedure, the planners want to highlight that it is more important to plan a city by transforming brownfield sites than to expand the city on the remaining non-built sites.

There are certain rules that can be identified if we analyse these seven examples, and they are:

- all of the sites are owned or are in the possession of the City of Zagreb or *Zagrebački holding* (an institution specialising in the management of city property and the city utilities system). Only one of the sites is owned both by the city and the state, and it is the complex of *Gredelj*, a former factory for manufacturing coaches, with a smaller part consisting of machinery and tracks of *Hrvatske željeznice* (Croatian Railways);
- except for one site (*Zagreb International Fair*), all the other sites are derelict and out of use (6/7), and the Fair itself is under temporary ownership (Jukić and Vukić 2015);
- for each of the locations (7/7), it is envisaged to preserve the existing structure both in full (3/7) or partially (4/7);
- these are mainly protected buildings or protected urban complexes (6/7).

In the case of only one of seven sites (1/7) is it envisaged to partially preserve the existing function (*Zagreb International Fair*), which is normal given the century-old tradition of organising large industrial shows or industrial and technological shows.

- even though all of the sites once had an industrial-production purpose, in most cases the planned transformation is substituted with a mixed and mainly business function (M2 according to GUP) which includes housing in the planned areas, but in a smaller scope;
- on only two sites (2/7) is the industrial function preserved.

Table 18.3 Zagreb—planned transformation, strategic city projects in compliance with the Development Strategy of Zagreb 2020

Site	Ownership	Utilisation (status)	Heritage protection	Planned preservation of the structure	Planned preservation of function	Planned function GUP	Model (Table 18.1)
Badel	City/private	Disused/temporary utilisation	Yes	Partially	No	Mixed business function M2	2B
Gredelj	Zagrebački holding/state railways	Disused	Protected complex	Partially	No	Mixed business function M2	2B
Paromlin	City	Disused	Yes	Yes	No	Mixed business M2	1C
Zagreb international fair	Zagrebački holding/ZV doo	Temporary utilisation	Protected complex	Partially	Partially	Business function K1	2A
Zagrepčanka	Zagrebački holding	Disused/temporary utilisation	Yes	Partially	No	Business function K1	2B
Military hospital	City of Zagreb	Disused	Yes	Yes	No	Social function D	1C
Blato hospital	City of Zagreb	Disused	No	Yes	No	Mixed business function M2	1C

18.5.2 London Brownfield Transformation Completed and Planned

Post-war plans focused on the decentralisation both of people and industry in designated new town locations under the Town and Country Planning Act of 1946 (Butina Watson et al. 2006). Despite these measures, London continued to grow quickly as a result of its national and international position, putting pressure on the available land for development that was shrinking as a result of population growth and the restrictions placed by the greenbelt.

Some of the structural changes of industry as well as changes in international maritime trade led to the closure of the London docks which closed one by one as a result of the introduction of larger ships that could no longer be serviced by the outdated infrastructure. This provided a vast area of brownfield land and buildings for development and regeneration.

18.5.2.1 Economic and Legal Foundation of Brownfield Transformation

Major political, economic, technological, planning and other changes took place during the 1980s, which brought about large-scale regeneration initiatives of brownfield land and buildings. Political changes came as a result of the Conservative government under Margaret Thatcher that promoted private capital investments rather than Local Authority and Central Government funding that were linked to the former Labour government policies. This turn around in policy led to other changes, of which a significant shift was the deregulation of the London stock exchange market in October 1986 which allowed foreign banks to trade in London. This deregulation and the need for large (100 m $\times$ 100 m) trading floors linked to the operational requirements of the banks created a great need for new types of buildings (Butina Watson 1993). The City and its land became the most expensive property in London, and in the world. Instead of demolishing historic buildings, a political decision was made to regenerate large areas of the discussed docklands and buildings. In addition, any remaining brownfield land in the historic areas was also designated for brownfield regeneration.

In parallel with the regeneration of historic city sites, and to meet the demand for international banking as a result of financial deregulation, under government initiatives legislation was passed in 1981 (City of London 1980) to set up the London Development Corporation with powers to regenerate the disused docks and land associated with maritime trade, amounting to some 5100 acres of development land. In order to deliver new developments quickly, a number of innovative planning mechanisms were introduced. The setting up of the London Dockland Development Corporation gave the newly founded body similar powers to those of the New Town Development Corporations, which included land acquisition, creating public/private partnerships, planning matters facilitated by the introduction of simplified planning zones (bypassing the complex planning procedures of existing local authority planning practices) and enterprise zones with favourable tax reductions to attract private developers.

Regeneration of brownfield sites and buildings in England is regulated by the Town and Country Planning (Brownfield Land Register) Regulations 2017 (Country of England 2017) which requires each local planning authority to prepare, maintain and publish a register of previously developed (brownfield) land suitable for housing as well as other uses. This is also backed by the National Planning Policy Framework (NPPF) (Country of England 2012) which provides national guidelines (for England) for delivering sustainable communities. The NPPF national guidelines are translated by local planning authorities into Local Development Plans and Frameworks that are subject to community engagement and approval under the Localism Act of 2011 (Country of England 2011). The use of brownfield sites is also supported by sustainable cities and compact city theories (Jenks et al. 1996), emphasising the densification of certain urban areas, mixed-use developments, different models of housing and employment provisions, as well as access to integrated transport networks to reduce the need for car travel. As many of the brownfield sites are within the greenbelt, they also provide a valuable resource for a variety of mixed-use developments.

18.5.2.2 Transformation of Derelict Docks of London

The early re-use of buildings such as St Katherine Docks was the beginning of heritage-led regeneration, using brownfield sites. St Katherine Docks were built on the former St Katherine Hospital site, so they were already utilising brownfield land. The docks closed in 1968 after being heavily damaged during the Second World War air attacks. The regeneration started in the 1970s under the leadership of the Greater London Council and by the 1990s it had turned into a leisure and heritage basin containing leisure (marina and hotel), retail, office and housing uses. It was a great success, funded by public and private money, and showed the way forward for much larger and more ambitious plans.

Under the umbrella of the London Dockland Corporation, many disused docks and surrounding land with industrial buildings were scheduled for redevelopment: East India Docks, the Royal Docks, the Victoria and Albert Docks, the Limehouse Basin and other areas and buildings all directed London's urban growth towards the east part of the city. The government also helped in terms of various financial subsidies for cleaning up the contaminated land as well as committing itself to improving the public infrastructure—the road and rail systems—thus incorporating previous dockland into the historic part of the city (Fig. 18.2).

Various specific projects were defined, of which Canary Wharf, built on the land and buildings that used to bring in fruit and food from the Canary Islands, and the Isle of Dogs; the rest was part of the redevelopment of the East India Docks. The initial development was carried out by a Canadian firm, Olympia and York, from 1987 to 1992 when the firm went into bankruptcy and the UK government stepped

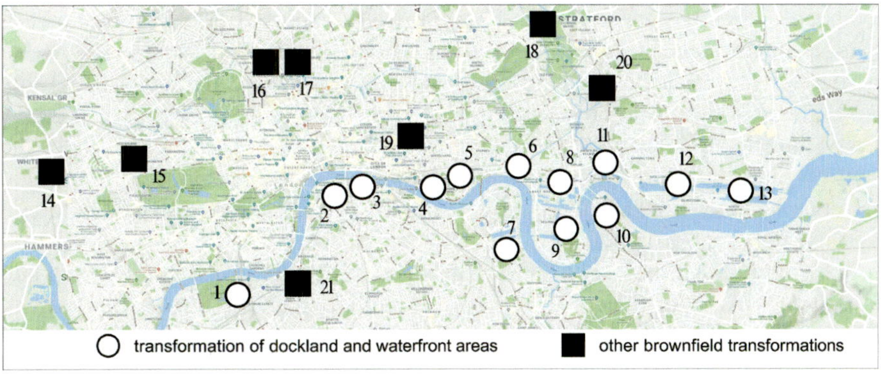

Fig. 18.2 Analysed brownfield sites in London: 1. Battersea Power Station; 2. Coin Street and OXO Tower Wharf; 3. Tate Group; 4. St Katherine Docks; 5. Tobacco Docks; 6. East India Docks; 7. Limehouse Basin; 8. Canary Wharf; 9. Isle of Dogs; 10. Millennium Dome; 11. Surrey Docks; 12. Royal Victoria Dock; 13. Albert Docks; 14. White City; 15. Paddington Basin; 16. St Pancras station; 17. King's Cross railway station; 18. The Queen Elizabeth Olympic sites; 19. Broad Street and Liverpool Street railway stations; 20. Stratford City; 21. Nine Elms (*Source* Google maps)

into save the project. First banks moved there in 1991, followed by various media and finance organisations, as well as residents seeking new types of accommodation. This highly controversial project was criticised for importing foreign (North American) architectural and urban design ideas that did not fit into London's historic townscape and for a lack of proper public consultation. However, it did generate government investments into improving public transport (Docklands Light Railway), roads and underground and rail extensions. The LDDC completed its work in 1998 and other projects continued as part of specific initiatives (Butina Watson et al. 2006).

Other brownfield regeneration projects followed, including the regeneration of the Limehouse Basin, Tobacco Docks (Wapping), the Royal and Surrey Docks and the Victoria and Albert Docks. Some of the regeneration schemes also include the Millennium Dome and the Queen Elizabeth Olympic sites under the umbrella of the Olympic Legacy, with the creation of sports facilities used during the Olympic Games of 2012, Linear Park along the Thames, and much-needed housing and other facilities. Other areas such as Stratford City, as well as the new Ebbsfleet Garden City are also spin-offs of the original London Docklands Development initiatives.

18.5.2.3 Examples of Transforming Central Brownfield Sites

In parallel to the brownfield regeneration work of the docklands, there were also other notable schemes, of which the Tate Modern and the Coin Street redevelopment stand out for their cultural and social value. What is today known as the Tate Modern (part of the Tate Group of museums and galleries) is the former Bankside Power Station built in 2 phases between 1947 and 1963 (closed in 1981) located on the

Southbank, directly opposite St Paul's Cathedral. It is linked to the historic part of London and its financial core in the City. The Millennium Bridge, designed by ARUP and the Foster Architects group links the two river banks. The building is listed and, like Battersea Power Station, was designed by Sir Giles Gilbert Scott. It represents a very innovative conversion from an industrial to a high art building, under the redesign intervention of Swiss architects Herzog and de Meuron. The conversion was completed in 2000 and later extended in 2016, funded by a variety of cultural grants and charitable donations.

It is today one of the most visited cultural buildings that are well integrated into the wider regeneration area of Southbank, particularly the Coin Street development. The whole regeneration area of brownfield buildings and land benefited from broad public consultation involving local community groups, the Coin Street Community Builders and the Coin Street Housing Cooperative groups (Ritchie and Thomas 2009). With a strong social mission to provide affordable housing, using creative cross-subsidy mechanisms between public, private and not-for profit investments, it is today known as one of the very successful examples of a sustainable and affordable housing area that also benefits from other facilities such as a craft centre (the former OXO Tower Wharf), a nursery, a community centre, arts and sports facilities and much needed open space. Housing is managed by five housing cooperative organisations. There is a good balance of private and affordable housing, which has created a strong sense of community. Both schemes, the Tate Modern and the Coin Street Community Builders work well together and have become an example of the social and cultural transformation of industrial areas and buildings, nesting between the National Theatre, the Hayward Gallery, the Royal Festival Hall and the Ballet Rambert. The whole stretch has become an important cultural quarter incorporating both high art and community projects.

18.5.2.4 Regeneration of Disused Railway Corridors

One of the largest schemes in the City of London during the 1980s was the regeneration of disused railway land of Broad Street and Liverpool Street railway stations, known as the Broadgate development (Butina Watson 1993). This transformation was also the result of the British Rail initiative to consolidate and redevelop/dispose of some of its brownfield sites as a result of the modernisation of the rail system as well as for the purpose of consolidating its operations. Located on the boundary of the City of London and the London Borough of Hackney it was a public/private partnership between British Rail and Rosehaugh Plc and Stanhope Properties, remodelled in 1982. The redeveloped scheme on some 12 hectares is a mixed-use development of ground-floor retail space and offices above, and includes the refurbishment of the heritage of Liverpool Street station. The design brief was prepared by the City Corporation of London Planning Department which increased the plot ratio from 3:1 to 5:1, using the morphological and typological reference to Finsbury Circus, creatively executed by Peter Foggo and ARUP Associates in the shape of a central open space used for various events and performances. Some of the profits from the development went into social housing improvements in the area. This was not an ad hoc intervention, but a strategic intervention that used the existing context of the area to repair the city's fragmented urban framework.

Within the closer parameters of the historic city, much of the brownfield railway land was also developed. Three distinctive projects were earmarked: King's Cross railway station and surrounding areas; St Pancras station and Paddington Basin (the last one is still being constructed). The modernisation of rail operations created surplus land and buildings ready to be utilised in the regeneration of the transport infrastructure, assisted also by the decision made by Transport for London to relocate the Eurotunnel link from Waterloo station (south of the River Thames) to the newly refurbished St Pancras and King's Cross complex of stations. As King's Cross station was damaged by fire, it also offered opportunities to completely update and refurbish the buildings and infrastructure. St Pancras, a listed heritage building, was empty between 1990 until its re-opening in 2007 when it became St Pancras International. St Pancras is also a listed building, designed by George Gilbert Scott (Bishop and Williams 2016).

The whole complex that includes both stations and their surrounding areas covers some 27 hectares of brownfield land and buildings. It was a high-profile public/private initiative, involving strong public participation. The regeneration area today covers some 20 new streets, 10 new public parks, 10.5 ha of open space, 1900 new homes (50% affordable) and 50 mixed-use buildings. In 2015 the stations were sold to an Australian pension fund as part of their portfolio of investments.

18.5.2.5 Other Brownfield Regeneration Projects

Other brownfield regeneration projects such as Paddington Basin (Paddington railway station disused land), White City, Nine Elms, Battersea Power Station, Ebbsfleet New Garden City, and the rest of the Thames Gateway such as Barking East, continue to shape the structure of London, using brownfield sites. The most ambitious of them all, the Crossrail Project (costing £15.5 billion) and Thames Link are cutting right through most parts of London in order to modernise the infrastructure and provide housing, employment, and other services so much needed in a constantly growing city.

18.6 Brownfield Land Register Drafting: London/Zagreb

London as the nation's capital has some 40,000 registered brownfield sites located within the greenbelt that is within the Greater London Authority (GLA). Other large sites are owned by Transport for London, National Health Service providers, and other public bodies. The Brownfield Land Register is held by the Greater London Authority and the Ministry for Housing, Communities and Local Government

18.8 Conclusion

The results of the implemented research should contribute to the better organisation and realisation of brownfield transformations in broader city centres, taking into account implemented protection measures, both of individual buildings and entire complexes of brownfield sites, and for the purpose of achieving a higher quality of life in the city. There are various procedures and models for revitalising the preserved heritage—vacant industrial premises—and one of them is a so-called 'city project', that is, a project of special importance for the city. Organised by the city, the procedure of preparing and monitoring the projects' realisation (the example of London) according to transparent procedures should enable systematic and controlled transformation of brownfield areas whose size and vicinity to the urban centres offer great opportunities for public or social functions and services. As expected, a uniform answer was given to the question of whether a unique brownfield transformation model is applicable and whether the procedure is unique, but implementation models vary; the question was also raised about how far the procedure is specific in relation to other locations and examples of brownfield sites of no particular importance for the city. However, the two cities are wholly different (Zagreb and London), based both on their size and experience acquired through the implementation of the brownfield transformation processes. For decades, London has had a developed legal and financial system which follows the transformation process, and Zagreb is only at the start of organising the processes and setting up a comprehensive brownfield land register.

Despite the fact that the two cities discussed in this paper have different political, historical and developmental profiles, and that the scale of interventions is much larger in London, the chapter has been able to derive some key principles that could lead to the transferability of theoretical knowledge as well as to practical interventions and solutions.

Acknowledgements The research is a part of the scientific project "Heritage Urbanism—Urban and Spatial Planning Models for Revival and Enhancement of Cultural Heritage". It is financed by the Croatian Science Foundation [HRZZ-2032] and carried out at the University of Zagreb, Faculty of Architecture.

References

Bishop P, Williams L (2016) Planning, politics and city making: a case study of King's Cross. RIBA Publishing, London

Butina Watson G (1993) The art of building cities: urban structuring and restructuring. In: Hayward R, McGlynn S (eds) Making better places—urban design now. Butterworth Architecture, Oxford, pp 64–72

Butina G, Brownill S, Carpenter J et al (2006) Thames Gateway evidence review. ODPM, London

Dixon T (2006) The role of the UK development industry in brownfield regeneration. CL:AIRE, SUBR:IM, London

Đokić I, Sumpor M (2010) Brownfield redevelopment: issues in Croatia. Privredna kretanja i ekonomska politika 123:57–86

CABERNET (2006) Sustainable brownfield regeneration. CABERNET network report, University of Nottingham. Retrieved from http://www.palgo.org/files/CABERNET%20Network%20Report%202006.pdf. Accessed 10 Jun 2017

City of London (1980) Local government, planning and land act 1980. Retrieved from http://www.legislation.gov.uk/ukpga/1980/65/pdfs/ukpga_19800065_en.pdf. Accessed on 10 Jun 2017

City of Zagreb (2017) City of Zagreb development strategy up to 2020. Available via Official Site of the City of Zagreb. Retrieved from https://www.zagreb.hr/razvojna-strategija-grada-zagreba/47768. Accessed on 10 Jun 2017

City of Zagreb (2016) Zagreb city master plan. Available via Official Site of the City of Zagreb, Retrieved from https://www.zagreb.hr/generalni-urbanisticki-plan-grada-zagreba-gup/89066. Accessed on 10 Jun 2017

Country of England, Town and Country Planning (2017) Town and country planning (Brownfield Land register) regulations 2017. Retrieved from http://www.legislation.gov.uk/uksi/2017/403/pdfs/uksi_20170403_en.pdf. Accessed on 10 Jun 2017

Country of England (2012) National planning policy framework (NPPF). Retrieved from https://assets.publishing.service.gov.uk/government/uploads/system/uploads/attachment_data/file/6077/2116950.pdf. Accessed on 10 Jun 2017

Country of England (2011) Localism act of 2011. Retrieved from http://www.legislation.gov.uk/ukpga/2011/20/pdfs/ukpga_20110020_en.pdf. Accessed 10 Jun 2017

Jenks M, Burton E, Williams K (eds) (1996) The compact city: a sustainable urban form?. E & FN Spon, London

Jukić T, Smode Cvitanović M (2011) Zagreb – Gradski projekti u postupku planiranja grada. Sveučilište u Zagrebu, Arhitektonski fakultet. Zagreb. ISBN 978-953-6229-79-6(AF)

Jukić T, Vukić F (2015) Zagreb fair—spatial model of transformation of single-purpose industrial heritage into a multi-functional urban area. In: Obad Šćitaroci M (ed) International scientific conference "Cultural heritage—possibilities for special and economic development—proceedings". Faculty of Architecture, University of Zagreb, Zagreb, pp 520–525

Jukić T (2015) Urban potential of Zagreb brownfield regeneration network. In: Ognen M, Alessandro A (eds) Projects for an inclusive city—social integration through urban growth strategies, SINERGI, pp 74–85. Available via City of Skopje. Retrieved from http://www.skopje.gov.mk/Uploads/Sinergi/SINERGI_BOOK_ONE-Projects%20for%20an%20Inclusive%20City.pdf. Accessed on 7 Jun 2018

Jukić T (2016) Urbana regeneracija, posljedica ili preventiva/Urban regeneration—consequence or prevention. In: Korlaet A (ed) Urban regeneration strategies—proceedings. Croatian Institute for Spatial Development, Zagreb, pp 50–57

Kojaković V (2016) Sustainable redevelopment of cities—EBRD as a catalyst for successful urban regeneration projects. Public lecture presented at the panel discussion Zagreb može oživjeti zapuštene gradske prostore, Arcotel Allegra, Zagreb, 10 Nov 2016

Matković I (2016) Strategija urbane regeneracije—urban regeneration strategy. In: Korlaet A (ed) Urban regeneration strategies—proceedings. Croatian Institute for Spatial Development, Zagreb, pp 8–13

Ritchie A, Thomas R (2009) Sustainable urban design—an environmental approach. Taylor & Francis, London

Smith N (2002) New globalism, new urbanism: gentrification as global urban strategy. Antipode 34(3):427–450. https://doi.org/10.1111/1467-8330.00249

Spirić A (2015) Spatial criteria in urban renewal of industrial brownfield sites. Građevinar 67(9):865–877

Svirčić Gotovac A (2010) Aktualni regeneracijski i gentrifikacijski procesi na primjeru Zagreba. Sociologija i prostor 48(2):197–221

Chapter 19
Walkspace as Cultural Heritage Within Urban Landscape

Tamara Zaninović, Garyfalia Palaiologou and Bojana Bojanić Obad Šćitaroci

Abstract The focus of this research is to explore the urban landscape through the promenade notion and contemporary walkability strategies to connect open spaces into a *walkspace system*. The research on urban public spaces seen as a walkspace system points out specific models of alternative urban heritage. These models are characterised by the presence of diverse cultural heritage and pedestrian connections which should be recognised in current design and city planning procedures. The aim is to create awareness of heritage values in practices of everyday life using public space as a mediator and spatial networking as a planning criterion. The identified walkspace models came out of case study comparisons in five cities: London, Barcelona, Budapest, New York and Madrid. The case studies represent diverse urban landscapes as pedestrian streets, boulevards and linear urbanscapes. These examples confirm that streets are not just traffic corridors and show ways in which streetscapes form walkspace systems in different scales. Pedestrianisations, landscaped streets, historic park streets and urbanscape parks are strategies which interconnect cultural heritage and create new heritage of contemporary promenades through public space design. Walkspace systems are the basis for heritage urbanism approach as means of achieving vitality and quality of public space in heritage revitalisation.

Keywords Walkspace system · Urban landscape · Public space · Contemporary promenades · Streetscape

T. Zaninović (✉) · B. Bojanić Obad Šćitaroci
Faculty of Architecture, University of Zagreb, Zagreb, Croatia
e-mail: tmaric@arhitekt.hr

B. Bojanić Obad Šćitaroci
e-mail: bobojanic@yahoo.com; bbojanic@arhitekt.hr

G. Palaiologou
School of Architecture, Building and Civil Engineering, Loughborough University, Loughborough, Leicestershire, UK
e-mail: g.palaiologou@lboro.ac.uk

M. Obad Šćitaroci et al. (eds.), *Cultural Urban Heritage*,
The Urban Book Series, https://doi.org/10.1007/978-3-030-10612-6_19

19.1 Introduction

This chapter discusses the concept of *walkspace systems* in relation to the heritage urbanism approach and international timely efforts which aim to promote heritage as a driver of urban sustainability.[1] The walkspace concept was introduced by Marić and Bojanić Obad Šćitaroci (2012) to highlight the cultural potential in understanding, planning and designing walkable spaces as systems of interconnected and accessible public spaces within the urban landscape. It treats public spaces, and specifically promenades, as cultural sites of significance in terms of both authorised and alternative heritage values (Smith 2006). It aims to equally address planned/designed qualities of cultural and mixed (cultural and natural) landscapes, as well as the way pedestrians experience and use the urban landscape.

Walkspace (two-dimensional) refers to the spatial features of a walkscape (three-dimensional). In contemporary landscape theory and for the purpose of this research, walkspaces are considered as areas with the possibility for public pedestrian movement, while promenades are types of walkspaces which are designed to support the dominance of pedestrians in urban space and have the visual characteristics of 'walkscapes' or areas in which we can walk and perceive the scenery.

The walkspace concept emphasises properties pertaining to spatial systems, such as street networks (connectivity, accessibility, permeability, configuration) and spatial layout (scale, street width, building and urban block footprint size, spatial arrangement). As such, it aims to acknowledge walkscapes as a method and tool for creating systems within the urban landscape, rather than disconnected locations. It follows that walkspace is the subject of spatial design and is preoccupied with spatial relationships, while walkscape addresses the overall physical setting of walking spaces within urban fabrics.

This walkspace research looks at urban streets and promenades which can be considered as a type of 'ordinary' heritage—a term explained in Chap. Preface. Streets, as public places of connecting the urban tissue, are recognised cultural heritage as either historical patterns in urban cores or as public places around architectural heritage. Streets constitute the dominant percentage of urban landscapes and mobility systems and are important at both local and regional levels. The main research questions are: (1) how and when are the streetscapes and promenades a type of cultural heritage? (2) In which ways walkspace and walkability strategies change urban tissue and cultural heritage?

The hypothesis is that public space designed and planned as a 'walkspace system' can form the base for sustainable urban landscape and heritage urbanism approach by raising the quality of urban life and vitality of public space and cultural heritage. The chapter examines the ways in which public space, specifically streets, can be used as urban heritage mediator and spatial networks as a planning criterion. It argues

[1] UN 2030 Agenda for Sustainable Development was adopted at the UN Summit in September 2015 with 17 Sustainable Development Goals among which is goal 11 to make cities inclusive, safe, resilient and sustainable with the 11.4 target to strengthen efforts to protect and safeguard the world's cultural and natural heritage.

that *walkspace systems* should be recognised in current design methods and city planning procedures. The research methods address the complexity of urban network by recognising multiscalar characteristics of urban landscapes. The results contribute to reaffirmation, redefinition and revision of the 'promenade' notion through the concept of a *walkspace system* by:

- confirming and further contributing to the movement of planning streets as public places for people, where streets are considered urban landscapes rather than motorised corridors;
- pointing out specific models of alternative urban heritage which come out of the case studies of walkspace systems characterised by the presence of cultural heritage with promenade experience and characteristics.

There are numerous conferences[2] (UN Habitat 2013; WALK21 2006), books (Table 19.1) and projects[3] suggesting that cities overwhelmed by cars need to raise the SOS awareness for enhancing public walkspaces in human scale (Mehta 2013). 'Place-making' strategies and studies are examples of this warning (Table 19.1). They use more 'bottom-up' than 'top-down' strategies for revitalisation of existing urban landscapes with a goal of turning public space into public places (similar to turning a house into a home). Cultural heritage issues in these revitalisation processes are delicate and often bypassed, but walkspaces and cultural heritage should be intrinsic to each other. Historical urban cores in cities, especially in the European context, consist of what used to be defined walled areas which gradually developed into pedestrian zones or shared spaces with main cultural and tourist sites (Gehl and Gemzøe 2004). These zones are a reaction to traffic problems and together with the tradition of promenade culture from the rise of the nineteenth-century boulevards (MacDonald et al. 2001) and public gardens represent the starting points for this research about walkability strategies and models.

19.2 Theoretical Background

The walkspace is a basic concept in the heritage urbanism approach since the focus of walkspace research is on public space enhancement where landscape urbanism, infrastructural urbanism and space syntax theory intersect. Unlike walkability models, which are used to predict how easy it is for people to reach the places by walking (Dhanani et al. 2017), the walkspace and walkscape concepts are related to spatial forms of pedestrian public places, which play a core functional role in the city as a whole. The main open public places, such as streets, squares and parks, are arranged into different combinations within urban fabrics. These combinations are

[2]Conferences such as Walk21 which is held annually since 2000, Future of Places Conference which is held annually since 2013 or City Street Conference which is held annually since 2016.

[3]List of projects can be followed online on landscape magazines such as Landezine or Project for Public Spaces websites.

Table 19.1 Systematisation of streetscape research and approaches

Research topics and aspects	References (publication title, author, year; ordered by significance in relation to the research topic/aspect)	Conclusion on streets as walkspace from heritage perspective
Issues, overview and dimensions of streets in urban design and cities with evident multidisciplinary character	• On Street, Anderson, S., 1978 • Urban Design Reader—Street and Squares, Moughtin, C., 2003 • Street and Patterns, Marshall, S., 2005	Urban characteristics and complexity of streets
Perception and urban research correlating cities and streets	• The image of the City, Lynch, K., 1960 • View from the Road, Appleyard, D., Lynch, K., Myer J., 1966 • Space is a Machine, Hiller, B., 2000	Streets as paths of connection characterised by vista properties
Sociological aspects of streets and other public places in urban studies and development	• The Life and Death of Great American Cities, Jacobs, J., 1961 • Social Life of Small Urban Spaces, Whyte, W., 1980 • Streets as Public Spaces and Drivers of Urban Prosperity, UN-Habitat, 2013 • Place Making, Bohl, C., Schwanke, D., 2002 • Placemaking, Thomas, D., 2016 • Cities for People, Gehl, J., 2010 • Life Between Buildings, Gehl, J., 2011 • The City Shaped and The City Assembled, Kostof, S., 1999 • Public Places Urban Spaces, Carmona, M., Heath, T., Oc, T., Tiesdell, S., 2003	Streets as the main urban public places with mixed-use functions

(continued)

Table 19.1 (continued)

Research topics and aspects	References (publication title, author, year; ordered by significance in relation to the research topic/aspect)	Conclusion on streets as walkspace from heritage perspective
Streets from pedestrian perspective within traffic issues	• Walkable City, Speck, J., 2012 • International Charter for Walking, Walk21 conferences (2000-2018), 2006 • Fußgeher- und Fahrradverkehr, Knoflacher, H., 1995 • Street-Forming|Re-Forming, City street conference proceedings, 2016 • Street Design Manual, New York City Department of Transport, 2013 • Architecture and Pedestrians, Echavarri, J., Dauden, F., Schettino, M., 2013 • Landscape Design Promenades, Krauel, J., 2008	Planning streets as walkspace and a promenade system with pedestrian-friendly approach
Specific street typologies	• Great Streets, Jacobs, A., 1993 • The Boulevard Book, MacDonald, E., Jacobs, A., Rofe, Y., 2001	Examples of streets which are heritage

explored in this chapter through the case studies where urban public spaces are read as *streetscapes*, since they are a part of the street network, and therefore as also *walkspace systems* or spatial sequence with various types of walkable areas. The body of the literature on streets is extensive and can be categorised from general topics and overviews to research on specific streets typologies (Table 19.1).

Also, numerous publications mention streets as a part of other topics but still leaving the research gap in linking streets and urban heritage. The recognition of heritage values as an important aspect of a street is clearer in boulevard and avenue studies (Jacobs 1993; MacDonald et al. 2001) where historic components are explicitly pronounced. Therefore, the starting points in choosing the case studies to look at heritage values associated with various walkspaces are examples of boulevards and promenades. The case studies were part of field research during the four years of 'Heritage Urbanism' research project[4] and of its precedent project titled 'Urban and Landscape Heritage of Croatia as a Part of European Culture'.[5]

19.2.1 Redefining the Notion of Promenade—Culture of Wandering

Contemporary promenades are redefinitions of the original promenade idea. The promenade is a concept in close connection to the notion of pleasure as '*locus amoenus*' in landscape architecture[6] (Bojanić Obad Šćitaroci 2018)—meaning a place where people can walk easily (Schäfer 2002) without dirty or bumpy paths and without distractions from other traffic forms, characterised by absence of stress, for relaxing and fun recreation of strolling (Smets 2002). In general, the idea that public spaces need to be pleasant is not a new one. In his Book 5, IX Chapter Vitruvius (Vitruvius Pollio and Krohn 1997) indirectly mentions the usefulness of promenades, porches and vegetation when he writes about building porches around public buildings and the ecological and health reasons for it. Nevertheless, the landscape theory agrees that, in their origins, promenades were not public. They were elite areas for pleasure or hunting. The Topos, European Landscape Magazine, number 41 deals with the theory of promenade topics. It asks how promenades are used to revive spaces. Smets (2002) explains that promenades combine two significances: the parade as a ritual of ceremonial display shared with others and the trail as an introspective path of self-discovery. The inspiration for promenades comes from ceremonial walks and processions which Le Notre turned into leisure parades with his gardens, and they soon became the models for public parks, at first solely for citizens who could afford it (Smets 2002). A need for redefinition evolved naturally from the function and usage of those types of spaces in urban and natural contexts. Similarly, the concept of boulevards developed and adapted over time from specific wide streets on the periphery, where the fortifications used to be, usually curved and full of public life—a mediator between new and old urban areas—into a multiway

[4]Field research on case studies in London (2016 and 2017), Budapest (2016), Barcelona (2015) and Madrid (2015) was done under the thematic unit called 'Urban public spaces of connecting the city' by the researchers Prof. Bojana Bojanić Obad Šćitaroci, Ph.D., and Tamara Zaninović, Ph.D. student, and it was partially financed by Croatian Science Foundation under the project 'Heritage Urbanism (HERU)—Urban and Spatial Models for Revival and Enhancement of Cultural Heritage' (period 2014–2018, project number HRZZ-2032 and leading investigator Prof. Mladen Obad Šćitaroci, Ph.D.).

[5]Field research on case studies Barcelona (2012) and New York (2013) was done by Tamara Zaninović, Ph.D. student, under the Croatian scientific project called 'Urban and Landscape Heritage of Croatia as a Part of European culture' (period 2006–2013, number 054-0543089-2967 and the head of the project Prof. Mladen Obad Šćitaroci, Ph.D.).

[6]*Locus amoenus* is a Latin for 'pleasant place', and it refers to literary *topos* which is idealized place of safety and comfort with three basic elements: trees, grass and water.

boulevard as a specific type of street described in The Boulevard Book (MacDonald et al. 2001).

'To walk' comes out of old English words for 'to throw' *wealcan* and 'to go' *gewealcan*. In German, it is *spazieren* similar to Italian *spaziare* coming from latin *spatium* (Schäfer 2002), which has multiple meanings ('space', 'distance', 'stroll', 'interval') all connected to *-scape* as in the word *landscape*. The -scape suffix denotes 'an extensive view, scenery' or 'a picture or representation' of such a view, as specified by the initial element (*cityscape*, *moonscape*, *seascape*). Subsequently, the promenade phenomenon can be translated as *walkscapes*. In slavic language, *hodati* ('walking') comes from word *hod* meaning 'a step', while the notion of 'to stroll' *šetati se* comes from an old Slavic word 'to move around' or *gibati se*. Altogether, these terms describe the concept of walkspaces and walkscapes as areas for movement which are pleasant. Pleasant places can be intended for recreational and health purposes or as a pleasant scenography and landscape—'pleasant to the eye'. Whereas walkscape emphasises an aesthetic view of heritage, walkspace acknowledges integration and accessibility of heritage within the urban landscape. *Spaziergang*, *paseo*, *passegiata* and *šetnja* are specific because they are not prefigured, unlike *march* which is a journey on foot with an objective, and unlike *exploration* which aims to discover undisclosed paths (Smets 2002). Walkscapes are places for gathering, wandering and individual contemplation—whether they are tourist heritage sites or just coastal line areas with vistas (known as waterfronts).

By walking, we create a path and the path is, therefore, a creative tool for organising and knowing the space and landscape. It is a symbolical act of transforming landscape (Careri 2002). Nodes at crossroads usually indicate a specific place, meaning and importance—a concentration, a square, a trade or a settlement. When landscape is filled with heritage—natural or urban—walkspaces become the criteria and mediators for the understanding, integration and evaluation of heritage because they enable its discovery by moving through streets. In the words of Dietrich (1996):

> ...we discover a landscape only by moving through it along the path: landscape without a path remains hidden to us. Perhaps it exists, but we lack the key to experience it.

19.2.2 Walkspace System in Contemporary Life

From the origins in parades and trails which were previously mentioned (Baecker 2002), in contemporary terms and projects promenades are redefined as: urban 'waterfronts' (seafronts and riverfronts) which refer to the ninetieth-century model of a boardwalk or as 'shopping malls' which originate in commercial streets (Smets 2002). They are also being interpreted and integrated into contemporary typologies such as airport and intermodal stations, multiplex malls, libraries and campuses, fitness and recreational areas, golf centres, entertainment areas (Baecker 2002). These specific typologies are not the topic of this heritage research because they represent contemporary urban and architectural typologies. Therefore, the chosen case studies are streetscapes which have specific pedestrian characteristics and are located in global metropolises. These streetscapes meet either one or more important criteria: they have significant historic and planning development connected to historic urban transformations and redevelopments; they have been the subject of contemporary redesign; they are important tourist or shopping pedestrian streets that are prominent within the urban fabric in terms of accessibility and socio-economic activity (Hillier 1999). Planning areas for the benefit of pedestrian movement are more crucial in the twenty-first century than before because of the various transport modes which need to be crosslinked in cities (Frank et al. 2006). With urban growth, there has been a need for areas to be rethought and reimagined for the purpose of new age and users.

19.3 Evaluation of Contemporary Promenades Through Walkspace Models in the Heritage Urbanism Approach

19.3.1 Research Background

The walkspace and walkscape relation in the urban analysis presented in journal 'Prostor' (Marić and Bojanić Obad Šćitaroci 2012) and at the conference about 'Green' in the city of Zagreb (Marić and Bojanić Obad Šćitaroci 2013) resulted in three models in the context of urban landscape theory. The first model (A) is 'linear movement' or linear promenades as a *streetscape model* which forms and connects an urban network. Linear promenade examples (Fig. 19.1) are pedestrian streets, historical routes and pathways, alleys, boulevards, avenues in various forms of mixed-use and specific (shopping and touristic) streets. The second model (B) is 'free movement' or promenade areas as a *landscape model* which has the size and scale associated with park and recreational areas. Promenade areas with free movement options are pedestrian zones with squares, urban waterfronts and urban parks and gardens, especially if they give panoramic views and scenes through walks. The third model (C) is 'planned movement' as a part of urban regeneration and development projects or an *urbanscape model* with superimposed urban and promenade networks so that both are still evident. Urbanscape models are strategic concepts in development which enable contemporaneity and vitality of both explained models and examples (A + B = C) such as Madrid Rio, New York High Line, Grand Paris, Milano Green Rays or La Sagrera Linear Park.

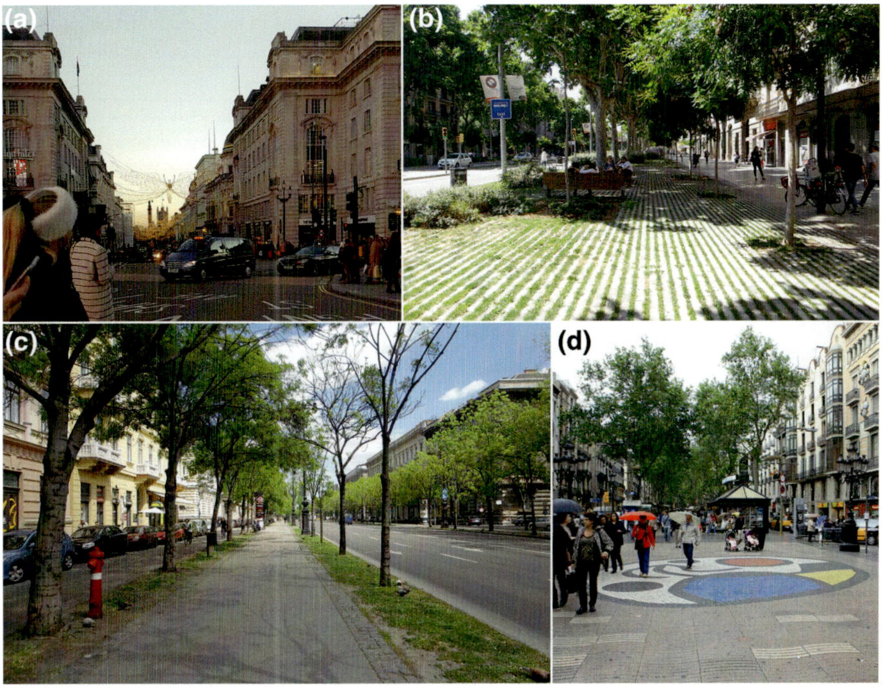

Fig. 19.1 Collage of streetscape case studies representing: **a** Regents Street Saint James's with vista on Waterloo Place as the southern part of the walkspace system between Regents Park and St. James's Park (©Tamara Zaninović, 2016). **b** Passeig de Sant Joan in Barcelona after contemporary redesign (©Tamara Zaninović, 2015). **c** Andrassy Avenue in Budapest (©Tamara Zaninović, 2016). **d** La Rambla Street in Barcelona (©Ana Sopina, 2012), landscaped street

This research builds upon the framework of these three models through case study comparison to fill in the gaps of typological identification in two ways: by giving criteria on how to identify models in specific urban settings (Table 19.2) and by analysing heritage and connectivity elements of the walkspace system (Table 19.3).

19.3.2 Walkspace Spatial Models at the City Level of Case Study Analysis

The case studies, located in London, Barcelona, Budapest, New York and Madrid, are chosen to explain the reinterpretation of the promenade notion. They all have promenade character but with different historical background, urban settings and planning concepts. They are either historically recognised urban landscapes or contemporary projects and plans with pedestrian-oriented strategies.

Table 19.2 Spatial and structural analysis of the walkspace systems in the case studies of London, Barcelona, Budapest, New York and Madrid

Cities and places Case study name, street length	Walkspace categorisation						
	Profile of linear movement/streetscape model of walkspace		Types of free movement/landscape model of walkspace encountered along or connected with the case study street			Urbanscape plans and projects applied on the case study street	
	A		A/B	B	B/C	C	
	Route	Axis	Squares and parks along the way	Historic urban park connection	Connection with waterfront	Waterfront	Contemp. linear park
London							
Regent Street Walkspace (Waterloo Pl.—St. Mark Sq.), 4 km	+		+	+			
Millennium Bridge Walkspace (St. Paul-Tate Modern), 600 m		+	+		+		
Exhibition Road, 800 m		+		+			
Oxford St., 2 km		+		+			
Barcelona							
La Rambla Walkspace, (La Rambla, Placa de Catalunya, Rambla Catalunya), 3 km	+		+		+		
Passeig de Sant Joan, 2 km		+		+			
La Sagrera Linear Park master plan, 8 km	+	+	+	+	+		+

(continued)

Table 19.2 (continued)

Cities and places Case study name, street length	Walkspace categorisation						
	Profile of linear movement/streetscape model of walkspace		Types of free movement/landscape model of walkspace encountered along or connected with the case study street			Urbanscape plans and projects applied on the case study street	
	A		A/B	B	B/C	C	
	Route	Axis	Squares and parks along the way	Historic urban park connection	Connection with waterfront	Waterfront	Contemp. linear park
Budapest							
Andrassy ut, 2.3 km		+	+	+			
Vaci u., 1.5 km	+		+				
Danube Banks Project, 3 km	+	+	+		+	+	
New York							
Broadway, 21 km	+		+	+			
42nd Street Project, 3.2 km		+	+		+		
High Line, 2.3 km	+		+				+
Madrid							
Madrid Rio, 5 km	+	+	+			+	+
Passeo Walkspace (P. del Prado, P. del Recoletos, P. de Castellana), 6 km	+	+	+	+			
Gran Via, 1.5 km	+		+				

The landscape models (B) are the promenades in the most general sense, but the study and analysis showed that streetscape models (A) are combined with landscapes models (B) while urbanscape models (C) contain both streetscapes and landscapes (A and B) altogether. The spatial profile of the streetscape model (A) is defined as either a route or an axis, depending on the spatial layout where a straight visual street line is an axis while curved or irregular layout lines are classified as routes. When this character changes significantly along the analysed streetscape area, it is marked as both axis and route character. Landscape model types are small-scale squares and parks along or at the ends of streetscape and linear park connections; historic public parks and recreational areas significant at the city scale; and waterfront areas (which can also be a part of streetscape or urbanscape models depending on

Table 19.3 Walkspace systems in London, Barcelona, Budapest, New York and Madrid analysed according to urban heritage and urban connection characteristics in correlation to walkspace models (A = linear street model of movement; B = free landscape model of movement; C = urbanscape model)

Cities and places case study, length	Model Table 19.2			Urban heritage Types and depth				Urban connection Types and depth				
	A	B	C	Urban	Landscape	Architectural	Intangible	Usage	Visual	Crossing	Height	Memory
London												
Regent Street Walkspace (Waterloo Pl.—St. Mark Sq.), 4 km	+	+		+	+	+	+	+	+	+	–	+
Millennium Bridge Walkspace (St. Paul-Tate Modern), 600 m	+	+		+	+	+	+	+	+	+	+	+
Exhibition Road, 800 m	+	+		+	+	+	–	+	+	–	+	+
Oxford St., 2 km	+	+		+	+	+	+	+	+	+	–	+
Barcelona												
La Rambla Walkspace (La Rambla, Placa de Catalunya, Rambla Catalunya), 3 km	+	+		+	+	+	+	+	+	+	–	+

(continued)

Table 19.3 (continued)

Cities and places case study, length	Model Table 19.2			Urban heritage Types and depth				Urban connection Types and depth				
	A	B	C	Urban	Landscape	Architectural	Intangible	Usage	Visual	Crossing	Height	Memory
Passeig de Sant Joan, 2 km	+	+		+	+	+	–	+	+	–	–	+
La Sagrera Linear Park, 8 km			+	+	+	+	+	+	+	+	+	+
Budapest												
Andrassy ut, 2.3 km	+	+		+	+	+	+	+	+	–	+	+
Vaci u., 1.5 km	+			+	–	+	+	+	–	+	–	+
Danube Banks Project, 3 km			+	+	+	+	+	+	+	+	+	+
New York												
Broadway, 21 km	+	+		+	+	+	+	+	+	+	–	+
42nd Street Project, 3.2 km	+	+		+	+	+	–	+	+	+	–	+
High Line, 2.3 km			+	–	+	+	+	+	+	+	+	+

(continued)

Table 19.3 (continued)

Cities and places case study, length	Model Table 19.2			Urban heritage Types and depth				Urban connection Types and depth				
	A	B	C	Urban	Landscape	Architectural	Intangible	Usage	Visual	Crossing	Height	Memory
Madrid												
Madrid Rio, 5 km			+	+	+	–	–	+	+	+	+	+
Passeo Walkspace (P. del Prado, P. del Recoletos, P.de Castellana), 6 km	+	+		+	+	+	+	+	+	+	–	+
Gran Via, 1.5 km	+			+	–	+	+	+	+	+	–	+

their primary function and design). Finally, the analysis identified that in the studied cities urbanscape models are large-scale urban interventions, i.e. new waterfronts, riverfronts or contemporary linear parks.

Altogether, the results from Table 19.2 confirm the hypothesis that open public spaces are interconnected by streets, and therefore, they should always be considered in the context of their scale and a wider system to which they belong. The results show that analysed examples are either streetscapes (A) or urbanscapes (C) in the cities with importance of urban landscape values (A/B are combined with A and C models in 13 out of 16 case studies, B are part of 7 out of 16 case studies, and B/C spatial models are part of 4 out of 16 case studies in Table 19.2).

19.3.3 Walkspace Evaluation Framework

Heritage along the promenades within walkspace can be: urban (streets and squares within urban areas), landscape (parks, gardens, trees and urban forests, sculptures and landscape design elements with historical significance), architectural (facades, details, materials, buildings and groups of buildings) and intangible heritage (patterns, vistas, historical events and memories associated with places). All these types can be present in walkspace areas and could be measured. It is argued here that this opens the possibility for these diverse heritage types to be included in planning evaluation and development procedures. This work proposes to call this the 'heritage depth' criterion. As a part of this criterion, we may also consider layers or strata of the built environment as a fabric genesis of the city which includes historical data on formative processes of both streets and buildings.

The mode of how the spatial areas and heritage are connected within walkspace depends on the case study and subject. There are: connections of usages and settlements (by public places such as squares, parks, paths and streets or by roads and traffic infrastructure network); visual connections (such as vistas, viewpoints, axis with landmarks, lighting design); connecting barriers (bridges and overpasses over barriers, tunnels and passages through it); connecting heights (stairs, ramps, escalators, lifts, cable car); connecting with the past events (memorials). All these elements could constitute separate research subjects (Bojanić Obad Šćitaroci et al. 2017; Marić and Bojanić Obad Šćitaroci 2015). This study focuses on the type of connection between walkspace area and heritage types, which we call the 'connection depth' criterion.

Both heritage and connection depth refer and allude to the 'time depth'[7] defined by historic landscape characterisation formed in Britain (Bandarin and van Oers 2012; Careri 2002). This approach suggests that historic layers of urban landscape

[7] Historic landscape characterisation is a programme developed in the 1990s by English Heritage for historic environment conservation and spatial planning where the time depth of cultural and historical processes plays the key role in understanding and managing landscapes.

can be determined in detail and be converted into precise and easy to use data through mapping technology and GIS tools.

19.3.4 Application of the Walkspace Evaluation on Case Studies at the City Level

In London, the case studies include two quite different streets. The first one is Regent Street, forming its unique system by changing directions and character. It is defined by urban parks in the north and south, which it connects as a Regent Street walkspace system (Fig. 19.1a) from Waterloo Place to Saint Mark Square, along which there are transversals and pedestrian zones, shopping areas, churches and monument sculptures. The second is Oxford Street, characterised by shopping and crowds which is why new proposals lead to the direction of its pedestrianisation. Two already pedestrian-friendly contemporary renewed areas in London have integrated different pedestrianisation strategies. The first is a pedestrian bridge connecting Thames shores and pedestrian zones around St. Paul and Tate Modern, enabling the exploration of both cultural and touristic sites with new and various vistas and walkspace experiences. The second is the shared space application in Exhibition Road. Both distances are pedestrian-friendly, unlike in Oxford and Regent streets with dominated traffic system.

In Barcelona, there are three types of streets which are found to be dominated by landscaping elements. They are represented here with three examples: the famous and historic La Rambla street; the newly redesigned Passeig de Sant Joan which connects urban areas and the public park at the coast; and the project of the La Sagrera Linear Park, a new biodiverse diagonal of connecting parks above the redeveloped traffic infrastructure and with the memory of Rec Comtal water infrastructure.

La Rambla (Fig. 19.1d) is a specific type of promenade where pedestrians dominate in the middle area within the tree line on each side. Tree lanes are a type of a border between the esplanade and two flanking narrow service car lanes. Car lanes are one way, each in different directions, and they are close to the building facades in front of which there are narrow pedestrian walkways. With different widths, these types of streets in cross sections can be found in other places in Barcelona and in other cities as well. La Rambla itself came into being by combining five Rambla streets in a 1.2-km-long[8] linear walkspace system. Its character continues further after Placa de Catalunya through the Rambla de Catalunya, which is 1.3-km long and runs west and parallel to the famous Passeig de Gracia.

The Barcelona case study is an example of how important the pleasant pedestrian areas are for the city. After the redevelopments of waterfront areas and designs for new public parks, the most recent projects deal with street issues such as redesigning

[8]Here, the lengths are given for the parts of the walkspace system La Rambla 1.2 km and Rambla Catalunya 1.3 km, while in the table length is the sum of its parts; therefore, with squares it is a 3-km walkspace system for this case study.

the southern part of the avenue Passeig de Sant Joan (Fig. 19.1b) creating 1.2-km-long contemporary walkways on each side of the traffic avenue with sitting areas in front of the building facades and a cycling area in the middle of the avenue. After the node between Passeig de Sant Joan and Avinguda Diagonal with the monument as spatial and visual focus, the Passeig becomes rambla type in section for 800 m. Also, the Superblock concept is an innovation for the Barcelona block system where new traffic regulation is transforming them into a combination of nine blocks, three by three, with pedestrian shared space system inside and city traffic outside. On the other hand, La Sagrera Linear Park is a combination of everything mentioned, from new public parks to new infrastructure. It includes history into design for identity and memory by redeveloping urban parts along the railway infrastructure and around important historic water canal. These types of projects are long-term master plans with the green strategy that can influence the development of the whole city that reflects on an even wider agglomeration area, with new stations and better public and regional traffic connectivity.

Andrassy Avenue in Budapest, or Andrássy út in Hungarian, is UNESCO protected cultural heritage (Fig. 19.1c). Its three parts start and end at the main urban nodes marked with squares or with crossroads that have wider distance between buildings, creating octagonal layout at one place and round layout composition on the other. Three parts have three densities and architectural typologies. The street starts and ends with wide and large open public areas: Park Erzébet tér in the southwest and Hősök square northeast as the entrance to the museums and the main city park Városliget. Vaci utca is a pedestrian tourist street, as in many European cities, placed in the historic centre. It connects two squares, one is the Fővam tér just in front of the Szabadság híd, a bridge on the first traffic ring where Pest fortifications used to be, and the other is Vörösmarty tér, which is placed central to assemblage of surrounding squares and parks in close proximity.[9] Szabadság Bridge is planned to be redesigned in the new Smart City Strategies which aim to introduce pedestrian changes, for the Danube Banks in Pest. In addition to these changes, a new pedestrian bridge will be added. For the city with traffic system of rings and avenues, these pedestrian waterfront and historic promenade areas such as in Buda part create the balance.

New York City initiatives for planning streets as public places which started with the Times Square pedestrianisation are now changing the city part by part, from waterfront to the well-published High Line project reusing the abandoned elevated rail line. New projects and visions continue with the competition for 42nd street redesign vision with adding the public light way. The American block regulation is characteristic of the Manhattan planning system and its grid, but one historic

[9]Váci Street is therefore through Vörösmarty tér in the north connected into a larger walkspace system which consists of: Vigadó tér, which is from the Vörösmarty tér located to the west on the coast and part of Danube Banks system as well; József Nádor tér and Hild tér located both to the north; Erzsébet tér in the northeast at the beginning of Andrássy út; Városháza park to the east; Szervita tér to the southeast; Széchenyi István tér placed at the coast northwest from the Vörösmarty tér and connected to the Szent Istvan tér with Szent Istvan Basilica through pedestrian street called Zrínyi utca.

pathway remained visible in that grid: Broadway Avenue. Its deviation from the regular orthogonal 1811 grid creates public open places such as the aforementioned Times Square, Madison Square Park and Union Square Park, as well as the Columbus Circle and the west entrance to Central park. The Broadway streetscape connects the main and important pedestrian and landscape areas of the city. Due to that and the fact that it is 21-km long (while the length of Manhattan is 53 km altogether), it can be recognised as a unique walkspace system with even regional significance.

The urban landscape of Madrid is dominated by avenues and wide car lanes. Similar to the case of Barcelona, there have been planning efforts to redefine the urban landscape by thinking about pedestrians and cycling. Madrid Rio is a built concept for the new urban landscape in the city over the highway. Its linearity is characteristic to the Madrid planning, from Soria y Mata projects to the streets Passeo del Prado, Passeo del Recoletos, Passeo de Castellana as well. This Passeo system is therefore combined out of three long multilane boulevards which have multiple tree lines in cross section. After the Madrid Rio and Avenida de Portugal which connect peripheral parts of the city, current strategies mainly deal with the central areas such as the plans to pedestrianise the main shopping street Gran Via.

The analysed case studies are characterised by diverse heritage and connectivity values in terms of walkspace models, and they are, therefore, significant in planning from the cultural heritage point of view inasmuch as they are urban pedestrian and public space connections. Evaluation of these factors in urban streetscapes is the basis for pointing out the specific streets in urban networks which possess identity factors for becoming walkspace systems and new urban landscapes in green strategies and infrastructures.

19.4 Discussion Through Case Study Comparison

19.4.1 Multiscale Character of Walkspace Systems

Urban streetscapes with promenade values were analysed as the research subject in five cities on 16 case studies. They differ in scales, and according to pedestrian movement and speed, there is the local pedestrian scale, local cycling scale, city and regional scales as follows:

- shared space projects and promenades around a kilometre in length (Millennium Bridge and Exhibition Road in London, Vaci Street in Budapest and Gran Via in Madrid);
- streets and projects from two to four kilometres in length (Regents Street and Oxford Street in London, La Rambla and Passeig de Sant Joan in Barcelona, Andrassy Avenue and Danube Banks Project in Budapest, 42nd Street Project and High Line in New York);
- linear landscape projects from 5 km and more and with large surface parks, squares and traffic stations (La Sagrera Linear Park in Barcelona and Madrid Rio);

- historical route in a city more than 20 km in length (Broadway Street in New York).

The case studies which include contemporary interventions and design concepts are a part of strategies for more sustainable cities through raising biodiversity, reducing traffic and pollution, stimulating recreation and pedestrian movement while preserving the genius loci. It could be said that the sustainability goal makes them new heritage candidates and cultural sites for present and future generations.

19.4.2 Urban Connections as Walkspace Spatial Models

The analysis of spatial models from Table 19.2 confirms that streets have a tendency to form spatial systems in cities. Both route and axis streetscape characters are not just sidewalks and traffic lanes but mediums for connecting main pedestrian places such as squares and parks. Therefore, the number of local squares and small parks (A/B column in Table 19.2) is not the factor for determining if the case study is a street (A) or landscape (B) spatial model. Nevertheless, this factor obviously is the heritage basis and initial criterion for a street to be considered a vivid public space and a walkspace system. Landscape models (B) are examples which include either historic large-scale urban parks or waterfront urban areas (B/C). Waterfronts themselves are contemporary urbanscapes (C) because of the promenade intention with the specific complexity in their planning almost as a 'city within a city' concept.

Observing the scale, heritage and connection elements within the urban context on case studies from Tables 19.2 and 19.3, there are subcategories based on the types of walkspace strategies in connecting public spaces into walkspaces which can be described as: (1) historic park street; (2) landscaped urban streets; (3) contemporary pedestrianisation and place-making methods; (4) contemporary urban landscape projects (Table 19.4).

While it is acknowledged that pedestrianisation is present to some degree in all walkability strategies, the categorisation here considers examples which specifically aim at pedestrian-friendly environments as a part of the third category (3). This category shows various methods of pedestrianisation which are implemented in either streetscape models (A) or in combination of streetscape and landscape model (A and B): new pedestrian bridges, shared space, new pedestrian zones and place-making interventions (Fig. 19.2). Pedestrian zones are strategies in use since the 1970s. Today, pedestrian zones are planned for shopping streets and areas which have high index of pedestrian crowds such as Oxford Street in London (Lancefield 2017)[10] and Gran Via in Madrid (Lyne 2017)[11] or nodes on Broadway Route in New York (Raphael

[10]The plans are there but have not been implemented yet, and it is uncertain if they will.

[11]The media announced plans to pedestrianize the main shopping street in Madrid, and this is currently in testing phase of analysing the last traffic restrictions from December 2017.

Table 19.4 Walkspace models and typology of walking-based strategies: 1. historic park street; 2. landscaped urban streets; 3. contemporary pedestrianisation and place-making methods; 4. urbanscapes contemporary pedestrianisation

Cities and places case study, length	Model			Walking-based strategy	
	A	B	C		
London					
Regent Street Walkspace (Waterloo Pl.—St. Mark Sq.), 4 km	x	x		Historic park street	1.
Millennium bridge Walkspace (St. Paul-Tate Modern), 600 m	x	x		Pedestrian bridge	3.
Exhibition Road, 800 m	x	x		Shared space	3.
Oxford Street, 2 km	x			Pedestrian zone plan	3.
Barcelona					
La Rambla Walkspace (La Rambla, Placa de Catalunya, Rambla Catalunya), 3 km	x	x		Landscaped street	2.
Passeig de Sant Joan, 2 km	x	x		Historic park street	1.
La Sagrera Linear Park master plan, 8 km			x	Parkscape with canalscape	4.
Budapest					
Andrassy ut, 2.3 km	x	x		Historic park street	1.
Vaci u., 1.5 km	x			Pedestrian zone	3.
Danube Banks Project, 3 km			x	Riverfront	4.
New York					
Broadway, 21 km	x	x		Historic route with place-making interventions	3.
42nd Street Project, 3.2 km	x	x		Landscaped street plan	2.
High Line, 2.3 km			x	Elevated reuse in park	4.
Madrid					
Madrid Rio, 5 km			x	Riverfront	4.
Passeo Walkspace (P. del Prado, P. del Recoletos, P. de Castellana), 6 km	x	x		Landscaped street	2.
Gran Via, 1.5 km	x			Pedestrian zone plan	3.

2009).[12] Shared space projects and new pedestrian bridges are more contemporary and specific examples. Bridges can become an urban attraction, like Millennium

[12]NYC Department of Transportation has been working on the pedestrianisation of Broadway Boulevard since 2009 through a series of projects which have been advocated by the Project for Public Space and its partners in the New York City Streets Renaissance since 2006.

Fig. 19.2 Contemporary pedestrianisation interventions in London: **a** Millennium pedestrian bridge (©Tamara Zaninović, 2016). **b** Shared space of Exhibition Road (©Tamara Zaninović, 2016)

Bridge in London,[13] or be a part of a bigger urban renewal, like in the Danube Banks Project in Budapest (Dorka 2018).

Historic park street (1) is a relation between important streets and the main public parks in cities. These park street connections were planned and conceptualised within the new public parks idea in the eighteenth/nineteenth century (Bojanic Obad Šcitaroci et al. 2004) and subsequent street network plans. In terms of this study, historic park streets are considered as hybrid walkspace models, being both streetscape and landscape models (A and B). The case study examples which represent the historic park street types are (a) Regents Street in London with Regents Park in continuation of route, (b) Andrassy Avenue in Budapest and (c) Passeig de Sant Joan in Barcelona, all three in different cities, countries and climates. Landscape streets (2) are also hybrid models (A and B) of streets which have a boulevard character with dominant tree-lined landscape, but without large surfaces of urban park areas, such as La Rambla in Barcelona and Passeo del Prado in Madrid.

Overall, out of 16 case studies of analysed walkspace systems the most numerous are examples which spatially combine both streetscape and landscape models (A and B). Nevertheless, these walkspace models have implemented diverse walkability strategies from historic planning and design to urban landscape redesigns and pedestrianisation. Beside the described case studies, there is also the 42nd Street Redevelopment Project in New York which has a contemporary redesign vision to turn the street into a landscape pedestrian boulevard with a low-floor light rail line. Opposite to that is the Broadway Avenue which, as discussed, is a long historic and diagonal route in the grid city. Diagonals have an inclination towards becoming systems the same as examples with waterfront areas, such as Madrid Rio and Danube Banks riverfronts and La Sagrera Linear Park.

[13]Foster+Partners, authors of Millennium Bridge in London, cited on their website the Evening Standard newspapers as a positive fact that the new bridge is a combination of design and construction pleasant as a view and for walking: "Something wonderful has happened. A British architect and a British engineer have built a confident, intelligent, striking, graceful, even playful structure in the shadow of St Paul's. Whether you're walking on it or simply looking at it from the banks, it gives deep satisfying pleasure".

Considered together, the spatial models and evaluation indicators can serve as a framework for characterisation and assessment of new public space strategies for connecting walkspaces into a system which is cultural heritage itself as an urban landscapes type.

19.4.3 Comparison of Heritage Thematic in Streetscapes and Urbanscapes

In streetscapes (Fig. 19.1), two themes are compared: historic park street with Andrassy Avenue as an example and landscaped street with La Rambla as an example. They are walkspace systems which are categorised in the same scale category, local cycling.[14] La Rambla's walkspace system extends from its one side to Rambla de Catalunya which has the same design and streetscape type and from its other side to the more famous Passeig de Gracia. Both these streets are unique in the urban network of Barcelona since they are positioned on what used to be a historical path. Passeig de Gracia stayed in inclination from the Barcelona strict gridiron because it was a road connection to the peripheral settlement in Garcia area. This difference is noticeable at the block scale change which is caused by the street irregularities. The change in scale is marked by passing through a park, Jardins de Salvador Espriu, which acts as a mediator between the grand avenue and the historic urban morphology. Rambla de Catalunya is also not perfectly aligned in parallel to the network gird. Again, this is the result of topography which in the past was not an empty space but an area with paths and streams. These route systems have preserved a heritage of references to their own development which is partially manifested through their surrounding urban form and building design. Andrassy Avenue is a product of planned regulations and covers a wider urban area with public buildings and prominent palaces and houses, a large-scaled square and a central park. This avenue is also partially influenced by the planning of the first underground line in Continental Europe. Notably, this is one of the reasons for including it in UNESCO heritage under 'ii' and 'iv' criteria.[15] Both Andrassy and La Rambla feature buildings and historic design features along the way with active mixed-use public places. They enable access and connection to main public places in the studied cities: La Rambla connects the waterfront and Catalunya Square, while Andrassy Avenue connects Erzébet Square and Városliget Park. Although La Rambla is not listed in the UNESCO World Heritage List, its architectural and archaeological heritage is recognised and protected by the city's planning legislation. In comparison with these two examples, Passeig de Sant Joan

[14]Andrassy Avenue is 2.3-km long, and La Rambla's walkspace system is 3-km long (including Rambla de Catalunya and Passeig de Gracia).

[15]Selection criteria: (ii) to exhibit an important interchange of human values, over a span of time or within a cultural area of the world, on developments in architecture or technology, monumental arts, town-planning or landscape design; (iv) to be an outstanding example of a type of building, architectural or technological ensemble or landscape which illustrates (a) significant stage(s) in human history.

could also be considered as historic street category because it is a planned avenue from the urban plan of Ildefons Cerdà Ensanche project (approved in 1859). Passeig de Sant Joan has a historic public park in its axis on the site where the Ciutadella fortress once stood before it was demolished according to the Josep Fontseré plan from 1872. Due to its urban design and regeneration project from 2008 (construction took place 2010–2011), it is landscaped street category. The last redesign and revitalisation project aimed to make this historic boulevard a more pedestrian-friendly area and a new green zone in the city (Domenech 2011). Therefore, Passeig de Sant Joan is an example whereby revival and redesign projects combine explored walkspace categories and promenade definition in new ways. These pedestrian strategies and design combinations eventually enrich the promenade culture and urban heritage.

In urbanscapes, two contemporary projects are completed as revitalisation projects (Fig. 19.3). These show differences in terms of character, setting and scale. The New York High Line (Fig. 19.3a) is a reuse project of a former rail line and performs at local cycling scale in terms of its length but without cycling possibility and characteristics since it is elevated from ground level and connected to it only with numerous stairs. Being a pedestrian greenway and elevated urban park, Madrid Rio (Fig. 19.3b) is an urban intervention which operates in between the local and regional scales. It defines a 5-km-long line of recreation and cultural areas alongside the riverfront and over the main highway traffic. As contemporary landscape projects, these places are designed cultural landscapes which embody the memories of their previous states and usages via integrated memorial and monumental elements. Unlike streetscapes, these examples are newly created and redefined promenade heritage. Their innovating concepts and large-scale intervention make these projects significant landmarks which may contribute to a new image and identity of the city. The High Line has already significantly influenced the revitalisation of the surrounding areas and nearby development followed by gentrification processes and population displacement—issues that should be considered in advance and addressed by planning policy. On the other hand, the effects of Madrid Rio projects are yet in the process of assimilation given the project's recent implementation.

Fig. 19.3 Urbanscape case studies: **a** New York High Line (©Tamara Zaninović, 2013). **b** Madrid Rio (©Ana Mrđa, 2015)

19.5 Conclusion

Initial historic terminologies are constantly being reused in a new context and with that redefined through new concepts, design ideas and situations. This research points out redefinitions in the notion of promenade. From a theoretical point of view, this redefinition remained focused on pedestrian areas which differ in terms of urban form and scale but they all include landscape features such as central parks and squares or different waterfronts with public buildings and memorial values.

The chapter develops walkspace theory in the context of urban and landscape design out of -scape-related terminology: walkscape, streetscape and urbanscape. Walkscape is defined as the entire landscape and environment which can be perceived through pedestrian motion. Streetscape is used for public space and urban fabric which forms streets or linear public space in the scale of an urban street. Urbanscape is a wider concept for newly planned urban landscapes which consist of various types of public spaces and urban morphologies. These concepts form three models of walkspace systems in cities differing in scale and, subsequently, in planning procedures: streetscape, landscape and urbanscape models. In the literature and planning, walkability is a measure of how easy it is to walk, which is investigated through various factors (density, land use mix, connectivity, urban morphology) which are not related to measuring pleasure. Walkscape, walkspace system and models have the basic relation to the pleasure moment because they are contemporary interpretations of the promenade notion being the place for pleasure and leisure.

The analysed case studies in five cities are generally characterised as either streetscape or urbanscape models. In both walkspace systems of streetscapes and urbanscapes, the analysis has identified landscape models being included or linked to the systems. Since there are no correlations between street factors (e.g. scale, usage, development) and spatial models (e.g. route, axis), the comparative discussion focused on the characteristics of scale, position, topography and heritage connectivity of the derived thematic subcategories which were identified within the streetscape and urbanscape models.

By reducing traffic and promoting active travel modes and public transport networks, walkspace models can support the green agenda and urban sustainability. Considering that streetscapes connect both architectural heritage and cultural landscapes of various public places, it is important that they are rethought and planned in the logic of walkspace systems, where promenades are just one of the elements. Streetscapes can be promoted as heritage not only for promenade use and landscape values, but also because of their historic spatial and socio-economic role in the urban landscape which is evident in landscaped street models, park street connections and historic routes and axes such as Broadway in New York and Oxford Street in London. Urbanscapes include streets and promenades together with landscaped areas—for example, urban parks, squares and waterfronts—all combined in such ways so as to form new parts in cities or reconfigure the way those cities work (e.g. Madrid Rio).

Both streetscape and urbanscape interventions are methods for ensuring vitality of heritage in the heritage urbanism approach which promote an integrated creation and protection of 'ordinary' everyday cultural urban heritage. Walkspace systems as urban landscapes are formed by cultural heritage; they form public places around heritage and give access to cultural heritage. Therefore, walkspace systems enable a reconceptualisation of urban identity and have the potential to upgrade pedestrian infrastructures, increase connectivity and urban integration in cities and their parts, and eventually to raise the overall quality of urban life. Altogether, the design and planning of walkspace models as heritage-integrated systems can bring new awareness of the urban landscape as lived experience and has the potential to regenerate urban parts (Obad Šćitaroci and Bojanić Obad Šćitaroci 2013)[16] within a city. Walkspace systems should be recognised in current design methods and city planning procedures because of their potential to support the identities of urban places and to integrate heritage values in the practices of everyday life.

Acknowledgements The research is a part of the scientific project 'Heritage Urbanism—Urban and Spatial Planning Models for Revival and Enhancement of Cultural Heritage'. It is partially financed by the Croatian Science Foundation [HRZZ-2032] and carried out at the University of Zagreb, Faculty of Architecture.
This research was also part of the research entitled 'Urbanscape Emanation' financed by the University of Zagreb and carried out at the Faculty of Architecture.

References

Baecker D (2002) Neue Promenaden braucht die Stadt/The city need new promenades. Topos 41:46–51
Bandarin F, van Oers R (2012) The historic urban landscape: managing heritage in an urban century. Wiley Blackwell, Oxford and West Sussex
Bojanić Obad Šćitaroci (2018) Aluzije i luzije u perivojnoj arhitekturi – kompozicija i oblikovanje/Allusion and illusions in landscape architecture—composition and design. In: Landscape architecture lectures, second year of bachelor studies of architecture and urban planning. University of Zagreb, Faculty of Architecture. http://scitaroci.hr/bbos/kolegiji-preddiplomskoga-i-diplomskoga-studija-arhitekture-i-urbanizma/. Accessed 19 June 2018
Bojanić Obad Šćitaroci B, Obad Šćitaroci M, Hajós G (2004) Public parks and gardens of Croatian towns in 19th century in European context. Šćitaroci d.o.o. and Faculty of Architecture, Zagreb
Bojanić Obad Šćitaroci B, Zaninović T, Sargolini M (2017) Design of memorials—the art of remembering–method of place regeneration. Prostor 25(2):306–315
Careri F (2002) Walkscapes—walking as an aesthetic practice. Editorial Gustavo Gili, SL, Barcelona
Dhanani A, Stockton J, Vaughan L (2017) Street mobility project: walkability models. Digital scholarly resource. http://www.ucl.ac.uk/street-mobility/toolkit. Accessed 16 June 2018
Dietrich L (1996) Wege/Pathways—editorial. Topos 15:4

[16]Power to re-generate urban parts meaning to give and create urban values, producing a city, fabricating a city, modelling a city, constructing a city, to mould a city is a notion translated from Croatian term which is both adjective and noun: „*gradotvornost*"—defined by authors prof. Mladen Obad Šćitaroci, Ph.D., and prof. Bojana Bojanić Obad Šćitaroci, Ph.D., in the paper called "*Gradotvornost perivoja i pejsaža - Re-interpretacija perivoja - konstelacija suvremenih tema*/Re-interpretation of Urban Gardens and Landscape".

Domenech L (2011) Passeig De St Joan Boulevard. http://www.landezine.com/index.php/2012/07/passeig-de-st-joan-boulevard-by-lola-domenech/. Accessed 16 June 2018
Dorka B (2018) Budapest to build new Danube bridge for pedestrians and cyclists. https://welovebudapest.com/en/2018/02/20/budapest-to-build-new-danube-bridge-for-pedestrians-and-cyclists/. Accessed 16 June 2018
Frank DL, Sallis FJ, Conway LT, Chapman EJ, Saelens EB, Bachman W (2006) Many pathways from land use to health: associations between neighbourhood walkability and active transportation, body mass index, and air quality. J Am Plann Assoc 72(1):75–87. https://doi.org/10.1080/01944360608976725
Gehl J, Gemzøe L (2004) Public spaces, public life. The Danish Architectural Press, Copenhagen
Hillier B (1999) Centrality as a process: accounting for attraction inequalities in deformed grids. Urban Des Int 4:107–127
Jacobs BA (1993) Great streets. The MIT Press, Cambridge
Lancefield D (2017) Mayor Sadiq Khan unveils plans for pedestrianisation of London's Oxford Street. https://www.independent.co.uk/news/uk/home-news/oxford-street-pedestrianisation-london-mayor-sadiq-khan-traffic-shopping-retail-a8039736.html. Accessed 16 June 2018
Lyne N (2017) Madrid mayor pledges to make iconic Gran Vía a pedestrian zone. https://elpais.com/elpais/2017/01/06/inenglish/1483700264_292073.html. Accessed 16 June 2018
MacDonald E, Jacobs AB, Rofe Y (2001) The boulevard book—history, evolution, design of multiway boulevards. The MIT Press, Cambridge
Marić T, Bojanić Obad Šćitaroci B (2012) Walkspace, Linearno povezivanje prostora na primjeru Splita/Walkspace/Linear space—motion in the city of split. Prostor 20(1):119–131
Marić T, Bojanić Obad Šćitaroci B (2013) Odnos prostora kretanja i prostora doživljaja 'zelenoga grada Zagreba'/The relationship between walkspace and walkscape in the 'Green City of Zagreb'. In: Božičević J, Nikšić M, Mlinarević TJ, Missoni E (eds) Zelenilo grada Zagreba. Croatian Academy of Sciences and Arts, Zagreb, pp 124–132
Marić T, Bojanić Obad Šćitaroci B (2015) Memorials as cultural landscapes. In: Obad Šćitaroci M (ed) International scientific conference cultural heritage—possibilities for spatial and economic development, proceedings. Faculty of Architecture, Zagreb, pp 244–249
Mehta V (2013) The street: a quintessential social public space. Routledge, New York
Obad Šćitaroci M, Bojanić Obad Šćitaroci B (2013) Gradotvornost perivoja i pejsaža - Re-interpretacija perivoja - konstelacija suvremenih tema/Re-interpretation of urban gardens and landscape. In: Božičević J, Nikšić M, Mlinarević TJ, Missoni E (eds) Zelenilo grada Zagreba. Croatian Academy of Sciences and Arts, Zagreb, pp 56–67
Raphael C (2009) Broadway dreams to be realized. https://www.pps.org/article/broadway-dreams-to-be-realized. Accessed 16 June 2018
Schäfer R (2002) Promenaden/Promenades—editorial. Topos 41:4
Smets M (2002) Promenade - eins und jetzt/Promenade, past and present. Topos 41:6–17
Smith L (2006) Uses of heritage. Routledge, Abingdon
UN Habitat (2013) Streets as public spaces and drivers of urban prosperity, Nairobi. https://unhabitat.org/books/streets-as-public-spaces-and-drivers-of-urban-prosperity/. Accessed 20 June 2018
Vitruvius Pollio M, Krohn F (1997) Vitruvius De Architectura Libri Decem. Translated to Croatian by Matija Lopac and Vladimir Bedenko from original Vitruvii DeArchitectura Libri Decem Iterum edidit Valentinus Rose, Lipsiae, B.G. Teubneri, MDCCCXCIX. Golden Marketing, Institut građevinarstva Hrvatske, Zagreb
WALK21 (2006) The International Charter for Walking—Creating healthy, efficient and sustainable communities where people choose to walk. In: WALK21 international conference series. https://www.walk21.com/charter. Accessed 12 June 2018

Chapter 20
Landmark Phenomenology of Sacred Architecture as Cultural Heritage

Zorana Sokol Gojnik and Igor Gojnik

Abstract Multisensory experience is an important part of the perception of sacred architecture. It is important to understand that the content of sacred architecture as cultural heritage, in the complex way of understanding, is not only its historical, artistic or cultural value, but also its symbolic value. One of its symbolic aspects is the multisensory experience it creates. This assertion is based on the theoretical framework of the phenomenology of architecture, an aspect of architectural theory that explores the experience of the built. Every experience of architecture is multisensory, and in the case of sacred space this aspect particularly influences the experience of the believer and of the visitor of cultural heritage because architecture relates, mediates and projects meanings anchored in a complex experience of space. Different phenomena such as light, sound, smell and touch directly affect the experience of sacred architecture. They create specific atmospheres that are intrinsic to the experience of the sacred, such as the atmosphere of the transcendent, the atmosphere of eternity, the atmosphere of belonging to the community and the atmosphere of an out-of-the-ordinary character of space. This approach changes the viewpoint on sacred architecture as cultural heritage and highlights numerous aspects of a building that should be considered in the process of the protection of cultural heritage.

Keywords Sacred architecture · Multisensory experience · Phenomenology · Cultural heritage

Z. Sokol Gojnik (✉)
Faculty of Architecture, University of Zagreb, Zagreb, Croatia
e-mail: zorana.sokol@arhitekt.hr; zorana.sokol@siloueta.com

I. Gojnik
„Siloueta arhitektura“, Zagreb, Croatia
e-mail: igor.gojnik@siloueta.com; igor.gojnik@gmail.com

M. Obad Šćitaroci et al. (eds.), *Cultural Urban Heritage*,
The Urban Book Series, https://doi.org/10.1007/978-3-030-10612-6_20

20.1 Introduction

The research in this chapter is the result of exploring the topic of preservation, presentation and enhancement of sacred architectural heritage. The research was inspired by the realization that plenty of sacred architectural heritage is in need of renovation and revitalization, as well as by a wish to avoid making conservational decisions without a complex insight into the specific purpose of sacred buildings which, apart from historical, cultural and artistic values also possess a symbolic value. That symbolic value is reflected not only in the liturgical and iconographical layer of space, which is not the topic of this paper, but also in the multisensory experiential aspects of space analysed in this paper.

In all religions, sacred space is a place of meeting with the phenomenon of the sacred. It is possible to experience sacred architecture through various forms of human perception, whose task is to mediate and to complete the experience of the sacred. The aim of this paper is to draw attention to different perspectives on the experience of sacred architecture through which sacred architecture achieves its task, in order to direct new conservational practices towards a complex understanding of sacred space and its complete protection. In the effort to regard sacred architecture as a landmark and a part of cultural heritage, it is necessary to include its reality in the totality of experience and to point out that protecting sacred buildings as a part of cultural heritage means protecting everything—material and nonmaterial aspects of their values and their primary task.

The research of the criteria for the enhancement of sacred architectural heritage is significant and internationally relevant because sacred architectural heritage is the topic of numerous documents regarding the revitalization of cultural heritage. However, the most common starting point of all documents is historical, cultural and artistic value of sacred architectural heritage, while its symbolic value is neglected.

At the beginning of the research, it is necessary to single out several important starting points:

- Sacred architectural heritage cannot be preserved from the historical, cultural and artistic aspect alone. The preservation needs to include its symbolic layer too, which is reflected in many layers of space (liturgical, iconographical and multisensory).
- New interventions in sacred architectural heritage are possible and reasonable, but through the preservation of the value of authenticity and integrity of heritage.

Through the research of sacred architecture as cultural heritage, we wish to confirm the formulated hypothesis:

- Apart from the historical, cultural and artistic criteria for the protection of sacred architectural heritage, it is also possible to establish additional criteria for the protection of sacred architectural heritage by relying on the theoretical framework of the phenomenology of architecture.

The aim of this research is to propose relevant and scientifically grounded criteria for the protection of the symbolic in sacred architectural heritage, which is reflected in the multisensory experiential layer of space, in order to preserve the originality and authenticity of sacred heritage.

In this paper, the phenomenology of sacred space is set within the theoretical framework of the phenomenology of architecture. Many authors have achieved considerable insights into the multisensory experiences of space in the field of the phenomenology of architecture.

The contribution of authors such as Juhani Pallasmaa, Karsten Harries, Steven Hall, Alberto Pérez-Gómez and Merleau-Ponty particularly stands out. Juhani Pallasmaa examines the role of the body and the senses in authentic architectural experience (Pallasmaa 2005). Karsten Harries allows a link between architecture and philosophy (Harries 1997). Steven Holl emphasizes the relationship between the building and the place (Holl 1996, 2000), Alberto Pérez-Gómez reaffirms the value of memory in architecture (Pérez-Gómez 1985) and Merleau-Ponty returns the body to the forefront of philosophy (Merleau-Ponty 2013).

Some authors have dealt with multisensory experience in religions or with certain aspects of the multisensory experience of architecture and their works have also become a valuable starting point for this research. Among them, Thomas Barrie stands out by emphasizing that in order to understand sacred architecture, it is necessary to understand that it has a mediating role as an 'in-between' place (Barrie 2010). De Wietse Boer and Christine Göttler examine the role of sensation in the religious transformations of early modern Europe (Boer and Göttler 2012). Gardner Rust talks about the importance of music and dance in religions (Gardner Rust 1996). Susan Elisabeth Hale talks about an inextricable link between sacred space and sound (Hale 2007). Victoria Henshaw warns about the importance of smell in the perception of urban space (Henshaw 2013), Tin Oberman, Bojana Bojanić Obad Šćitaroci i Kristian Jambrošić deal with the issue of enhancing the soundscape of city spaces (Oberman et al. 2014).

Based on the previous knowledge and by using the method of analysis and synthesis, in this paper we have presented a specific application of the set theoretical framework of the phenomenology of architecture to the most important multisensory experiential aspects of sacred space, since an integral discussion about the phenomenology of sacred space from this aspect has not been done before in the existing literature.

20.2 The Phenomenology of Sacred Space

The phenomenology of architecture is a field in the theory of architecture that deals not with the question of designing space, but designing the experience of space. The aim of this theory is to integrally include all the senses of perception into the experience of architectural form. The starting point is the thesis that since the modern era to these days ocularcentrism has dominated the experience of architecture (Pallasmaa 2005). The concept of ocularcentrism is based on the assumption that since the modern era, the cognition in western civilization has been determined by the visual paradigm through the separation of the subject (who comprehends) and the object (what is being comprehended).

The era of technology relies heavily on visual perception through photography, and the contemporary period in particular emphasizes the virtual, not the real experience of architecture, which strengthens the ocular-centric position. The origin of the phenomenon of ocularcentrism can be found in the rationalist assumptions of the modern era in which the main criteria for the truth is intellectual, thus significantly neglecting the other ways of understanding the truth such as sensory perception, faith and sensibility (Gadamer 2004; Heidegger 1996, 2004; Johnson 1994).

As a reaction to ocularcentrism, the phenomenology of architecture, as a theoretical movement from the end of twentieth and the beginning of twenty-first century, shifts the focus from the rationalist paradigm which emphasizes visual perception and perceives architecture as a primarily aesthetic 'subject', to the holistic paradigm in which all our senses make up the integrality of perception, which means that perception is a priori multisensory. It tries to 'reaffirm its role as the theatre of memory and metaphor that there is no such thing as a meaningless structure' (Pérez-Gómez 1985). The human body 'is the locus of all formulations about the world; it not only occupies space and time but consists of spatiality and temporality' (Pérez-Gómez 1985). Such experientiality of the body involves the entire person in the experience of architecture.

The phenomenological approach offers a new view on sacred architecture which has the role of an 'intermediary' (Barrie 2010). It connects two realities: the earthly—experiential reality and the transcendent reality—the phenomenon of the sacred which it is never possible to experience fully because neither space, nor music, image, words, gestures nor rite can express it fully. They can only make it closer and give us a 'foretaste' through the atmosphere created by space, music and artistic content. That is why sacred architecture in particular is one of the privileged places of multisensory experience which brings the believer closer to the substance of faith. In that way, sacred architecture brings us into contact with the numinous through different experiential aspects.

20.3 Multisensory Experiential Aspect in the Context of Sacred Space

If we look at sacred architecture of different religions, it can be seen that sacred architecture, more than any other architectural topic, has carefully preserved many aspects of multisensory experience. In different religions, the emphasis on involving the body in the rite is different, and with that involving the body in the experience of the substance of faith. By that, the body becomes an active factor of the experience of all the phenomena that bring it closer to the substance of faith, and thus to the substance of space, music and rites. In Hinduism, the believer is involved through a dynamic state (movement before entering the temple) and a static state (standing in the temple). In Buddhism, similarly to Hinduism, dynamic (walking) and static (standing) elements of the rite interchange. In Judaism, the temple rite had a dynamic character (offering sacrifice, ritual washing, etc.), whereas the rite in synagogues has a dominantly static character. Islamic space, although static in its structure in the sanctuary, cherishes exceptionally dynamic rite elements (ritual washing, kneeling, standing up, hand gestures). In the Christian rite, dynamic (entering the church, procession, standing up, kneeling, moving as a part of the rite) and static elements (keeping still and sitting) also interchange.

It is obvious that in different religions believers do not participate in the rite only with sober thoughts but also with the body, by changing gestures, movement of the body stance and ritual walk. Through their bodies, believers enter an interaction with space. Space, in all aspects of multisensory experience, thus becomes one of the important elements of the experience of faith.

20.3.1 Light

The role of light in architecture is of vital importance in the experience of space. Studying the daylighting concept as a part of the complete architectural concept is very rare in contemporary architectural solutions (Gojnik et al. 2015). Thus, in contemporary practices light is often a 'by-product' of the architectural concept which basically relies on building with surfaces and masses. The phenomenological approach warns us that light is one of the important elements of the identity of space and one of its basic structural elements. Light in architecture defines shapes. It makes shapes enter our visual perception.

Light in sacred space is a particularly important topic because it has symbolic meanings which differ among religions. In the architecture of the Jewish temple in Jerusalem, the topic of light is interpreted through the contrast to light, through the topic of the dark Holy of Holies which is illuminated by candles from the outside. The candles symbolize the divine light which lightens up the dark. In Islamic sacred architecture, we can discern the idea of God who belongs to the other ontological reality and who never speaks directly to the man, but rather touches him, although never in totality and clarity but through a 'veil' of human experience. God's divine touch of the man in Islam is like scattered light that permeates through perforated walls and domes of Islamic sacred architecture. In Buddhism, God is a processual reality, and the aim of the life journey is to reach the *enlightenment*, the spiritual awakening. The walk to the brightest illuminated part of space with the sculpture of Buddha (the awakened) is a common feature of Buddhist sacred architecture. In the architecture of Christian churches, light is the symbol of Christ who transforms the human life in the Eucharist. That is why in Christian churches light accentuation is the strongest in presbytery, the area where the Eucharist takes place. The symbol of light is also associated with the area of baptistery in Christian sacred architecture. Through baptism, the believer joins the community of believers and is lightened by 'the light of Christ' (Sokol Gojnik et al. 2015).

Fig. 20.1 The church of St. Anne in Rijeka, competition work

The task of sacred architecture is to transfer theological concepts to space. It is necessary to design the daylighting concept carefully in new solutions, bearing in mind its symbolic character which differs among religions. Through the intensity of light and contrast, it is possible to emphasize the theological concept, and also to create dynamics and rhythm through light gradation. That is why it is important to design with light and not to design the volume not taking into account the effects of light that create the experience of indoor space. Moreover, it is important to recognize the value of the daylighting concept in the created buildings (Fig. 20.1).

20.3.2 Sound

Sound is an important element of space. Architectural space has its own 'music', that is, sounds which we experience by entering space and being in it. In contrast to the eye which is 'captivated' by space, sound 'captivates' the man. It reaches the man and creates a special atmosphere of being in space.

The meaning of sound is especially significant for sacred architecture. Anthropologists believe that the first rites in the history of mankind were not sacrifices, but rather dance and music (Gardner Rust 1996). Through music, that is, sounds, the man participates in the totality of religious experience. The sound of sacred space, just like the music inside it, is a sonorous icon of the sacred, its basic tonality. 'Sight is the sense of the solitary observer, whereas hearing creates sense of connection… our look wanders lonesomely in the dark depths of a cathedral, but the sound of the organ makes us immediately experience our affinity with the space' (Pallasmaa 2005). Sacred space therefore has its own soundscape. Landscape architecture theoretician Hedfors with his concept of sonotope (a layer of information about soundscape of a certain space) differentiates between: *resonating landscape* (landscape/architecture with its acoustic properties) and *generating landscape* (sources of sound connected with the subject landscape/architecture) (Hedfors 2008; Oberman et al. 2014). This theory is applicable to the architecture of sacred space as well. Sacred space integrates the resonating and generating aspect of architecture. It is the complete experience of soundscape, in which its resonating aspect is the sound of architecture itself, whereas the generating aspect of sacred space is the music made in liturgy which has the function of the rite. In that way, the sonotope of sacred space is an experience that involves the human experience and through which realities that open the believer to the phenomenon of the sacred intertwine and permeate (Hale 2007).

The experience of silence of sacred space is the basic characteristic of the resonating aspect of sacred architecture. The silence of sacred space can manifest itself in various forms. The silence that can be silent, quiet, cacophonous, shrill, sibilant—absence of sound (Ballou 2015). When planning sacred buildings, acousticians need to be especially sensitive to the specificity of the sound of sacred space. Sacred space cannot be planned according to the same criteria as, for example, a concert hall. Carefully moderated sound of sacred space has a symbolic value. It maintains a certain level of echo which contributes to creating sacred atmosphere, the atmosphere of divinity and awe.

Church towers as an element of sacred architecture have special importance in landscape. Except with their visual presence, they also become an important element of wider environment with their sound because they shape the soundscape of a city. They become present at a particular time of the day, inviting to prayer, reminding the man of his ontological groundedness. Through sound, certain landscapes become etched in our memory, and through the same sound, they come alive in our memory. In big, contemporary cities, in which the generating landscape has almost muffled the resonating landscape, the sound of a bell has lost its intensity in the soundscape of a certain part of the city. However, in smaller environments, in which soundscape has been preserved, the sound of a bell provides the place and space with an identity and keeps the genius loci of a certain space.

Church bells are a 'significant part of how local people define who they are and their relationship to where they are. These buildings incorporate remembrance as active memorialising, as part of enriching present-day existence' (Kelly et al. 2018). The sounds of sacred architecture in a wider space give a cultural character to a certain area.

20.3.3 Smell

The element of smell in architecture began to be acknowledged as an important factor of architectural identity in the last ten years or so. Jonathan Foyle says 'smell is intrinsic to the personality of a place' (Foyle 2012). The importance of smell in architecture is also emphasized by Chandler Burr, the curator of the department of olfactory art at the Museum of Arts and Design in New York, the first museum to exhibit scent as a historic art form. This recently awoken substantial interest in the significance of smell for the perception of architecture (Henshaw 2013) can be considered as a reaction to the dominant ocular-centric view on architecture.

It is precisely sacred architecture that has always acknowledged and cherished that perceptive viewpoint and has given it ritual meaning. That is why in many cultures smell is associated with the sacred. In many religions, smell used to be a part of the rite, and it was often accompanied by visual images of smoke rising into the sky, permeating the earthly and the transcendent reality. In ancient Egypt, smell had a particularly sacred meaning, especially the smell of myrtle and incense which was considered the sweat of gods. The Greeks and Romans used smells to please gods and to consecrate a certain place. In the Vedas, the smells of rose, jasmine, sandalwood and cedar are mentioned and they are an important part of Hinduist rites in which scented gifts are offered to gods. In Judaism and Christianity, the use of incense is particularly present since it was considered 'God's resin' and it was highly esteemed. In Judaism, the rite of offering sacrifice was accompanied by burning incense. In Christianity, incense occupies a special symbolic place. In the Adoration of the Magi (Matt. 2:11), incense is one of the gifts that the Magi lay before the newborn Messiah—Jesus Christ. The smell of incense rises into the sky and symbolizes human aspiration towards the divine. The smell of incense in Christian liturgy symbolizes prayers rising to the Heaven. It is ceremonially burned during entering processions, circling the altar, in front of the Evangeliary, within the rite it is swung around the priest celebrating the Mass, etc. The smell of incense has permeated the experience of the entire Christian sacred architecture. That makes it an important element of religious identity because 'it is not only a collective or individual experience, but forms a memory and connects past, present and future as an element of identity' (Reichling 2010).

Apart from the use of smell in the rite, the characteristics of certain materials that have their distinctive smell also give character to a certain building. For example, the smell of wood gives a different character to architecture compared to the smell of brick or concrete. Smells complete the experience of architecture and give it a specific character which, in synergy with other characters, contributes to the totality of the experience of architecture.

20.3.4 Tactile Elements of Architecture

The haptic aspect of space is another important factor of the identity of space. Human skin is our biggest organ, and touch is a complex type of sense. Through touch we discover whether an object is soft or hard, coarse or smooth, warm or cold, dry or wet, porous or dense, etc. Through touch we can feel shape. 'All the senses, including vision, can be regarded as extensions of the sense of touch—as specialisations of the skin. They define the interface between the skin and the environment—between the opaque interiority of the body and the exteriority of the world' (Pallasmaa 2005). The touch of hands (handshake) is the first contact we make with a person before we get to know them. The character of handshake reveals the character of a person (Chaplin et al. 2000). The man can be 'touched' by architecture too. The verb 'touch' in many languages means to 'emotionally affect or move someone or something'. Being touched by architecture means to experience architecture in the totality of the multisensory experience. The verb 'touch' here speaks of multiple aspects of the experience of architecture, but it also emphasizes the importance of the haptic sense.

The phenomenologists of architecture emphasize the loss of haptic perception in contemporary architecture (Pallasmaa 2005; Bloomer and Moore 1977), and thus the loss of the entire experience of space. To experience space through touch immediately evokes an association with different experiences of wooden, concrete, stone, mortar, plastic and similar surfaces of walls or floors. Different materials create different atmospheres of space. Walking on a wooden and a stone floor are two different realities. Leaning your body against a concrete, wooden or, for example, stone wall are completely different experiences. Experience can also be 'tactile experience from distance' (Holl 1996, 2000). Haptic experience is significant for sacred architecture too. The dignity of stone, the intimacy of wood, the lifelessness of plastic, the crudeness of concrete, etc. are the characteristics of material which can be used to emphasize the desired characters of sacred architecture. The sense of touch also becomes apparent in the conduct towards certain elements of sacred space (Boer and Göttler 2012). That is why in comprehensive consideration of sacred space, it is necessary to look at sculpture, paintings, relics, holy objects, etc., and architecture integrally. These elements of sacred space are susceptible to touch and through touch they complete the experience of faith. In touch, the harmony of visible and invisible, and of form and content is achieved. Through touch, we participate in the transcendent. Touch and physical experience are especially immanent in Christianity where God himself was embodied, and after the resurrection he revealed himself to the apostles and through touch confirmed the truth about the resurrection (St. Thomas). Touch is also an important element of the Christian rite. Believers touch (shake hands) when they offer each other the sign of peace; in the Holy week, the cross is touched and the culmination is receiving the host in the Eucharist when the bread is touched and tasted as the body of Christ.

Touch completes the totality of the experience of the rite and architecture which, together, help the believer to experience the meeting with the transcendent (Fig. 20.2).

Fig. 20.2 The church of Our Lady of Lourdes in Zagreb

20.4 Research Results—The Criteria for Evaluating Sacred Architecture from the Phenomenological Aspect in the Context of Built Heritage

The aim of this paper was to point out the importance of the preservation of relevant elements of multisensory experience of sacred buildings which are the components of the total experience of space. They create specific atmosphere characteristics of sacred architecture, such as the atmosphere of the transcendent, the atmosphere of eternity, the atmosphere of belonging to a community, the atmosphere of an out-of-the-ordinary character of space and so on. Because of that, in the protection of sacred buildings as cultural heritage it is necessary to acknowledge, evaluate and predict the preservation criteria:

- *The criteria of space light*—the daylight of a space creates a special lighting atmosphere. It is necessary to acknowledge the symbolic role of a certain daylighting concept and to preserve all elements of space contributing to such a lighting atmosphere. The absence of light can also be a certain daylighting concept, and it should be acknowledged as such. Materials and colours also contribute to the character of light in space. It is important to consider a sacred building in interaction with a wider urban tissue and to protect the daylighting concept of a building, not disrupting it with a new building in close surroundings.
- *The criteria of space sound*—the sound of space is defined by a certain spatial disposition of a sacred building and the materials it is made of. In sacred buildings, the resonating component of architecture is important, the component which

maintains the originality of sound of a sacred building and which contributes to the total experience of architecture. Each space has a characteristic degree of silence and echo which needs to be acknowledged. In the protection of built heritage soundscape, whose specific character is provided by the bells of church towers of sacred buildings, is also important. The soundscape of an urban tissue in which a certain sacred building is situated gives a cultural character to that space.

- *The criteria of space smell*—the smell of space participates in the total experience of sacred space. It is emphasized through the use of certain smells in space, but it is architecturally defined by the materials a building is made of. Certain materials have their own specific smell which gives a certain character to space. When intervening in buildings that original character may be disrupted.
- *The criteria of the haptic aspect of space*—the haptic aspect of space is defined by materials and surface finishing. Integral care for all elements of space and for the inventory such as sculptures, paintings, relics and holy objects susceptible to touch is essential in sacred space. In the process of protection, it is necessary to acknowledge the important elements of space contributing to haptic experience and to preserve them in subsequent interventions.

20.5 Conclusion

The value of sacred buildings in the previous forms of protection is most commonly viewed from an ocular-centric position which has dominated the theory of architecture and the protection of built heritage. From that position, historical, cultural and artistic values have been emphasized, while there has been no evaluation of the symbolic value of sacred architecture. This paper confirms the hypothesis set at the beginning of the research that it is possible to establish additional criteria of the protection of sacred architectural heritage by relying on the theoretical framework of the phenomenology of architecture.

It is possible to recognize the carriers of symbolic meaning in multisensory experiential elements of architecture covered in this paper. They are relevant and scientifically grounded criteria for the protection of the symbolic layer of sacred architectural heritage. They can be recognized in the specificity of space light, sound, smell and tactile aspects of architecture. These elements of architecture affect the creation of certain atmospheres of sacred spaces such as the transcendent, the atmosphere of eternity, the atmosphere of belonging to the community and the atmosphere of an out-of-the-ordinary character of space. That is why, when evaluating sacred built heritage in need of renovation and revitalization, it is necessary to acknowledge these elements of space too and to maintain their originality with the aim of preserving and enhancing heritage.

A joyful encounter

I met a piper. He was blind.
But he said: 'Life is beautiful, my son!'

I was confused for a while: 'Beautiful you say?'

'You will see, my son, when you are blind!'

Gustav Krklec

Acknowledgements The research is a part of the scientific project 'Heritage Urbanism—Urban and Spatial Planning Models for Revival and Enhancement of Cultural Heritage'. It is financed by the Croatian Science Foundation [HRZZ-2032] and carried out at the University of Zagreb, Faculty of Architecture.

References

Ballou G (2015) Handbook for sound engineers. Focal Press, Waltham

Barrie T (2010) The sacred in-between: the mediating roles of architecture. Routledge, Abingdon

Bloomer KC, Moore CW (1977) Body, memory, and architecture. Yale University Press, New Haven

Boer W, Göttler C (2012) Religion and the senses in early modern Europe. Brill, Leiden

Chaplin WF, Phillips JB, Brown JD, Clanton NR, Stein JL (2000) Handshaking, gender, personality and first impressions. J Pers Soc Psychol 79(1):110–117

Foyle J (2012) Scent of the building. Financial Times, November 30, 2012. https://www.ft.com/content/8db9b640-34c2-11e2-8986-00144feabdc0. Accessed 6 June 2018

Gadamer HG (2004) Truth and method. Crossroad, New York

Gardner Rust E (1996) The music and dance of the world's religions. Greenwood, Santa Barbara

Gojnik I, Sokol Gojnik Z, Bojanić Obad Šćitaroci B, Veršić Z (2015) Dnevno svjetlo u sakralnom prostoru kao doživljajni element naslijeđa. In: Obad Šćitaroci M (ed) Cultural Heritage—possibilities for spatial and economic development. Faculty of Architecture University of Zagreb, Zagreb, pp 262–268. ISBN: 978-953-8042-11-9

Holl S (1996) Intertwining. Princeton Architectural Press, New York

Holl S (2000) Parallax. Princeton Architectural Press, New York

Hale SE (2007) Sacred space, sacred sound: the acoustic mysteries of holy places. Quest Books, Wheaton

Harries K (1997) The ethical function of architecture. MIT Press, London

Hedfors P (2008) Site soundscapes: landscape architecture in the light of sound—sonotope design strategies. VDM Verlag, Saarbrücken

Heidegger M (1996) Being and time. State University of New York Press, Albany

Heidegger M (2004) Phenomenology of the religious life. Indiana University Press, Bloomington

Henshaw V (2013) Urban smellscapes: understanding and designing city smell environments. Routledge, Abingdon

Johnson GA (1994) The Merleau-Ponty aesthetics reader: philosophy and painting. Northwestern University Press, Evanston

Kelly A, McCrum S, Sammon P, Stalling D (2018) Memorialising the sacred. Architecture Gallery of the Irish Architectural Archive, Dublin

Merleau-Ponty M (2013) Phenomenology of perception. Routledge, Abingdon

Oberman T, Bojanić Obad Šćitaroci B, Jambrošić K (2014) Unaprjeđenje zvučnog okoliša gradskih prostora - utjecaj na urbanizam i pejsažnu arhitekturu. Prostor 22(48):200–211. ISSN 1330-0652

Pallasmaa J (2005) The eyes of the skin—architecture and the senses. Wiley, Chichester

Pérez-Gómez A (1985) Architecture and the crisis of modern science. The MIT Press, Cambridge

Reichling P (2010) About odoriferousness and malodorousness in the golden legend. Religion for the senses. Part III: religious meanings of taste, smell and touch in ancient and medieval Asia and Europe. Käte Hamburger Kolleg (KHK), Ruhr-Universität Bochum, 21.04.2010–22.04.2010
Sokol Gojnik Z, Gojnik I, Bojanić Obad Šćitaroci B (2015) Light as a symbol in sacred architecture. In: ARCHTHEO'15: DAKAM (Eastern Mediterranean Academic Research Center), Istanbul, pp 609–617

Chapter 21
Heritage Urbanism and Landscape with the Sense and Limitations of the "Place"

Ester Higueras Garcia

Abstract Since the European Landscape Convention, where everything is culture and everything is landscape, the need to provide strategies of integrated action between both realities is a highly desirable goal. The structure to log it adopts the following sequence:

1. Knowledge of **the place**, as physical-spatial, landscape and historical interaction. We propose an initial analytical matrix as a starting point.
2. Determining **carrying capacity**. The physical and environmental physical load according to the characteristics of the place, using a simplified method of calculation.
3. Determining the **attraction capacity** of the place, by its patrimonial heritage (physical or immaterial).

Overarching these three realities, we can summarise the following aspects:

- An array of compatibilities of use, according to the thresholds of the impacts on air, soil and water, to mitigate the negative impacts.
- The result of the positive impacts, through the generation of jobs and wealth for the local population, and addressing an interesting experience for the tourist as well.

Architects, ecologists, urban planners and specialists in landscape sciences and ecology need new tools, hence the importance of these strategies, written in the form of a handbook for their determination. Finally, the case study of Easter Island is explained, thanks to the singularity of its location, the existence of a rich cultural heritage and with a continuous increase in tourism demand, where a new balance is required. This proposal could be replicated in other places and contexts, with similar characteristics, within the environmental protection of cultural landscapes and heritage.

Keywords Place · Carrying capacity · Landscape impacts · Attraction capacity · Cultural landscapes

E. Higueras Garcia (✉)
Departamento de Urbanística y Ordenación del Territorio, Universidad Politécnica de Madrid, Madrid, Spain
e-mail: ester.higueras@upm.es

M. Obad Šćitaroci et al. (eds.), *Cultural Urban Heritage*,
The Urban Book Series, https://doi.org/10.1007/978-3-030-10612-6_21

21.1 General Framework

Maurice Strong (UN 1972) started using the term eco-development. Since then, it has been associated with a concept based on the following: it considers the human being as the most valuable resource. Use of local resources must be selected from the point of view of solidarity with future generations; however, the authorities must promote the participation of communities to reach the objectives of eco-development and should use existing local resources and create suitable techniques for their best use. Besides, they will try to minimise aggression on the local environment and use waste through recycling. Education is a prerequisite for proper environmental management. It seems clear that tourism and eco-development must go hand in hand since they complement each other perfectly. This means, first, that it is imperative to have a new plan based on an intimate knowledge of the environmental local conditions, with methodologies and objectives adapted to each place, and to rule out the use of imported solutions used in different cultural or ecological realities.

The European Landscape Convention (Council of Europe 2000) adds a new application to what since 1973 has been no more than a wish. Now, it is necessary to analyse the interactions between the urban, the lived, and the environmental physical frame, since all these factors configure a system that needs to be protected from actions, seeking interaction and project mechanisms that achieve a dynamic long-term balance between both systems.

From this need, a methodology of action is presented which is not completely new in its structure, but is so in relation to the integrated objectives that are pursued in the face of the current challenges of cultural patrimonial landscapes and heritage urbanism. The following parts are distinguished in the methodology (Table 21.1).

Table 21.1 Heritage urbanism and landscape with the sense and limitations of the "place"

Heritage urbanism and landscape with the sense and limitations of the "place"	
Analytical part	Analysis of the components of the "place" Carrying capacity calculation The attractiveness of a place
Conclusive	The values of the patrimonial place in landscape heritage urbanism Thresholds of the impacts of activities An integrated matrix of an evaluation of the impacts on heritage and landscape Temporary verification of the list of indicators for patrimonial uses in singular landscapes

21.2 Heritage Urbanism in Landscapes: Concept

Since the earliest times, the selection of a place for urban settlements, agricultural livestock or extractive activities has been decisive. The existence of resources (mainly raw materials and water), the search for moderate thermal conditions over a large number of months per year, the absence of adverse risks (seismic, erosive, etc.), and even the character ratios of the symbolic, esoteric, religious or magical have determined this selection. While it is true that there are many settlements that have been demolished, radically transformed or abandoned, we find another large number where this relationship of empathy and harmony between the place and the built-up area still today overwhelms us. We find patrimonial settlements in balance with their environment, establishing new synergies which it is necessary to understand and not to substantially alter. This involves heritage urbanism in landscapes.

Undoubtedly, we must go back to Greek culture to seek the root of this relationship, through what was termed as the "*genius loci*", or the spirit of the place.

According to Bote (1990), there are two major models of strategic approaches in tourism development. The first calls for large-scale, demand-driven development which can generate negative effects on both the tourist and the receiving localities. We understand that it arises from fundamentally economic assumptions. In addition, negative effects are particularly noticeable in the devaluation of the natural attractions of the landscape and the environment in general. Above all, some areas have a certain fragility regarding this type of resource, such as islands and water lands. Faced with this approach, the cited author indicates that there are other strategies, which he calls "artisanal", or handmade, in which the positive aspects of both realities can prevail. With regard to heritage urbanism in landscapes, it is undoubtedly a question of this second path, one that strengthens both visitors and local people, but requires a detailed analysis, synthesis and adequate planning effort.

The aim of heritage urbanism in landscapes is to achieve a balance between tourism and the physical-environmental-cultural aspects in which it is located. The patrimonial attractions of the place must be associated with a certain capacity of use, the visitors must be established, as must their derived actions, so that the place does not reflect symptoms of deterioration in the long term.

According to Casasola (1990):

> The rate of use of tourism ecosystems and the degree of disruption and pollution of these ecosystems depends not only on the volume of tourism activity, but also on their modalities, styles, content, space distribution and proposed objectives. If the raw material of tourism is the natural and cultural attractions, then eco-development must design, plan and carry out the appropriate strategies for its rational use, profitability, conservation and protection.

21.3 Knowledge of the Place: Analysis of Components

Currently, we find numerous methodologies to carry out a holistic analysis of the physical-environmental conditions of a territory. Interactions with the landscape are not so frequently added to this matrix, and it is even rarer to find how the cultural and patrimonial conditions join the process to determine homogeneous zones, with all the integrated variables. The disciplinary limitations and the exhaustiveness of all sciences which can follow these processes do nothing more than to condition negatively the intention of a useful, synthetic, clear and operative plan for decision-making.

In this sense, it seems very opportune to revise the methodology of the spatial analysis of territorial determinants, as proposed by McHarg (1969), and which was a determinant for the studies of environmental impact, environmental planning and, decades later, all the sustainability objectives that would arrive. McHarg leads us to the essence of the main goals to be achieved and which can be applied to landscape heritage.

To study something complex and dynamic, it seems appropriate to divide it into sections or components, which could be first analysed one by one. The next step is to study the interaction between all the previous components to establish the territorial dynamics of the past, which will help us to know the present and to project the future. The first basic division of the variables is that of biotic and abiotic components. Among the biotic components are flora and fauna. Among the abiotic, we can enumerate the soil, water, subsoil, climate and landscape. This first division can be extended or reduced in each case, and it must always be based on the specific characteristics of each place. We look for a base plane, where at least all the relevant conditions of the territory will appear and serve to determine the homogeneous analytical zones. At least it will contain those listed in Table 21.2.

Table 21.2 Analysis of the components of the "place"

Analysis of the components of the "place"
(i) Abiotic components
The relief. Relationship between hypsometry and the clinometric components. High and low points of the territory. Plain and depressed areas, high and flat areas. Slopes with high or moderate ratios
Hydrological structure, rivers, streams. Determination of natural runoff. Aquifer recharge zones, and water ponds and wetlands
(ii) Biotic components
Diversity index. Flora and fauna by areas (Tuomisto 2010)
(iii) Landscape as a cultural process
Determination of landscape assessment by homogeneous and perceptive zones
Landscape fragility determined by light territorial colours, absence of slopes and lack of tall vegetation
The tangible and intangible assets of the place

21.3.1 Integrated Plan of the Analysis of the "Place"

The plan must collect all these variables in order to be able to encompass them. The synthesis plans between the project actions and the environmental values will be the key, not only for the correct understanding of the interactions, but they will greatly help to describe the corrective measures, as a determining element, in an appropriate management plan for landscape heritage urbanism. There are numerous possibilities to graph the results, among which it could be useful to provide:

1. Punctual graphic information: suitable for data on the presence of cultural resources or singular values of the territory, such as the location of relevant landmarks (natural or man-made), hermitages and historical-artistic heritage elements.
2. Linear graphic information: for territorial linear elements such as riverbeds, roads, electricity lines, fencing, all supply networks and infrastructures.
3. Net graphic information: usually with information in cells suitable for landscape studies, visibility analysis or ecotypes areas.
4. Isoclines for collecting information according to some gradient functions, where each line represents an equal value. These plans are suitable for information on water table heights, soil depth, precipitation, altitudes, temperatures or even time schedules to access a site.
5. Surface graphic areas: to represent homogeneous zones according to any criteria or sum of criteria, such as maps of vegetation units, faunal habitats, soil types, lithology, watersheds and morphological urban structure.

21.4 Carrying Capacity Calculation

There are two basic parameters that directly affect the relationship between conservation and development in landscape heritage urbanism. Firstly, the carrying capacity and secondly, the attractiveness of the place. The first establishes the limits of use of the resources based on the structure of the reception of tourists and balance between the resident and floating population and the natural environment. The World Tourism Organization (UNWTO 2010) defines carrying capacity "as the maximum number of people who can visit a tourist place at the same time, without harming the physical, economic or sociocultural environment, without appreciably reducing the quality of the visitor's experience". One of the debates facing territorial planning is precisely how to put into practice the harmonious inclusion of tourist activities in each place, which necessarily involves limiting the occupation of areas with the most attractive landscape. One of the most useful tools is the carrying capacity, a positive concept for the regulation and growth of tourism in heritage areas.

The concept of carrying capacity arose in the mid-1930s to regulate the intensity and frequency and sustained recreational use of resources and to maintain the natural conditions of the environmental protection areas of the USA (Robert 2002). Then, in the 1960s, its application was centred on the social aspects of recreational experience, defining levels of use and satisfaction of visitors, since it is recognised that the intensity of use of the resource alters the tourist experience (LaPage 1963). In the 1970s, research continued to focus on the social aspects of capacity, leading to concepts such as "saturation", a negative density assessment, or encounter numbers generated in a given area (Desor 1972) and "degrees of sensitivity between activities", as a factor of sensation of saturation or discontent of the tourist (Castellani and Sala 2012). At the end of the 1970s, the ecological aspects, detected in some areas, and the need to conserve their natural characteristics, were recaptured, and Thurot (1973, cited in Sharpley and Tefler 2002) argues that the impacts of the tourism system imply economic, social and ecological dimensions. In the 1980s, research sought to set thresholds, as discussion pointed to a lack of definition of "precise limit". In this direction, Cooper and Jackson (1989) attempt to establish ecological and behavioural thresholds, beyond which the physical environment deteriorates, and the tourist experience diminishes, developing elements of design of wild areas, based on criteria of preservation of the area and outdoor recreation. However, it was Getz (1983) who proposed the identification of variables and indicators to obtain a load capacity model. In addition, integration of cultural development into the concept arises (Weaver 2000), as does the definition of recreational activities based on uses in ecologically sensitive areas (Cortés Chavez 2009).

Carrying capacity, based on the environmental, social and economic sustainability of tourist activity, was delimited in the 1980s, with numerous specific definitions, among which the following stand out: (i) "the ecological load capacity" related to the negative results of environmental impacts (Pigram 1983); (ii) "economic carrying capacity", where the impacts of tourism activity are linked to other activities (Mathieson and Wall 1982); (iii) "tourist capacity" understood as the development of tourism without affecting the economic, social, cultural and environmental structure of the place (Tburot 1980); (iv) "the physical carrying capacity" (Mathieson and Wall 1982), as the limit of tourist infrastructure and the regulation of the influx of visitors; (v) "the capacity of psychological load" (Pigram 1983), where it is proposed to limit the amount of tourists and activities in order to have a beneficial effect on the quality of the visitor's recreational experience.

In the 1990s, research was linked to the concept of sustainability, and in Latin America, a new method of calculating the management capacity of protected areas arose. It allows setting boundaries and establishing strategies for organising and controlling visits in protected areas (Cifuentes 1992). This proposal was complemented by a tool that allows the re-evaluation of load capacity on a periodic basis called "selection and monitoring of impact indicators" (Amador et al. 1996). This was complemented with a methodology for the determination of tourist cargo capacity, which was applied to the coastal areas of the Mediterranean, developed by the Priority Action Programme (1998), from where development scenarios are calculated.

In this period, the conceptualisation of "ecological load capacity" defines the limits, based on expert advice, of the level of tourism development or recreational activity (Watson and Kopachevsky 1996). "Social capacity" (UNWTO 2010) can be determined from its value and the visitor's behaviour and finally the approach between the local population and tourists.

In India, a group of researchers (Khanna 1999) developed a methodology that incorporates carrying capacity in regional and territory planning, defining the concept of the "ecological load of the territory" and the "intensity of development".

In sum, it is possible to indicate that carrying capacity presents a wide spectrum of uses that more precisely contributes not only to defining thresholds of a determined territory, but also to generating policies, plans and strategies in balance with the environment, and it is suitable to control and monitor impacts.

For heritage urbanism in landscapes, the carrying capacity contributes to:

- Reconciling local urban development with the ecological characteristics of the place, at the same time reducing environmental risks in heritage areas;
- Overcoming the deficiency in the environmental assessment of the diagnostic stage, providing elements of analysis which allow the state of the site to be determined and its limits of use defined;
- Integrating environmental assessment factors in the analysis of development alternatives, based on the limitations and fragility of the place according to its intrinsic characteristics and cultural landscape.

The new directions of landscape heritage urbanism in tourism capacity management can point to and connect with local development and sustainability. The fixing of the highest number of visitors cannot avoid considering the forms of sustainable heritage tourism, the preservation of the quality of resources, the maintenance of the local character and the development of products based on natural and cultural heritage. In this way, the integration of the load capacity must be part of development management and its planning, indicating different cases according to each period. One way to succeed is to identify the environmental, social and economic conditions desired by a community, and to set up development strategies that allow the tourist load capacity to be managed and for the necessary social and local goals to be achieved (Table 21.3).

21.5 The Attractiveness of a Place

The attraction capacity of a place in landscape heritage urbanism is based on the identification of the heritage values intrinsic to it as poles of attraction, and on a consideration of polyvalent resources, where it is possible to combine patrimonial aspects, historical-cultural research and conventional tourism.

These two parameters, carrying capacity and attraction, are decisive when setting the objectives of the balance of each place. In the case of small municipalities that base their tourism development on the territory, it has already been found that they are inversely related: the greater the reception capacity, the lower the attraction capacity and the negative repercussions on their carrying capacity. It is essential to reach a certain balance, in other words to offer good reception capacity based on the

Table 21.3 Simplified determination of carrying capacity in landscape heritage urbanism (According to Cortés Chavez 2009 and Higueras 2006)

Simplified determination of carrying capacity In landscape heritage urbanism
(A) Inventory and characterisation of the place
1. Delimitation of the patrimonial and environmental study area (the patrimonial city in its place)
2. Biotic, abiotic and cultural landscape components (Table 21.1)
3. Synthesis plan, covering sectoral plans of previous limiting constraints Determination of homogeneous environmental units: physical-environmental-cultural-landscape
4. Identification of: weaknesses, strengths, limiting factors (fixed factors, flexible or dynamic factors) impacts (effects of intensity and type of use in each homogeneous area) levels of air, soil and water pollution
(B) Analysis and evaluation of the capacity of the place (use of soil, ecological and tourist demands)
1. Assessment of the state of the components of the territory:
– assessment of the carrying capacity
– determination of assimilative capacity
– evaluation of homogeneous units
– evaluation of conditions for the quality of life
– definition of patrimonial and landscape development keys of the place
2. Current scenario modelling
3. Definition of the load capacity for urban, ecological and tourist use
4. Definition of environmental management units
5. Definition of the desired objective image, and direction of priority actions

reasonable management of the patrimonial territory, without exceeding the carrying capacity of each area of the place.

On the other hand, it is essential to diversify activity to avoid both seasonality and intensive use of the territory. From this perspective, the need arises to overcome specialisation in a specific tourist activity. In particular, tourism activities based on consumption of the natural environment should include, wherever possible, elements of varied activities. In this way, it will be possible to remove pressure on the tourist use of the territory. In addition, the wealth and jobs generated may revert to the local population. For this to happen, it is necessary to detect values of interest for the tourist and include them as part of what patrimonial tourism has to offer.

21.6 Case Study: Easter Island

For most island territories, tourism represents the main economic activity in terms of the generation of income and jobs. However, due to their small size and isolation, islands are quite vulnerable to negative environmental and social impacts that tourism development can entail. Therefore, it is important to plan and manage a tourism model to achieve sustainable objectives according to the criteria of landscape heritage urbanism.

The increase of tourism leads to threats to the island's ecosystem, even accelerating the deterioration of the archaeological, cultural and social heritage, with problems that affect the disposal of waste, water and air quality, or the loss of biodiversity. The Easter Island analysis is established through three large blocks: tourist attractions; environmental values; and economic and population factors. By synthesising the three blocks, the environmental units of the island were determined, which has allowed for a diagnosis presenting the most critical areas or those exposed to the greatest tourism pressure from 1990 to 2017.

The methodology to evaluate the transformation of Easter Island is based on the three key characteristics of the island territories (UNWTO 2012) (Fig. 21.1). The adaptation of the three key characteristics to the context and situation of the island is shown in Fig. 21.2.

Each block should be analysed separately to unify and compare the data obtained by a SWOT analysis to determine graphically the environmental and functional current situation of Eastern Island in terms of carrying capacity and landscape heritage demand. Several maps illustrate this. The first (Fig. 21.3) shows the tourism attraction for landscape heritage urbanism, with four factors: (i) natural landscape; (ii) historical and monumental resources; (iii) crafts and gastronomy; and (iv) folklore and scheduled events.

An environmental sectoral analysis is shown in Fig. 21.4 with four variables, such as: (i) relief, hypsometry, altitude (from 0 to 550 m); (ii) hydrological analysis, with lakes and surface drainage; (iii) vegetation, such as forest, farming areas, wetlands or

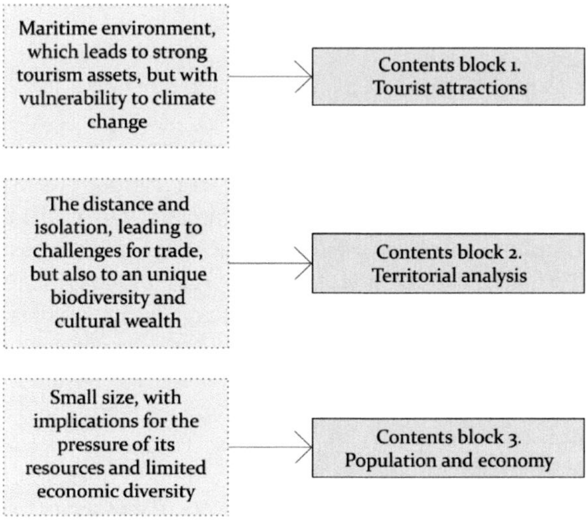

Fig. 21.1 Three key characteristics presented by island territories to evaluate the transformation of Easter Island. *Source* Alba Roque (2018) Research work, Universidad Politécnica de Madrid

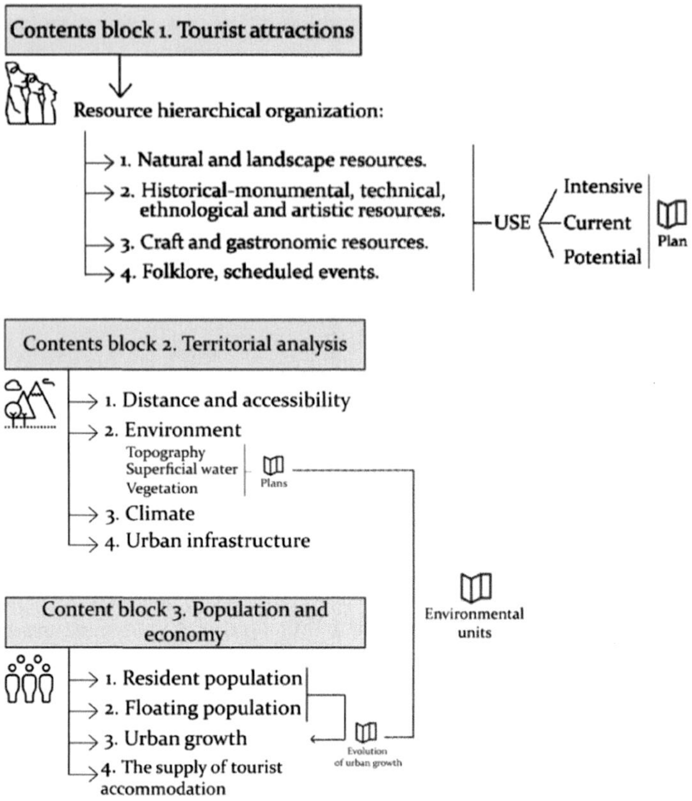

Fig. 21.2 Three characteristics adapted to the Easter Island context. *Source* Alba Roque (2018) Research work, Universidad Politécnica de Madrid

national protected areas; (iv) urban and regional infrastructures areas: dirt, cobbled, asphalt or paved roads.

The analysis presented in Fig. 21.4 was integrated to determine environmental landscape heritage areas (Fig. 21.5). Five groups were delimited (i): EU-slope terrain areas (I–II–III); (ii) EU-protected areas (IV–V); (iii) EU-forest plain area (VI–VII); (iv) EU-farming areas (VIII); and (v) EU-urban areas (IX).

Finally, natural landscape resources and historical-monumental ones were overlaid with the environmental landscape heritage areas, and the final map was ready (Fig. 21.6). Each area has a different intrinsic pressure which determines the proposed land uses to preserve the balance. (i) Subarea I—the North-North-East Anakena sector has the greater pressure, so it must be a protected area, with controlled intensive uses; (ii) Subarea II—the South-east Rano Raraku and Tongariki sector has great pressure, thus it should be protected and intensive uses should be controlled; (iii) Subarea III—the North-west Ana Te Pahu sector presents moderate pressure, so other alternatives could be considered; (iv) Subarea IV—the South-West Tahai sec-

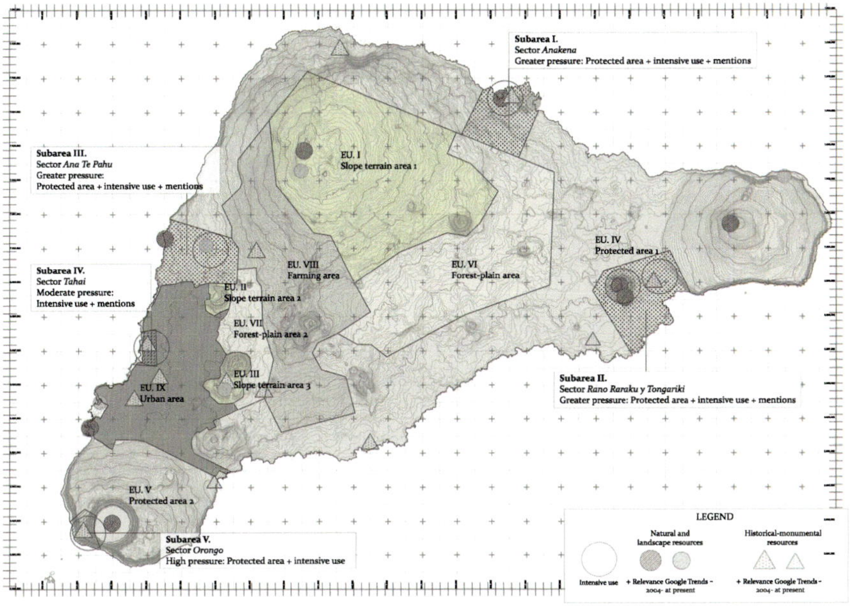

Fig. 21.3 Tourism attraction for landscape heritage urbanism, four factors: natural landscape; historical and monumental resources; crafts and gastronomy; and folklore and scheduled events. *Source* Alba Roque (2018). Research work, Universidad Politécnica de Madrid

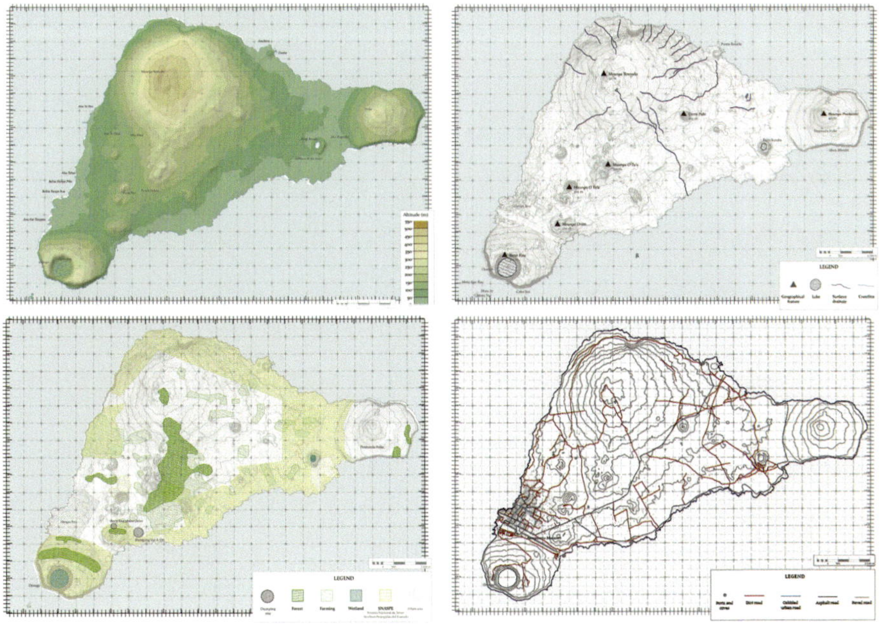

Fig. 21.4 Territorial and environmental analysis with four variables (upper left to lower right): relief; hydrological analysis; vegetation and regional infrastructures. *Source* Alba Roque (2018). Research work, Universidad Politécnica de Madrid

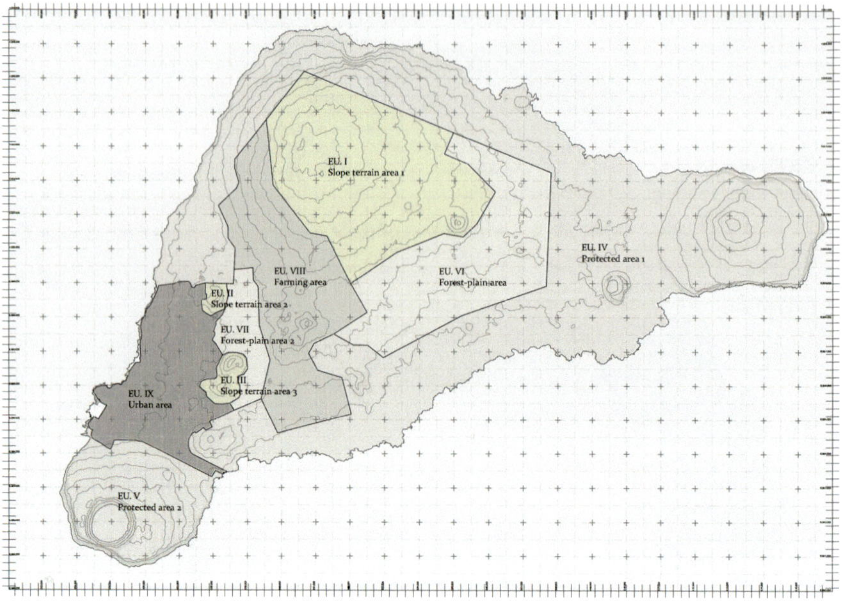

Fig. 21.5 Environmental landscape heritage areas. Five groups: EU-slope terrain areas (I–II–III); EU-protected areas (IV–V); EU-forest plain area (VI–VII); EU-farming areas (VIII); and EU-urban areas (IX). Source: Alba Roque (2018) Research work, Universidad Politécnica de Madrid

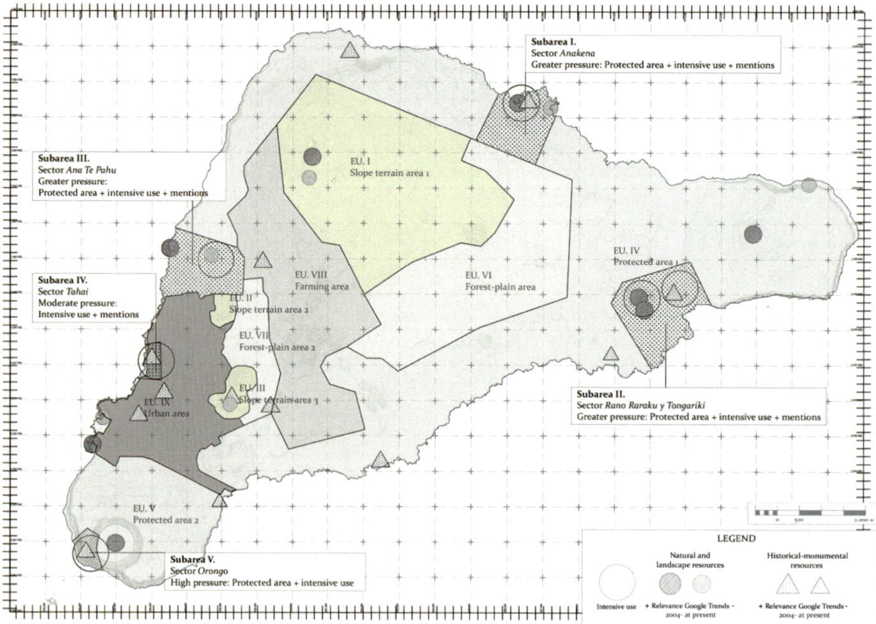

Fig. 21.6 Heritage landscape areas. Subarea I—North–North–East Anakena sector; Subarea II—-South–East Rano Raraku and Tongariki sector; Subarea III—North–West Ana Te; Subarea IV—-South–West Tahai sector; Subarea V—South Orongo Sector. *Source* Alba Roque (2018) Research work, Universidad Politécnica de Madrid

tor has moderate pressure as well; (v) and finally Subarea V—the South Orongo sector presents high pressure, so a protected area with non-extensive uses would be the most suitable proposal.

According to the previous layer of the environmental and heritage analysis, the landscape resources on Eastern Island are not the same in each of the five subareas determined, so this indicates how new actions must be carried out regarding the impacts on the physical, biotic, landscape and heritage components listed in the integrated matrix shown in Table 21.4.

Finally, Table 21.5 shows the temporary verification list, in a period of one and five years, on the Eastern Island case study as a landscape heritage urbanism plan.

21.7 Conclusion

Models of enhancement and revitalisation of the cultural heritage are connected to the holistic conceptualisation of the sense of the place. According to the conceptual background and the case study, some tools are summarised to determine the components of landscape cultural heritage.

Generally, tangible (physical) and intangible (sociocultural, ethnographic, etc.) values can be differentiated (Table 21.6). Physical heritage values are usually grouped into two major categories: the urban (historical heritage, popular architecture, quality public spaces, traditional urban morphology, unique architectural typology, painting, sculpture, etc.) and the natural (quality of the environment, the landscape, the diversity of fauna, native vegetation, territorial hydrography, etc.). Intangible sociocultural values can be understood as manifestations of the traditional culture of the society: religion, folklore, gastronomy, local handcrafts and so on.

The proposed uses and the matrix of compatibilities according to thresholds of impacts on air, soil and water for landscape heritage urbanism are described in Table 21.7, and as an integrated matrix in Table 21.8.

Finally, the temporary verification list (Table 21.9) will help to establish the action and management plan on the landscape heritage urbanism site.

The heritage landscape urbanism approach, through an integrated spatial and management plan, can promote compatible uses of territory, reduce negative impacts on air, soil and water and raise the quality of life for residents and visitors alike.

Table 21.4 Integrated matrix on the evaluation of the impacts on heritage and landscape

	Integrated matrix on the evaluation of the impacts on heritage and landscape Easter Island case study										
		Homogeneous areas Low or high impact (⊗)									
	Effects on the environment Physical, biotic, landscape and heritage components	[Subarea 1] Anakena Sector		[Subarea 2] Rano Rarak and Tongariki Sector		[Subarea 3] Ana Te Pahu Sector		[Subarea 4] Tahai Sector		[Subarea 5] Orongo Sector	
		L	H	L	H	L	H	L	H	L	H
Negative impacts	Direct or indirect effects on fauna		⊗		⊗	⊗		⊗			⊗
	Cumulative and synergistic effects on fauna		⊗		⊗	⊗		⊗			⊗
	Direct or indirect effects on biodiversity		⊗	⊗		⊗		⊗			⊗
	Cumulative and synergistic effects on biodiversity		⊗	⊗		⊗		⊗			⊗
	Direct or indirect effects on soil (contamination)		⊗	⊗		⊗		⊗			⊗
	Cumulative and synergistic effects on the soil		⊗	⊗		⊗		⊗			⊗
	Direct or indirect effects on air (pollution)	⊗			⊗		⊗	⊗			⊗
	Cumulative and synergistic effects on the air	⊗			⊗		⊗	⊗			⊗
	Direct or indirect effects on surface water		⊗		⊗	⊗		⊗			⊗
	Cumulative and synergistic effects on surface water		⊗		⊗	⊗		⊗			⊗
	Direct or indirect effects on airborne noise		⊗			⊗			⊗		⊗
	Cumulative and synergistic effects on airborne noise		⊗		⊗	⊗			⊗		⊗

(continued)

Table 21.4 (continued)

	Integrated matrix on the evaluation of the impacts on heritage and landscape Easter Island case study										
		Homogeneous areas Low or high impact (⊗)									
	Effects on the environment Physical, biotic, landscape and heritage components	[Subarea 1] Anakena Sector		[Subarea 2] Rano Rarak and Tongariki Sector		[Subarea 3] Ana Te Pahu Sector		[Subarea 4] Tahai Sector		[Subarea 5] Orongo Sector	
		L	H	L	H	L	H	L	H	L	H
	Direct or indirect effects on aquifer contamination		⊗		⊗	⊗			⊗		⊗
	Cumulative and synergic effects on aquifer contamination		⊗		⊗	⊗			⊗		⊗
	Direct or indirect effects on the landscape		⊗		⊗	⊗			⊗		⊗
	Cumulative and synergistic effects on the landscape		⊗		⊗				⊗		⊗
Evaluation	Total impacts of tourist uses	Low 02/16 HIGH 14/16		Low 04/16 HIGH 12/16		Low 14/16 HIGH 02/16		Low 10/16 HIGH 06/16		Low 00/16 HIGH 16/16	
	Total patrimonial operation in each place	Greater pressure Uses controlled		Greater pressure Preservation		Moderate pressure Compatible uses		Moderate pressure Compatible use		High pressure Protected area + intensive use	
Positive effects	Direct or indirect effects on heritage landscape urbanism		⊗		⊗	⊗			⊗		⊗
	Accumulative and synergistic effects on heritage		⊗		⊗	⊗			⊗		⊗
	Direct or indirect effects on employment generation	⊗		⊗			⊗	⊗			⊗

Table 21.5 Temporary verification of indicators list of the patrimonial uses in singular landscapes

Landscape heritage urbanism Temporary verification indicators list Eastern Island		
List of selected indicators	Checklist	
	1 year	5 years
First preventive and corrective measures to protect the atmospheric resources		
Verify that there are no new sources of emission of pollutants into the atmosphere	✓	✓
Verify that there are no unauthorized uses that could increase pollutant emissions	✓	✓
Second preventive and corrective measures to protect geological and geomorphological resources		
Verify that there are no clearings or landfills	✓	
Verify that the original geomorphology of the relief is respected	✓	✓
Check that there is no material removal	✓	
Third preventive and corrective measures to protect edaphic resources		
Verify that materials are not accumulated in areas with a slope of more than 5%	✓	
Verify maintenance of the current runoff rate	✓	✓
Check that the ground is not compacted by the passage of animals or heavy machinery	✓	
Verify that there is no contamination of the subsoil, by accident of materials or liquids		✓
Fourth preventive and corrective measures to protect water resources		
Verify the level of contamination of surface watercourses	✓	
To verify the good hydrological functioning of the channels and their aquifers		✓
Preventive and corrective measures to protect water resources	✓	✓
Check that there are no extraordinary risks during events		
Verify that no alien species are planted		✓
Verify that no current species are started, extracted, rooted, etc.		✓
Verify that pruning, cleaning and thinning are in accordance with the regulations	✓	
Sixth preventive and corrective measures to protect fauna		
Verify that there are no intrusive species in the area		✓
Verify that the breeding seasons of animals are not disturbed in breeding and nesting areas		✓
Verify that live specimens of any wild species are not captured		✓
Verify that the quality of the ecosystems is maintained		✓

(continued)

Table 21.5 (continued)

Landscape heritage urbanism Temporary verification indicators list Eastern Island		
List of selected indicators	Checklist	
	1 year	5 years
Seventh preventive and corrective measures to protect the landscape		
Verify that no posters are installed	✓	✓
Check that there are no bright or white shading elements. Awnings, tarpaulins, etc. should harmonise with the colour of the landscape: brown, ochre, sienna, dark green, mole green, or similar		✓
Eighth preventive and corrective measures of use. Terms of use		
Check that no other activities are carried out other than those regulated		✓
Verify that no inscriptions or signs are made on trees, stones or other natural elements		✓
Check that the decibel level does not exceed 55dBA at night, or 65dBA during the day, on no day of the year	✓	✓
Verify that megaphones are not used	✓	✓
Verify that wildlife is not disturbed	✓	✓
Ninth preventive and corrective measures for tourist uses		
Verify that there is no wild camping	✓	✓
Check that there are no bathing activities or nautical activities without control in surface water channels	✓	✓
Tenth preventive and corrective measures to protect waste discharges		
Verify that no landfill sites appear		✓
Check that there are no uncontrolled residues	✓	✓
Check the purification of liquid spills	✓	✓
Actions to be implemented		

Table 21.6 Values of the patrimonial place in landscape heritage

Values of the patrimonial place in landscape heritage urbanism
(i) Tangible values
The urban place (historical heritage, popular architecture, quality public spaces, traditional urban morphology, unique architectural typology, painting, sculpture, etc.)
The natural place (quality of the environment, singularity of the landscape, diversity of fauna, native vegetation, territorial hydrography and other similar resources)
(ii) Intangible sociocultural values
Traditional culture of the settlement from the religious, folkloric, gastronomic, local handicraft aspects, etc. The views, the landscape, the intervisibility

Table 21.7 Thresholds of impacts of activities

Values Landscape heritage urbanism	Thresholds of impacts of activities High, medium or low impacts											
	Air impacts			Soil impacts			Water impacts			Community impacts		
	L	M	H	L	M	H	L	M	H	L	M	H
(i) Tangible values												
Historical heritage												
Popular architecture												
Quality public spaces												
Traditional urban morphology												
Unique architectural typology												
Painting												
Sculpture												
Quality of the environment												
Uniqueness of the landscape												
Diversity of fauna												
Native vegetation												
Territorial hydrography												
Landmark elements												
(ii) Intangible sociocultural values												
Traditional culture												
Religious values												
Folklore												
Gastronomy												
Ethnographic values												
Local crafts												

Table 21.8 Integrated matrix of the evaluation of the effects on heritage and landscape

	Integrated matrix of the evaluation of the effects on heritage and landscape						
		Homogeneous areas Low or high impact					
	Effects on the environment Physical, biotic, landscape and heritage components	[AREA 1]		[AREA 2]		[AREA3]	
		L	H	L	H	L	H
Negative impacts	Direct or indirect effects on fauna						
	Cumulative and synergistic effects on fauna						
	Direct or indirect effects on biodiversity						
	Cumulative and synergistic effects on biodiversity						
	Direct or indirect effects on soil (contamination)						
	Cumulative and synergistic effects on soil						
	Direct or indirect effects on air (pollution)						
	Cumulative and synergistic effects on the air						
	Direct or indirect effects on surface water						
	Cumulative and synergistic effects on surface water						
	Direct or indirect effects on airborne noise						
	Cumulative and synergistic effects on airborne noise						
	Direct or indirect effects on aquifer contamination						
	Cumulative and synergic effects on aquifer contamination						
	Direct or indirect effects on the landscape						
	Cumulative and synergistic effects on the landscape						

(continued)

Table 21.8 (continued)

	Integrated matrix of the evaluation of the effects on heritage and landscape						
		Homogeneous areas Low or high impact					
	Effects on the environment Physical, biotic, landscape and heritage components	[AREA 1]		[AREA 2]		[AREA3]	
		L	H	L	H	L	H
Evaluation	Total impacts of tourist uses						
	Total patrimonial operation in each place						
Positive effects	Direct or indirect effects on heritage landscape urbanism						
	Accumulative and synergistic effects on heritage						
	Direct or indirect effects on employment generation						

Table 21.9 Temporary verification indicators list of patrimonial uses in singular landscapes

Landscape heritage urbanism Temporary verification indicators list Patrimonial uses in singular landscapes		
List of selected indicators	Checklist	
	1 year	5 years
First preventive and corrective measures to protect atmospheric resources		
Verify that there are no new sources of emission of pollutants into the atmosphere		
Verify that there are no unauthorised uses that could increase pollutant emissions		
Partial evaluation		
Second preventive and corrective measures to protect geological and geomorphological resources		
Verify that there are no clearings or landfills		
Verify that the original geomorphology of the relief is respected		
Check that there is no material removal		
Partial evaluation		
Third preventive and corrective measures to protect edaphic resources		
Verify that materials are not accumulated in areas with a slope of more than 5%		

(continued)

Table 21.9 (continued)

Landscape heritage urbanism Temporary verification indicators list Patrimonial uses in singular landscapes		
List of selected indicators	Checklist	
	1 year	5 years
Verify maintenance of the current runoff rate		
Check that the ground is not compacted by the passage of animals or heavy machinery		
Verify that there is no contamination of the subsoil, by accident of materials or liquids		
Partial evaluation		
Fourth preventive and corrective measures to protect water resources		
Verify the level of contamination of surface watercourses		
Verify the good hydrological functioning of the channels and their aquifers		
Preventive and corrective measures to protect water resources		
Check that there are no extraordinary risks during events		
Partial evaluation		
Fifth preventive and corrective measures to protect flora and vegetation		
Verify that no alien species are planted		
Verify that no current species are started, extracted, rooted, etc.		
Verify that pruning, cleaning and thinning are in accordance with the regulations		
Partial evaluation		
Sixth preventive and corrective measures to protect fauna		
Verify that there are no intrusive species in the area		
Verify that the breeding seasons of animals are not disturbed in breeding and nesting areas		
Verify that live specimens of any wild species are not captured		
Verify that the quality of ecosystems is maintained		
Partial evaluation		
Seventh preventive and corrective measures to protect the landscape		
Verify that no posters are installed		
Check that there are no bright or white shading elements. Awnings, tarpaulins, etc., should harmonise with the colour of the landscape: brown, ochre, sienna, dark green, mole green or similar		
Partial evaluation		

(continued)

Table 21.9 (continued)

Landscape heritage urbanism Temporary verification indicators list Patrimonial uses in singular landscapes		
List of selected indicators	Checklist	
	1 year	5 years
Eighth preventive and corrective measures of use. Terms of use		
Check that no other activities are carried out than those regulated		
Verify that no inscriptions or signs are made on trees, stones, or other natural elements		
Check that the decibel level does not exceed 55dBA at night, or 65dBA during the day, on no day of the year		
Verify that megaphones are not used		
Verify that wildlife is not disturbed		
Partial evaluation		
Ninth preventive and corrective measures for tourist uses		
Verify that there is no wild camping		
Check that there are no bathing activities or nautical activities, without control in surface water channels		
Partial evaluation		
Tenth preventive and corrective measures to protect waste discharges		
Verify that no landfill sites appear		
Check that there are no uncontrolled residues		
Check the purification of liquid spills		
Partial evaluation		
Total evaluation		

References

Alba Roque AP (2018) Isla de Pascua: visualización de la acción del turismo. Research work E.T.S. Arquitectura Universidad Politécnica de Madrid, Madrid. http://oa.upm.es/49733/. Accessed 21 June 2018

Amador E, Cayot L, Cifuentes M, Cruz E, Cruz F (1996) Determinación de la Capacidad de Carga Turística en los sitios de visita del Parque Nacional Galápagos. Servicio Parque Nacional Galápagos, Instituto Ecuatoriano Forestal y de Áreas Naturales y Vida Silvestre, Puerto Ayora. http://81.47.175.201/stodomingo/attachments/article/205/CCT_Galapagos.pdf. Accessed 21 June 2018

Bote V (1990) Planificación económica del turismo, de una estrategia masiva a una artesanal. Trillas, México

Casasola L (1990) Turismo y ambiente. Trillas, México

Castellani V, Sala S (2012) Carrying capacity of tourism system: assessment of environmental and management constraints towards sustainability. In: Kasimoglu M (ed) Visions for global tourism industry—creating and sustaining competitive strategies. IntechOpen, London, pp 296–316. ISBN 978-953-51-0520-6

Cifuentes M (1992) Determinación de Capacidad de Carga Turística en Áreas Protegidas. Centro Agronómico Tropical de Investigación y Ensenanza, CATIE, Turrialba
Coccossis H (1998) Assessment of integrated coastal management in Africa. Government of Finland, United Nations Environment Programme, Priority Actions Programme, Split. https://www.researchgate.net/publication/277135951_Assessment_of_Integrated_Coastal_Management_in_Africa. Accessed 21 June 2018
Cooper C, Jackson S (1989) Destination life cycle: the Isle of Man case. Ann Tourism Res 16:377–398
Cortés Chavez S (2009) La capacidad de carga como herramienta para laordenación sostenible delterritorio. Cuad Invest Urbanística 65:35–55
Council of Europe (2000) European landscape convention. Available via Council of Europe. https://www.coe.int/en/web/conventions/full-list/-/conventions/rms/0900001680080611. Accessed 21 June 2018
Desor JA (1972) Toward a psychological theory of crowding. J Pers Soc Psychol 21(1):79–83
Getz D (1983) Capacity to absorb tourism—concepts and implications for strategic planning. Ann Tourism Res 10(2):239–263
Higueras E (2006) Urbanismo bioclimático. Impreso en Espagna, Barcelona
Khanna P (1999) Carrying capacity as a basis for sustainable development: a case study of national capital region in India. Prog Plann 52(2):101–166
Lapage WF (1963) Some sociological aspects of forest recreation. J Forest 61:32–66
Mathieson A, Wall G (1982) Tourism: economic, physical and social impacts. Longman House Press, New York
McHarg IL (1969) Design with nature. Doubleday/Natural History Press, New York. ISBN 0471-11460-X
Pigram P (1983) Outdoor recreation and resource management. St. Martin's Press, New York
Robert E (2002) How much is too much? Carrying capacity of national parks and protected areas. In: Arnberger A, Brandenburg C, Muhar A (eds) Monitoring and management of visitor flows in recreational and protected areas—conference proceedings. Bodenkulur University, Vienna, pp 306–313
Sharpley R, Telfer DJ (eds) (2002) Tousim and development: concepts and issues. Aspects of tourism, no. 5. ISBN-1-873150-34-2
Tburot JM (1980) Capacite de Charge et production touristique. CHET, Aix-en-Province
Tuomisto H (2010) A consistent terminology for quantifying species diversity? Yes, it does exist. Oecologia 164(4):853–860. https://doi.org/10.1007/s00442-010-1812-0
UN (1972) Declaration of the United Nations conference on the human environment. Available via UN. http://www.un-documents.net/unchedec.htm. Accessed 21 June 2018
UNWTO (2010) UNWTO annual report—a year of recovery. Available via UNWTO. http://media.unwto.org/sites/all/files/pdf/finalannualreportpdf.pdf. Accessed 21 June 2018
UNWTO (2012) UNWTO annual report 2012. http://cf.cdn.unwto.org/sites/all/files/pdf/annual_report_2012.pdf. Accessed 21 June 2018
Watson GL, Kopachevsky JP (1996) Tourist carrying capacity: a critical look at the discursive dimension. Prog Tourism Hospitality Res 2(2):169–179. https://doi.org/10.1002/pth.6070020205
Weaver DB (2000) A broad context model of development scenarios. Tourist Manage 21(3):217–224

Chapter 22
Regeneration of Historical Urban Landscapes in the Hinterland of Marche Region

Flavio Stimilli and Massimo Sargolini

Abstract The hinterland of Marche region has been declining for a long time, mainly due to the process of *littoralization*, ongoing phenomenon worldwide that represents a great challenge for planners and policy-makers. Increasing urbanization, seasonal mass tourism and overall growing of human pressure are just some of the issues affecting the coast. Vice versa, dramatic depopulation, economic depression and abandonment of rural activities are just some of those affecting the hinterland. In many places, this trend has become chronic, almost irreversible. Few models of regional development have barely stopped it. Most frequently, the historical villages, towns and the related landscape heritage, so peculiar and precious in the hinterland of Marche region, just further decay. After the earthquakes of 2016, the Marche Regional Council has commissioned a research team of five Italian universities to draft a strategic programme aimed at regenerating the hinterland. Supported by local authorities, scholars and experts in different fields, however related to regional and spatial planning, have made a joint effort to conceive possible solutions to the long-lasting crisis of these areas, which started in fact long before the recent earthquakes. The programme highlights the unexplored potential and resources of the hinterland, promoting a polycentric, integrated and synergistic model of development. In this framework and spirit outlined in the first part of the paper, and from *Heritage Urbanism* perspective, the focus shifts onto the case study of Camerino, outstanding example of urban heritage, exploring issues and options for different approaches and intervention models to regenerate the historical urban landscapes.

Keywords Camerino · Historical village · Historical town · Intervention model · Strategic planning · Urban heritage

---

F. Stimilli (✉)
International School of Advanced Studies, University of Camerino, Camerino, Italy
e-mail: flavio.stimilli@unicam.it

M. Sargolini
School of Architecture and Design, University of Camerino, Ascoli Piceno, Italy
e-mail: massimo.sargolini@unicam.it

M. Obad Šćitaroci et al. (eds.), *Cultural Urban Heritage*,
The Urban Book Series, https://doi.org/10.1007/978-3-030-10612-6_22

22.1 Introduction

The series of earthquakes occurred in central Italy between the end of 2016 and the beginning of 2017 has struck a huge hilly and mountainous area, stretching across four regions—Lazio, Marche, Umbria and Abruzzo—and affecting 140 municipalities, more or less damaged by the earthquakes.

Lazio and Marche have been the hardest hit regions in terms of direct victims—249 and 49, respectively—and in terms of damages to the historical and cultural heritage. In this regard, Lazio has been facing the most striking aftermaths, with the old towns of Amatrice and Accumoli razed to the ground, while Marche holds the largest affected area, with 87 municipalities and countless works of art and architecture forever compromised.

The profile and image of the historical urban landscapes (Sonkoly 2017) have been changing dramatically in Marche region, due to the high number of old dwellings and towns either destroyed or strongly damaged by the earthquakes of 2016.[1] All around the Sibillini Mountains—especially in the upper part of the Tronto Valley to the south-east and in the upper area of the province of Macerata to the north-west—most of the historical centres in the old villages and towns are now empty and abandoned, the entrance to these present-time *red zones* being actually forbidden for security reasons.[2]

Throughout nearly one year after the earthquakes, mayors and city administrations have worked hard to secure the historical centres, often surrounded by medieval defensive walls, and preserve them from further collapse of the buildings. In the meanwhile, they have also managed to provide temporary housing and alternative accommodations to their citizens, thus allowing for the start of the reconstruction projects and activities, which are still at the very beginning.

However, what has been changing is not only the skyline, the physical appearance and the conditions of the urban settlements within the historical walls—with total or partial collapse of the buildings and the progress of spontaneous vegetation in the gardens, squares and streets—but also the whole landscape outside and the overall social and economic structures of the hinterland. Many people indeed have chosen or had to move from the countryside too. They resettled somewhere else, most frequently on the coast, and often stopped their rural and forest activities, thus letting the natural processes take back the lead in the evolutionary dynamics of the landscape. This trend has actually started long before the earthquakes of 2016, but the last disasters have undoubtedly deepened and accelerated the crisis of the hinterland. The Italian government and the regional authorities have taken immediate and strong countermeasures in the short run and are now rethinking globally the development strategies for these areas in the long term (Esposito et al. 2017).

As difficult as it is to restore historical buildings and medieval towns in compliance with modern standards and anti-seismic regulations, although with the great help of the most advanced earthquake-proof technologies, much harder in fact is to conceive

[1]Especially the three main quakes of 24 August, and 26 and 30 October 2016.

[2]For example, Arquata del Tronto, Castelsantangelo sul Nera, Visso and Camerino.

and implement a feasible strategy and vision for the overall future and the sustainable development of the hinterland.

The case of Nocera Umbra, lovely medieval town heavily damaged by another earthquake in 1997 and very well reconstructed, has taught planners and decision-makers that rebuilding houses, though in the best way possible, is not enough. The process of reconstruction lasted almost twenty years, and people are still far from having comeback to repopulate the town, which is empty for more than a half.

With this in mind, the Marche Regional Council has funded an interuniversity research project that, after having studied in depth the overall situation of the hinterland before and after the earthquakes, has pointed out priorities of intervention and new paths of development for the Apennine area of Marche region.

22.2 The Regional Approach and the Model of Transformation and Development

Capitalizing on the *Italian national strategy for inland areas*,[3] the research "New paths of development for the Marche Apennines after the earthquake of 2016"[4] has tried to provide policy-makers with an innovative tool that could help and guide them in the elaboration of the regional plan for the reconstruction and development of the hinterland.

What is now on the agenda of decision-makers at a regional scale, although with a strong delay and urgency, is indeed the choice of the best and most needed projects and interventions to pursue and carry out in the following years, as well as the draft of a consistent and balanced framework, within which to place them harmonically.

On the ground of these premises, the programme outlined by the joint team of some forty researchers is composed of eleven "new paths of development", covering the main strategical topics and issues, from smart mobility to innovative management of natural resources and cultural heritage. Some of them relate in particular to the regeneration of the historical urban settlements of the hinterland, suggesting a general model of intervention to apply with slight adjustments for each homogeneous area and for similar cases.

To understand thoroughly the issues affecting each area and to take into account the views of local communities, the research team has also carried out a series of in-depth interviews with the mayors of the 87 municipalities that fall within the seismic crater of Marche region.

Then, by merging this "local knowledge and perspective" with their scientific studies and expertise, researchers have outlined a strategic interpretation of each analysed topic, identifying critical issues and potentialities, and providing possible

[3]Strategia Nazionale per le Aree Interne (SNAI), promoted by the national Agency for Territorial Cohesion: http://www.agenziacoesione.gov.it/it/arint.

[4]Nuovi Sentieri di Sviluppo per l'Appennino Marchigiano dopo il Sisma del 2016: http://www.consiglio.marche.it/iniziative/appennino/index.php.

Table 22.1 (continued)

Topic	Issue	Potential	Main guidelines/actions
Disaster risk reduction	Excessive **building** and **population density**	Enhancement of **public open spaces** and **quality of life**	**Selective demolitions** for new public gardens and squares, gathering places and escape routes
Territorial cohesion and **critical mass**	**Dispersion of villages** and **towns** in a large area	**Complementarity** of services and heritage	**Unification of municipalities**
			Shared management of services and **heritage** by unions/networks of municipalities
Castled position of villages and towns	**Risk of *musealization*** of the historical centres	**Modern** and **lively downstream settlements**	**Pilot projects** for reuse of architectural heritage in strategic position between upper and lower towns and between towns

These typical factors are:

- Historical centre on the ridge of a hill, with medieval and compact urban fabric, rich in architectural heritage and enclosed by (high) defensive walls.
- Increasing depopulation and ageing of people in the old town (started decades ago), resulting in potential availability of many houses and buildings, already empty or abandoned for a long time before the earthquakes.
- Modern neighbourhoods downstream quite disconnected and far from the historical settlement—which is physically and perceptually isolated inside the walls—with strong differences in height (Fig. 22.1).
- Hydrogeological fragility of the spatial context (e.g. steep slopes, landslide risk).
- "Centrifugal thrusts" towards urban expansion and abandonment of the old town, included those on behalf of the university—which is playing in fact a crucial role for the future of the city, as in the paradigmatic case of Konuralp (Tanriverdi Aysegul 2016).
- Severe damage to the architectural heritage (especially churches) and access forbidden to most of the historical centre ("red zone").
- Access points to the old town too narrow and compromised by the (risk of) collapse of bell towers and/or other structures and buildings.
- Explicit and shared desire of reducing building density by demolishing a few of "out-of-place" buildings and making room to new public open spaces.

Fig. 22.1 Free interpretation of the spatial arrangement of Camerino and the relationship between the upper and the lower towns (Drawing by Martina Alessandrini, 2017)

- Temporary houses[5] built downstream, quite far and with no connection or relationship with the historical centre, but with a strong landscape impact and extraordinary land take.

Most evident in Camerino is the widening gap between the upper (old) town and the lower (modern) settlements—already enlarged with new constructions, whether temporary or not—and the change in the historical urban landscape and skyline. This is mainly due to the (partial) collapse of a few churches, bell towers and buildings (especially some erected on the city walls), as well as by the impressive support and reinforcement structures built up throughout the town.

However, the most striking impression and proof of change after the earthquakes is the image of Camerino by night: a huge black hole has taken the place of the bright upper town, and the overall night landscape has completely turned to strong darkness.

As for the gap between the upper and the lower towns, actually it has quite distant and very deep origins, with roots that sink ideally into the dawn of modernism, which implied a radical break in urban conservation paradigms, having separated the historical city from the main paths of urban development (Bandarin and van Oers 2012). In Camerino, where the modern expansion started mainly after the Second World War, a comparison between two different urban plans of 1965 and 1967,

[5]S.A.E. Soluzioni Abitative d'Emergenza ("emergency housing solutions").

about the development of the university, clarifies well the issue (Figs. 22.2 and 22.3). These are the urban development plan drawn up by Ezio Mariani in 1965, which finally prevailed, and the plan drafted in 1967 by famous architect and urban planner Ludovico Quaroni, who tried to conceive an expansion of the town very close to the historical centre and in consonance with the historical layout and the intrinsic logic of its urban fabric. As opposed to this principle of historical continuity and physical consistency, Mariani suggested a linear expansion along a main road stretching directly and straight out of the town (Mariani 1986). In later years, the City Hall chose and implemented in fact this rather than that idea.

Following this trend of "taking distances" from the historical centre, the recently built-up emergency housing solutions are just on the edges of modern districts, deepening the gap between these and the upper town. If it is true in fact that an even worse option on the desk of decision-makers could have been to place these temporary houses completely out of Camerino, down the valley towards Castelraimondo, another possibility was actually to use and place them to bridge the gap between the old town and the modern neighbourhoods, filling the spaces in between. Considering the pseudo-temporariness of the emergency housing—which has required heavy urbanization and infrastructure works that will last undoubtedly for a long time—this is a missed opportunity for Camerino.

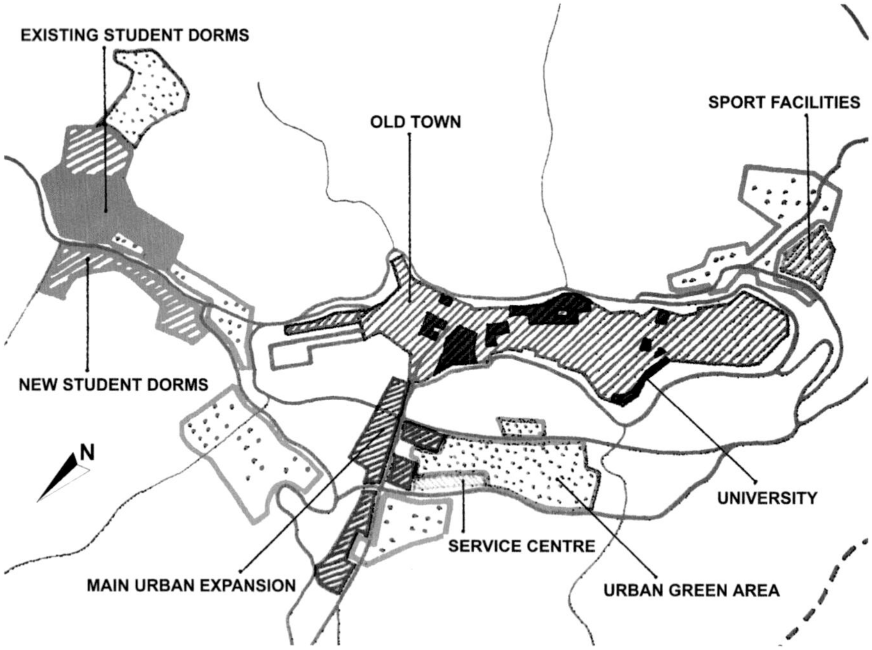

Fig. 22.2 Interpretation of urban development plans for the city of Camerino, by architect Mariani

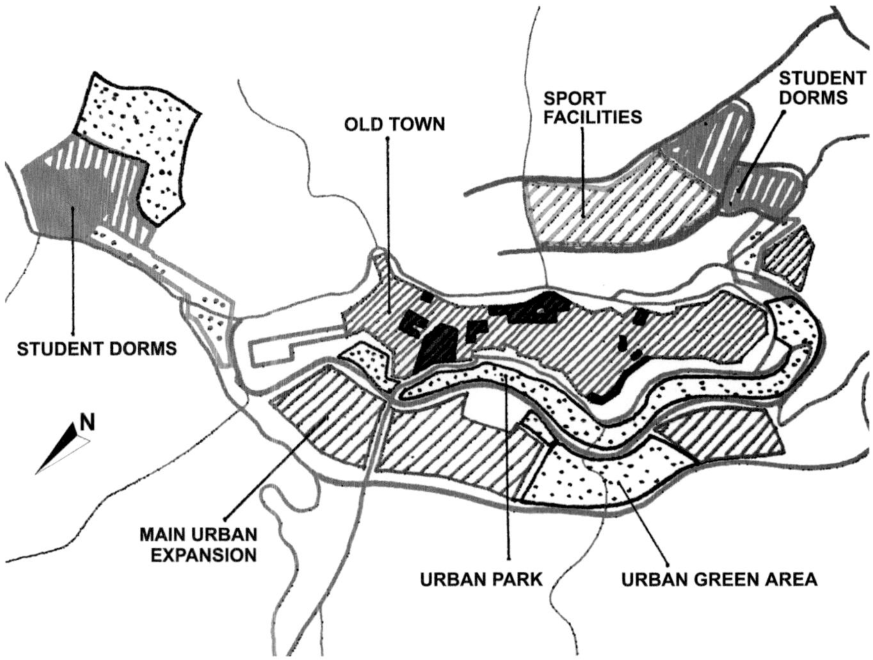

Fig. 22.3 Interpretation of urban development plans for the city of Camerino, by architect Quaroni

Quite meaningful, in a sense revelatory, is also that the new commercial centre, to be soon built up in the major of the modern neighbourhoods, is arranged around a square that features precisely the same shape of the main square in the old town, recalling even the same proportions and outward appearance of the facades. The interpretation of this resemblance, specifically required and commissioned to the architects by the City Hall, is actually twofold. On the one hand, it could testify the bond and affection of citizens for their historical centre, with the hope of coming back soon to live and experience it once again. On the other hand, considering that the new commercial centre is not temporary at all, not even on the paper, it might represent the "last farewell" to the old town, "identically" reproduced somewhere else.

Although one would tend most likely towards the second hypothesis, a reconnection between the historical centre and the modern satellite neighbourhoods is nonetheless something that any sensible urban planner would try at least to address (Zucconi Galli Fonseca 1994). The map in Fig. 22.4 is an attempt in this sense, exploiting the upcoming new highway that will run through the valley, providing a new gate to the town of Camerino at the bottom of the hill. From this point, a mechanized ascent—such as a small cable railroad, whether suspended or on the ground—could link the lower settlements to the upper town, bridging a linear distance of just over two kilometres but covering an overall height difference of more

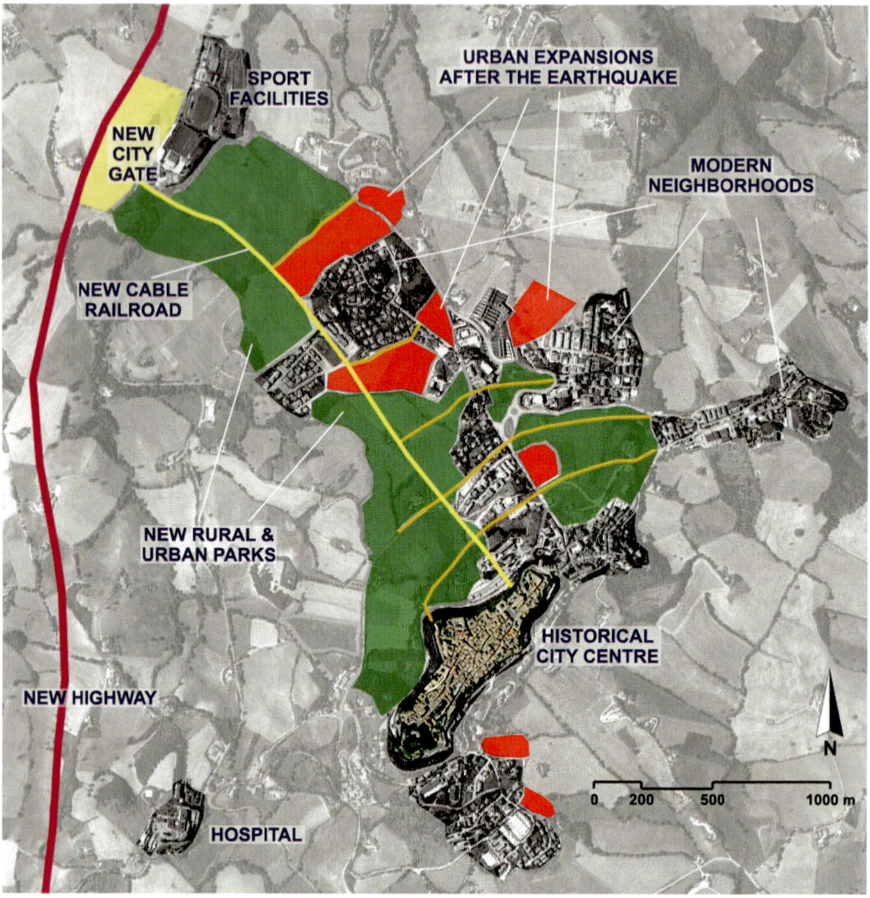

Fig. 22.4 Possible new layout of Camerino, after the construction of the highway

than 300 m. With some stops on the way, it could provide an easy access to the main districts of Camerino, reconnecting them all in a new spatial configuration, by redeveloping the empty fields between the parts through landscape architecture interventions for new urban and rural parks.

Unfortunately, the guidelines commissioned to famous architect Mario Cucinella by the City Hall and drawn up by him and his workgroup in support of the upcoming reconstruction and redevelopment plan of Camerino just focus on the method and phases to restore and reopen the historical centre. They do not take into account, altogether, what is going on outside the city walls, with the high risk of retracing the somehow negative experience of Nocera Umbra.

22.4 Models of Intervention on the Historical Urban Heritage

Drawing from the current public debate, either among specialists or among common people, and from the above case study—in particular from the curious reproduction of the old city square in a modern neighbourhood and from possible demolitions imagined in the upper town, a fact that in Italy is not obvious at all—four slogans describe the basic models of intervention:

- Where it was, how it was.
- Where it was, how it will be.
- Where it will be, how it was.
- Where it will be, how it will be.[6]

The first model, namely to restore or reconstruct the built heritage where it was, in the same way and with the same appearance, is feasible only for those cases whether so precious from the historical–cultural point of view or not too much compromised, provided that modern technologies allow for keeping the old construction method and materials with minor updates. Actually, this model was more in use and popular in the past—and often invoked by politicians to ensure people that the *status quo* will be set up again—but nowadays is quite old-fashioned and not very sought after, at least in principle.

The second instead is probably the most quoted and promoted today. It states the will of not giving up the loss of the built heritage, especially the place where it stood up, which is probably the main factor and driver of identity (Wang and Prominski 2016). According to this model, a building or an architectural complex could stay on the same site where it was, but with significant changes in terms of construction techniques and/or of aesthetic appearance.

In some cases, however, it will not be possible to restore or reconstruct the built heritage on the same site. For some reasons—e.g. seismic micro-zonation revealing a fragmented subsoil that accelerates the seismic waves too much and socio-economic opportunities prevailing on other assessments—that heritage will have to move to another, more appropriate, location. At this point, the choice will be between transferring or rebuilding it in the same way as it was, or building it up again but differently, with the use of new techniques and/or in a different, likely modern style.

Sometimes, it will be impossible to reconstruct on their own sites even entire settlements or some relevant parts of them, as in the case of the upper town of Arquata del Tronto and of many hamlets scattered throughout the seismic crater. In the latter case, the reason is usually of socio-economic and strategic nature, whereas in the first instance the choice is mainly due to the excessively bad conditions of the subsoil.

---

[6]"Dov'era, com'era", "Dov'era, come sarà", "Dove sarà, com'era", Dove sarà, come sarà".

22.5 Conclusion

However different from each other, all four models of intervention presuppose the possibility of rebuilding or restoring the lost or damaged heritage, somehow and somewhere. Nevertheless, this is not always true and does not apply in fact to everything. For better or worse, it will not be possible altogether to restore or rebuild many cases, especially in very small and remote hamlets, where the owners may not have any interest to come back or to reconstruct anything, even though in a different site, which should fall, however, within the same municipal area.

Apart from that, the four models cover all the rest of instances that are nowadays under discussion. As simple as they are, the models can be valid and helpful not only in Italy—where they have already managed to break at least the taboo of "*everything must remain/return as it is/was*"—but potentially abroad too.

Croatia, for example, is as rich in historical villages and small towns as Italy, especially in the hinterland. Likewise, the process of *littoralization*, also named *coastalization* (Terkenli et al. 2015), has been affecting this country too, all along the coastline and in the hinterland. Moreover, the last Balkan War had a strong impact—on the hinterland of Dalmatia in particular—with effects on the historical urban heritage comparable to the ones produced in Italy by the earthquakes of 2016. Therefore, starting from the present research, it can be extremely interesting and promising to carry out a comparative study between these two cases and the corresponding models of intervention.

Acknowledgements The research is a part of the scientific project "Heritage Urbanism—Urban and Spatial Planning Models for Revival and Enhancement of Cultural Heritage". It is financed by the Croatian Science Foundation [HRZZ-2032] and carried out at the University of Zagreb, Faculty of Architecture.

References

Bandarin F, van Oers R (2012) The historic urban landscape: managing heritage in an urban century. Wiley-Blackwell, Chichester. https://doi.org/10.1002/9781119968115. ISBN: 9780470655740

Bittarelli AA (1992) Camerino: Le sue vie, le sue frazioni. Mierma, Pieve Torina

Esposito F, Russo M, Sargolini M, Sartori L, Virgili V (2017) Buiding Back Better: idee e percorsi per la costruzione di comunità resilienti. Carocci, Rome. ISBN: 9788843090600

Mariani E (1986) Università degli Studi di Camerino: Aggiornamento del piano di sviluppo edilizio. Università di Camerino, Camerino

Sonkoly G (2017) Historical urban landscape. Palgrave Macmillan, Cham. https://doi.org/10.1007/9783319491660. ISBN: 9783319491653

Tanriverdi Aysegul K (2016) Method for assessment of the historical urban landscape. Procedia Eng 161:1697–1703. https://doi.org/10.1016/j.proeng.2016.08.648

Terkenli TS, Douguédroit A, Cassar LF (2015) Connections, mobilities, urban prospects and environmental threats: the Mediterranean in transition. Cambridge Scholars Publishing, Cambridge. ISBN: 9781443872461

Wang F, Prominski M (2016) Urbanization and locality: strengthening identity and sustainability by site-specific planning and design. Springer, Heidelberg. ISBN 9783662484920
Zucconi Galli Fonseca C (1994) Le mura di Camerino. Università degli Studi di Camerino, Camerino

Chapter 23
Models of Terraced Landscape Regeneration in the Case of Slovenia

Lucija Ažman Momirski

Abstract Terraced landscapes in Slovenia are a feature that has only been partly inventoried and has not been properly recognized. However, it is possible to find many examples of terraced landscapes in the Register of Slovenian Cultural Heritage, which is a central repository of data on heritage maintained by the Slovenian Ministry of Culture. There are 220 immovable cultural heritage units registered as cultural landscapes. The register defines landscape heritage as a heritage site that is an open space with natural and artificial components in its structure, the development and use of which are chiefly determined by human processes and activities. All types of agricultural terraces can be found protected as cultural heritage areas. A comparison of terraced landscapes in Slovenia shows that terraces have been dramatically decreasing over the last decade. Models for renewing terraced landscape heritage in Slovenia have an agronomic basis. There is no model similar to terraced landscape management, and none of Slovenia's terraced landscapes are protected as a UNESCO World Heritage Site. This study presents new approaches to terraced landscape renewal based on strategic spatial planning, although the models are patterns whose idea can be understood but cannot be copied because the meaning of the context of individual examples varies. The main fear of people living in terraced landscapes, which are defined as cultural heritage sites, is that protective measures could pose many obstacles to their lives, work, and residence. Every regeneration of terraced landscapes must be based on the participation of the local community.

Keywords Terraced landscapes · Slovenia · Cultural heritage · Spatial planning · Topolò

L. Ažman Momirski (✉)
Faculty of Architecture, University of Ljubljana, Ljubljana, Slovenia
e-mail: lucija.azman@fa.uni-lj.si

M. Obad Šćitaroci et al. (eds.), *Cultural Urban Heritage*,
The Urban Book Series, https://doi.org/10.1007/978-3-030-10612-6_23

23.1 Introduction

Terraced landscapes are anthropogenic landscapes that people have created to grow crops on the sides of hills or mountains by planting on graduated platforms. The features of terraced landscapes are outstanding considering their diversity and appearance. They have recognizable cultural and symbolic meaning that is important at the local, national, and international levels. Terraced landscapes in Slovenia have only been partly inventoried in detail so far, and they have not been properly recognized by the legislative framework at various territorial levels or by various professions such as spatial planning. Nonetheless, considering all Slovenian territory, only nineteen municipalities (out of 211)—corresponding to 3.3% of the country's entire territory—have no terraced landscapes at all. In other words, terraced areas that form part of the cultural landscape in Slovenia can be found in more than 90% of the country's municipalities, corresponding to a little less than 97% of Slovenian territory. The presence of terraced landscapes in these municipalities is not uniform: In some municipalities, where most of the territory is flat, there may be only a few terraces on slopes that are not very steep at the edge of the municipality's territory (borderline cases were included in the category of municipalities with terraced landscapes). In other municipalities, terraced landscapes may be the dominant landscape feature. Both active (i.e., cultivated) and abandoned terraces were considered in the review (Ažman Momirski and Berčič 2016). Further studies showed that 104 municipalities in Slovenia have terraced vineyards (Ažman Momirski 2019).

In the Register of Slovenian Cultural Heritage, which is a central repository of data on heritage maintained by the Slovenian Ministry of Culture, it is also possible to find examples of terraced landscapes among cultural landscapes. There are 220 immovable cultural heritage units registered as cultural landscapes in the register, which defines landscape heritage as a heritage site that is an open space with natural and artificial components in its structure, and the development and use of which are chiefly determined by human processes and activities. All types of agricultural terraces can be found protected as cultural heritage areas, such as the Jeruzalem Hills near the village of Jeruzalem (Republika Slovenija Ministrstvo za kulturo 2018a).

The list of outstanding landscapes in Slovenia includes almost one hundred small landscape areas (Republika Slovenija Ministrstvo za okolje in prostor 2018). They are divided into field landscapes, landscapes of grasslands, landscapes of vineyards and orchards, and landscapes with exceptional natural structure. For each type of outstanding landscape, the actual problems, desired state, and guidelines for preserving or maintaining these landscapes have been defined. Some terraced landscapes have also been recognized as outstanding landscapes, such as the terraced areas in the Izola "amphitheater," in Rut in the Bače Gorge in the Alpine Mountains, in Jeruzalem in the Pannonian low hills, and in Velika Slevica in the Dinaric valleys and corrosion plains.

The Franciscan Cadaster prepared between 1823 and 1869, which was also prepared for tax regulation, shows historical terraced landscapes in Slovenian territory (Republika Slovenija Ministrstvo za kulturo 2018b). The surveying in this historical

Fig. 23.1 Cultural terraces as depicted in the Franciscan Cadaster: 1. Rut in the Bače Gorge in the Alpine Mountains; 2. Ostrožno Brdo in the Brkini Hills; 3. Dečja Vas in the Dinaric plateaus; 4. Smoleva in the Alpine Hills (Source: Republika Slovenija Ministrstvo za kulturo 2018b)

source is extremely precise considering the surveying conditions and instruments at the time. Most of the original cadastral maps and sketches were made at a scale of 1:2880. The basic units of the Franciscan Cadaster were the cadastral communities formed during the preparation of the Josephinian Cadastre between 1784 and 1789 (Republika Slovenija Ministrstvo za kulturo 2018c). There is a very clear depiction of terraced land in some plans, and the accuracy of the measurements is even greater than the accuracy of today's maps (Fig. 23.1). On the other hand, the plans do not contain any data for identifying the form of the terrain.

23.2 Terraced Landscapes: Landscapes of Abandonment and Ignorance

A review of terraced landscapes in Slovenia shows that terraces have been dramatically decreasing over the last decade, which is best represented by data on terraced vineyards. In 2007, 37% of vineyards were still located on terraces (Štabuc and Hauptman 2008), but in 2011 this share was only 35%, and in 2016 only 31% (Republika Slovenija Ministrstvo za kmetijstvo, gozdarstvo in prehrano 2018). The physical condition of abandoned lands shows that landowners gave up the basic responsibilities of ownership, such as routine maintenance. A landscape that has been abandoned is among the most visible outward signs of a community's decay, and depopulation is one of the major threats to rural landscapes (ICOMOS-IFLA 2018). Not only the loss of population but also a decline in the number of households and low income of the inhabitants have negative effects that exacerbate land decay. Often a concentration of abandoned and unmaintained land affects adjacent plots, where similar processes start. It is also important to know how long the land has been in this particular condition, although the degree of land degradation depends on many factors such as soil structure, relief characteristics, insolation, and so on. There is a connection between the mode of decay of terraced landscapes and their construction. Trees and forest overgrow agricultural terraces that are no longer cultivated. The irregularly maintained terrace slopes deteriorate due to movements of the earth. Only a small number of terraces are built in such a stable way that terraces can remain intact for decades or even centuries. In Slovenia, an example of an exceptional terraced landscape is in Ostrožno Brdo in the Brkini Hills. Because grass-covered terrace slopes are much more common in the Slovenian cultural landscape than dry stone wall terrace construction, the disintegration of such a terraced landscape is seen as an irregular slope characterized by the remnants of elements of the terraced landscape. The construction of dry walls also requires constant repair and clearing of weeds, and walls often collapse because of erosion (Fig. 23.2).

Fig. 23.2 Abandoned terraced landscapes in Slovenia: 1. Collapsed dry stone walls in the Koper Hills; 2. Due to the abandonment of agriculture, the terraced landscape in Izola was transformed and eroded; 3. Forest overgrowing terraces in the Brkini Hills (*Photograph* Lučka Ažman Momirski); 4. In Haloze, in eastern Slovenia, trees have expanded onto agricultural terraces (*Photograph* Matevž Lenarčič)

Terraced landscapes cannot only be called a landscape of abandonment (Varotto 2015), but also a landscape of ignorance (Ažman Momirski and Berčič 2016). This definition refers to three levels of observation: regarding information on the extension of cultural terraces in Slovenia, recognizing terraced regions as a unique landscape system, and accepting newly developed and improved methods of constructing terraces. Although a variety of methods for analytical observation of terraced landscapes applied in the past, it is necessary to inventory terraced areas following the methodology of different territorial levels.

In some regions in Slovenia, terraced landscapes are increasing: An updated plan of terraced landscapes in the Gorizia Hills shows that the terraced area has grown by almost one hundred hectares in the last ten years (Ažman Momirski 2019). The share of land designated for olive cultivation, with terrace construction, also continues to increase despite high investment costs. Models for regenerating terraced landscapes obviously exist. Terrace construction ranks among the major land improvement measures. Slovenia's Agricultural Land Act (Republika Slovenija 2011) mentions terraces in the paragraph on improving agricultural land, which also includes leveling land, and creating and removing terraces. A successful model to regenerate terraced landscapes in Slovenia can be found in the southern Gorizia Hills (Fig. 23.3). The lower slopes in the western and southern parts of the Gorizia Hills are the most favorable areas for agriculture. There are fluctuations in the volume of terraced agricultural land from year to year. Certainly, the greatest impact on the ups and downs of renewed terraced areas and a key condition for the preservation of agricultural

Fig. 23.3 Renewed terraced slopes in the southern Gorizia Hills (*Photograph* Lučka Ažman Momirski and Matevž Lenarčič)

production is when cultivation allows farmers to achieve a reasonable income for the work invested. The favorable circumstance is that modern and mechanized farming has been established in the Gorizia Hills.

Agriculture has a decisive influence on the development of a terraced cultural landscape. Today, the slopes that have been transformed by people for centuries and even millennia are presented as a heritage landscape, but at the same time this recognition becomes an obstacle to the further development of these landscapes. Therefore, the direction of the development of terraced landscapes must be derived from both protection and development motifs. The main fear of the people that live in terraced landscapes defined as heritage sites is that the protection measures could pose many obstacles to their lives, work, and residence.

23.3 Theoretical Background

The literature concentrates on identifying terraced systems (Berčič 2016), their technologies (Konold and Petit 2013), and their social organization (Watson 2009), on research about natural hazards (Brancucci and Masetti 2008), land degradation (Inbar and Zagier 2016), and conservation (Laureano 2012), on studies of the development of terraced landscapes (Harfouche 2007), their role in agricultural production (Salas 2015), and food security (Junchao 2012), on tourism development and promotion (Sun et al. 2010), and on policies (Tillmann and Salas 2016), regulations (Fontanari 2008), and management for preserving terraced landscapes and their functioning (Yamaoka 2012). In the past, studies in Slovenia have concentrated on reporting the extensive abandonment of cultivated terraces in the Koper Hills and Gorizia Hills (Vrišer 1954; Melik 1960; Titl 1965), on terrace construction methods following the development of agricultural technology and terrace construction methods using agricultural machinery (Colnarič 1971; Colnarič et al. 1985; Colnarič and Vrabl 1991; Škvarč 1999; Vršič and Lešnik 2005; Škvarč and Kodrič 2007), and on terminology connected with terraces and old viticulture techniques in villages in the countryside around Koper from the mid-nineteenth century to the 1950s (Presl 1995).

The literature refers to several models of terrace renewal. The project Adopt a Terrace (Varotto and Lodatti 2014) relies on negotiations between landowners that are abroad or unable to take care of their terraces and mountain enthusiasts interested in supporting terrace conservation. Terraced landscapes are degraded by the loss of functionality. Despite the fact that the appearance of terraces is increasingly being studied, the problem of maintaining and managing abandoned terraces remains. The authors propose that it is possible to establish new models to preserve the terraced landscape through new forms of social governance, in particular through a partnership between the residents of cities and mountain areas, thus surpassing twentieth-century models based on tourism and nature conservation. In this way, families that care for the mountain landscapes on a daily basis can return to mountain regions. The pilot project was carried out in the Brenta Valley, which has been marked by tobacco cultivation since the seventeenth century. There, negative demographic development also influenced the abandonment of traditional agricultural systems. The strategic model of terrace renewal was based on the following steps: identifying abandoned land, establishing contacts with owners of potentially usable land, preparing lease agreements with owners, and ensuring compliance with a number of good administrative rules.

Such a project is not the only adoption initiative, although there are some characteristic issues developed only by this project: an established institutional partnership among academia, local governments, and associations; using wasteland and overcoming the limitations of private property; and referring to and using adoption as a tool for enhancing a non-profit and multifunctional approach to the land. Land stewardship is an innovative and useful strategy for managing and conserving the land through the involvement of stakeholders (LandLife n.d.) and their commitment. It encourages the involvement of people that work in and know their landscape and through voluntary agreements between landowners/users and land stewardship organizations. One of the strengths of the stewardship approach is that all social groups can use it or participate in some way.

The term cultural heritage encompasses several main categories of heritage (UNESCO 2018):

- Cultural heritage
  - Tangible cultural heritage:
  - movable cultural heritage (paintings, sculptures, coins, manuscripts)
  - immovable cultural heritage (monuments, archaeological sites, and so on)
  - underwater cultural heritage (shipwrecks, underwater ruins and cities)
- Intangible cultural heritage: oral traditions, performing arts, and rituals
- Natural heritage: natural sites with cultural aspects such as cultural landscapes, physical, biological, or geological formations
- Heritage in the event of armed conflict.

Since the new cultural landscape category was introduced in 1992 for potential world heritage sites, terraced landscapes have been listed as UNESCO World Heritage Sites (Peters 2015). The representatives of the cultural landscapes that endeavor to secure the status of a UNESCO World Heritage Site can expect that the maintenance of the terraced landscape would not only be performed by landowners and local and national authorities. With UNESCO's financial support, terraces can be preserved, thereby preventing all other processes, such as the danger of landslides. Such areas of exceptional history, aesthetic value, and grandeur usually attract many people, and also an increased inflow of tourists, investors/entrepreneurs, artisans, and jobseekers. "The Rice Terraces of the Philippine Cordilleras is an outstanding example of an evolved, living cultural landscape that can be traced as far back as two millennia ago in the pre-colonial Philippines" and was added to the list in 1995 (UNESCO 2015a). The cultural landscape of the province of Bali in Indonesia consists of rice terraces and their water temples and is a result of the *subak* system as a manifestation of the *Tri Hita Karana* philosophy (UNESCO 2015b). It was added to the list in 2012. Spectacular terraces of the cultural landscape of the Honghe Hani rice terraces were acknowledged in 2013 (UNESCO 2015c). Recently, the cultural landscape of southern Jerusalem in Battir, a land of olives and vineyards, was identified as a representative of an outstanding example of a landscape (UNESCO 2015d). Two European terraced areas can be found on the UNESCO list. In 1997, "Portovenere, Cinque Terre, and the Islands (Palmaria, Tino, and Tinetto)" was added to the UNESCO World Heritage List (UNESCO 2015e); Cinque Terre is a belt on the northeastern coast of the Ligurian Sea in Italy. The Lavaux vineyard terraces along the south-facing northern shores of Lake Geneva were added to the list in 2007 (UNESCO 2015f). In these terraced landscapes, not only the terraces themselves are protected and safeguarded, but also the "intangible culture and knowledge of the people that create them" (Peters 2015). Critical views on the application to gain the status of a UNESCO World Heritage site demonstrate that there is a kind of production of heritage landscapes (Puleo 2012). A comparison of the representation of terraces found in applications for UNESCO World Heritage sites to the alternative representations based on archival and field research demonstrates that many parts of the landscapes do not fit the UNESCO World Heritage selection criteria (UNESCO 2018). Puleo's (2012) research shows that a reduced dimension of the area is proposed to create a new form of property that suits its nomination for UNESCO World Heritage. This manipulation method is a common feature of UNESCO World Heritage proposals, and so this kind of approach does not challenge the rules that regulate a property's authenticity and integrity. The author of this study shows that, in fact, UNESCO World Heritage easily accepts such a transformation of a candidate property.

23.4 Research Aim and Methodology

In addition to agricultural models, land stewardship models, and UNESCO World Heritage Sites, there are also other models of terraced landscape regeneration. The motivation to explore them is to achieve multiple answers to how to change passive heritage into active heritage, how to foster the longevity of terraced landscapes, and how to contribute to the local community as part of this goal. This study applies the methodological principles of strategic spatial planning in the case of heritage urbanism. Strategic spatial planning is increasingly becoming an alternative to traditional spatial planning as the process of designing visions, strategic goals, and directions, steps toward achieving defined goals, and implementation measures. Vision refers to an image of possible or alternative futures. These futures are constructed in a specific context: a place with a certain scale or scales and a specific moment (Van den Broeck 2008). The creation of a vision must be conceived as a conscious and purposeful action to represent values and meanings for the future that a particular place is committed to (Ogilvy 2002). Strategic goals identify the endpoint that participants want to reach. Participatory actions and negotiations are decisive steps for strategic spatial planning to become successful. Strategic projects are an active force in making changes possible; they relate to action and to implementation because steps must be achieved (Albrechts 2006).

23.5 Selected Case Study: Topolò Station

The village of Topolò (Sln. *Topolove*) in the Natisone valleys in the municipality of Grimacco (Garmak), Italy, is located in a mountainous region called Slavia Friulana between the town of Cividale del Friuli and the Slovenian border. The village's houses stand on terraces cut into a slope, and the village, which consists of several hundred buildings, has the shape of an equilateral triangle (Fig. 23.4). The Province of Udine is a minority area because most people in Slavia Friulana are Slovenians speaking four different dialects, named after the major valleys that form the territory. After the mid-nineteenth century, the village of Topolò no longer developed and many residents emigrated. The entire area was also largely depopulated after 1961, when many of the villages lost more than two-thirds of their residents. Two earthquakes that hit Friuli in May and September 1976 caused extensive damage, adding to land abandonment. Social circumstances in the area were difficult because of restrictions near the border, fear, and intense nationalism.

Fig. 23.4 1 Orthophoto of the settlement of Topolò in 2006 (*Source* Google Earth). 2. View of Topolò from the southwest. 3 and 4. The terraced landscapes, dry stone walls, and houses of Topolò have deteriorated and are overgrown by trees (*Photograph* Lučka Ažman Momirski)

After the early 1980s, (Slovenian) cultural activities again started taking place in the region. One of them, Slovenian Multimedia Window (Sln. *SMO_Slovensko Multimedialno Okno*; the acronym *SMO* also means 'we are') is a museum of landscapes and stories in San Pietro al Natisone dedicated to the cultural landscape that runs from the Julian Alps to the sea, from Mount Mangart to the Gulf of Trieste. The museum was conceived as an active, dynamic, welcoming, and stimulating space, founded on communication. The story of the locations is told through thematizing aspects that characterize the culture of the territory.

Another now internationally renowned cultural event called Topolò Station (2018; Ital. *Stazione di Topolò*, Sln. *Postaja Topolove*) has taken place every July for the last twenty-four years and has significantly helped the new development of the village and its landscape. In 2016, the event Art and Culture against Abandonment devoted particular attention to learning about and reacquiring terraced landscapes. The terraced landscapes started being created in the sixteenth century, when two families in Topolò built stone walls to drain the water on the steep slopes. Today, the terraced landscapes, dry stone walls, and houses have deteriorated and are overgrown by trees.

23.6 Model for Improving Abandoned Terraced Landscapes

The vision of Topolò Station is to become a place of artistic searching and experimentation. Its strategic goal is for participants to draw inspiration from the village, its inhabitants, the landscape, past events, and so on. Therefore, in all respects, the village with its characteristics is a protagonist and not simply a background for artistic activities. To achieve this vision, Topolò Station allows the greatest possible informality, the use of what is available, the complete absence of any haste, and the total absence of schedules because events take place "after sunset," "at dawn," and "until dawn." Only the opening of the main event has a more rigid timing, "around six pm" (Associazione Topolò-Topoluove, n.d.). The strategic approach has remained the same since the beginning. The main people involved are individuals (Donatella Ruttar), residents of Topolò (today, there are thirty-nine people in the village, who are united in the Topolò-Topoluove Association), the Venetian Artists Association (Ital. *Associazione artisti della Benecia*, Sln. *Društvo beneških likovnih umetnikov*), artists, travelers, nonconventional researchers, and those that are simply curious. The artistic, musical, and performance-oriented events of Topolò Station attract numerous visitors every year. Whoever comes to Topolò can also bring a reproduction of a painting to donate to the gallery. In addition to building a collective property capital, the collection also accumulates information for an experiment in cultural topology defined by places in geographical space and historical time from which the reproductions and models of the individual works come. The task and responsibility of everyone involved are also to articulate meaning, visions, and other ideas about access to and reuse of abandoned land and heritage. Creative, organic, and sometimes temporary use of abandoned land emerges when participants act. Their actions precede those of city governments, land banks, or developers. Choosing the right participants ensures the durability and persistence of the model.

The originality and uniqueness of Topolò Station lie in the fact that all events and artistic creations arise in Topolò and as part of the village and its residents. An event held for the Third World Conference on Terraced Landscapes in October 2016 by the village was the thematic session Awareness through Art, proposed by Italian section of the International Terraced Landscapes Alliance and organized by the Topolò-Topoluove Association (International Terraced Landscapes Alliance 2016).

Several activities were carried out, the most important of which was the video contest Terraced Landscapes—The Past, the Future, which addressed artists, documentary-film directors, farmers, builders of dry stone walls, researchers, and supporters of terraced landscapes at the international level in order to bring their experience to a world audience. The expectations and intentions of presenting videos and documentaries were to show the best practices of terraced landscapes with the aim of releasing terraces from the prejudice of marginality and ignorance. Part of the event was a walk in the abandoned terraced landscape in Topolò during the full-day program with the meaningful title *Kamani govorijo* (The Stones Speak), which heightened the perception of the archaeological remains of houses and terraces through musical passages and brief texts or excerpts. The participants in the event also visited *Kozolec* (The Hayrack), a monument to Slovenian rural architecture (Fig. 23.5).

Fig. 23.5 The event Art and Culture against Abandonment and the program *Kamani govorijo* (The Stones Speak): 1. The restored Mlin house among the terraces; 2. *Kozolec* (The Hayrack), a monument to Slovenian rural architecture; 3. A reconstructed dry stone wall along the route; 4. Terraces and their elements as a scene for an artistic performance (*Photograph* Lučka Ažman Momirski)

23.7 Conclusion

The methodological principles of strategic spatial planning for regenerating terraced landscapes link these with society and promote the innovative use of cultural heritage. This is in line with the New Urban Agenda of Habitat III (The United Nations Conference on Housing and Sustainable Urban Development 2017), which emphasizes promoting the use of cultural heritage for sustainable spatial development and recognizes its role in extending participation and civic responsibility. Models for renewing terraced landscape heritage in Slovenia have an agronomic basis. There is no model similar to terraced landscape management, and none of Slovenia's terraced landscapes are protected as a UNESCO World Heritage Site. However, the model observed here (Table 23.1) can be interpreted as a general model that can be applied to future cases, although the models are patterns whose idea can be understood but not copied because the meaning of the context of the individual examples varies.

Strategic spatial planning is a process by which visions, actions, and means of implementation have defined that shape a place and frame what it could become. Strategic spatial planning has prevailed in determining the development of cities, but it is not sufficiently used for planning the development and renewal of landscapes. In the future, a heritage-led sustainability model will require a management structure for access to and reuse of abandoned lands that distinguishes a high and medium extent of abandonment and measures for further development (Table 23.2). The next step toward managing terraced landscapes could be systematic training of participants in managing the innovative reuse of abandoned landscapes through e-learning and

Table 23.1 Model for regenerating terraced landscapes

Methodological principles of strategic spatial planning	Brief definition	Case study
Vision	Possible or alternative futures	Topolò Station: a place of return
Strategy	Strategic goals and directions	The village of Topolò with its residents and surrounding landscape is the central protagonist of the strategic goals
Tactics	Steps to achieve defined goals	Commitment of participants and their active involvement, open dialogue, accountability, collaboration, inhabitants, local associations, and devoted individuals
Operative	Implementation: projects and events	Awareness through Art: the video contest Terraced Landscapes—The Past, the Future

Table 23.2 Management measures for further development of abandoned terraced landscapes

Level of area abandonment	Management steps
Unstructured: abandonment is high	Good knowledge of the location Good knowledge of activities at the site Defining common goals of participants at the site Network of participants Promoting the territory and activities
Structured: abandonment is medium	Selecting forms of management by skilled participants Participatory process (involvement of all participants) Innovative services Planning, design, and regeneration of abandoned areas

workshops in order to make a participatory model possible. However, to refer to the case study of Topolò Station, terraces will only survive in their complexity and uniqueness if they satisfy the needs of people in a living cultural landscape.

References

Albrechts L (2006) Bridge the gap: From spatial planning to strategic projects. Eur Plan Stud 14(10):1487–1500
Associazione Topolò-Topoluove (n.d.) Stazione Topolò—Postaja Topolove. http://www.stazioneditopolo.it/
Ažman Momirski L (2019) Slovenian terraced landscapes. In: Varotto M, Bonardi L, Tarolli P (eds) World terraced landscapes: History, environment, quality of life. Environ Hist 9: 45–62.
Ažman Momirski L, Berčič T (2016) Ignored regions: Slovenian terraced landscapes. Ann Ser Historia et Sociologia 26(3):399–418. http://dx.doi.org/10.19233/ASHS.2016.37
Berčič T (2016) Discovering terraced areas in Slovenia: reliable detection with Lidar. Ann Ser Historia et Sociologia 26(3):449–468. http://dx.doi.org/10.19233/ASHS.2016.35
Brancucci G, Masetti M (2008) Terraced systems: heritage and risk. In: Scaramelini G, Varotto M (eds) Terraced landscapes of the Alps—Atlas. Marsilio, Venice, pp 46–54
Colnarič J (1971) Ureditev vinograda. Kmečki glas, Ljubljana
Colnarič J, Vrabl S (1991) Vinogradništvo. Kmečki glas, Ljubljana
Colnarič J, Gregorič J, Hrček L, Korošec Koruza Z (1985) Posebno vinogradništvo. Biotehniška fakulteta, VTOZD za agronomijo, Ljubljana
Fontanari E (2008) Laws and policies. In: Fontanari E, Patassini D (eds) Terraced landscapes of the Alps: projects in progress. Marsilio, Venezia, pp 102–103
Harfouche R (2007) Histoire des paysages méditerranéens terrassés: aménagements et agriculture. Archaeopress, Oxford
ICOMOS-IFLA (2018) ICOMOS-IFLA principles concerning rural landscapes as heritage. Available via ICOMOS. https://www.icomos.org/images/DOCUMENTS/General_Assemblies/19th_Delhi_2017/Working_Documents-First_Batch-August_2017/GA2017_6-3-1_RuralLandscapesPrinciples_EN_final20170730.pdf. Accessed on 10 January 2018
Inbar M, Zagier A (2016) Physical and social aspects of land degradation in mediterranean highland terraces: a geodiversity approach. Ann, Ser Historia et Sociologia 26–3:419–432. http://dx.doi.org/10.19233/ASHS.2016.34
International Terraced Landscapes Alliance (2016) 3rd World Meeting on Terraced Landscapes. http://www.terracedlandscapes2016.it/en/. Accessed on 7 February 2018

Junchao S (2012) Exploring the orientation of farming civilisation growth in low-carbon times from Honghe Hani ethnic terraced fields. In: Peters HA, Junchao S (eds) First terraced landscapes conference (Honghe—China) paper collection. Yunnan People's Publishing House, Kunming, pp 323–338

Konold W, Petit C (2013) Historische Terrassenweinberge. Bristol-Stiftung, Zürich, Baugeschichte, Wahrnehmung, Erhaltung

LandLife (n.d.) European policies and land stewardship. http://www.landstewardship.eu/land-stewardship/european-policies. Accessed on 27 March 2018

Laureano P (2012) The Tourist Impact of the UNESCO Inscription: from the Urban Exodus to Mass Tourism in the Abandoned Towns of Matera in Italy and Petra in Jordan. In: Peters HA, Junchao S (eds) First terraced landscapes conference (Honghe—China) paper collection. Yunnan People's Publishing House, Kunming, pp 436–450

Melik A (1960) Slovensko Primorje. Slovenska matica, Ljubljana

Ogilvy JA (2002) Creating better futures. Oxford University Press, Oxford

Peters H (2015) Disappearing terraces: can international tools support safeguarding terraced landscapes and their traditional knowledge? In: Tillmann T, Bueno de Mezquita M (eds) II Congreso Internacional de Terrazas. Encuentro de culturas y saberes de terrazas del mundo. CORDESAN, CBC, Cuzco, pp 345–355

Presl I (1995) Vinogradništvo v vaseh koprskega zaledja od sredine 19. stoletja do petdesetih let 20. stoletja. In: Darovec D (ed) Dežela refoška: vinogradništvo in vinarstvo slovenske Istre. Koper, Zgodovinsko društvo za južno Primorsko, pp 177–196

Puleo TJ (2012) Parasitizing landscape for UNESCO World Heritage. Geoforum 45:37–345

Republika Slovenija (2011) The Agricultural Land Act (Zakon o kmetijskih zemljiščih). http://www.uradni-list.si/1/objava.jsp?sop=2011-01-3086. Accessed on 12 December 2017

Republika Slovenija Ministrstvo za kulturo (2018a) Register nepremične kulturne dediščine (RKD). http://www.mk.gov.si/si/storitve/razvidi_evidence_in_registri/register_nepremicne_kulturne_dediscine/. Accessed on 4 January 2018

Republika Slovenija Ministrstvo za kulturo (2018b) SI AS 176 Franciscejski kataster za Kranjsko, 1823–1869 (Fond). http://arsq.gov.si/Query/detail.aspx?id=23253. Accessed on 15 March 2018

Republika Slovenija Ministrstvo za kulturo (2018c) SI AS 175 Jožefinski kataster za Kranjsko, 1784–1790 (Fond). http://arsq.gov.si/Query/detail.aspx?ID=23252. Accessed on 15 March 2018

Republika Slovenija Ministrstvo za kmetijstvo, gozdarstvo in prehrano (2018) Vinogradništvo in vinarstvo. http://www.mkgp.gov.si/si/delovna_podrocja/kmetijstvo/kmetijski_trgi/vino/. Accessed on 4 January 2018

Republika Slovenija Ministarstvo za okolje in prostor (2018) Outstanding landscapes. http://www.ppz.mop.gov.si/. Accessed on 6 January 2018

Salas MA (2015) Agrobiodiversidad, soberania y seguridad alimentaria y nutricional. In: Congreso II (ed) Tillmann T, Bueno de Mezquita M. Internacional de Terrazas. Encuentro de culturas y saberes de terrazas del mundo. CORDESAN, CBC, Cuzco, pp 105–118

Škvarč A (1999) Obnova vinogradov. In: Stanek K, Srebrenič M, Radikon B (eds) Znanje kot ključni dejavnik razvoja kmetijstva. Kmetijska svetovalna služba KZ Goriška Brda, Združenje vinogradnikov in vinarjev Brda, Občina Brda, pp 1–5

Škvarč A, Kodrič I (2007) Nature and regulation: management of vineyards and orchards on terraces. Urbani izziv 17(1–2):78–84

Štabuc R, Hauptman S (2008) The possibilities of revitalisation of vineyards on the terraces in Podravje. In: Ažman Momirski L, Černič Mali B (eds) Living Terraced Landscapes: Perspectives and strategies to revitalise the abandoned regions. Fakulteta za arhitekturo, Ljubljana

Sun Y, Min Q, Li W (2010) Study on the agricultural heritage tourism development in minority areas. In: Peters HA, Junchao S (eds) First terraced landscapes conference (Honghe—China) paper collection. Yunnan People's Publishing House, Kunming, pp 244–251

The United Nations Conference on Housing and Sustainable Urban Development (2017) New Urban Agenda. Available via HABITAT III http://habitat3.org/wp-content/uploads/NUA-English.pdf. Accessed on 30 March 2018

Tillmann HJ, Salas MA (2016) The Mountain/coastal sea farmers and the stone walls of the terraces resist the threats to terraced landscapes and cultures: ITLA—The International Terraced Landscapes Alliance. Ann, Ser Historia et Sociologia 26(3):375–388. http://dx.doi.org/10.19233/ASHS.2016.36

Titl J (1965) Socialnogeografski problemi na koprskem podeželju. Lipa, Koper

UNESCO (2015a) Rice Terraces of the Philippine Cordilleras. http://whc.unesco.org/en/list/722. Accessed on 27 March 2018

UNESCO (2015b) Cultural Landscape of Bali Province. http://whc.unesco.org/en/list/1194. Accessed on 27 March 2018

UNESCO (2015c) Cultural Landscape of Honghe Hani Rice Terraces. http://whc.unesco.org/en/list/1111. Accessed on 27 March 2018

UNESCO (2015d) Cultural Landscape of Southern Jerusalem, Battir, Palastine. http://whc.unesco.org/en/list/1492/. Accessed on 27 March 2018

UNESCO (2015e) Cultural Landscape of Portovenere, Cinque Terre, and the Islands. (Palmaria, Tino and Tinetto). http://whc.unesco.org/en/list/826. Accessed on 27 March 2018

UNESCO (2015f) Cultural Landscape of The Lavaux Vineyard Terraces. http://whc.unesco.org/en/list/1243. Accessed on 27 March 2018

UNESCO (2018) What is meant by "cultural heritage"? http://www.unesco.org/new/en/culture/themes/illicit-trafficking-of-cultural-/unesco-database-of-national-cultural-heritage-laws/frequently-asked-questions/definition-of-the-cultural-heritage/. Accessed on 27 March 2018

Van den Broeck P (2008) The changing position of strategic spatial planning in Flanders. A socio-political and instrument-based perspective. Int Plann Studies 13(3):261–283

Varotto M (2015) Terraced landscapes of the Alps: decay, rediscovery, revitalization. In: Tillmann T, Bueno de Mezquita M (eds) II Congreso Internacional de Terrazas. Encuentro de culturas y saberes de terrazas del mundo. CORDESAN, CBC, Cuzco, pp 38–48

Varotto M, Lodatti L (2014) New family farmers for abandoned lands. Mt Res Dev 34(4):315–325. https://doi.org/10.1659/MRD-JOURNAL-D-14-00012.1

Vrišer I (1954) Goriška brda: gospodarska geografija. Geografski zbornik 2:51–113

Vršič S, Lešnik M (2005) Vinogradništvo. Kmečki glas, Ljubljana

Watson EE (2009) Living terraces in ethiopia—konso landscape. Culture and Development. Boydell & Brewer, Suffolk

Yamaoka K (2012) Collaboration between Public Policies and Private Activities for Preserving and Making Good Use of Rice Terraces in Japan. In: Peters H A, Junchao S (eds) First Terraced Landscapes Conference (Honghe–China) paper collection. Yunnan People's Publishing House, Kunming, pp 517-526

Chapter 24
Application of MCDM Methods to Tourism Evaluation of Cultural Sites

Ivana Stević, Stevan R. Stević and Zélia Maria de Jesus Breda

Abstract This paper explores the usage of two specific multi-criteria decision -making (MCDM) methods—Simple Additive Weighting (SAW) method and Technique for Order Preference by Similarity to Ideal Solution (TOPSIS)—in the tourism field, where their usage is practically non-existent. Multi-criteria analysis methods use objective or subjective criteria to rank different variables of interest, the purpose of ranking being to facilitate different strategic decision-making and problem-solving processes. By using these methods, the paper assesses the attractiveness of six cultural heritage sites in Porto, Portugal, namely some of the city's architectural masterpieces, and ranks them based on the following set of criteria: (i) historic value; (ii) aesthetic/artistic value; (iii) representativeness; (iv) state of preservation; (v) infrastructure and accessibility and (vi) social significance. The purpose of the study is to understand different levels of attractiveness of cultural sites and to discuss the possible reasons why ones are given higher importance than the others. The two methods give almost identical outcomes, showing parity between their usages and affirming the integrity of the results.

Keywords Cultural sites · Tourism assessment · Multi-criteria analysis · Strategic decision-making · Problem-solving processes

---

I. Stević (✉) · Z. M. de Jesus Breda
Department of Management, Economy, Industrial Engineering and Tourism, University of Aveiro, Aveiro, Portugal
e-mail: ivana.stevic87@ua.pt

Z. M. de Jesus Breda
e-mail: zelia@ua.pt

S. R. Stević
Faculty of Economics, University of East Sarajevo, Sarajvo, Bosnia and Herzegovina
e-mail: stevan.stevic.efb@gmail.com

M. Obad Šćitaroci et al. (eds.), *Cultural Urban Heritage*, The Urban Book Series, https://doi.org/10.1007/978-3-030-10612-6_24

24.1 Scope and Objectives

The main objective of the paper is to evaluate the attractiveness of some of the main cultural sites/tourism attractions in Porto, Portugal, based on the opinions of experts in the tourism field, from both the academia and the industry. The assessment of the sites, i.e. the alternatives, is done through a set of criteria or attributes, which results in a best-to-worst ranking of the alternatives in question. The intention is to withdraw conclusions that are in line with the need for heritage enhancement, preservation and sustainability of its space values and sociocultural identity. The understanding of the concept of culture has significantly evolved, broadened and deepened since its beginnings when it used to refer to essentially tangible objects and sites. Presently, it is thought to have a fundamental importance for the construction of social identities, relating not only to material, but also to immaterial expressions like language and spoken tradition, social practices, rituals and different kinds of events (Robinson and Picard 2006). So, nowadays, culture and heritage are much more than just buildings and architectural styles. They represent meanings, identities, narratives, feelings. A heritage site does not only refer to a monument anymore, but the monument with its surroundings—both natural and social. By exploring the MCDM methods, we intend to, as well, analyse how the evaluators that participated in the study see heritage and which values they give the most importance to.

The approach focuses on sites of different typologies, enabling a discussion of the importance of different values in the tourism context, and the ways in which tourism could contribute to the improvement of inferiorly ranked sites. The study uses two particular multi-criteria decision-making methods, so as to investigate and compare the implications of their usage in decision-making processes related to heritage. By exploring these methods, we tend to provide an example of a framework that could assist in planning and decision-making related, but not limited, to cultural sites. The same methods could be used in different contexts, e.g. for evaluation and ranking of different aspects of tourism destinations and assistance in regulation and legislation of spatial planning.

24.2 Introduction

Heritage is the embodiment of such diverse cultural identities, traditions and collective memories, resulting from complex historical development processes (ICOMOS 2002). In today's globalized world, it is both impacting and being continuously impacted by its setting, where the need for communicating and raising awareness about its importance, whether natural or cultural, tangible or intangible, has become imperative. Heritage conservation has become beneficial for cities' and communities' social, economic and cultural development. Apart from its cultural value, heritage is increasingly becoming an instrument of economic diversification and growth, particularly due to tourism.

There is a growing dependence between tourism and cultural heritage, due to cultural heritage creating a basis for tourism growth, and tourism having the power to generate funds that make conservation of cultural heritage possible (NWHO 1999). Tourism and cultural sectors have been joining forces in the rehabilitation of urban cores, supporting the preservation and sustainability of traditional and contemporary cultural values, identities and traditions (UNWTO nd). However, tourism also contributes to the decline of these very values, overcrowding many attractive urban cultural destinations, with the city under study—Porto—being one of them.

Portugal, generally, and Porto, in particular, are becoming increasingly trendy tourism destinations. Portugal was awarded the title of Europe's and world's leading tourism destination by the World Travel Awards (nd) in 2017 and was ranked world's 14th most competitive travel destination in 2016 (World Economic Forum 2017), 11th in tourism revenue at the EU level in 2015 and 2016 (Eurostat 2017a, b) and 24th in tourism revenue in the world ranking in 2016 (The World Bank 2018), where the revenue more than tripled over the last two decades, growing from 5646.00 to 17,185.00 million US$, representing a 5% average annual growth rate. Not falling behind, Porto was elected best European destination by the travellers in 2012, 2014 and 2017 (Europe's Best Destinations nd) and is nominated for Europe's leading city break destination in 2018 by the World Travel Awards. It was also awarded the title of "The Best Start-up Friendly City of Europe" by the World Business Angels Investment Forum (O portal de notícias do Porto 2018). Overall, the entire country is experiencing a booming tourism demand that, more than ever, calls for the protection of its cultural—both constructed and intangible—and natural heritage.

Uncontrolled tourism growth and massification can cause irreversible damage over time and affect not only the built environment and tangible resources, but intangible heritage as well, in terms of loss of cultural values, identity, social exclusion and depopulation of touristically attractive areas, alongside the impacts on the natural environment, natural resources and wildlife (Mason 2016). This is why the decision-making processes and policies related to tourism growth ought to be subordinate to the ones concerning conservation and protection of cultural sites.

Tourism policy and planning are nowadays far beyond dealing with a single set of impacts, due to the complexity of interconnections with several other areas, culture being one of the most dominant ones. Neither tourism nor culture can be dealt with alone, without being integrated into wider policy frameworks (Robinson and Picard 2006). This must be the reason why there is a considerable academic attention to the tourism–culture relationship, though there are still quite a few gaps to be bridged. This study focuses on the methodological one, which regards quantitative approaches to cultural heritage assessment, from a tourism perspective.

For being a set of approaches that aids the decision-making processes, multiple criteria decision analysis (MCDA) or multi-criteria decision-making (MCDM) was selected to be explored in this paper, in order to study, on a concrete example, its possible usages in the tourism field, which are practically non-existent. These methods use objective or subjective criteria to rank different variables of interest, in order to facilitate different strategic decision-making and problem-solving processes. By using these methods, the paper assesses the attractiveness of six cultural heritage sites in Porto, Portugal, namely some of the city's architectural masterpieces, by ranking them based on a set of criteria, which will be explained in the following sections.

24.3 Cultural Heritage Assessment

Almost two decades ago, in 1999, ICOMOS (2002) has defined communication of heritage values and the need for its preservation to host communities and to visitors as the primary objective of heritage management. In order for cultural heritage to be valued, and consequently preserved, its value and significance have to be determined; i.e. it has to be assessed, where the evaluation involves different approaches, depending on what wishes to be achieved. There is no single, universally accepted approach to cultural heritage evaluation, particularly when it comes to tourism contexts. It can be done from the economic, social, cultural or environmental perspective, employing qualitative and/or quantitative methodologies.

When it comes to cultural heritage evaluation, cultural economics literature often shows the distinction between economic and cultural values of cultural goods (Hutter and Frey 2010; Throsby 2001), arguing they should be evaluated separately. Klamer (2004, 2016) categorizes social or societal values in separate, claiming art as social, due to its values being developed in a social setting. On the other hand, Throsby (2001), for example, includes the social strand as a subcategory of cultural values, nonetheless separated from the economic. Ultimately, Klamer (2017) goes a step further and suggests the distinction between personal, social, societal and transcendental/cultural values, leaving out the economic value. Mazzanti (2002) classifies cultural heritage as a multidimensional, multi-attribute and multi-value economic resource, with numerous other works in the field of economics, both classical and cultural, comparing or referring to heritage as (cultural) capital, asset or resource (Dalmas et al. 2015; Necissa 2011; Noonan, 2003; Throsby 2001) that has a use value.

However, heritage differentiates itself from the mere notion of capital because its cultural dimensions go beyond the use values and also include the intangible non-use values, for which the monetary value is quite difficult to assign. The two sets of values together form the total economic value (Choi et al. 2010; Dalmas et al. 2015; Ferretti et al. 2014). There is a strong linkage between cultural goods that are available in a certain area and the active production of material culture, which is of particular importance to economic development (Ferretti et al. 2014). However, it is not only the material culture that is being produced in relation to cultural goods, but the immaterial one, as well. The evaluation of a site should consider both its intrinsic values and the opportunities for value creation it provides to the community, since both represent the basis of the site's identity and the culture it embodies (Ferretti et al. 2014). It is inevitable that, due to their significant contribution to economic growth, culture and cultural sites are discussed in the economic context and seen as economic means. This, however, calls upon the need for greater efforts to emphasize the sociocultural values of both tangible and intangible heritage, as, without these, there can be no economic value either.

All things considered, we believe it is a matter of author's personal perspective and, moreover, of what intends to be evaluated, since all categorizations cannot be reconciled into a single one. This paper focuses on a set of values of cultural heritage sites, which can be subcategorized within the main sustainability pillars and the four dimensions of capital—cultural, social, environmental and economic. Sets of criteria were adopted from various different sources and adapted into one, but initially building upon Throsby's (2001) categorization. The author distinguishes the aesthetic, social, spiritual, historic and symbolic values, as subcategories of cultural values. We, however, do not agree with social values being a subcategory of the cultural ones, but believe they should be categorized separately, following Klamer's (2004) logic and adding the environmental and economic subcategories to the evaluation criteria.

24.3.1 Cultural Heritage Sites and Tourism Impacts

Tourism has become one of the major economic forces of modern time, triggering numerous direct and indirect social, cultural and environmental impacts, and having a strong multiplier effect on the countries' economies (Costa 2006; Massieu 2006; Mathouraparsad and Maurin 2017). Tourism expenditures have direct, indirect and induced effects on production, income, employment and import outflows (Khan et al. 1990), supporting further job creation, boosting local businesses, (re)investment and collaboration between different stakeholders (Rusu 2011). The complexity of tourism lays in the fact that it is entirely dependent on the destination's resources and their sustainability, them being not only natural resources and tangible cultural heritage but also the land physiognomy—e.g. geographical position, morphological features and agricultural products—as well as intangible sociocultural resources—such as tradition, habits, hospitality and so on (Di Martino and Petrillo 2006).

Cultural sites are playing a leading role in policy-making and development strategies, as well as in efforts to promote sustainable development (Richards 2001) that, to a great extent, is empowered by tourism. But, just like tourism supports enhancing and protecting cultural heritage, heritage supports sustainable tourism and socio-economic growth. By contributing to the dispersal of the tourism activity more uniformly across the territory, it allows a more even distribution of its economic impacts and protection of social and environmental aspects of destinations pressured by tourism (Matarasso 2001).

Culture and cultural heritage also influence tourism actors and stakeholders. In domestic tourism, they stimulate national pride in one's history, while in international tourism they inspire respect and understanding of other cultures and, as a consequence, promote peace and understanding (UNWTO nd). The importance of heritage encourages tourism actors, from both the supply and the demand sides, to communicate a greater respect, appreciation and concern for culture, a superior cultural orientation within the tourism context, and a stronger spirit for protection and conservation of material and immaterial heritage.

Tourism–culture co-dependency and interconnectedness are, hence, so solid that both impact and enhance the other. Still, there is no such thing as negative impacts of culture or heritage on tourism. Tourism, on the other hand, even though it generates many positive impacts in terms of enhancing the revitalization, conservation and protection of cultural attractions and heritage sites, it has also been linked to a decline in cultural values, trivialization of culture and loss of authenticity, local distinctiveness, identity, customs and traditions (Dyer et al. 2003; Mason 2016; Richards and Wilson 2004; Smith 2007; Swarbrooke 1999), caused by unethical, short-term led and/or unqualified tourism planning and management. Some of these problems are: (i) vulnerability of cultural sites; (ii) over-commercialization and over-exploitation of culture through tourism; and (iii) lack of local control, that is, little or no participation of local communities and governments in the decision-making, planning, and organizational and managerial processes concerning cultural sites, due to most power resting in hands of non-local investors (Swarbrooke 1999).

Heritage conservation should provide responsible and well-managed opportunities, for both the members of host communities and the visitors, to experience and understand heritage and culture at first hand (ICOMOS 2002). Tourism ought to be socially inclusive and receptive to local residents' participation and active involvement in decision-making concerning its planning and organization. Heritage and cultural specificities of places allow tourism to grow and, hence, should tourism promotion programmes be committed to the protection and enhancement of heritage singularities (ICOMOS 2002). Without culture, after all, there can be no tourism (Jafari 1996).

24.3.2 MCDM and Cultural Heritage

Multi-criteria decision-making methods are used for supporting groups or individuals to rank, select and/or compare different alternatives, by allowing several criteria to be taken into account simultaneously (Belton and Stewart 2002, 2010). Real-life decision problems are rarely mono-criterion-based and coherent. They rather incorporate a variety of, often contradictory, criteria (Botti and Peypoch 2013). The ultimate objective of their application is to assist the decision-makers (DMs) in integrating different options (which reflect the opinions of the evaluators), in a prospective or reflective framework (Ferretti et al. 2014). Multi-criteria framework is an efficient tool to implement within the multi-/inter-/trans-disciplinary approaches, where experts come from different backgrounds, like in case of the present study. This is a scenario where the communication process can be hampered, and a multi-criteria framework helps bridge the communication gap (Munda 2005, 2016).

According to Belton and Stewart (2002, 2010), the starting point for any multi-criteria analysis is a well-structured problem, consisting of three clearly stated components: (i) the set of alternatives or decision space from which a decision has to be made; (ii) the set of criteria against which the alternatives are to be evaluated; and (iii) the method to be used to affect that evaluation. The first two components are explained in the methodology section, whereas here we are focusing on discussing the third component, that is, the evaluation methods.

The MCDM (MCDA) techniques were first introduced in the mid-1960s, from when on a great variety of them was developed (Yoon and Hwang 1995). Among most commonly used methods in practice today are: Weighted Sum Method (WSM), Analytic Hierarchy Process (AHP), Analytic Network process (ANP), Technique for Order of Preference by Similarity to Ideal Solution (TOPSIS), *Elimination Et Choix Traduisant la Realité* (ELECTRE), Preference Ranking Organization Method for Enrichment Evaluations (PROMETHEE), Multi-Attribute Value/Utility Theory (MAVT/MAUT) and *VIšeKriterijumska Optimizacija I Kompromisno Rešenje* (VIKOR—from Serbian: Multi-criteria Optimization and Compromise Solution) (Lakshmi and Venkatesan 2014; Triantaphyllou et al. 1998; Yoon and Hwang 1995; Zardari et al. 2015).

When it comes to the application of MCDA methods on decision-making regarding cultural heritage, AHP is among the most frequently used methods, either alone (e.g. Thórhallsdóttir 2007) or in combination with Geographical Information Systems (GIS) (e.g. Fusco Girard and De Toro 2007). Among the research topics that stand out, one of the most addressed ones is certainly the reuse of historical buildings (e.g. Dutta and Husain 2009; Ferretti et al. 2014; Ferretti and Comino 2015; Fuentes 2010; Giove et al. 2011; Kim et al. 2010; Wang and Zeng 2010). Dutta and Husain (2009) use a linear additive multi-attribute model, i.e. a WSM, for grading heritage buildings in Calcutta, aiming to assist urban planners in decision-making that either concerns heritage renovation/maintenance/protection or defends a more efficient land use by their demolition. Fuentes (2010) uses the same method for the assessment of abandoned traditional farms in Central Spain, so as to advise on the preservation of the most valuable ones, where he resorts to GIS for information management and integration.

Ferretti et al. (2014) use MAVT for assessment of reuse of industrial heritage buildings in the metropolitan area of Torino. Ferretti and Comino (2015) propose an integrated approach to determine the suitability of recovering a set of abandoned farms inside a natural park in Northern Italy for touristic purposes, where they combine MAVT with GIS and SWOT analysis. Giove et al. (2011) use a mathematically founded MAVT—the Non-Additive Measures (NAM)—and the Choquet integral as an aggregation operator, for the evaluation of the sustainability of potential projects for the economic reuse of historical buildings in Venice. Kim et al. (2010) assess the restoration priorities for 14 Korean cultural heritage buildings, by using stochastic AHP and knowledge-based experience curve (EC). Wang and Zeng (2010) combine ANP with the fuzzy Delphi method for the reuse selection of historical buildings in Taiwan. These methods are, hence, used from Iceland, Italy and Spain to India, Korea and Taiwan.

24.3.3 MCDM and Tourism

There are very few studies that use MCDA methods in the tourism field, leaving a lot of space for their exploration. The most frequently explored are variations of ELECTRE, and other MCDM techniques in combination with GIS, mainly for the evaluation of tourism destination competitiveness (TDC). Botti and Peypoch (2013) explore ELECTRE-I in combination with the Weighted Sum Method for the evaluation of TDC. Zhou et al. (2015) also assess DC but use AHP for the matter. Zhang et al. (2011) evaluate TDC in the Yangtze River Delta in China, by the means of TOPSIS in combination with Information Entropy Weight (IEW). Göksu and Kaya (2014) assess and rank tourist destinations in Bosnia and Herzegovina by combining fuzzy AHP and TOPSIS. The most recent study in the field develops a socio-technical evaluation index, again, for TDC, by using cognitive mapping and MCDA (Carayannis et al. 2018). The authors combine cognitive maps with Measuring Attractiveness by a Categorical-Based Evaluation Technique (MACBETH), which results in ranking of the most competitive tourism destinations in Portugal, with Porto being ranked the 9th most competitive, together with Coimbra and Évora.

There are also studies that apply more advanced methods, such as ELECTRE-III-H, which combines mathematical models, social science theories and artificial intelligence techniques and is based on optimization algorithms. Del Vasto-Terrientes et al. (2015) use this method in combination with the Web Quality Index (WQI) to assess communication of tourism brands, by assessing the official tourist destination websites. Similarly, Mailly et al. (2013) use a variation of ELECTRE for identifying favourable climates for light tourism, aiming to address some of the methodological problems in composite indices, such as the Tourism Climatic Index. Michailidou et al. (2016) explore ELECTRE-III for the assessment of mitigation and adaptation options in tourism areas, in the environmental/climate change context.

A single study was found that compares the Weighted Sum Method with TOPSIS, in addition to PROMETHEE, for a real-life decision-making problem of finding the most suitable area for the construction of a new casino, while considering profitability, on one hand, and social benefits, on the other (Ishizaka et al. 2013). Furthermore, a study that uses somewhat similar logics to this paper explores the WSM with the support of SWOT and GIS to measure the potential for tourism development at heritage sites (Yan et al. 2017). The authors use two sets of indicators (criteria), these being (i) resource values and (ii) state of development, which have some common points with the indicators used in this paper. For example, they also assess the historic value, aesthetic value, ambiance or setting, accessibility and tourist facilities, the last three being comparable to representativeness, and infrastructure and accessibility criteria used in this study. Another paper that is somewhat comparable to the present evaluates destination's attractions from the experts' points of view, although it uses AHP for the matter and assesses both man-made and natural tourist attractions (Emir et al. 2016). Aminu et al. (2014) also integrate GIS with a MCDA method, the ANP in this case, to assess the priorities for a sustainable tourism development in a Malaysian district.

24.4 Methodology and Case Study Characterization

24.4.1 Rationale

This is an experimental study that uses a quantitative methodological approach. To the knowledge of the authors, this is the first and only study that explores the usage of the Simple Additive Weighting (SAW) method and the Technique for Order Preference by Similarity to Ideal Solution (TOPSIS) combined, and in the context of cultural heritage evaluation/valorization/assessment. Thus, there is a clear contribution of the paper to the state of the art in tourism, in terms of methodological design, where an uncommon approach is explored.

The paper ranks six cultural sites, using six value criteria and the opinions of seven experts from the tourism field, five of them being from the academia and two being the decision-makers associated with public institutions/governing bodies. The six monuments were chosen for their exceptionality, all of them being composite sets of different elements, architectural styles, design and/or purpose, apart from being among the main tourism landmarks of the city. The choices were based on the intention to discover how aforesaid exceptionality is seen and evaluated by the experts from the tourism field and how attractive the sites' features are in the tourism context.

24.4.2 Alternatives

The set of alternatives, i.e. cultural sites chosen for evaluation, is given in Table 24.1, followed by a detailed description of each alternative, aiming to justify the selection by portraying sites' exceptionality.

Alternative A1 is the first building built in Portugal exclusively dedicated to music, whether for public performances and displays or for artistic creation and training. It is both a concert house and a cultural institution/foundation. The project is a result of an international architectural competition, whose objective was to mark and celebrate the year of 2001 when Porto was the European Capital of Culture, though it was

Table 24.1 Alternatives (A_n) whose values in the context of tourism attractiveness are evaluated by the experts from the field

Alternatives	A_n
Casa da Música concert house	A_1
Serralves Foundation	A_2
Monument Church of S. Francisco de Assis	A_3
Churches of Carmo and Carmelitas	A_4
Clérigos Tower	A_5
Cathedral	A_6

only completed and inaugurated in 2005 (Casa da Música nd). The solution that was ultimately chosen as the winning one was presented by Rem Koolhaas, winner of the Pritzker Prize for architecture for this very project. The construction of the house was quite controversial at the time, since it was about to take place where the former Tram Terminal was located. Among the project Evaluation Committee, there were Álvaro Siza Vieira and Eduardo Souto de Moura (Casa da Música nd), the most distinguished Portuguese architects, both Pritzker winners themselves.

Alternative A2 is a privately owned foundation of public utility, created in 1989, consisting of a contemporary art museum, an art deco mansion and a vast park. Its creation marked the beginning of an innovative partnership between the state and the civil society. This remarkable cultural site is today considered one of the most successful and well-established cultural projects in the country, being classified as a National Monument in 2012 (Serralves 2018). The museum is a project of Álvaro Siza Vieira, winner of the Pritzker Prize for architecture in 1992. The Serralves House is a unique example of art deco architecture, and the park is a complex ensemble of green spaces: a garden influenced by the sixteenth and seventeenth centuries' French gardens, designed by the French architect Jacques Gréber, awarded with the Henry Ford Prize for the preservation of the environment in 1997, a woodland area and a farmstead. The park is a home to significant tree and shrub heritage, composed of approximately 200 species and a variety of autochthonous and exotic ornamental plants, at the same time being an extension of the museum, where some of the sculptures from the foundation's collection are exhibited permanently (Serralves 2018). Additionally, it is an institution committed to education, developing a series of events and activities for schools, children, youth, adults and families throughout the year.

Alternative A_3 is the city's most important gothic temple, originating from the fourteenth century. It is also one of the country's most important baroque works of art, with baroque and rococo elements being added posteriorly in the seventeenth and eighteenth centuries, with an exuberance in gild carved woodwork. It is a National Monument since 1910 and a World Cultural Heritage by UNESCO since 1996 (Ordem São Francisco do Porto nd). It is an institution consisted of a hospital, a senior residence, a church, a convent and a museum that bets on technological innovation and a quality development through culture, diffusion of knowledge, art and history, human rights and social responsibility. Therefore, it is not only a work of art and a remarkable architectural site, but a complex and constantly growing community.

Alternative A_4 is a twinned assemblage of two churches—Carmelitas on the left and Carmo on the right—built in a setting of the seventeenth and eighteenth centuries' spirit of the illuminist urbanism. The churches are separated by what must be one of the world's narrowest houses, supposedly built to prevent the contact between the nuns and the monks. It is a space that provides access to the bell tower, and it belongs to the church of Carmo but, curiously, shares a lateral wall with the other church, resulting in the amalgamation of the buildings and adding to their distinctiveness. Carmelitas church was the first one built as a part of a seventeenth-century convent, characterized by a classical façade with a single bell tower and a rich gilded wood carved interior. The church was used as barracks during the French Invasion of Porto in the early nineteenth century (Gomes nd) and now serves as headquarters of the Portuguese National Guard (De Sousa nd, Direção-Geral do Património Cultural nd). Carmo Church is the more recent one, built in the eighteenth century in baroque/rococo style. The most prominent feature of the edifice is not the interior, but the lateral façade entirely covered with traditional Portuguese blue and white tiles, forming a magnificent panel that depicts allusive scenes of the foundation of the Carmelite Order and the Mount Carmel, where the Order originated. Due to their unique illustrative value in terms of historical evolution of art in Portugal, the pair was classified as a National Monument in 2013 (De Sousa nd).

Alternative A_5 is part of the architectural complex of Clerics, classified as a National Monument in 1910. The complex consists of a tower, a museum and a church. In focus of the paper is, however, only the tower, for being among the city's main landmarks and a must-see tourism attraction. It is an eighteenth-century baroque bell tower, built at the request of the Brotherhood of Clerics by the brilliant Nicolau Nasoni (Torre dos Clérigos 2018), whose remains were buried in the very church. The tower was built on the "hill of the hanged men", where the executed prisoners were buried (Civitatis Porto nd). With its 75 m elevation, it is the country's tallest bell tower and, at the time of its construction, the tallest building overall. It had been one of the most important public buildings that, for a long time, served as a lighthouse to sailors navigating along the Douro River, while its clock showed the city's official time (Descobrir Portugal 2010).

Alternative A_6 is the Cathedral of Porto, built between the first half of the twelfth and the beginning of the thirteenth centuries. Its origins probably date back to the early Middle Ages and Swabian times, due to the evidence that points out to the existence of a bishop of Porto during this period (Direção Regional de Cultura do Norte nd). It is a remarkable building that still preserves traces of the original Romanesque structure, with its fortified façade flanked by two towers and a rosette window at the front. It underwent modifications in the Mannerist and Baroque periods and was added a gothic cloister in the late fourteenth century, during the reign of King John I (Direção Regional de Cultura do Norte nd). Today, it is a dynamic urban organism, with many ongoing restoration schemes and projects it makes part of, such as the Route of the Cathedrals, done in partnership with the Ministry of Culture, or the official website of the Diocese of Porto (Diocese do Porto nd).

24.4.3 Evaluation Criteria

As aforementioned, this paper assesses a group of cultural heritage sites, based on a set of values with cultural, social, environmental and economic dimensions. Some of them are difficult to categorize under only one dimension, particularly the social one that easily overlaps with the cultural and the economic one. This is how we opt for evaluating the sites for their: (i) historic and (ii) aesthetic/artistic values—perceived as cultural dimensions; (iii) representativeness as an environment subcategory; (iv)

Table 24.2 Criteria (C_m) used for evaluation of the alternatives (A_n)

Criteria (C_m)	Description
Historic value	Evolution or pattern of local history
Aesthetic/artistic value	Aesthetic and/or artistic features of local significance
Representativeness	Typical features of cultural and environmental space where they are located
State of preservation	Preservation of integrity and authenticity, and management that follows the sustainability standards
Infrastructure and accessibility	Ease of access to any age group, with any sort of needs
Social significance	Association with the local community or a particular group for social, cultural, educational or spiritual reasons

state of preservation as an economic dimension; (v) infrastructure and accessibility as socio-economic dimension; and (vi) social significance as a sociocultural subcategory. Table 24.2 summarizes the evaluation criteria.

Historic value criterion does not necessarily imply the age of the building, but rather the association with important historic events or person(s). Aesthetic/artistic value concerns architectural design features, purity and/or rarity of the construction, and/or the prominence of the architect. Representativeness regards the harmony and continuity of the site with its environment and the importance of it as a landmark. State of preservation is related to the current structural condition of the building and the sustainable practices implemented at the site. Infrastructure and accessibility regard urban planning around the site, the existence and proximity of different facilities, and the attention to social inclusion, i.e. the access for local residents and people with special needs. Lastly, social significance has to do with the site's connection with the culture and identity of the local community.

24.4.4 Evaluation Process

The evaluation process starts with the elaboration of a decision matrix (Eq. 24.1) that typically has the following form:

$$
\begin{array}{c}
\qquad\qquad\quad C_1 \;\; C_2 \;\cdots\; C_m \\
\qquad\qquad\quad w_1 \;\; w_2 \;\cdots\; w_m \\
DM = \begin{array}{c} A_1 \\ A_2 \\ \vdots \\ A_n \end{array}
\begin{bmatrix}
x_{11} & x_{12} & \cdots & x_{1m} \\
x_{21} & x_{22} & \cdots & x_{2m} \\
\vdots & \vdots & \vdots & \vdots \\
x_{n1} & x_{n2} & \cdots & x_{nm}
\end{bmatrix}
\end{array}
\tag{24.1}
$$

where A_n are the alternatives to be evaluated, C_m are the criteria against which the alternatives are evaluated, w_m are the weighting coefficients of the criteria that determine their importance, and x_{ij} are the parametric values, i.e. the scores assigned by the experts to the alternatives for each of the criteria. The first step in the assessment is determining the criteria weights, where $w_j \geq 0$; $\sum w_j = 1$. This means that the decision on the importance of the criteria needs to be made, since all the criteria cannot be equally significant. The second step involves the actual grading of alternatives by the experts, based on the criteria (attributes), on a determined scale that, in our case, is the 5-point Likert scale, where: 1—weak, 2—moderate, 3—strong, 4—very strong and 5—exceptional. Lastly, the evaluations for each alternative are summed up as follows:

$$\sum_{i=1;j=1}^{n;m} x_{ij} \tag{24.2}$$

Table 24.3 shows the average weight values attributed to each criterion by the experts, based on their opinions on the importance of the criteria, in a way that the sum of the weights gives a total of 1, as given in Eq. 24.2.

The sums of the assessments of each cultural site (A_n) are given in Table 24.4.

Once this stage is completed, it is proceeded with the subsequent steps of the methods, as follows. We first complete the procedure of the Simple Additive Weighting method, followed by the Technique for Order Preference by Similarity to Ideal

Table 24.3 Criteria (C_m) used for evaluation of cultural sites and their corresponding weights (w_m)

Evaluation criteria (C_m)	C_1 Historic value	C_2 Aesthetic /artistic value	C_3 Representativeness	C_4 State of preservation	C_5 Infrastructure and accessibility	C_6 Social significance	
Criteria weights (w_m)	w_1	w_2	w_3	w_4	w_5	w_6	Total
Values	0.2043	0.2086	0.1343	0.1614	0.1414	0.15	1

Table 24.4 Evaluation matrix. Sums of experts' evaluations of the alternatives (A_n), based on criteria (C_m)

Evaluation matrix	C_1	C_2	C_3	C_4	C_5	C_6
A_1	12	25	16	30	27	19
A_2	18	23	16	28	26	21
A_3	26	30	23	24	16	25
A_4	25	30	22	23	19	23
A_5	27	29	26	27	19	29
A_6	29	28	26	22	20	27

Solution, the latter being slightly more complex than the former. Ultimately, the outcomes of the two methods are compared and discussed.

24.4.5 Methods

SAW

Simple Additive Weighting, or Weighted Sum Method, or Scoring Method (SM) is the earliest and possibly the most widely used MCDM method for its simplicity and quickness to form a comprehensive judgement on the performance of alternatives in question (Afshari et al. 2010; Ishizaka et al. 2013; Mela et al. 2012; Triantaphyllou et al. 1998). The method is based on the weighted average, where the ranks are obtained by multiplying the normalized values of the decision matrix with the criteria weights, assigned by the decision-makers to each criterion. It is the simplest form of multi-attribute utility analysis that applies a linear relationship, which involves standardization of the scores across all criteria by: (i) assigning the preference weights; (ii) multiplying these by the scores; (iii) adding up the resulting scores to obtain the total weighted scores for each alternative and (iv) ranking the total weighted scores (Zardari et al. 2015).

The two simple steps of the procedure are the following, sequentially applied to the case study in focus of the paper.

(1) Normalization of decision matrix (criteria) →

$$r_{ij} = \frac{x_{ij} - x_j^{**}}{x_j^{*} - x_j^{**}} \tag{24.3}$$

(2) Construction of weighted normalized decision matrix by multiplying normalized matrix (Table 24.5) by the corresponding criteria weights, given in Table 24.3 →

Table 24.5 Normalized decision matrix after applying Eq. 24.3 to the evaluation matrix, given in Table 24.4

Matrix normalization	C_1	C_2	C_3	C_4	C_5	C_6
A_1	0	0.285714286	0	1	1	0
A_2	0.352941176	0	0	0.75	0.909090909	0.2
A_3	0.823529412	1	0.7	0.25	0	0.6
A_4	0.764705882	1	0.6	0.125	0.272727273	0.4
A_5	0.882352941	0.857142857	1	0.625	0.272727273	1
A_6	1	0.714285714	1	0	0.363636364	0.8

Table 24.6 Weighted normalized decision matrix, based on Eq. 24.4

Alternatives (A_n)	Values
A_1	0.3624
A_2	0.351701337
A_3	0.601207059
A_4	0.564148048
A_5	0.782803342
A_6	0.659018182

$$\begin{bmatrix} r_{11} & r_{12} & \cdots & r_{1m} \\ r_{21} & r_{22} & \cdots & r_{2m} \\ \vdots & \vdots & \vdots & \vdots \\ r_{n1} & r_{n2} & \cdots & r_{nm} \end{bmatrix} = \begin{bmatrix} r_{11}w_1 + & r_{12}w_2 + & \cdots & + r_{1m}w_m \\ r_{21}w_1 + & r_{22}w_2 + & \cdots & + r_{2m}w_m \\ \vdots & \vdots & \vdots & \vdots \\ r_{n1}w_1 + & r_{n2}w_2 + & \cdots & + r_{nm}w_m \end{bmatrix} \tag{24.4}$$

(3) Ranking of results, based on the weighted decision matrix values, given in Table 24.6

Ranked alternatives	Rankings
Casa da Música concert house	V
Serralves Foundation	VI
Monument Church of S. Francisco de Assis	III
Churches of Carmo and Carmelitas	IV
Clérigos Tower	I
Cathedral	II

TOPSIS

Technique for Order Preference by Similarity to Ideal Solution was originally proposed by Hwang and Yoon (1981) to help select the best alternative against a fixed number of criteria. It is a simple MCDA method that aids in ranking of a set of alternatives that simultaneously have the shortest Euclidean distance from the positive ideal solution and the furthest distance from the negative ideal solution (Behzadian et al. 2012; Lakshmi and Venkatesan 2014; Zavadskas et al. 2016). The method is applied through five successive steps (Hwang and Yoon 1981): (i) normalization of the initial decision matrix; (ii) building of the weighted normalized decision matrix by multiplying the normalized matrix by criteria weights; (iii) determining the positive ideal (A^*) and the negative ideal (A^-) solutions; (iv) calculating the separation measures for each alternative, i.e. their distance from the positive ideal and the negative ideal solutions; and (v) calculating the relative closeness coefficient (C^*), i.e. the closeness of the alternatives to the ideal solution. The method results in ranking of the alternatives in question, based on the values of the closeness coefficients for each alternative.

The procedure is explained subsequently and applied to the case study in focus of the paper. The results of both methods are compared and discussed in the following section.

(1) Normalization of decision matrix →

$$r_{ij} = \frac{x_{ij}}{\sqrt{\sum_{k=1}^{n} x_{kj}^2}} \tag{24.5}$$

(2) Construction of weighted normalized decision matrix by multiplying normalized decision matrix (Table 24.7) by the corresponding criteria weights, given in Table 24.3 →

$$\begin{bmatrix} r_{11} & r_{12} & \dots & r_{1m} \\ r_{21} & r_{22} & \dots & r_{2m} \\ \vdots & \vdots & \vdots & \vdots \\ r_{n1} & r_{n2} & \dots & r_{nm} \end{bmatrix} \cdot \begin{bmatrix} w_1 \\ w_2 \\ \vdots \\ w_m \end{bmatrix} = \begin{bmatrix} n_{11}+ & n_{12}+ & \cdots & +n_{1m} \\ n_{21}+ & n_{22}+ & \cdots & +n_{2m} \\ \vdots & \vdots & \vdots & \vdots \\ n_{n1}+ & n_{n1}+ & \cdots & +n_{nm} \end{bmatrix}. \tag{24.6}$$

Table 24.7 Normalized decision matrix after applying Eq. 24.5 to the evaluation matrix, given in Table 24.4

Matrix normalization	C_1	C_2	C_3	C_4	C_5	C_6
A_1	0.207669653	0.369449163	0.298297801	0.474223108	0.511808107	0.319972347
A_2	0.311504479	0.33989323	0.298297801	0.442608234	0.492852251	0.353653647
A_3	0.449950914	0.443338996	0.428803089	0.379378486	0.303293693	0.421016246
A_4	0.43264511	0.443338996	0.410159476	0.36357105	0.360161261	0.387334946
A_5	0.467256718	0.428561029	0.484733926	0.426800797	0.360161261	0.488378846
A_6	0.501868327	0.413783062	0.484733926	0.347763613	0.379117116	0.454697546

Table 24.8 Weighted normalized decision matrix, based on Eq. 24.6

Normalized matrix weighting	C_1	C_2	C_3	C_4	C_5	C_6
A_1	0.04242691	0.077067095	0.040061395	0.07653961	0.072369666	0.047995852
A_2	0.063640365	0.070901728	0.040061395	0.071436969	0.069689308	0.053048047
A_3	0.091924972	0.092480514	0.057588255	0.061231688	0.042885728	0.063152437
A_4	0.088389396	0.092480514	0.055084418	0.058680367	0.050926802	0.058100242
A_5	0.095460548	0.089397831	0.065099766	0.068885649	0.050926802	0.073256827
A_6	0.102531699	0.086315147	0.065099766	0.056129047	0.05360716	0.068204632

Table 24.9 Ideal solutions for each criterion, based on values from Table 24.8

Ideal solutions	C_1	C_2	C_3	C_4	C_5	C_6
A^* (positive ideal solution)	0.102531699	0.092480514	0.065099766	0.07653961	0.072369666	0.073256827
A^- (negative ideal solution)	0.04242691	0.070901728	0.040061395	0.056129047	0.042885728	0.047995852

(3) Determining positive ideal (A^*) and negative ideal (A^-) solutions from weighted normalized decision matrix, where →

$A^* = (\max n_{ij} \mid j \in V), (\min n_{ij} \mid j \in V');$
$A^- = (\min n_{ij} \mid j \in V), (\max n_{ij} \mid j \in V');$
$V = (j = 1, 2, \ldots, m \mid j \in \text{max. criteria}$
$V' = (j = 1, 2, \ldots, m \mid j \in \text{min. criteria}$

(4) Calculating separation (S_i) of each alternative from the positive ideal (*) and the negative ideal ($^-$) solutions →

$$S_i^* = \sqrt{\sum_{j=1}^{m} \left(n_{ij} - n_j^*\right)^2} \tag{24.7}$$

and

$$S_i^- = \sqrt{\sum_{j=1}^{m} \left(n_{ij} - n_j^-\right)^2} \tag{24.8}$$

(5) Calculating relative closeness (C^*) to ideal solution (Table 24.10) →

$$C_i^* = \frac{S_i^-}{S_i^* + S_i^-}, \text{ where } 0 \le C_i^* \le 1. \tag{24.9}$$

where $0 \le C_i^* \le 1$, meaning that the alternative is closer to ideal solution if its closeness value is nearer to 1 and farther from 0.

Table 24.10 Separation of alternatives from ideal solutions (see Table 24.9), resulting from Eqs. 24.7 and 24.8

Distance of alternatives (A_n) from ideal solutions	S^* (positive)	S^- (negative)
A_1	0.071520599	0.036385511
A_2	0.055197065	0.037792853
A_3	0.037076397	0.058980098
A_4	0.036177106	0.054563286
A_5	0.024039211	0.068169635
A_6	0.028847169	0.07071367

Table 24.11 Closeness of alternatives to ideal solution, resulting from Eq. 24.9

Closeness to ideal solution (C^*)	Values
A_1	0.337196023
A_2	0.406418821
A_3	0.614014677
A_4	0.601311995
A_5	0.739296044
A_6	0.71025587

(6) Ranking of results, based on the values of closeness to ideal solution, given in Table 24.11 →

Ranked alternatives	Rankings
Casa da Música concert house	VI
Serralves Foundation	V
Monument Church of S. Francisco de Assis	III
Churches of Carmo and Carmelitas	IV
Clérigos Tower	I
Cathedral	II

24.5 Discussion of Findings

The two MCDA methods, applied to the case study in focus of this paper, produce practically identical outcomes, the only difference being that the last two ranked alternatives change places. Therefore, their usage is perfectly comparable in this instance, which is a fact that differs from the outcomes of various other case studies.

For example, in research done by Ishizaka et al. (2013), WSM and PROMETHEE methods offer a consensual recommendation, while TOPSIS gives a different one, showing a disparity in their application, on the contrary to what this study reveals. Similarly, the application of ELECTRE and WSM to the evaluation of destination competitiveness by Botti and Peypoch (2013) produces essentially different outcomes. Therefore, the outcomes of a simultaneous application of various methods to a single problem are not always consensual. They might even be in accordance in a certain context but vary in another.

The MCMA results in practically clustering two groups of alternatives—the religious and the nonspiritual ones. The two bottom-most-ranked cultural sites are the most recent, secular structures, one of them being classified as a National Monument (The Serralves Foundation), whereas the other (Casa da Música) does not yet have this designation, although the process of its classification is ongoing. All the alternatives are complex *objets d'art*, attractive to visitors of diverse profiles for their great intrinsic values, this being the main reason behind the decision to study them. The study, however, appoints to the contemporary, secular sites being considered inferior to the religious ones that are, simultaneously, of older date. Tables 24.12 and 24.13 summarize the rankings of the alternatives, resulting from SAW and TOPSIS, respectively.

When it comes to criteria values, the aesthetic/artistic value is scored highest, followed by the historic value, state of preservation, social significance, infrastructure and accessibility and, lastly, the representativeness. Therefore, aesthetic and historic values were given the highest importance. Here, we find pertinent to emphasize that the historic value is not to be equalized with historical, meaning that the said value refers to the buildings' importance throughout the history, not the fact that they

Table 24.12 SAW rankings

Ranked alternatives	Rankings
Clérigos Tower	I
Cathedral	II
Monument Church of S. Francisco de Assis	III
Churches of Carmo and Carmelitas	IV
Casa da Música concert house	V
Serralves Foundation	VI

Table 24.13 TOPSIS rankings

Ranked alternatives	Rankings
Clérigos Tower	I
Cathedral	II
Monument Church of S. Francisco de Assis	III
Churches of Carmo and Carmelitas	IV
Serralves Foundation	V
Casa da Música concert house	VI

were built in the past, i.e. not their oldness. State of preservation of the structure and its social significance are thought of as being of medium importance, while infrastructure/accessibility and building's representativeness were thought to be the least important values.

It is interesting to compare these results with the research by Yan et al. (2017) where, within the set of resource values, historic value comes first, followed by authenticity of the traditional style, while the aesthetic value and ambience or setting are of medium importance, whereas this case study results in the latter being the most important criteria. Apart from the resource values, the same authors evaluate the state of development indicators, such as the factor of accessibility, tourist facilities, interpretation in situ, tourist information, among others. This paper, however, explores a single set of indicators, where the intrinsic values and the state/condition of cultural sites were not separated. Therefore, it would be interesting to apply different methodologies in the future and evaluate multiple sets of indicators, so as to see whether the findings would differ from the present ones.

24.6 Word of Conclusion

This paper explores the usage of two MCDM methods—SAW and TOPSIS—for the assessment of six particular cultural heritage sites from a tourism perspective. By exploring these methods, we aimed at providing a framework able to assist in planning and decision-making related, but not limited, to cultural sites. It was opted for the assessment of sites of different typologies, so as to focus on their different intrinsic values, independently of their nature, purpose or age. It was, hence, opted for the assessment of: (i) a contemporary concert house; (ii) an art foundation consisting of a contemporary art museum, an art deco mansion, and vast terrains; (iii) a gothic temple; (iv) a complex of two baroque/rococo churches; (v) a baroque church bell tower and (vi) the Romanesque/mannerist/baroque/gothic cathedral. The differences of the sites in question are more than obvious—in terms of architectural styles, age, purpose, dimension, etc. However, they have one thing in common, which is that they all represent the city's landmarks and the must-see tourism attractions. This was the logic that was followed upon deciding on the alternatives to assess in this paper.

The alternatives were ranked on the basis of their socio-economic, sociocultural and environmental values, as seen by the experts from tourism and the related fields. The rankings reveal that the aesthetic/artistic value of cultural heritage buildings is perceived as most important by the experts, followed by the historic value, state of preservation, social significance, infrastructure and accessibility, and representativeness. However, it seems that the age of the buildings has actually played the central role in both the criteria weighting and the alternative evaluation decisions, since both methods result in the "youngest" two buildings being ranked last. Even though the lastly ranked Casa da Música is among the city's must-see tourism attractions—aside from being one-of-a-kind masterpiece—this result was somehow to be expected.

The literature reveals that MCDA is quite underexplored in the tourism field, when compared to its application to cultural heritage. When it comes to its engagement in joint contexts of cultural heritage and tourism, it is practically inexistent. This is where this paper's contribution lays, it being twofolded. Firstly, it adds theoretically to the state of the art in tourism, where it explores: (i) application of MCDA in tourism in general and (ii) its usage for cultural heritage evaluation in the tourism context. Practical contribution rests in the usage of two MCDA methods in a single case study, where their applicability is shown on a concrete empirical example. This serves to suggest that their area of application can go beyond the context of the present case study.

Cultural heritage is getting each time more affected by tourism dynamics, thus calling for planning and decision-making regarding tourism to be in accordance with heritage enhancement, preservation and sustainability of its space values and identity. Tourism's interdisciplinarity is one of its most transparent features, which additionally calls for resorting to methodologies from other fields, offering vast research possibilities. In case of the MCDM techniques, a great number of different methods can be explored in different contexts and combined with other types of analysis, such as GIS or SWOT analysis. Also, same methods could be applied in different contexts for support in regulation/legislation of spatial planning. In tourism, particularly, they can be successfully applied to comparing/ranking/evaluating of the best/worst aspects of both tangible and intangible capital, such as destinations, attractions, events, services, human resources and so on.

References

Afshari A, Mojahed M, Yusuff RM (2010) Simple additive weighting approach to personnel selection problem. Int J Innovation Technol. Manage 1(5):511–515

Aminu M, Matori AN, Yusof KW, Zainol RB (2014) Application of geographic information system (GIS) and analytic network process (ANP) for sustainable tourism planning in Cameron Highlands, Malaysia. Appl Mech Mater 567:769–774. https://doi.org/10.4028/www.scientific.net/AMM.567.769

Behzadian M, Otaghsara SK, Yazdani M, Ignatius J (2012) A state-of the-art survey of TOPSIS applications. Expert Syst Appl 39(17):13051–13069. https://doi.org/10.1016/j.eswa.2012.05.056

Belton V, Stewart TJ (2002) Multiple criteria decision analysis: an integrated approach. Kluwer Academic Publishers, Boston

Belton V, Stewart TJ (2010) Problem structuring and multiple criteria decision analysis. In: Ehrgott M, Figueira JR, Greco S (eds) Trends in multiple criteria decision analysis, Springer, New York, pp 209–239. http://dx.doi.org/10.1007/978-1-4419-5904-1

Botti L, Peypoch N (2013) Multi-criteria ELECTRE method and destination competitiveness. Tourism Manage. Perspect. 6:108–113. https://doi.org/10.1016/j.tmp.2013.01.001

Carayannis EG, Ferreira FAF, Bento P, Ferreira JJM, Jalali MS, Fernandes BMQ (2018) Developing a socio-technical evaluation index for tourist destination competitiveness using cognitive mapping and MCDA. Technol Forecast Soc Chang 131:147–158. https://doi.org/10.1016/j.techfore.2018.01.015

Casa da Música (nd) A obra. http://www.casadamusica.com/en/a-casa-da-musica/a-obra/. Accessed 5 Feb 2018

Choi AS, Ritchie BW, Papandrea F, Bennett J (2010) Economic valuation of cultural heritage sites: a choice modelling approach. Tour Manag 31(2):213–220. https://doi.org/10.1016/j.tourman.2009.02.014
Civitatis Porto (nd) Igreja e Torre dos Clérigos. https://www.tudosobreporto.com/igreja-torre-dos-clerigos. Accessed 7 Feb 2018
Costa C (2006) Tourism planning, development and the territory. In: Buhalis B, Costa C (eds) Tourism management dynamics: trends, management and tools. Elsevier Butterworth-Heinemann, Oxford, pp 236–243
Dalmas L, Geronimi V, Noël JF, Tsang King Sang J (2015) Economic evaluation of urban heritage: an inclusive approach under a sustainability perspective. J Cult Heritage 16(5):681–687. https://doi.org/10.1016/j.culher.2015.01.009
De Sousa M (nd) Igrejas do Carmo e dos Carmelitas. http://portoby.livrarialello.pt/igrejas-do-carmo-dos-carmelitas/. Accessed 5 Feb 2018
Del Vasto-Terrientes L, Fernández-Cavia J, Huertas A, Moreno A, Valls A (2015) Official tourist destination websites: Hierarchical analysis and assessment with ELECTRE-III-H. Tourism Manag. Perspect. 15:16–28. https://doi.org/10.1016/j.tmp.2015.03.004
Descobrir Portugal (2010) Torre dos Clérigos, o mais conhecido monumento do Porto. https://descobrir-portugal.com/2010/09/torre-dos-clerigos/. Accessed 7 Feb 2018
Di Martino P, Petrillo C (2006) An international project to develop networking for promoting a specific destination: emigration as a tool to enhance tourism in Cilento area. In: Lazzeretti L, Petrillo C (eds) Tourism local systems and networking. Elsevier, Oxford, pp 219–233
Diocese do Porto (nd) Sé do Porto. http://www.diocese-porto.pt/index.php?option=com_content&view=article&id=877&Itemid=60. Accessed 7 Feb 2018
Direção Regional de Cultura do Norte (nd) Sé do Porto. http://www.culturanorte.pt/pt/patrimonio/se-do-porto/. Accessed 7 Feb 2018
Direção-Geral do Património Cultural (nd) Conjunto constituído pela Igreja dos Carmelitas Descalços e Igreja de Nossa Senhora do Monte do Carmo. http://www.patrimoniocultural.gov.pt/pt/patrimonio/patrimonio-imovel/pesquisa-do-patrimonio/classificado-ou-em-vias-de-classificacao/geral/view/155835. Accessed 5 Feb 2018
Dutta M, Husain Z (2009) An application of Multicriteria Decision Making to built heritage: the case of Calcutta. J Cult Heritage 10(2):237–243. https://doi.org/10.1016/j.culher.2008.09.007
Dyer P, Aberdeen L, Schuler S (2003) Tourism impacts on an Australian indigenous community: a Djabugay case study. Tour Manag 24:83–95. https://doi.org/10.1016/S0261-5177(02)00049-3
Emir O, Bayer RU, Erdoğan NK, Karamaşa Ç (2016) Evaluating the destination attractions from the point of experts' view: an application in Eskişehir. TURIZAM 20(2):92–104. https://doi.org/10.18421/TRZ20.02-04
Europe's best destinations (nd) European best destinations 2017. https://www.europeanbestdestinations.com/best-of-europe/european-best-destinations-2017/. Accessed 30 Jan 2018
Eurostat (2017a) Estatísticas do turismo. http://ec.europa.eu/eurostat/statistics-explained/index.php?title=Tourism_statistics/pt&oldid=351465. Accessed 18 Feb 2018
Eurostat (2017b) Tourism statistics. http://ec.europa.eu/eurostat/statistics-explained/index.php/Tourism_statistics. Accessed 3 Feb 2018
Ferretti V, Comino E (2015) An integrated framework to assess complex cultural and natural heritage systems with multi-attribute value theory. J Cult Heritage 16(5):688–697. https://doi.org/10.1016/j.culher.2015.01.007
Ferretti V, Bottero M, Mondini G (2014) Decision making and cultural heritage: an application of the multi-attribute value theory for the reuse of historical buildings. J Cult Heritage 15(6):644–655. https://doi.org/10.1016/j.culher.2013.12.007
Fuentes JM (2010) Methodological bases for documenting and reusing vernacular farm architecture. J Cult Heritage 11(2):119–129. https://doi.org/10.1016/j.culher.2009.03.004

Fusco Girard L, De Toro P (2007) Integrated spatial assessment: a multicriteria approach to sustainable development of cultural and environmental heritage in San Marco dei Cavoti, Italy. CEJOR 15(3):281–299. https://doi.org/10.1007/s10100-007-0031-1
Giove S, Rosato P, Breil M (2011) An application of multicriteria decision making to built heritage: the redevelopment of Venice Arsenale. J Multi-Criteria Decision Anal 17(3–4):85–99. https://doi.org/10.1002/mcda.455
Göksu A, Kaya SE (2014) ranking of tourist destinations with multi-criteria decision making methods in Bosnia and Herzegovina. Econ Rev—J Econ Bus 12(2):91–103
Gomes M (nd) Carmo and Carmelitas' Churches. http://www.localporto.com/carmo-camelitas-church/. Accessed 5 Feb 2018
Hutter M, Frey BS (2010) On the influence of cultural value on economic value. Revue d'économie politique 120(1):35–46. https://doi.org/10.3917/redp.201.0035
Hwang CL, Yoon KP (1981) Multiple attribute decision making: methods and applications. Springer-Verlag, New York
ICOMOS (2002) International cultural tourism charter: principles and guidelines for managing tourism at places of cultural and heritage significance. http://www.frh-europe.org/cms/wp-content/uploads/2017/12/ICOMOS-International-Cultural-Tourism-Charter-English1.pdf. Accessed 12 Mar 2018
Ishizaka A, Nemery P, Lidouh K (2013) Location selection for the construction of a casino in the Greater London region: a triple multi-criteria approach. Tour Manag 34:211–220. https://doi.org/10.1016/j.tourman.2012.05.003
Jafari J (1996) Tourism and culture: an inquiry into paradoxes. In: Proceedings of a round table. Culture, tourism, development: Crucial issues for the XXIst century, UNESCO Headquarters, Paris, 26–27 June 1996. http://unesdoc.unesco.org/images/0010/001085/108539Eo.pdf. Accessed 23 Feb 2018
Khan H, Seng CF, Cheong WK (1990) Tourism multiplier effects on Singapore. Ann Tourism Res 17(3):408–418
Kim C-J, Yoo WS, Lee U-K, Song K-J, Kang K-I, Cho H (2010) An experience curve-based decision support model for prioritizing restoration needs of cultural heritage. J Cult Heritage 11(4):430–437. https://doi.org/10.1016/j.culher.2010.03.004
Klamer A (2004) Cultural goods are good for more than their economic value. In: Rao V, Walton M (eds) Culture and Public Action. Stanford University Press, Stanford, CA, pp 138–162
Klamer A (2016) The value-based approach to cultural economics. J Cult Econ 40(4):365–373. https://doi.org/10.1007/s10824-016-9283-8
Klamer A (ed) (2017) Doing the right thing: a value based economy, 2nd edn. Ubiquity Press, London
Lakshmi TM, Venkatesan VP (2014) A comparison of various normalization in techniques for order performance by Similarity to ideal solution (TOPSIS). Int J Comput Algorithm 3:882–888
Mailly D, Abi-Zeid I, Pepin S (2013) A multi-criteria classification approach for identifying favourable climates for Tourism. J Multi-Criteria Decision Anal 21(1–2):65–75. https://doi.org/10.1002/mcda.1507
Mason P (ed) (2016) Tourism impacts, planning and management, 3rd edn. Routledge, London
Massieu A (2006) Managing economic impacts, tourism satellite accounts and observatories. In: Buhalis B, Costa C (eds) Tourism management dynamics: trends, management and tools. Elsevier Butterworth-Heinemann, Oxford, pp 229–235
Matarasso F (ed) (2001) Recognising Culture: a series of briefing papers on culture and development, Da Costa Print: The Book Factory, London. http://www.culturenet.cz/res/data/004/000567.pdf. Accessed 16 Mar 2018
Mathouraparsad S, Maurin A (2017) Measuring the multiplier effects of tourism industry to the economy. Adv Management & Applied Economics 7(2):123–157
Mazzanti M (2002) Cultural heritage as multi-dimensional, multi-value and multi-attribute economic good: toward a new framework for economic analysis and valuation. J Socio-Econ 31(5):529–558. https://doi.org/10.1016/S1053-5357(02)00133-6

Mela K, Tiainen T, Heinisuo M (2012) Comparative study of multiple criteria decision making methods for building design. Adv Eng Inform 26:716–726. https://doi.org/10.1016/j.aei.2012.03.001

Michailidou AV, Vlachokostas C, Moussiopoulos N (2016) Interactions between climate change and the tourism sector: Multiple-criteria decision analysis to assess mitigation and adaptation options in tourism areas. Tour Manag 55:1–12. https://doi.org/10.1016/j.tourman.2016.01.010

Munda G. (2005) Multiple Criteria Decision Analysis and Sustainable Development. In: Figueira JR, Greco S, Ehrgott M (eds) Multiple criteria decision analysis: state of the art surveys. international series in operations research and management science, vol 78. Springer, New York, pp 953–986. https://doi.org/10.1007/0-387-23081-5_23

Munda G. (2016) Multiple criteria decision analysis and sustainable development. In: Greco S, Ehrgott M, Figueira JR (eds) Multiple criteria decision analysis: state of the art surveys, vol 1/2, 2nd edn. International series in operations research and management science, vol 233. Springer, New York, pp 1235–1268. https://doi.org/10.1007/978-1-4939-3094-4

Necissa Y (2011) Cultural heritage as a resource: its role in the sustainability of urban developments. The case of Tlemcen Algeria. Procedia Eng 21:874–882. https://doi.org/10.1016/j.proeng.2011.11.2089

Noonan DS (2003) Contingent valuation and cultural resources: a meta-analytic review of the literature. J Cult Econ 27:159–176. https://doi.org/10.1023/A:1026371110799

NWHO (1999) Sustainable tourism and cultural heritage: a review of development assistance and its potential to promote sustainability. http://www.msu.ac.zw/elearning/material/1329113217culture_fulltext%5B1%5D.pdf. Accessed 13 Mar 2018

Ordem São Francisco do Porto (nd) http://ordemsaofrancisco.pt/instituicao/. Accessed 6 Feb 2018

O portal de notícias do Porto (2018) Porto vence o prestigiado world excellence award para a cidade europeia mais amiga daschoi startups. http://www.porto.pt/noticias/porto-vence-o-prestigiado-world-excellence-award-para-a-cidade-europeia-mais-amiga-das-startups-). Accessed 24 Feb 2018

Richards G (2001) The development of cultural tourism in Europe. In: Richards G (ed) Cultural attractions and European Tourism. CABI, Wallingford, pp 3–30

Richards G, Wilson J (2004) Social networks, culture and tourism in Catalunya. Paper presented at the ATLAS Annual Conference: Networking and partnerships in destination development and management, Naples, 4–6 April 2004

Robinson M, Picard D (2006) Tourism, culture and sustainable development. Société Édition Provence, Nîmes. http://unesdoc.unesco.org/images/0014/001475/147578e.pdf. Accessed 9 Mar 2018

Rusu S (2011) Tourism multiplier effect. J Econ Bus Res 17(1):70–76

Serralves (2018) The foundation. https://www.serralves.pt/en/foundation/the-foundation/. Accessed 5 Feb 2018

Smith MK (2007) Space, place and placelessness in the culturally regenerated city. In: Richards G (ed) Cultural tourism: global and local perspectives. Haworth Press, New York, pp 91–108

Swarbrooke J (1999) Sustainable tourism management. CABI, Wallingford

The World Bank (2018) International tourism, receipts (current US$). https://data.worldbank.org/indicator/ST.INT.RCPT.CD. Accessed 25 Mar 2018

Thórhallsdóttir TE (2007) Environment and energy in Iceland: a comparative analysis of values and impacts. Environ Impact Assess Rev 27(6):522–544. https://doi.org/10.1016/j.eiar.2006.12.004

Throsby D (2001) Economics and culture. Cambridge University Press, Cambridge

Torre dos Clérigos (2018) History and architecture. Tower, museum, church. http://www.torredosclerigos.pt/en/history-and-architecture/tower-brotherhood-museum-church/. Accessed 7 Feb 2018

Triantaphyllou E, Shu B, Nieto Sanchez S, Ray T (1998) Multi-criteria decision making: an operations research approach. In: Webster JG (ed) Encyclopedia of electrical and electronics engineering, vol 15. Wiley. New York, NY, pp 175–186

UNWTO (nd). Cultural Heritage. http://sdt.unwto.org/en/content/cultural-heritage-1. Accessed 21 Feb 2018

Wang H-J, Zeng Z-T (2010) A multi-objective decision-making process for reuse selection of historic buildings. Expert Syst Appl 37(2):1241–1249. https://doi.org/10.1016/j.eswa.2009.06.034

World Economic Forum (2017) The travel and tourism competitiveness report 2017: paving the way for a more sustainable and inclusive future. http://www3.weforum.org/docs/WEF_TTCR_2017_web_0401.pdf. Accessed 2 Feb 2018

World Travel Awards (nd) World Winners 2017. https://www.worldtravelawards.com/winners/2017/world. Accessed 30 Jan 2018

Yan L, Gao BW, Zhang M (2017) A mathematical model for tourism potential assessment. Tour Manag 63:355–365. https://doi.org/10.1016/j.tourman.2017.07.003

Yoon KP, Hwang CL (eds) (1995) Multiple attribute decision making: an introduction. Sage University paper series on quantitative applications in the social sciences, Sage publications, Thousand Oaks

Zardari NH, Ahmed K, Shirazi SM, Yusop ZB (2015) Weighting methods and their effects on multi-criteria decision making model outcomes in water resources management. Springer Briefs in Water Science and Technology, Springer, New York. http://dx.doi.org/10.1007/978-3-319-12586-2

Zavadskas EK, Mardani A, Turskis Z, Jusoh A, Nor KMD (2016) Development of TOPSIS method to solve complicated decision-making problems—an overview on developments from 2000 to 2015. Int J Inf Technol Decis Making 15(3):645–682. https://doi.org/10.1142/S0219622016300019

Zhang H, Gu C, Gu L, Zhang Y (2011) The evaluation of tourism destination competitiveness by TOPSIS & information entropy—a case in the Yangtze River Delta of China. Tour Manag 32(2):443–451. https://doi.org/10.1016/j.tourman.2010.02.007

Zhou Y, Maumbe K, Deng J, Selin SW (2015) Resource-based destination competitiveness evaluation using a hybrid analytic hierarchy process (AHP): the case study of West Virginia. Tourism Manag Perspect 15:72–80. https://doi.org/10.1016/j.tmp.2015.03.007

Chapter 25
Reactivation of Functionally Derelict Areas with Cultural Heritage Sites

Mojca Foški, Gašper Mrak, Alma Zavodnik Lamovšek and Barbara Lampič

Abstract Functionally derelict areas (FDAs) refer to not fully utilised land or abandoned land with the impact of its previous use still visible and being of lower utility value, which can have potential for further sustainable spatial development and also for protecting undeveloped natural or agricultural land, i.e. greenfield land. This chapter addresses FDAs where there is immovable cultural heritage (ICH) on wholly or partially abandoned sites with substantially reduced functioning. The purpose of this study is to show the possible regulatory solutions at national and local levels for the regeneration/reactivation of FDAs with cultural heritage sites. Analysis of FDAs in Slovenian territory has revealed that problems associated with their regeneration originate from the applicable provisions concerning cultural heritage protection. The relevance of the investigated topic is supported by the fact that in more than a third of all FDAs in Slovenia we recorded various cultural heritage units and legal regimes for their protection, which are included in the national register of immovable cultural heritage. FDAs with the presence of ICH were classified into three groups (A, B, and C) depending on the intensity of the ICH phenomenon in the FDAs. Based on the cases presented in this chapter, we propose various instruments for the reactivation of FDAs with the presence of ICH.

Keywords Derelict areas · Cultural heritage · Methodological approach · Reactivation

M. Foški · G. Mrak · A. Zavodnik Lamovšek (✉)
Faculty of Civil and Geodetic Engineering, University of Ljubljana, Ljubljana, Slovenia
e-mail: alma.azvodni@fgg.uni-lj.si

M. Foški
e-mail: mojca.foski@fgg.uni-lj.si

G. Mrak
e-mail: gasper.mrak@fgg.uni-lj.si

B. Lampič
Faculty of Arts, University of Ljubljana, Ljubljana, Slovenia
e-mail: barbara.Lampic@ff.uni-lj.si

M. Obad Šćitaroci et al. (eds.), *Cultural Urban Heritage*,
The Urban Book Series, https://doi.org/10.1007/978-3-030-10612-6_25

25.1 Introduction

Functionally derelict areas (hereinafter FDAs) refer to not fully utilised land or abandoned land with the impact of its previous use still visible and of lower utility value, which can have potential for further sustainable spatial development and also for protecting undeveloped natural or agricultural land, i.e. greenfield land (Lampič et al. 2017a). This chapter addresses FDAs where immovable cultural heritage (hereinafter ICH) is present on wholly or partially abandoned sites with substantially reduced functioning. The purpose of this study is to show the possible regulatory solutions at national and local levels for the reactivation of FDAs with the presence of cultural heritage.

The problem of brownfields has been identified in almost all European countries over the last few decades (Lampič et al. 2017b), while the growing share of abandoned areas and brownfields (particularly in Eastern European countries) has intensified as a result of transition from the socialist to the capitalist system, through the transformation of the economy and society, and through the global economic crisis in the recent decade. This problem has been also identified in Slovenia; nevertheless, efforts undertaken so far to address the problem of various types of spatial dereliction have been unsuccessful.

The basis for managing FDAs is to identify them in all their aspects and to establish a single record. In Slovenia, the register of functionally derelict areas was established in 2017 (Lampič et al. 2017a; Lampič and Bobovnik 2017), at the initiative of the Ministry of Economic Development and Technology of the Republic of Slovenia (hereinafter MEDT). MEDT supported the setting up of a register for steering economic investments in FDAs. In doing so, it recognised the importance of the study, which was methodologically designed to reflect a broader focus, both in terms of developing the FDA definition and in the manner of recording. The result of the entire study was the FDA record serving other purposes as well.

In terms of the usability of the FDA record for various purposes and further studies, it is important to note that each FDA was captured using a single inventory form and a physical field visit. The interviewer filled out the inventory form with a wide data set and interviewed representatives of professional municipal services (Lampič et al. 2017a). It appears from the local experts' responses that stakeholders (municipal staff, owners, developers, etc.) at the local level frequently underlined the presence of immovable cultural heritage as an obstacle for fast and efficient FDA revitalisation; moreover, immovable cultural heritage was not recognised as a development potential or an added value of the brownfield site either by municipal professional staff or by real property owners or developers. Such a wide-scale presence of ICH in FDAs also surprised the researchers, i.e. the FDA record developers.

"Heritage protection shall be provided for in the public in accordance with the cultural, educational, developmental, symbolic and identity significance of the heritage for the state, regions and municipalities" (Article 2 ZVKD-1 2008). The mere fact that there is any ICH present in an FDA is thus contrary to the principles of protecting cultural heritage sites as entities of national significance. What matters most from the protection perspective is the prevention of harmful effects on cultural heritage and integrated heritage conservation in the comprehensive reactivation and/or regeneration of FDAs.

This opposing perception of FDAs, as a negative spatial, economic, and social phenomenon, and the presence of ICH, as an important entity for designing spatial identity, provided the starting point for further studies. Furthermore, it was found that in the identified FDAs the number of ICH units was so large that rather than random anomalies occurring in space, they should be regarded as a large-scale problem. The presence of ICH cannot come in second place, after the FDA reactivation process, but requires an in-depth review of the situation in terms of the number, size, protection regimes, and ICH types present in the FDAs identified in Slovenia. In this study, we wanted to confirm the working hypothesis, i.e. that cultural heritage regeneration, when present in derelict areas, is an important factor of FDA reactivation, either by regenerating the former or establishing a new functionality of a site.

In this chapter we will: (1) analyse in detail the interaction of FDAs and ICH in Slovenia; (2) classify FDAs with the presence of ICH into various types and analyse their spatial distribution; (3) present two best practices; (4) based on the results gained, give orientations to help reduce the number of FDAs with ICH in Slovenia in FDA reactivation and regeneration. Given the universality of the problem, which was thoroughly addressed in a previous study, and given the establishment of a single FDA record (Lampič et al. 2017a), it is reasonable to expect that the results will be useful also on the broader European scale. Until now, we have not come across many studies that systematically and in a targeted way address the questions of FDA reactivation and regeneration with ICH present; for the most part, only case studies have been carried out.

25.2 Theoretical Background

In this subchapter, we provide a brief review of extensive professional and scientific, domestic and international literature which relates to FDAs in general and FDAs with immovable cultural heritage sites. The purpose of this review of the literature and other sources, including the review of good practices of already reactivated FDAs, is manifold. We seek to show not only the various approaches to defining FDAs in various (particularly European) countries, but we also focus on exploring their potentials, models, and FDA reactivation actions, particularly those where ICH is also present.

The definition of FDAs is generally intricately connected with the purpose and goals of addressing derelict areas and brownfields, while the related problems are often left to regional and/or national governance (Lampič et al. 2017). Nevertheless, as Alker et al. (2000) indicate, there are many definitions of derelict areas, depending on the study purpose, national policies, and even the individual developer's interests. *Brownfields handbook* (Bergatt Jackson et al. 2006) states that in Europe a derelict area is any site that is mostly abandoned, vacant or not fully utilised, where its former use has ceased and where it may also be contaminated, while the market has not been in the position to use it effectively (Alker et al. 2000; Bergatt Jackson et al. 2006; Špes et al. 2012). In her book *The death and life of great American cities* (1961), Jacobs already found that degradation starts with the lack of function and not vice versa; i.e. the function does not cease after a site becomes derelict. FDAs can be defined as sites where former activities (functions) have ceased, which was the main reason why other processes began, which even accelerated degradation (e.g. the physical deterioration of buildings, becoming overgrown, depopulation; Lampič et al. 2017). Depending on former uses, different typologies have been proposed (Vojvodíková et al. 2011; Bergatt Jackson et al. 2006; Gauchon 1997; Lampič et al. 2017a). Other proposals have come from various international projects dealing with the problem of derelict areas (Ferber and Grimski 2002; Ferber et al. 2006; COBRAMAN 2009a, b; TIMBRE 2011; RETINA 2012).

FDAs are mostly sites of abandoned factories and other industrial facilities, warehouses, abandoned mines, ports, abandoned farms, and military sites. Gauchon (1997) also draws attention to vacant sports and tourist facilities and other public services facilities, such as schools and hospitals. Lampič et al. (2017a) also classified as FDAs sites of unfinished investments (mostly residential and commercial constructions) that emerged after 2008, during the economic crisis.

There is potential for reactivation which can originate from the location, size, degree, and the extent of abandonment, even of the former, abandoned function. Transformations with reactivation and/or the regeneration of FDAs into industrial and residential areas or areas of service activities are the most common, and they also have a greening potential (Sanches and Pallegrino 2016; Kim 2018) and a potential for increasing the biodiversity of urban areas (Prentis and Northon 1992), a potential for energy generation (Donaldson and Lord 2018) and the development of ecotourism (Lemoine 2016). Navratil et al. (2018) find that abandoned areas (particularly industrial ones) can provide an excellent potential for establishing attractive tourist destinations, and they even suggest protecting abandoned industrial sites as the cultural heritage of a former (industrial) time. In practice, we can trace such cases in various European countries, as well as in Slovenia.

The presence of cultural heritage is often perceived as an obstacle to faster FDA reactivation (Berg and Stenbro 2015); however, the presence of cultural heritage must also be understood as an added value of a site, as it can increase public interest for reactivation and thus justify public engagement and public investments (Sable and Kling 2001). Since cultural heritage is an important social value, we are faced with a paradoxical situation in brownfield sites, as FDAs are perceived, contrary to ICH, as lacking in quality and value. The absence of functions and physical degradation in FDAs also usually has negative effects on the condition and quality of the protected ICH entity.

The identification of cultural heritage potential in FDAs often depends on the local community and individual developers (Rypkema 2009). A combination of remnants from the past coupled with new interventions allows for the development of unique sites, to which tourists and/or special social groups (young people, artists, etc.) are often attracted. Examples of good practices of developing entire abandoned industrial, military, and mining sites can be found in many European countries. Nevertheless, cases of including cultural heritage into site reactivation are less frequently addressed in the literature.

25.3 Methodology and Data

In the first part of this study, the descriptive method was used to explore the theoretical background. In the second part, a two-level empirical analysis in Slovenia was carried out: (1) first we classified FDAs into three groups, based on the presence of protected ICH as stipulated by the Cultural Heritage Protection Act (ZVKD-1 2008) and the Handbook of Legal Protection Regimes that Should Be Taken into Account in Spatial Planning and Interventions in Space in the Areas of Cultural Heritage (MoC 2018a); then (2) we also analysed the dependence of ICH on the FDA type and the spatial distribution in Slovenia (FDA record after Lampič et al. (2017a) and Lampič et al. (2017).

(1) The first part of the analysis was made based on selected criteria and indicators. FDAs were classified into three groups depending on the number of units of protected ICH (1; 2–5; more than 5 units of ICH), the proportion of the area covered by ICH in the FDAs (0–10; 11–50; 51–95, 96–100%), and the type of protection regime of the ICH according to the MoC (2011): registered archaeological site, cultural heritage site, and cultural monument site present in individual FDAs:

Group A: *FDAs with an accentuated presence of ICH satisfy the following combinations of indicators and criteria*:

- More than 5 units of ICH, regardless of the proportion of area that they occupy all types of ICH protection regimes.
- 2–5 units of ICH, proportion of area 96–100%, all types of ICH protection regimes.
- 1 unit of ICH, proportion of area 96–100%, all types of ICH protection regimes.

Group B: *FDAs with ICH, but satisfying, to a lesser degree, the following combinations of indicators and criteria:*

- 2–5 units of ICH, proportion of area 51–95%, all types of ICH protection regimes.
- 2–5 units of ICH, proportion of area 11–50%, all types of ICH protection regimes.
- 1 unit of ICH, proportion of area 51–95%, all types of ICH protection regimes.

Group C: *FDAs with a non-accentuated presence of ICH satisfy the following combinations of indicators and criteria*:

- 2–5 units of ICH, proportion of area 11–50%, all types of ICH protection regimes.
- 2–5 units of ICH, proportion of area 0–10%, all types of ICH protection regimes.
- 1 unit of ICH, proportion of area 11–50%, all types of ICH protection regimes.
- 1 unit of ICH, proportion of area 0–10%, all types of ICH protection regimes.

(2) In the second part of the analysis, the classified FDAs with the presence of ICH were addressed also depending on the FDA type and spatial distribution in

Slovenia. FDAs in Slovenia are divided into 9 basic types (Lampič et al. 2017a): (1) AGR—FDA of agricultural activities; (2) SERV—FDA of service activities; (3) TOUR—FDA of tourist, sports and recreation activities; (4) IND—FDA of industrial activities; (5) DEF—FDA of defence, protection and rescue services; (6); MIN—FDA of mineral extraction; (7) INF—FDA of infrastructures; (8) TRANS—FDA of transitional use; and (9) HOU—FDA for housing.

Based on the classification of FDAs with the presence of ICH into groups A, B, and C, according to the FDA types identified, we also analysed their spatial distribution in all Slovenian municipalities. Separately, we analysed urban municipalities and other municipalities, as specified in the Local Government Act (ZLS-UPB2 2007). We assumed that the number of ICH units in urban municipalities would be higher than in other municipalities.

25.3.1 The Data Used

As mentioned in the previous subchapter, in this analysis, we used the data from the FDA record (Lampič et al. 2017a; Lampič et al. 2017) and the cultural heritage register (MoC 2018a). For a better understanding of the study results, we also need to know the criteria for FDA capture, which took place using a common working method (Lampič et al. 2017a). All FDAs larger than 0.5 ha in open space and those larger than 0.2 ha in urban (built-up) areas were recorded. A recording criterion was also the complete or partial abandonment of activity (at least 10% of the site had to be abandoned), with visible signs of physical degradation (poor or lack of maintenance). In Slovenia, on the cross-sectional date of 30 September 2017, there were 1081 FDAs recorded in a total area of 3422.7 ha.

The cultural heritage register (MoC 2018a) is a publicly available viewer of immovable cultural heritage, while vector data were obtained from the Ministry of Culture of the Republic of Slovenia (hereinafter MoC). The register was first introduced in 1991, while its administration and use are specified in the Rules on the Cultural Heritage Register (2009). The heritage register lists all ICH by class, type, scope, ownership, and the unit's protection status. Cultural heritage sites are thus sites as well as buildings and parts of buildings protected under cultural heritage protection legislation (ZVKD-1 2008). The proposals for inscription in the heritage register are prepared by the Institute for the Protection of Cultural Heritage of Slovenia, while the minister for culture decides on including the unit into the register. A change in entry or deletion from the heritage register is done in the same way as the registration of a unit. The heritage register system of immovable cultural heritage is in line with international recommendations (MoC 2018a). Along with basic descriptive information, each cultural heritage unit also contains geolocation data, allowing for further spatial analyses. The heritage register is continuously updated. In January 2018, 32,408 ICH units were registered in Slovenia, which are, as to their type of protection regimes and size, very heterogeneous, of which cultural heritage sites with 18,777 units prevail (MoC 2018b).

Table 25.1 Number of Cultural Heritage (CH) sites by type and total number of CH sites in Slovenia from the Ministry of Culture data set

	(1) registered archaeological site	(2) cultural heritage site	(3) cultural monument site	Total no.
ICH sites in Slovenia	2494	18,777	8394	29,665
ICH in FDAs	137	617	523	1277

The number of CH sites that appear at least in one FDA in Slovenia

Table 25.2 Total number of FDAs in Slovenia is 1081 of which 415 FDAs have at least one ICH units present

	AGR	SERV	TOUR	IND	DEF	MIN	INF	TRANS	HOUS	Total
No. of FDAs by type	77	162	60	237	35	170	133	112	95	1081
No. of FDAwICHs	23	87	32	78	10	44	47	33	61	415

The data had to be appropriately organised for the needs of the presented study and its results. Based on an agreement, it was established that the total number of CH units can be reduced to 29,665 (Table 25.1), as 2743 units did not meet the criteria selected in this study. Then we made a clip with the ICH data set and FDAs data set. In total, 1331 ICH units were found within the FDAs. In the cross-sectional layer, we detected errors resulting from various levels of accuracy of data capture and typological discrepancies. During the data review, a number of border disputes were found (18 features) and some with an area less than 10 m^2 (36 features). All 54 features were deleted from the data set. The final number of ICH units in the FDAs is 1277 (Table 25.1).

Of 1081 FDAs in Slovenia, 415 were found that contain at least one unit of ICH (Table 25.2).

Below, we analysed 415 FDAs with 1277 ICH units present—these FDAs were named FDAs with ICH present (hereinafter FDAwICH).

25.4 Results

We find that cultural heritage sites are present in all types of FDAs, while the occurrence of ICH varies considerably by type. At least one ICH protection regime was found in 38.5% of all FDAs (Table 25.2; Fig. 25.1). In FDAs for housing, FDAs of service activities and FDAs of tourist, sports and recreation activities, ICH is present in more than half of the FDAs recorded (Fig. 25.1). ICH has a limited presence in

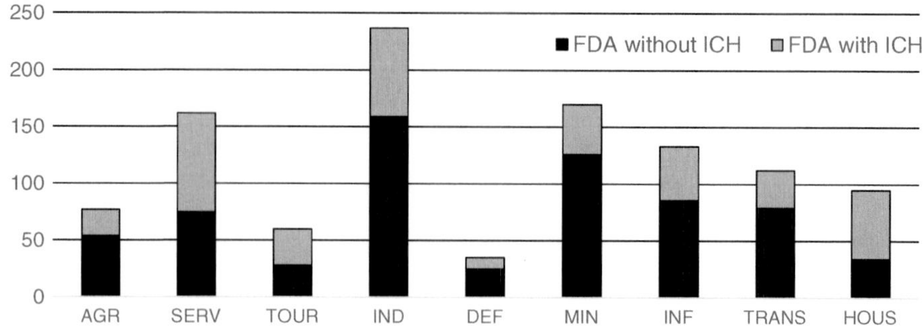

Fig. 25.1 FDAs with and without ICH

FDAs of mineral extraction, FDAs of defence, protection and rescue services, FDAs of agricultural activities, and FDAs of transitional use.

The largest number of ICH units (Fig. 25.1) is in FDAs of service activities (737 ICH units), followed by FDAs of industrial activities (125 ICH units) and FDAs for housing (120 ICH units).

The results of analysis by protection regimes of ICH reveal (Fig. 25.2) that cultural heritage sites are the most frequent (617 ICH units), followed by cultural monument sites (523 ICH units) and registered archaeological sites (137 ICH units). The total number of FDAwICH units is 1277, meaning that in an individual FDA there are several different ICH units present (Fig. 25.2). In 58% of FDAs, there is only one ICH unit present, in 38% there are 2–5 ICH units, and only in 4% of FDAs are there more than 5 ICH units. A more thorough consideration of the data shows that most of the FDAwICH units with service activities are found in historical town and village centres. The historical background and the age of these sites are usually the reason for the greater number of cultural heritage facilities present there. A detailed analysis of cultural heritage sites shows that settlement heritage sites and building heritage sites prevail, as classified by MoC (2011).

The FDAwICH analysis shows that, given the combination of the criteria of the proportion of area that ICH occupies in FDAs (0–10; 11–50; 51–95, 96–100%) and the ICH protection regime, 268 of 415 or 65% of FDAwCHs are completely (100%) covered in at least one protection regime of ICH (Table 25.3). In this area, there is a prevalence of FDAs of service activities (63 FDAwCHs), FDAs of industrial activities (42 FDAwCHs), and FDAs for housing (43 FDAwCHs). These are sites whose reactivation and regeneration are of key importance both for improving the quality of life as well as for the regeneration of sustainable economic development that must be based on the use of local resources and small private businesses (Rypkema 2009), which can contribute the most to the recovery of the local labour market and to the improvement in environmental, social, and particular cultural responsibility at the local level.

Table 25.3 Number of ICH units and proportion of area depending on the FDA type

FDA type	Total no. of FDAwICHs	Proportion of ICH area per FDA				Protection regime of ICH			
		0–10%	11–50%	51–95%	96–100%	Registered archaeological site	Cultural heritage site	Cultural monument site	Total no. of ICH units
IND	78	20	9	7	42	22	73	30	125
INF	47	7	5	3	32	21	36	16	73
AGR	23	2	2	1	18	3	21	12	36
DEF	10	2	2	0	6	5	5	4	14
TRANS	33	5	3	1	24	20	20	17	57
MIN	44	13	6	6	19	22	25	12	59
SERV	**87**	**6**	**6**	**12**	**63**	**17**	**361**	**359**	**737**
TOUR	32	3		8	21	12	27	17	56
HOUS	61	4	8	6	43	15	49	56	120
Total	415	62	41	44	268	137	617	523	1277

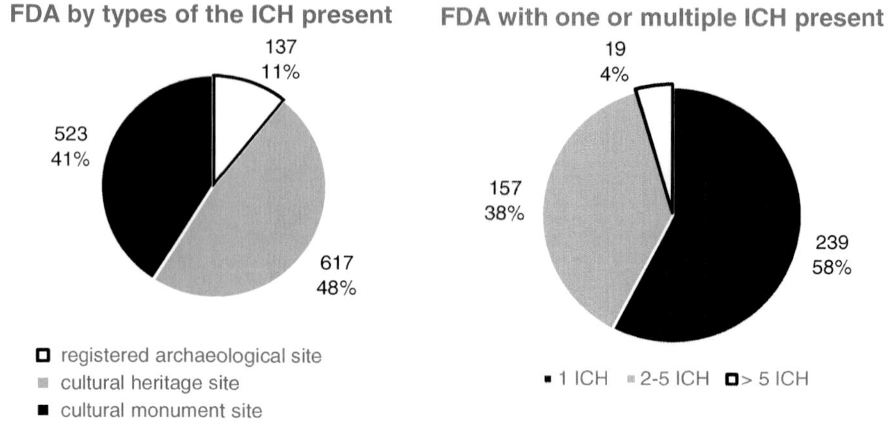

Fig. 25.2 Number of ICH units present in the FDA by types of protection regimes of the ICH (left) and the number of FDAs with one or multiple ICH units present (right)

25.4.1 *Classification of Functionally Derelict Areas with Immobile Cultural Heritage into Groups A, B, and C*

The results of graphical cross sections of the edited data of FDAs and ICH show that as many as 274 FDAwICHs can be classified into group A (more than half) with a total of 1102 ICH units (86.3% of all ICH units), 40 FDAwICHs into group B with a total of 45 ICH units, and 101 FDAwICHs into group C with a total of 105 ICH units (Table 25.4). The maximum presence of FDAwICHs in group A, where there is also the maximum number of CH units (1102 units), confirms our hypothesis, i.e. that cultural heritage regeneration is an important factor in reactivating FDAs. Along with ICH regeneration, the inclusion of intangible heritage into regenerating activities in FDAs plays an important role. Table 25.4 shows that 60 FDAwICHs have registered archaeological sites (42 FDAwICHs with 96–100% proportion of area per FDA, 4 FDAwICHs with a proportion of area of 51–95%, 11 FDAwICHs with 11–50%, and 14 FDAwICHs with 0–10% proportion of area per FDA). Registered archaeological sites are mostly sites below the Earth's surface, which are largely unexplored, and they are generally unaffected by the abandonment of activities above ground. This effect can be achieved only in the case of FDA reactivation and changes in activities, as in this case it makes sense to carry out "preliminary research (means the research of heritage or archaeological remains) that needs to be carried out in order to obtain the information necessary for the evaluation of heritage or archaeological remains prior to the works being carried out in an area or prior to construction" (Article 3, ZVKD-1 2008).

If we classify FDAwICHs also based on FDA types, we see that in group A FDAs of service activities, FDAs for housing, and FDAs of industrial activities are most

Table 25.4 Classification of FDAwICHs into three groups, A, B, and C, in terms of the number of ICH units, proportion of area, and the ICH protection regime

FDAwICH group	Combination of indicators and criteria	No. of FDAwICHs	No of units of ICH		
			Registered archaeological site	Cultural heritage site	Cultural monument site
Group A	More than 5 units according to the proportion of area	19	6	316	302
	2–5 units and 96–100%	136	53	167	139
	1 unit and 96–100%	119	42	43	34
	Total group A	274	101	526	475
Group B	2–5 units and 51–95%	10	1	16	11
	2–5 units and 11–50%	7	5	9	3
	1 unit and 51–95%	23	4	13	13
	Total group B	40	10	38	27
Group C	2–5 units and 0–10%	4	6	8	3
	1 unit and 11–50%	39	11	13	7
	1 unit and 0–10%	58	14	32	11
	Total group C	101	31	53	21
	Total	415	137	617	523

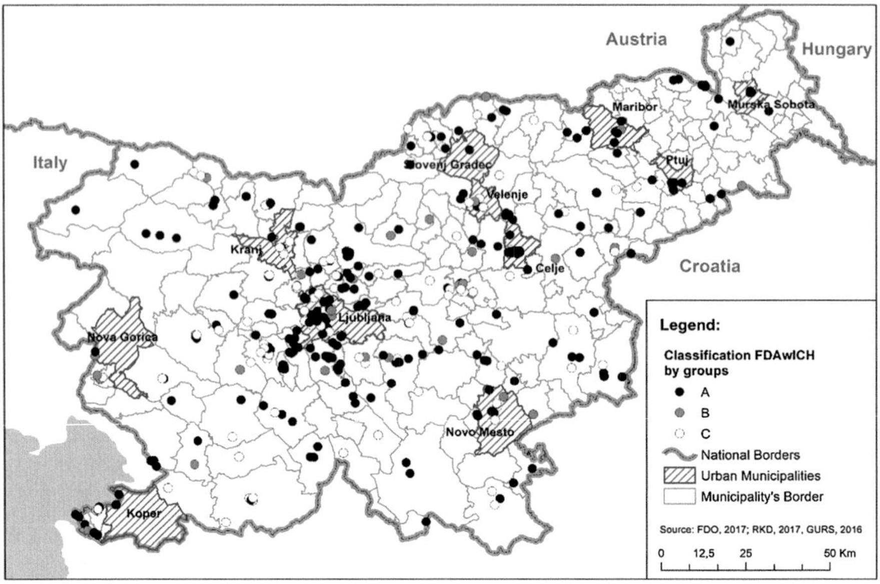

Fig. 25.3 Spatial distribution of FDAwICHs into groups A, B, and C by municipality

common (Table 25.5). Most ICH units are found in these areas, while the number of FDAs of service activities (737 CH units) stands out, which was also expected according to the first part of the analysis.

The spatial distribution of FDAwICHs by groups A, B, and C (Fig. 25.3) shows that all groups of FDAwICHs are found throughout Slovenia, while they are most densely located in bigger cities. The phenomenon of the presence of ICH in FDAs is thus not territorially distinct, i.e. concentrated in specific areas. FDAwICHs are represented in urban municipalities and other municipalities in Slovenia (according to ZLS-UPB2 2007), but a further analysis reveals that in urban municipalities (Table 25.6) there are approximately 25% of all FDAwICHs (103–415), which is almost half of all ICH units (59 out of 1277). We find that in urban municipalities, FDAwICHs do not prevail among FDAs, but the number of ICH units in FDAs is still large. The Urban Municipality of Celje stands out, where there are the most FDAwICHs (14 out of 22), with as many as 259 ICH units present. A detailed look at the data shows that in the FDAwICHs in the town of Celje there are as many as 226 ICH units (out of 259 recorded throughout the Urban Municipality of Celje). The FDAwICHs of Maribor city centre between Orožnova Street and the Drava River (77 ICH units) and the FDAwICHs of the historical city centre of Kranj (81 ICH units) similarly stand out. These cities saw major changes during the transition period after Slovenia's declaration of independence in 1990, but they were also greatly affected by the 2008 global economic crisis.

In contrast, the results in the city of Ljubljana reveal that ICH is present in almost half of the FDAs recorded (46 out of 103), but the number of ICH units is not

Table 25.5 FDAwCH groups by FDA type

FDA TYPE	No. of FDAwICHs group A	No. of FDAwICHs group B	No. of FDAwICHs group C	TOTAL No. of FDAwCHs	Registered archaeologi-cal site	Cultural heritage site	Cultural monument site	Total ICH units
IND	43	7	28	78	22	73	30	125
INF	32	3	12	47	21	36	16	73
AGR	18	1	4	23	3	21	12	36
DEF	6	0	4	10	5	5	4	14
TRANS	24	1	8	33	20	20	17	57
MIN	19	6	19	44	22	25	12	59
SERV	67	8	12	87	17	361	359	737
TOUR	21	8	3	32	12	27	17	56
HOUS	44	6	11	61	15	49	56	120
Total	274	40	101	415	137	617	523	1277

Table 25.6 Occurrence of FDAs and FDAwICHs in Slovenian urban municipalities

Urban municipalities	Total no. of FDAs	FDAwICHs	FDAwICH classification according to groups A, B, C in urban municipalities			Registered archaeological site	Cultural heritage site	Cultural monument site	Total no. of ICH units
			A	B	C				
Ljubljana	103	46	31	4	11	21	22	37	80
Maribor	19	7	6	1	0		18	69	87
Celje	22	14	12	1	1		179	80	259
Kranj	12	6	2	0	4	2	16	70	88
Novo mesto	21	8	4	1	3	4	2	4	10
Murska Sobota	13	5	5	0	0			14	14
Nova Gorica	10	2	1	1	0		2	1	3
Koper	16	2	2	0	0	2		2	4
Velenje	12	6	2	1	3		16	2	18
Ptuj	11	4	4	0	0		3	16	19
Slovenj Gradec	14	2	1	0	1		1	8	9
Total	253	102	70	9	23	29	259	303	591
Slovenia	1081	415	274	40	101	137	617	523	1277

high in individual FDAs. In general, there is a prevalence of cultural heritage sites (22 ICH units) and cultural monument sites (37 CH units), while there are fewer registered archaeological sites (21 ICH units) (Table 25.6). In urban municipalities, most FDAwICHs are classified in group A, as expected.

In conclusion, despite the presence of FDAwICHs throughout Slovenia, urban municipalities and urban centres have a greater potential for reactivating FDAs, particularly seen from the perspective of potential economic and tourist development, where cultural heritage must be seen as an asset rather than as an obstacle for further development. This is also confirmed by experience from many European cities which have managed to revitalise historical city cores through past investments in the restoration of cultural heritage (Rykpema 2009). Below, we show two examples of good practices in the revitalisation of historical city cores in Slovenia, which provide valuable lessons for future systemic measures for FDAwICH reactivation and regeneration.

25.4.2 Examples of Good Practices: Regeneration of the Old City Centres of Kranj and Celje

In terms of the presence of ICH in FDAs, the historical centres of Kranj and Celje are classified into group A, which is mostly because the entire historical centres of both towns are recognised as cultural heritage sites. In the FDAwICHs of the historical centres of Kranj and Celje, there are 81 and 226 ICH units, respectively. In both cases, there was a need to strengthen the collaboration of city authorities and the Institute for the Protection of Cultural Heritage of Slovenia in efforts to improve the condition and to ease the transition from an impoverished and degraded site into an area that would pursue sustainable economic development and provide a high-quality environment for its inhabitants and visitors. After five years of introducing the regeneration process in both towns, many favourable effects have been identified, and some of these measures have been adopted in other Slovenian towns as well.

25.4.2.1 Example of Gradual Reactivation of the Old City Centre of Celje

The rehabilitation of the old city centre of Celje started in April 2012. The municipality renovated 12,500 m^2 of its urban areas. The infrastructure, public areas and squares, and urban furniture were completely renovated, water fountains were provided, the traffic regime was reorganised to provide more room for pedestrians and cyclists, bus stops were renovated and bicycle racks added, etc. The extensive regeneration of the historic centre of Celje was partially covered by European funding (€2.5 million), while the total project value was €232.4 million. The renovations of the old centre were completed in 2015, and even earlier, in 2014, the Urban Municipality of Celje started the project of co-financing the restoration of street façades (on average €150,000 annually) to increase the town's attractiveness. This action was initiated by the European project EPOurban (2014), whose purpose was to include private condominium owners in the city restructuring process. Since 2014, the Institute for the Protection of Cultural Heritage of Slovenia, Regional Office Celje, has been involved by providing professional recommendations in the renovation of façades (Prenova stavbe… 2014). In 2017, the tender was extended to the restoration of religious (sacral) and cultural heritage in the cultural heritage site of the historic centre (Urban Municipality of Celje 2014). With the decree on the uses of public spaces (2017), the user charges for caterers and summer gardens in the historic centre were set at a symbolic amount of €0.01/m^2/day). In this way, attempts are being made to attract various activities to cities. All activities put in place in Celje since 2012 have significantly changed the image of the functionally abandoned and degraded area. Based on further activities by the Urban Municipality of Celje in some currently still abandoned and vacant areas (Urban Municipality of Celje 2017), it is reasonable to expect that these areas will in time be excluded from the FDA record.

25.4.2.2 Example of Gradual Reactivation of the Old City Centre of Kranj

The centre of Kranj was slowly emptying because of the abandonment of service activities, which have been relocated since 1990 to the outskirts, and the proximity of transport hubs, as well as due to changed consumption habits and way of life. In 2015, vacant and abandoned business premises were inventoried (Lampič et al. 2015) in the historic centre of Kranj. As many as 30% of all business premises (55 of 182) had been vacant for more than five years. Back in 2015, financial incentives were adopted, e.g. allocation of funds for encouraging small-scale trades, exemption from ground exploitation fees, exemption from rent payment for catering gardens, and exemption from municipal fees for promotion purposes (Delo 2014).

At the same time, since 2014, each year a tender has been published for co-financing rents for ground floor business premises in the historic centre of Kranj (Urban Municipality of Kranj 2017d) at 50% of the rent or up to €5 per m^2, while shops are allowed to stay open until 9.00 p.m. After introducing the measures, 19 new business premises bounced back and the city regained its energy. The Urban Municipality of Kranj also started to co-finance the restoration of buildings to encourage development and to regenerate the old centre through the maintenance and preservation of its cultural heritage (Urban Municipality of Kranj 2015). With projects such as Euro-Free Wednesday (Urban Municipality of Kranj 2017a), Thursdays at the Well (Urban Municipality of Kranj 2017b), Happy Friday (Urban Municipality of Kranj 2017c), and Downtown with Kranček (Urban Municipality of Kranj 2017d), Kranj wants to include intangible cultural heritage into its historic centre in an engaging and innovative way (MMC RTV SLO and STA 2018). The reactivation and regeneration of the historic centre of Kranj are not yet completed; in 2017, resources were allocated for the restoration of the market, square, new urban furniture, and an office for the integrated regeneration of Kranj (Urban Municipality of Kranj 2018).

25.5 Measures for the Reactivation and Regeneration of Functionally Derelict Areas with Immovable Cultural Heritage

FDA reactivation measures can be generally classified under (1) legislative, (2) spatial planning, (3) financial, and (4) support measures (more on the system of measures in Lampič et al. 2017a). Their implementation falls under the responsibility of various ministries and state and local services. The evaluation of FDAwICHs and their classification into groups A (FDAs with an accentuated presence of ICH), B (FDA with ICH, but satisfying, to a lesser degree a combination of indicators and criteria), and C (FDAs with a non-accentuated presence of ICH) allow for the preparation of targeted activities in FDAwICH reactivation and regeneration. In the case of FDAwICHs in group A, MoC measures prevail in many aspects, but in view of integrated FDA reactivation, collaboration with other sectors is paramount, particularly with the Ministry of the Environment and Spatial Planning, the Ministry of Economic Development and Technology of the Republic of Slovenia, and other institutions at national and regional levels, as well as local communities.

In the first step, we produced a proposal for a model with key activities for the reactivisation of FDAwICHs (after Bergatt Jackson et al. (2006) and Adams et al. (2010) with the authors' amendments), which is briefly presented here: (1) FDA identification; (2) recognition of potentials and threats for FDAwICH reactivation; (3) environmental remediation; (4) vision development; (5) reactivation and regeneration planning; (6) programming/reactivation implementation plan; (7) reactivation implementation (project management method); and (8) monitoring of implementation, situation, and the changes necessary (Table 25.8).

The time of reactivating FDAwICHs, as in any other project, depends on the quality of the project and project management. We need to ensure the inclusion of stakeholders in the early phases of project design and thus provide the basis for active collaboration. The procedures and activities are known, while appropriate planning is required.

In the presented case of the historic centre of Celje, the time necessary for archaeological investigations was taken into account in the project but due to exceptional archaeological finds the regeneration process took a few months more. Therefore it is possible to conclude that the presence of ICH does not necessarily have a decisive influence on the reactivation timeframe.

Detailed guidelines and measures for preserving ICH in FDAs must be the result of the professional support of responsible institutions. The two examples of good practices clearly illustrate the possibility of collaboration among responsible institutions (particularly the Ministry of Culture and Urban Municipality) for the purpose of FDAwICH reactivation and regeneration. We should search for new functional, programmatic, and design possibilities and think about ICH transformation and innovation in line with the needs and the times (Fakin Bajec 2013).

Based on the analysis of FDAs with the presence of ICH and the examples of good practices, we then developed a proposal of possible and mostly previously tested

Table 25.7 Proposal of appropriate measures to reactivate FDAwICHs of old city centres

Common characteristics of FDAwICHs of old city centres	Specific measures and incentives in the field of ICH protection	Related intersectoral activities
– The function of the entire area ceases because of deteriorated buildings and changed needs (relocation of activities to the outskirts into newly built areas, as a result of emptying of housing) – The presence of ICH is a drawback for investors to attempt to reactivate the area (conservation plan for the entire city centre should be produced) – As a rule there are few owners in the site; ownership is either public or private – Good potential for reactivation; there is a need to search for appropriate activities – Real property value decreases with higher degree of degradation	– Inclusion in various EU projects – Direct financial resources of MoC for ICH owned by local and religious communities are available every other year – Investments in public infrastructure, appearance of settlements, façade renovation, organisation of common public areas (improvement of city centre image). Co-financing of programmes (activities) of operational costs in city centres – Purchase of business premises with public resources and inclusion of public programmes – Financial incentives for development of small-scale trades – Buildings to be sold/donated to private owners or favourable rental agreements to be signed under the conditions of maintenance and inclusion of appropriate programmes (integration of intangible cultural heritage) – Promotion of good practices and awareness-raising among investors and developers – Appropriate promotion for more visibility in broader professional and lay community	– Continuous collaboration between various sectors (good practices under the Rural Development Programme 2007–2013 are presented in Hostnik (ed) (2015) – Preparation of materials and free-of-charge guidance on FDAwICH reactivation (procedures, accessible advisory services—project offices in the field, accessible financial mechanisms, etc.) – Site management with detailed urban plans coupled with conservation plans – Horizontal, intersectoral projects and financing from the Operational Programme for the Implementation of the EU Cohesion Policy (Delak Koželj 2016) – Indirect financial mechanisms and products (write-offs or reduction in national and local taxes, grants, soft loans, etc.)

solutions for reactivating the individual FDAwICH. Here we provide an example that proposes measures to reactivate FDAwICHs of old city centres (Table 25.7).

25.6 Discussion and Conclusions

The FDAwICH problems that we detected while studying and establishing the register (Lampič et al. 2017a) are also recognised elsewhere in Europe (Sable and Kling 2001; Berg and Stenbro 2015). Due to the extensiveness of this phenomenon and the negative attitude to FDAwICH reactivation, which we identified in the interviews with professionals at the local level, we decided to focus our further research precisely on producing a proposal of a model with key activities and actions to reactivate FDAwICHs.

The acquisition and organisation of additional data on the recorded ICH with types of protection regimes was demanding and required special discussions with MoC representatives. Firstly, it is necessary to underline the MoC's standpoint, i.e. that no distinction in the value of various types of protection regimes should be made, and even less so in terms of various ICH units inside the same protection regime; the MoC underlined that spatial degradation does not put at risk registered archaeological sites, as they are generally located below ground. The significance of these sites in FDAwICH reactivation has been previously explained, and we treated them equal to any other ICH unit throughout the study. This decision turned out to be correct as it enabled the production of an integrated study, which showed that ICH is present in 38% of all FDAs in Slovenia. FDAwICHs in group A (*FDAs with an accentuated presence of ICH*), which are mostly located in urban municipalities, particularly stand out. We can confirm that ICH regeneration, due to the large proportion of area and the number of units, is an important FDA reactivation factor. Indirectly, this was shown in examples of good practices concerned with reactivating the old centres of Celje in Kranj.

The underlined presence of ICH in FDAs for potential investors (owners or developers) generates fear of strict guidelines and measures to protect ICH, increased financial input, and a longer reactivation timeframe (Lampič et al. 2017a). Investors are often unable to recognise the potential and added value (Rypkema 2009) of cultural heritage in the integrated regeneration of a site and/or a building. FDAwICH reactivation and regeneration measures must focus on reducing and eliminating the negative connotation of the presence of ICH and the fear of potential investors. To this end, the model for the reactivation of FDAwICHs was even further elaborated and shown in Table 25.8.

Nevertheless, it is important to note that each derelict site should be addressed individually, regardless of the type or region. We believe that, apart from a general model with recommended key activities, it is not possible to form a single reactivation model that could apply to, let alone prescribe, FDAwICH reactivation. The activities and available or used measures depend on the FDAwICH type. Reactivation takes place depending on the FDAwICH characteristics and the needs identified in the

Table 25.8 Model for reactivation of FDAwICHs with key activities and phases (after Bergatt Jackson et al. (2006) and Adams et al. (2010) with authors' amendments)

Key activity	Phases in the activity
1. FDA identification (this activity does not apply when it is systematically undertaken at the state level)	– Identifying and understanding the problem of FDAs – FDA content specification (typology) – Definition of FDA capture criteria – Establishing FDA records
2. Recognition of potentials and threats for FDAwICH reactivation	– Analysis of problems and potentials – Status diagnostics (SWOT analysis, main environmental problems, economic and social aspects of regeneration) – Analysis of stakeholders and developing a framework for their potential collaboration (motivating the private sector for FDA reactivation)
3. Environmental remediation	– In the case of environmental contamination, environmental remediation is first necessary (given the environmental contamination type, provision of financial assets)
4. Vision development	– Setting goals and alternatives (including development of FDA reactivation scenarios)
5. Reactivation and regeneration planning	– Formation of a project group – Production of development plans, coordination of interests (common goals, priorities, definition of new programmes at the site, etc.)
6. Programming/ reactivation implementation plan	– Production of implementation plan for specific cases of FDA reactivation (definition of necessary activities, criteria, indicators, input data, including a communication plan and public involvement in the FDA reactivation process) – Definition of financial resources – Definition of possible institutional support by the state, local communities – regarding the accepted systemic measures (technical and other support for preparation and implementation of the project, such as local offices, funds)

(continued)

Table 25.8 (continued)

Key activity	Phases in the activity
7. Reactivation implementation (project management method)	– Implementation of planned activities, coordination of stakeholder activities in real time and space (depending on the purpose of regeneration, existing infrastructure and its condition, ownership type, new uses: private or public property; the nature of the site following the regeneration: housing, business, commercial, open space)
8. Monitoring of implementation, situation, and the changes necessary	– Monitoring the implementation of the adopted plans and action programme (feedback, including strategic assessment, ex-ante evaluation) – Continuous monitoring of progress/development (sustainable development assessment) – Adjustment of goals set, plans, and the programme in line with the results of monitoring the implementation and situation

municipality/region/state. It depends on economic and social trends, stakeholder involvement, and the active role of public institutions. We must follow the goal to restore the previous function of these areas, while the paths to reach this goal vary considerably among the individual FDAwICHs, even when the FDA type is the same (i.e. they have a similar former activity). The success of reactivation does not depend on FDA types, but rather on the ownership structure, the alignment of the restoration programme with the envisaged use of space, legal regimes, spatial restrictions, available financial resources, etc. The presence of ICH in FDAs cannot and must not pose a development constraint, but must be promoted as an opportunity to create high quality and culturally diverse spaces.

References

Adams D, De Sousa CH, Tiesdell S (2010) Brownfield development: a comparison of North American and British approaches. Urb Stu. 47(19):75–104

Alker S, Joy V, Roberts P, Smith N (2000) The definition of brownfield. J of Env Plan and Man 43(1):49–69. https://doi.org/10.1080/09640560010766

Berg SK, Stenbro R (2015) Densification or dilution? On cultural and economic value creation along the Aker River in Oslo Norway. Hist Env: Policy & Practice 6(3):197–213. https://doi.org/10.1080/17567505.2015.1100360

Bergatt Jackson J, Drobiec L, Ferber U, Gorski M, Nathanail P, Petríková D (2006) Brownfields handbook. Cross-disciplinary educational tool focused on the issue of brownfields regeneration. http://fast10.vsb.cz/lepob/index1/handbook_eng_screen.pdf. Accessed 1 Oct 2016

COBRAMAN (2009a) Brownfield types. http://database.cobraman-ce.eu/BrownfieldTypes.php. Accessed 3 Jul 2016

COBRAMAN (2009b) Manager coordinating brownfield redevelopment activities. http://www.cobraman-ce.eu/. Accessed 20 Feb 2017

Delak Koželj Z (2016) Finansiranje kulturne dediščine. Javno predavanja pametni razvoj na primeru vključevanja kulturne dediščine in ohranjanja narave. ZVKDS, Služba za razvoj in informatiko, Ljubljana. http://www.zvkds.si/sites/www.zvkds.si/files/u6/finansiranje_kulturne_dediscine.pdf. Accessed 14 Feb 2018

Delo (2014) V Kranju pripravili nabor ukrepov za oživljanje mestnega jedra. http://www.times.si/slovenija/v-kranju-pripravili-nabor-ukrepov-za-ozivljanje-mestnega-jedra–9e2c0703a91761d6a49a63dc4f1e9d110eba9d6e.html. Accessed 4 Oct 2017

Donaldson R, Lord R (2018) Can brownfield land be reused for ground source heating to alleviate fuel poverty? Ren Ene 116:344–355. https://doi.org/10.1016/j.renene.2017.09.037

EPOurban (2014) Omogočanje vključevanja zasebnih etažnih lastnikov stanovanjskih objektov v proces mestnega prestrukturiranja [Enabling private owners of residential buildings to integrate them into urban restructuring processes]. Urban Municipality of Celje, Celje. https://moc.celje.si/projekti/960-epourban. Accessed 18 Aug 2016

Fakin Bajec J (2013) Interpretacija dediščine med teorijo in prakso. Vloga stroke pri rabi kulturne dediščine v lokalnem okolju. In: Interpretacija kulturne dediščine. Posvet Slovenskega etnološkega društva, Ljubljana. https://www.etno-muzej.si/files/events/iterpretacijekulturnedediscine.pdf. Accessed 14 Feb 2018

Ferber U, Grimski D (2002) CLARINET: Brownfields and redevelopment of urban areas. Austrian Federal Environment Agency, p 147 http://www.commonforum.eu/Documents/DOC/Clarinet/brownfields.pdf. Accessed 21 Dec 2015

Ferber U, Grimski D, Millar K, Nathanail P (2006) CABERNET: Sustainable brownfield regeneration. CABERNET University of Nottingham network report, Nottingham. http://www.palgo.org/files/CABERNET%20Network%20Report%202006.pdf. Accessed 12 Feb 2017

Gauchon MC (1997) Anciennes remontées mécaniques dans les montagnes françaises: pour une géographie des friches touristiques. Bulletin de l'Association de géographes français, 74(3):296–310. https://www.persee.fr/doc/bagf_0004-5322_1997_num_74_3_1986. Accessed 3 Feb 2018

Hostnik B (ed) (2015) Obnova objektov kulturne dediščine v okviru Programa razvoj podeželja Republike Slovenije 2007–2013. Ministrstvo za kmetijstvo gozdarstvo in prehrano in Zavod za varstvo kulturne dediščine, Ljubljana. http://www.zvkds.si/sites/www.zvkds.si/files/uploads/files/publication/elaborat-digitalna-verzija_koncna_.pdf. Accessed 10 Feb 2018

Jacobs J (1961) The death and life of great American cities. Random House, New York

Kim G (2018) An integrated system of urban green infrastructure on different types of vacant land to provide multiple benefits for local communities. Sus Cit and Soc 36:116–130. https://doi.org/10.1016/j.scs.2017.10.022

Lampič B, Bobovnik N (2017) Dodatni popis funkcionalno degradiranih območij v petih statističnih regijah (Osrednjeslovenska. Končno poročilo. Univerza v Ljubljani, Filozofska fakulteta, Ljubljana, Savinjska, Primorsko-notranjska, Obalno-kraška, Koroška), p 29

Lampič B, Marot N, Gamse M, Jenko I, Kljun U, Mali K, Korošec T, Verlič L, Žabota B (2015) Vzpostavitev aktivnega registra prostorsko in funkcionalno degradiranih območij za Gorenjsko regijo: Končno poročilo. Oddelek za krajinsko arhitekturo Biotehniške fakultete, Ljubljana, Oddelek za geografijo Filozofske fakultete, p 120

Lampič B, Cigale D, Kušar S, Potočnik Slavič I, Foški M, Zavodnik Lamovšek A, Barborič B, Meža S, Radovan D (2017a) Celovita metodologija za popis in analizo degradiranih območij, izvedba pilotnega popisa in vzpostavitev ažurnega registra: Končno poročilo. Ljubljana, Filozofska fakulteta Univerze v Ljubljani, Fakulteta za gradbeništvo in geodezijo Univerze v Ljubljani, Geodetski inštitut Slovenije, p 192

Lampič B, Cigale D, Kušar S, Potočnik Slavič I, Foški M, Zavodnik Lamovšek A, Barborič B, Meža S, Radovan D (2017b) Evidentiranje in analiza funkcionalno degradiranih območij v izbranih statističnih regijah Slovenije. Urbani izziv. Strokovna izdaja 7:10–18

Lampič B, Kušar S, Zavodnik Lamovšek A (2017c) Model celovite obravnave funkcionalno degradiranih območij kot podpora trajnostnemu prostorskemu in razvojnemu načrtovanju v Sloveniji. [Model for a comprehensive assessment of derelict land as a support for sustainable spatial and developmental planning in Slovenia]. Dela 48:5–31. https://doi.org/10.4312/dela.48.1.5-31

Lemoine G (2016) Brownfield restoration as a smart economic growth option for promoting ecotourism, biodiversity, and leisure: two case studies in Nord-Pas De Calais. In: Prasad MNV (ed.) Bio and bio. Elsevier, pp 361–388. https://doi.org/10.1016/B978-0-12-802830-8.00015-0

MMC RTV SLO, STA (2018) V starem Kranju začenjajo srede brez evra. http://www.rtvslo.si/zabava/zanimivosti/v-starem-kranju-zacenjajo-projekt-srede-brez-evra/449453. Accessed 22 Mar 2018

MoC (2011) Priročnik pravnih režimov varstva, ki jih je treba upoštevati pri pripravi planov in posegih v območja kulturne dediščine. Ministry of Culture of the Republic of Slovenia, Ljubljana. http://giskd2s.situla.org/evrdd/P_09_04_03.htm. Accessed 18 Aug 2016

MoC (2018a) Register nepremične kulturne dediščine. Ministry of Culture of the Republic of Slovenia, Ljubljana. http://www.mk.gov.si/si/storitve/razvidi_evidence_in_registri/register_nepremicne_kulturne_dediscine/. Accessed 12 Feb 2018

MoC (2018b) Varstveni režimi kulturne dediščine. Ministry of Culture of the Republic of Slovenia, Ljubljana. http://gisCH6s.situla.org/evrd/. Accessed 20 oct 2017

Navratil J, Krejci T, Martinat S, Pasqualett MJ, Klusacek P, Frantal B, Tochackova K (2018) Brownfields do not "only live twice": the possibilities for heritage preservation and the enlargement of leisure time activities in Brno, the Czech Republic. Cities 74:52–63. https://doi.org/10.1016/j.cities.2017.11.003

Prenova stavbe v starem mestnem jedru (2014) Mestna občina Celje in Zavod za varstvo kulturne dediščine Slovenije. https://moc.celje.si/images/Projekti_v_teku/sofinanciranje/141204brosurazaweb.pdf. Accessed 18 Aug 2016

Prentis E, Norton GA (1992) MEADOWS — an expert system for the establishment of diverse wildflower grasslands on derelict land in urban areas in the UK. Eco Eng 1–3:213–228. https://doi.org/10.1016/0925-8574(92)90003-K

RETINA (2012) BRM Brownfield revitalisation methodology. Faculty of Arts of the University of Maribor, Maribor, p 90. https://southeast-europe.net/document.cmt?id=457. Accessed 21 Dec 2015

Rypkema DD (2009) The role of heritage conservation in a sustainable economy. European Heritage Heads Forum, Bratislava. http://ehhf.eu/sites/default/files/201407/Rypkema_pwp.pdf. Accessed 23 May 2016

Sable KA, Kling RW (2001) The double public good: a conceptual framework for "shared experience" values associated with heritage conservation. J of Cul Eco 25(2):77–89

Sanches PM, Pallegrino PRM (2016) Greening potential of derelict and vacant lands in urban areas. Urb For & Urb Gre 19(2):128–139. https://doi.org/10.1016/j.ufug.2016.07.002

Špes M, Krevs M, Lampič B et al (2012) Sonaravna sanacija okoljskih bremen kot trajnostno razvojna priložnost Slovenije, Degradirana območja: [Ciljni raziskovalni program (CRP) "Konkurenčnost Slovenije 2006–2013"]: zaključno poročilo. Oddelek za geografijo Filozofske fakultete Univerze v Ljubljani, Ljubljana, p 66

TIMBRE (2011) Tailored improvement of brownfield regeneration in Europe. EU FP7 project. Helmholtz-Centre for Environmental Research—UFZ, Leipzig. http://www.timbre-project.eu/. Accessed 12 Feb 2017

Urban Municipality of Celje (2014) Projekti v teku. Sofinanciranje obnove stavb v Mestni občini Celje, Celje. https://moc.celje.si/projekti/4280-sofinanciranje-obnove-stavb-v-mestni-obcini-celje. Accessed 4 Mar Oct 2017

Urban Municipality of Celje (2017) Drevored Mestnega parka Celje bo obogatilo 100 novih, rdečecvetnih kostanjev, Celje. https://moc.celje.si/novice-in-obvestila/5482-drevored-mestnega-parka-celje-bo-obogatilo-100-novih-rdececvetnih-kostanjevih. Accessed 4 Oct 2017

Urban Municipality of Kranj (2015) Javni razpis za sofinanciranje obnove stavb v starem Kranju. Razpisna dokumentacija, Kranj. https://www.kranj.si/KRANJ_SI.asp?lytoo=kranj_moje_mesto/aktualno/novice&showNews=NEWSYPXUKO615201591929. Accessed 15 Sep 2017

Urban Municipality of Kranj (2017) Javni razpis za sofinanciranje najemnin pritličnih poslovnih prostorov v starem Kranju. Javni razpis za sofinanciranje stavb v starem Kranju. Razpisna dokumentacija, Kranj. http://www.kranj.si/files/06_mestna_obcina/javni_razpisi_in_narocila/2017/javni-razpis.pdf. Accessed 15 Sep 2017

Urban Municipality of Kranj (2017a) Sreda brez evra. https://www.nakupujemvmestu.si/projekti?sreda-brez-evra. Accessed 15 Sep 2017

Urban Municipality of Kranj (2017b) Četrtki ob vodnjaku, Kranj. https://www.nakupujemvmestu.si/projekti?cetrtki-ob-vodnjaku-2. Accessed 15 Sep 2017

Urban Municipality of Kranj (2017c) Srečni petek, Kranj. https://www.nakupujemvmestu.si/projekti?srecni-petek. Accessed 15 Sep 2017

Urban Municipality of Kranj (2017d) S Krančkom v mestu, Kranj. https://www.nakupujemvmestu.si/projekti?s-kranckom-v-mestu-1. Accessed 15 Sep 2017

Urban Municipality of Kranj (2018) Poročilo seje občinskega sveta, Kranj. http://www.kranj.si/KRANJ_SI,medijsko_sredisce,sporocila_za_javnost.htm&showNews=NEWSUPJBRM1125201682118. Accessed 10 Feb 2018

Vojvodíková B, Potužník M, Bürgermeisterová R (2011) The database on brownfields in Ostrava (Czech Republic): some approaches to categorization. Moravian geographical reports, 19(4): 50–60. http://www.geonika.cz/EN/research/ENMgr/MGR_2011_04.pdf. Accessed 23 Jun 2016

ZLS-UPB2 (2007) Zakon o lokalni samoupravi (uradno prečiščeno besedilo). Official Gazette RS 94/07

ZVKD-1 (2008) ZVKD-1 Cultural Heritage Protection Act, Ministry of Culture of the Republic of Slovenia. Official Gazette RS 16/08, 123/08, 8/11, 90/12, 111/13 in 32/16

Chapter 26
Models of Bastion Fortifications Integration in Cities

Damir Krajnik and Lea Petrović Krajnik

Abstract This paper examines the models of contemporary integration of bastion fortifications into the urban tissue, either in case of preserved bastion fortifications or in case of valuable urban areas created by their transformation during the nineteenth and the twentieth centuries, and also in case of spaces at the place of former fortifications in which visible physical remains of fortifications have not been preserved nor has recognizable and historically valuable urban area been created by historical transformation. Bastion fortifications, which had significantly characterized the morphology of renaissance and baroque cities during the four centuries of their existence, lost their purpose at the end of the eighteenth century. Depending on specific strategic and political circumstances, fortifications in certain cities have been preserved in their original form, whereas in others they have been partially or completely removed and the spaces at their place have experienced urban transformation. Based on the analysis of historical integration models of bastion fortifications, the analysis of urban features of spaces at the place of fortifications at the beginning of the twenty-first century and the identification of contemporary identity factors of city centres of such genesis, this paper defines three possible models to enhance future development of spaces at the place of bastion fortifications.

Keywords Bastion fortifications · Urban transformation · City centres · Integration models

26.1 Introduction

Historical heritage of bastion fortifications has significantly marked urban development of European cities, both at the time of their active use for defence purposes from the sixteenth to the eighteenth century and after the loss of their purpose when,

D. Krajnik (✉) · L. Petrović Krajnik
Faculty of Architecture, University of Zagreb, Zagreb, Croatia
e-mail: dkrajnik@arhitekt.hr

L. Petrović Krajnik
e-mail: lea.petrovic@arhitekt.hr

M. Obad Šćitaroci et al. (eds.), *Cultural Urban Heritage*,
The Urban Book Series, https://doi.org/10.1007/978-3-030-10612-6_26

at the end of the nineteenth century, the spaces at their place in many cities experienced urban transformation and became integrated into the urban tissue. In many cases, recognizable spaces of high urban quality were created, whether they were preserved fortification complexes or central city spaces formed at the place of former fortifications.

The incentive for this research is the fact that the revitalization of bastion fortifications, as well as urban planning of unfinished city areas created by their inadequate or uncompleted transformation, is in danger of being left to the reckless initiatives of the market and private capital on one hand, and on the other hand, its development is 'halted' by the conservation approach of maintaining the 'status quo' because of the high heritage value. The question is how to solve those urban planning and architectural problems and how to reconcile them with or adjust them to conservation demands, at the same time enabling a long-term active protection and economically sustainable use in accordance with the value status of heritage. The answer to this question is provided by the analysis of cities in which the preserved bastion fortifications have an appropriate function and are actively used (with the adequate protection of historical heritage at the same time) and by the analysis of urban features of spaces created by the complete transformation of fortifications during the nineteenth and the beginning of the twentieth century, providing recognition of the historical urban concepts of transformation.

The primary aim of this chapter is to present the method of analysis of urban features of spaces at the place of bastion fortifications at the beginning of the twenty-first century and the application of the results of the analysis in determining the appropriate contemporary integration model of spaces created by partial or uncompleted transformation of bastion fortifications in city centres.

26.2 Construction of Bastion Fortifications in European Cities

At the time of their active use for defence purposes (from the fifteenth to the eighteenth century), bastion fortifications represented a morphologically and functionally clearly defined area which constituted the separation zone between the urban tissue of the city core and the suburb.

With the development of artillery in the fifteenth century Europe, medieval fortifications with high defensive walls and towers began to lose their purpose. The development of artillery, in which German military engineers took the lead at the beginning of the fifteenthth century, had crucial impact on the loss of their efficiency (Gutkind 1964). Such historical context had led to the development and construction of a new, bastion type of fortifications which represented the response to the new demands of efficient defence. Unlike medieval walls, bastion fortifications had a markedly horizontal dimension compared to the vertical one and a complex layout. The material fortifications were built of had also changed, since stone and brick were replaced or combined with earth in order to enable the absorption of the impact power of cannon balls. Earth material for the construction of walls was obtained by digging out moats, which also became an important part of the defence system. Given that the number of 'rings' of defensive walls could be expanded only to a certain degree, due to the constant improvement of artillery (the increase of range) the concept of defence soon changed. It changed from 'passive' to 'active' through the construction of bastions as platforms for artillery at strategic points (corners) and through the creation of *no-building zones* in the area between bastion fortifications and the suburb.

Although it is difficult to define the bastion type of fortifications precisely since they had varied significantly in function and form during the four centuries of their construction, we can conclude that the bastion type of fortifications is a defence system with a significant layout size, consisting of the main part of the fortification (bastions connected with walls) and the outer defensive belt (consisting of one or more outer fortification elements, depending on the time of construction and the complexity of the fortification conditioned by the size and strategic importance of a certain city).

26.3 Historical Integration Models of Bastion Fortifications

With the change of geopolitical situation in Europe and the change of methods of warfare due to the modernization of firearms at the end of the eighteenth century, bastion fortifications in most cities lost their purpose. The need for their maintenance had gradually faded; fortifications began to present an obstacle to urban development due to hygienic reasons and the difficulties with traffic connections between city core and the suburb which in many cities resulted in the need for their removal and integration into the urban tissue (Krajnik 2007).

At the beginning of the nineteenth century in European cities there was a variety of structural changes, from the development of new economic activities and the organization of transport infrastructure to strong demographic growth, with the burgeoning middle class which wanted to increase the quality of life by improving basic hygienic conditions and developing education, culture and art. The transformation of cities in the nineteenth century was greatly influenced by political reasons, too: from representational to military; from demonstrating power and glorifying the nation to preventing street barricades and social unrest, accompanied with the affirmation of humanistic principles adopted by the burgeoning middle class. The middle class became the initiator of the creation of urban public spaces, especially city parks and the construction of public buildings. Bastion fortifications—spaces without a real function in the very centre of the city—arose as the ideal location for the formation of these areas.

In such a socio-historical context, depending on the specific strategic, spatial and political situation during the nineteenth century, in most cities bastion fortifications became integrated into the urban tissue, with fortification structure being preserved in its original form in some cities, whereas in others fortifications were removed and a new urban structure was formed in their place.

General plans for the transformation of spaces at the place of fortifications were developed in the middle of the nineteenth century in numerous cities in which new spatial possibilities for the connection of the previously separated urban areas and the creation of new, highly valuable land in the city centre had been created by the removal of fortifications in the first half of the nineteenth century.

According to those plans, urban areas created by the removal of fortifications experienced transformation under which some of the most important urban planning solutions of that time were achieved (Knežević 1996), since most areas formed at the place of former fortifications represented spaces of new identity and new urban values—extensions of the city core dominated by public buildings, representative residential buildings and urban public spaces (squares, parks, avenues and boulevards). Among the cities which have undergone this type of transformation, Vienna, whose *ring* had become a paradigmatic example and had greatly influenced the formation of the European city of the nineteenth century, holds special importance since it has established a new 'conservation' approach to reconstruction of the city. It is an approach not based on radical removal (demolition) of the existing urban tissue but on the formation of morphologically and functionally separate area between the existing urban tissues (the city core and the suburb) whose form is mainly kept intact (Mollik et al. 1980) (Fig. 26.1).

Although in most cities the transformation of fortifications was carried out following a similar pattern and it resulted in a new type of valuable historical heritage, in some cities such as Kassel or Györ the transformation was carried out in a way that a new urban structure created at the place of former bastion fortifications negates the former zone of separation of the city core and the suburb with its geometry, and the newly created urban tissue is not characterized by the concentration of urban public spaces and public buildings (Krajnik 2011).

The information about the development of cities during the nineteenth and at the beginning of the twentieth century indicates the existence of two historical models of integration of bastion fortifications into urban tissue. The first model represents functional integration of the fortifications through the change of use and ownership, whereby fortifications lose their military function (they usually fall under the city administration) by keeping the morphology of construction (physical structure of fortification elements), while the second model represents a functional and morphological transformation of spaces at the place of bastion fortifications.

Representative examples of bastion fortifications systems in smaller urban centres were mostly integrated into the urban tissue based on the first model, whereby the preserved military buildings were given civil use, and the fortification belt was given the function of urban public space. This model implied smaller interventions in urban planning and architectural sense, while the emphasis was on finding a way of maintaining the buildings (change of purpose and sustainable use) and the newly formed public space of exceptionally large scale.

The model of functional and morphological transformation of spaces at the place of bastion fortifications was mostly used in urban areas in which there was a greater

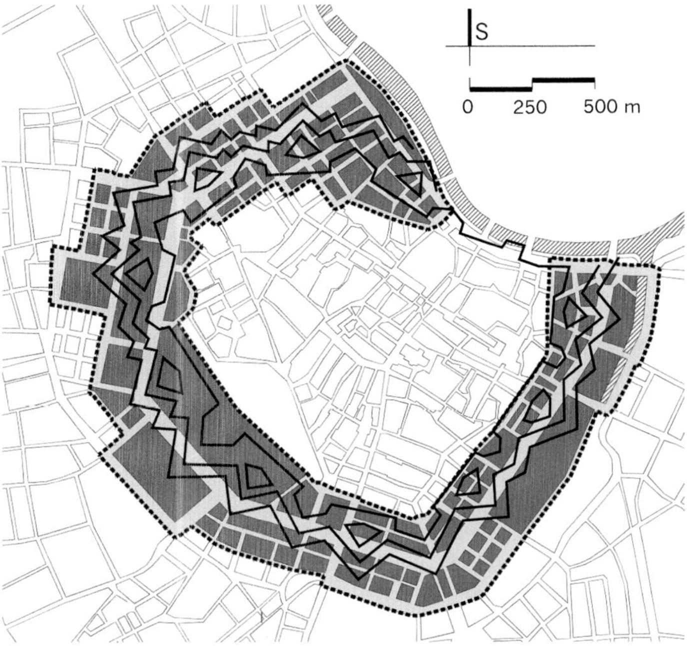

Fig. 26.1 Vienna (Austria)—overlap of the fortification layout with the urban matrix formed in the second half of the nineteenth century

need for the use of valuable space in the immediate contact with the city centre. In this model, we can differentiate between two sub-models/sub-types, the first one being the transformation of space by acknowledging historical heritage (preserving the layout geometry of the former fortification complex and preserving certain fortification elements), and the second one being the transformation which negates historical heritage (the geometry of the former fortification belt cannot be read in the layout of the city and the physical structure of fortification elements is not preserved).

These two models also differ in urban structure and land use of space formed by the transformation of fortifications, whereby the creation of central urban spaces of high urban quality and recognizability that have become valuable urban heritage in themselves was based on the first sub-model, whereas with the second sub-model that was not the case. In the literature about the historical development of cities, urban transformations of spaces at the place of fortifications during the nineteenth century which have created new urban values and spaces of new identity of the city are connected with the concept of the *ring* and the formation of the *inner fringe belt*. The concept of the *ring* represents a model of transformation, dominant in the planning of spaces at the place of bastion fortifications in European cities of the nineteenth century, which is characterized by very similar urban features. Spaces formed in this way at the same time represent recognizable morphological areas within the urban tissue (Conzen 1975) and clearly defined caesuras in the growth of residential parts of the city, which is why they can be defined as *fringe belts* (Whitehand 2005).

26.4 The Spaces of Bastion Fortifications at the Beginning of the twenty-first Century

At the beginning of the twenty-first century, spaces at the place of bastion fortifications in European cities can be divided into two basic groups based on their urban features (urban structure, land use and traffic scheme). The first group is made of spaces with preserved bastion fortifications, whereas the second group consists of spaces created by the removal (disintegration) of fortifications and the transformation of spaces at their place.

Unlike in cities with preserved bastion fortifications in which urban features of the subject spaces are almost identical, in cities in which fortifications have undergone transformation urban features of the spaces at their place vary considerably from case to case. This is due to the fact that their defining has been influenced by various factors, such as size and importance of a certain city at the time of transformation, constructional characteristics of bastion fortifications (the complexity of fortifications in the sense of type and number of elements of the outer defence belt which conditioned the total layout size of the fortification system), geopolitical position, cultural, historical, socio-political and geographical conditions in which the cities were developing.

Given that in certain cities the transformation of fortifications has not been carried out completely or in an appropriate way on the entire area of the former fortification structure, an analysis of urban features of spaces of such genesis at the beginning of the twenty-first century has been made for the purpose of defining contemporary models of integration of such spaces into the urban tissue. The analysis has been done on 15 examples of European cities (Vienna, Bremen, Brno, Frankfurt, Geneva, Graz, Hamburg, Klagenfurt, Copenhagen, Mainz, Riga, Wroclaw, Würzburg, Zittau and Zürich), chosen based on three main criteria: (1) the existence of bastion-type fortifications in the space between the city core and the suburb at the end of the eighteenth century; (2) the existence of (military)-civilian town in the city core and in the suburb at the time of the existence of bastion fortifications and (3) the readability of spaces of former fortifications in the city layout at the beginning of the twenty-first century.

By creating an analytical model (a graphical representation of basic urban elements in the analysed space which enables the definition of urban features—traffic system scheme, land use and urban structure and their mathematical and statistical commensurability) for each example, seven groups of data have been determined:

(1) the prevailing urban structure in insulas/urban blocks; (2) the prevailing land use of insulas; (3) the urban structure and land use in insulas with the biggest number of public buildings; (4) the urban structure and land use in insulas with the biggest

number of public spaces; (5) the urban structure and land use in insulas with the biggest number of *groups* of public buildings (buildings for public use and urban public spaces); (6) the urban structure and land use in insulas which form the biggest number of *series (uninterrupted rows) of groups* of public buildings and (7) the traffic system scheme of space at the place of (former) fortifications.

Comparative analysis of spatial data obtained from the analytical models of the chosen cities has indicated the existence of characteristic types of spaces at the place of bastion fortifications in European cities according to the criteria of the prevailing *urban structure* and the prevailing *land use* of insulas in the analysed space, and by combining characteristic types of *land use*, characteristic types of *urban structure* in insulas and the traffic system scheme we have determined five *integral types* of spaces with typical urban features.

The examined spaces with prevailing (mainly) unbuilt urban structure in combination with (mainly) public use in insulas belong to the *integral type 1-A*. The spaces formed at the place of bastion fortifications in Bremen, Frankfurt, Graz, Hamburg and Copenhagen belong to this type. In these cities, the space at the place of bastion fortifications is the most visible in today's urban matrix.

The spaces belonging to the *integral type 1-B* have very similar urban features, with the prevailing *mixed* urban structure in combination with *(mainly) public use* in insulas. In the *integral type 1-B* cities, there is a bypass street around the former city core, and clear urban reading of spaces at the place of former fortifications is possible in today's urban matrix. The spaces formed at the place of bastion fortifications in Vienna, Riga, Wroclaw and Würzburg belong to this type.

In the cities belonging to the *integral type 1-C (mainly) built urban structure* prevails, which indicates that in the spaces created by the transformation of bastion fortifications in these cities the biggest part of the area is not taken up by urban public spaces but rather by (mainly) public use building. In the cities belonging to this type, there is a bypass street following the outer edge of former fortifications which enables the readability of spaces at the place of fortifications in the urban matrix at the beginning of the twenty-first century. Brno and Klagenfurt belong to the *integral type 1-C*.

The examined spaces with the *prevailing (mainly) unbuilt structure* in combination with *mixed use* in insulas belong to the *integral type 2-A*, whereas in the spaces belonging to the *integral type 2-B* mixed urban structure and mixed land use prevail. Such spaces can be found at the place of bastion fortifications in Geneva, Zittau and Zürich (Figs. 26.2, 26.3, 26.4, 26.5 and 26.6).

26.5 The Contemporary Identity Factors

Spaces (at the place) of bastion fortifications in many European cities represent valuable cultural heritage and a recognizable element of urban identity.

Depending on whether bastion fortifications have been preserved in their original form or they have experienced a certain type of transformation, we can identify

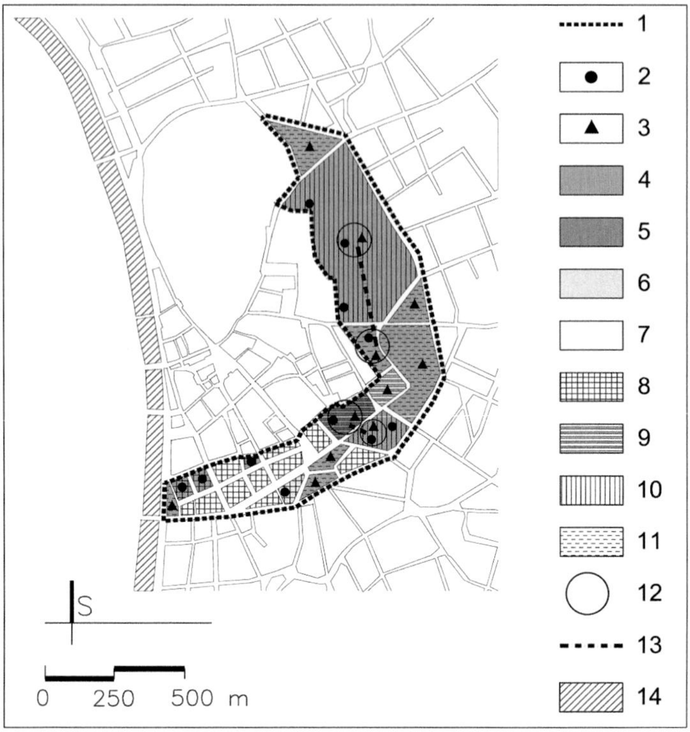

Fig. 26.2 Graz (Austria)—analytical model of the urban area at the place of bastion fortifications belonging to the *integral type 1-A* based on its urban features. 1—boundaries of analysed area; 2—public building; 3—public space; 4—public use; 5—mixed, mainly public use; 6—mixed, mainly *other* uses; 7—*other* uses; 8—block (unbroken building frontage along perimeter roads, with buildings occupying more than 50% of area); 9—mainly built (more than 50% of area built up, with discontinuous building frontage along perimeter roads); 10—mainly unbuilt (less than 50% of area built up and discontinuous building frontage along perimeter roads); 11—insula without buildings; 12—group (public building and public space within same insula); 13—stretch (two or more insulas with groups, separated with single road) and 14—water

several types of cultural heritage which, as factors of historical identity, can be viewed from urban planning, architectural, landscape, cultural, historical and sociological viewpoint.

In the cities in which the transformation of fortifications has not been carried out, urban identity factors are made of a fortification complex in the space between the city core and the suburb which is clearly readable in today's urban layout, such as Lucca and Palmanova in Italy (Fig. 26.7).

In the cities in which the transformation of bastion fortifications has been carried out, urban identity factors represent entirely formed central city areas which are characterised by the concentration of public buildings and urban public spaces, as well as by the readability of the former urban structure in today's urban layout,

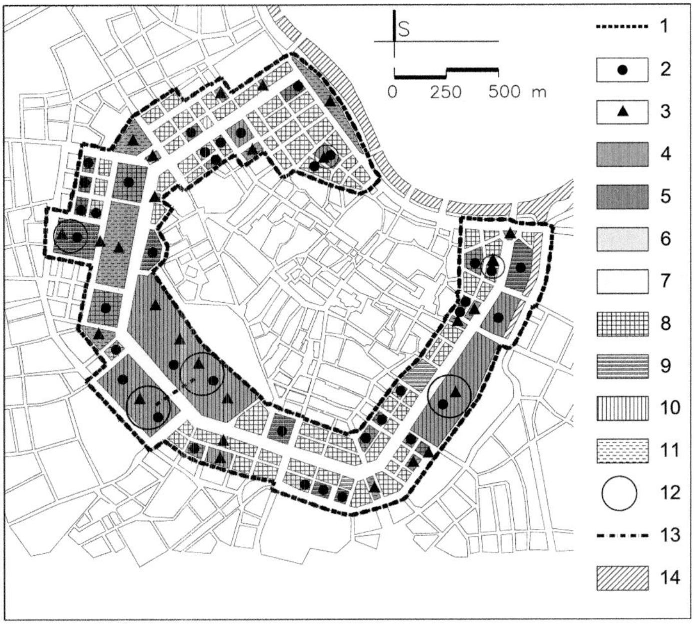

Fig. 26.3 Vienna (Austria)—Analytical model of the urban area at the place of bastion fortifications belonging to the integral type 1-B based on its urban features. For key, see Fig. 26.2

since their geometry has developed from the geometry of former fortifications. Such examples exist in cities like Vienna, Brno, Frankfurt, Riga, Würzburg and others (Fig. 26.8).

Depending on whether fortifications have been preserved or have undergone transformation, architectural identity factors consist of the preserved elements of fortifications (whether they are entirely preserved fortification belts such as those in Palmanova and Neuf-Brisach or only fragments of former impressive fortresses such as Eugene's bastion in Osijek) and/or the construction of public buildings (museums, theatres, city halls, libraries etc.) and representative residential buildings from the nineteenth and the beginning of the twentieth century.

An important element of identity are urban landscape areas too, whether they are landscape areas of entirely or partly preserved fortifications (Palmanova, Lucca, Bremen and Frankfurt) or representative squares, parks and boulevards created as a part of transformation of fortifications (in cities like Vienna, Brno, Osijek, etc.).

Historical identity factors are primarily made of a preserved complex or fragments of bastion fortifications connected with some events important for the historical development of a certain city.

In spaces created by transformation of fortifications we can clearly identify sociological identity factors too, since those areas have generally become centres of social life of the middle class soon after the transformation. That social life took place in public buildings and urban public spaces for which there was no available space

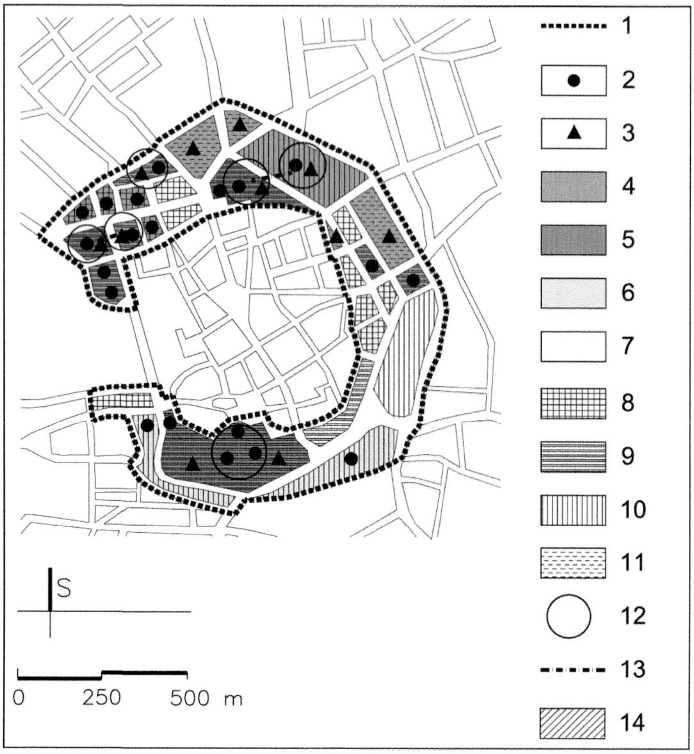

Fig. 26.4 Brno (The Czech Republic)—Analytical model of the urban area at the place of bastion fortifications belonging to the integral type 1-C based on its urban features. For key, see Fig. 26.2

in the city core before the transformation, while the considerable floor area created by demolishing fortifications was enough to accommodate all the necessary, new amenities at that time, and that function has been preserved to these days (Krajnik et al. 2008).

Identity features of spaces at the place of (former) bastion fortifications differ depending on urban genesis that took place in a certain city from the end of the eighteenth to the beginning of the twentieth century. In the context of the relationship towards heritage and the planning of future development of such city areas, it is important to recognize specific identity features of every individual example in order to ensure their affirmation in the context of future development, which has to be based on the respect of heritage, but also on creating possibilities for its modernization which will enable the continuity of heritage.

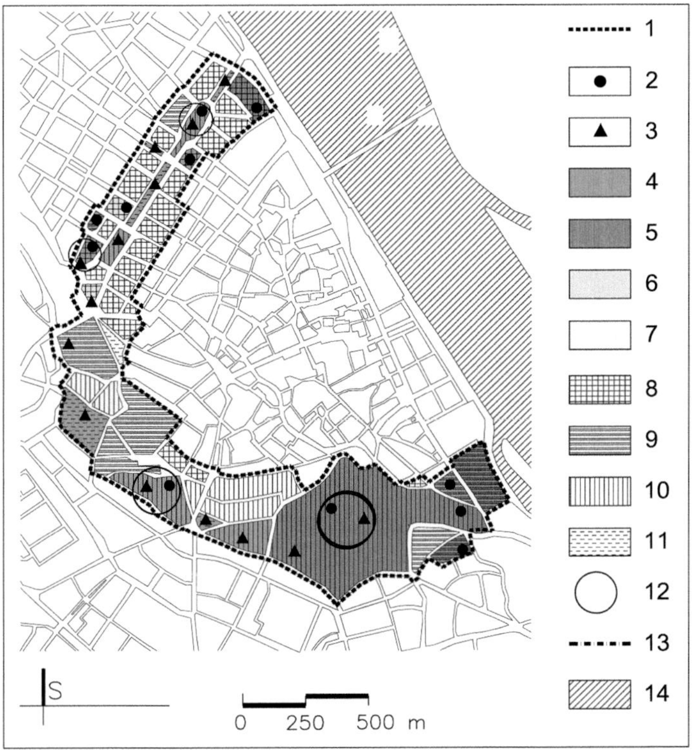

Fig. 26.5 Mainz (Germany)—Analytical model of the urban area at the place of bastion fortifications belonging to the integral type 2-A based on its urban features. For key, see Fig. 26.2

26.6 Contemporary Integration Models of Bastion Fortifications

The contemporary age requires the synthesis of historical and contemporary urban space, which enables quality management of the living environment and the restoration of cultural–historical values at the same time. That is why spaces (at the place) of bastion fortifications need to be appropriately integrated into the contemporary life of a city as a whole, in order to create conditions for their better protection and, at the same time, to contribute to the general quality of urban space.

The revitalization of spaces (at the place) of bastion fortifications requires a research approach based on the historical development of a city, with the identification of characteristic stages and their features in the urban planning, architectural, sociological, economic and every other sense.

In the examples in which fortifications have been preserved, it is necessary to analyse the existing condition of certain fortification elements and to examine the original purpose of the buildings in order to determine their typological characteris-

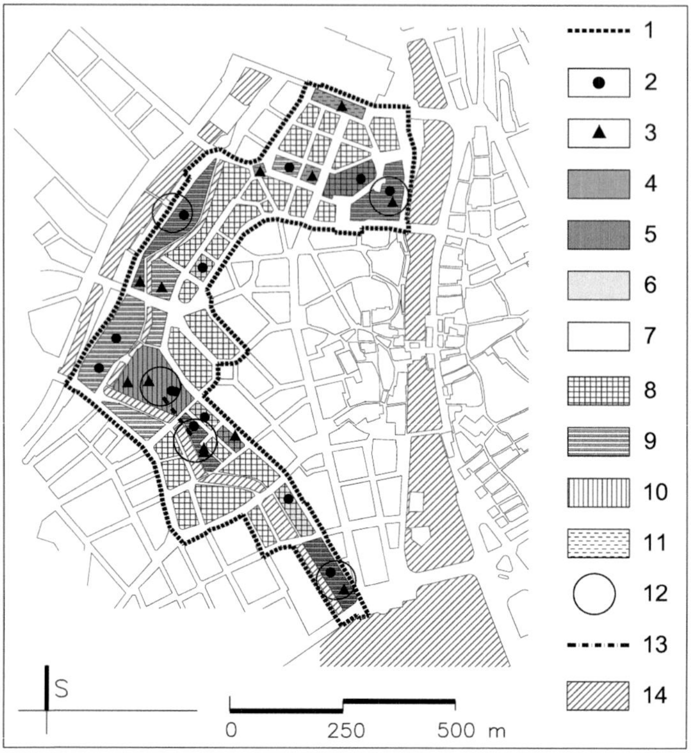

Fig. 26.6 Zürich (Switzerland)—Analytical model of the urban area at the place of bastion fortifications belonging to the integral type 2-B based on its urban features. For key, see Fig. 26.2

tics, the categorization of monuments and ambient values of space as a foundation for establishing or redefining the modality of protection. The research needs to ensure a quality foundation for the development of guidelines for restoration, reconstruction and conservation interventions if necessary, and for the creation of future amenities and functions in accordance with the inherited values, which enables active protection and sustainability of cultural heritage.

Apart from stronger integration into the everyday life of residents, contemporary integration of spaces (at the place) of bastion fortifications also requires their stronger integration into the existing tourist offer of a city, with the emphasis on cultural tourism (Božić et al. 2005).[1]

Depending on the identity features of a certain example, future development of bastion fortifications and/or of spaces created by their transformation can be based on three main models:

[1]Cultural tourism is increasingly replacing the mass tourism of the second half of the twentieth century. Programs that recognize and also activate and systematically connect typical localities or amenities in some area, offering culture as an individual attraction or as a network of localities connected into a cultural route by a certain topic, are of vital importance to this type of tourism.

Fig. 26.7 Lucca (Italy)—satellite image of the preserved bastion fortifications (*Source* Google Earth)

(1) the model of preservation and affirmation of fortifications; (2) the model of preservation and affirmation of the urban area created by the transformation of fortifications and (3) the model of creative interpretation.

The model of preservation and affirmation of bastion fortifications, whether it is about the entire system or only about fragments of a former fortification structure, implies the implementation of various conservation interventions and the enhancement of presentation of the preserved fortifications, but also the introduction of acceptable and sustainable functions which have to be in accordance with the original character of the fortification and must not disrupt urban, architectural and landscape values of the area.

Considering the character of space, in this model the integration into tourist offer is especially important for the affirmation of fortifications, whereby the presentation of fortifications needs to be particularly adapted to 'cultural tourists' who possess wider knowledge of specific topics, such as fortification architecture. When planning the presentation, it is important to integrate tangible and intangible cultural heritage related to the topic of fortifications in order to ensure a high level of experience, which can be used to justify charging fees for entrance to certain buildings, guided tours, workshops etc. (McKercher and Du Cros 2002). The presentation may include the exhibition of typical historical equipment (old cannons etc.), the scale model of the

Fig. 26.8 Vienna (Austria) satellite image of the urban area created by the transformation of bastion fortifications in the second half of the nineteenth century (*Source* Google Earth)

original fortification, setting up information panels and/or the rent of digital guides for visitors which makes the information about the history of the fortress available to them etc., but also the creation of different activities such as programs in which interesting historical events (like famous battles) and the former everyday life in the fortress are enacted (Krajnik and Petrović Krajnik 2011).

The model of preservation and affirmation of the urban area created by the transformation of fortifications needs to be applied in the cities in which the transformation of fortifications has been carried out, whereby, if the transformation has been carried out consistently and on the entire area of the former fortification belt, the emphasis should be on preservation and affirmation of historical buildings and public spaces created during the historical transformation. In case that in certain parts of the space the transformation has not been carried out consistently, that is, in accordance with the identified historical model of transformation, it is necessary to enable new construction and introduce new amenities in such spaces in order to ensure sustainable use. In this process, the geometry and use of new buildings, the articulation of public space and potential additions to the transport network need to be planned in such a way to be in accordance with the features of the *integral type* to which the area belongs. If there are any parts of space with preserved fortification elements, it is necessary to apply the elements of the first model in order to achieve their quality integration.

The model of creative interpretation primarily refers to spaces in which no visible physical remains of fortifications have been preserved, nor there is a recognizable and historically valuable urban area created by historical transformation. This model involves reconstruction of the existing and/or the construction of new buildings whose design refers to the former fortification structure, introduction of new amenities (primarily of public use), and the design of urban public spaces in whose articulation potentially preserved elements of the former fortifications should be used (archaeological presentation) and/or the former fortification structure should be creatively interpreted with new composition elements (topography, the use of water elements and so on).

In all three models, it is crucial to present that segment of cultural heritage through the suggested activities' scenarios both to the residents as well as to the future visitors (tourists), whereby the emphasis should be on the realization of interventions that do not require substantial funding (marking certain important localities, creating digital guides, presenting archaeological sites, organizing tourist trips, enacting important historical events etc.). Such interventions would further contribute to raising awareness of the value of both lost and preserved fortifications and their importance in the life of cities throughout history. With interventions implemented in such a way, bastion fortifications could be integrated into (re)defining the *brand* of a city much better, and with that more possibilities for the constructional restoration of the preserved but derelict and unused parts of the fortification structure would be created. Together with better integration of the fortification system into the *branding* of a city, the affirmation of fortifications can also be contributed to by connecting cities into a network through joint organization of tourist visits or seasonal cultural events, that is, by planning cultural routes with the topic of bastion fortifications.

26.7 Conclusion

The identified models of future development of bastion fortifications, which is of the spaces created by their transformation, enable a universal/general application in the process of planning the sustainable development and protection of specific types of cultural heritage in European cities. Exact urban planning and architectural interventions conceptually based on these models need to be elaborated and harmonised with the spatial context and identity features of every individual example. Such an approach can contribute to the protection of the preserved heritage, to its affirmation as a part of cultural–historical identity of a city, but also as a recognizable cultural tourism product. In that way, the approach to heritage as an active resource integrated into the contemporary life and as a driver of sustainable development is achieved.

Acknowledgements The research is a part of the scientific project 'Heritage Urbanism—Urban and Spatial Planning Models for Revival and Enhancement of Cultural Heritage'. It is financed by the Croatian Science Foundation [HRZZ-2032] and carried out at the University of Zagreb, Faculty of Architecture.

References

Božić N, Krajnik D, Ivanković V (2005) Manors of North-West Croatia—possible tourist itinerarires. In: Obad Šćitaroci M (ed) Contemporary use and creative management of Manors, Castles and Villas, University of Zagreb, Faculty of Architecture, Zagreb, pp 49–56 (ISBN 953-6229-40-4)

Conzen MRG (1975) Geography and townscape conservation. In: Uhlig H, Lienau C (eds) Anglo-german symposium in applied geography. GiessenWürzburg-München, Lenz, Giessen, pp 95–102

Gutkind EA (1964) International History of city development—Urban development in central Europe, vol I. Collier-Macmillian Limited, London

Knežević S (1996) Zagrebačka zelena potkova. Školska knjiga, Zagreb

Krajnik D (2007) Urbanistička preobrazba bastionskih utvrđenja. University of Zagreb, Faculty of Architecture, Doktorska disertacija

Krajnik D (2011) Preobrazba bastionskih utvrđenja u europskim i hrvatskim gradovima. University of Zagreb, Faculty of Architecture, Zagreb (ISBN 978-953-6229-78-9)

Krajnik D, Petrović Krajnik L (2011) Possible approaches to the reconstuction and presentation of bastion fortifications in Croatia. In: Dolkart A, Al-Gohari OM, Rab S (eds) Conservation of architecture, Urban areas, nature and landscape, vol II. CSAAR Press, Amman

Krajnik D, Obad Šćitaroci M, Šmit K (2008) Preobrazba gradskih utvrđenja u javne prostore urbane društvenosti. Društvena istraživanja 17:463–482

McKercher B, Du Cros H (2002) Cultural tourism: the partnership between tourism and cultural heritage management. Haworth Hospitality Press, Philadelphia (ISBN 0789011069, 9780789011060)

Mollik K, Reining H, Wurzer R (1980) Plannung und Verwicklichung der Wiener Ringstrassenzone. Franz Steiner Verlag GMBH, Wiesbaden

Whitehand JWR (2005) Urban Morphology. Urban Landscape Management and Fringe Belts, Urban Design 93:19–21

Chapter 27
Revitalisation of Historic Gardens—Sustainable Models of Renewal

Mladen Obad Šćitaroci, Mara Marić, Koraljka Vahtar-Jurković and Ksenija Radić Knežević

Abstract The main research aim of this paper is to determine appropriate and sustainable models of revitalisation of historic gardens and parks which must respect the gardens' authenticity and inherited identity features, but also allow for contemporary interventions and new facilities. The fundamental criterion for such interventions is to preserve the historical identity and revitalise the former atmosphere while affirming authenticity. The goal is to identify suitable models of revitalisation of gardens and parks. Three different examples are used in this research: (1) gardens of Renaissance summer villas in Dubrovnik (sixteenth to nineteenth century, Southern Adriatic); (2) historicist gardens in Opatija (late nineteenth and early twentieth centuries, Northern Adriatic); (3) historicist spa gardens (late nineteenth and early twentieth centuries, Northern Croatia). The models of renewal and revitalisation of the gardens and parks identified and examined are divided into two groups: basic models and thematic models. In relation to basic models, the following models are applicable: conservation, urban planning, architectural, functional, aesthetic and ambient models. Regarding thematic models, the following can be applied: activation, spatial (linear, networked or dispersed), economic, ecological, transformational and participative models. Research was conducted on selected examples of typologically diverse cultural heritage based on which historical and contemporary models were

M. Obad Šćitaroci (✉)
Faculty of Architecture, University of Zagreb, Zagreb, Croatia
e-mail: scitaroci@gmail.com

M. Marić
Department for Mediterranean Plants, University of Dubrovnik, Dubrovnik, Croatia
e-mail: mara.maric@unidu.hr

K. Vahtar-Jurković
Department for Spatial Planning, Faculty of Civil Engineering, Civil Engineering and Environmental Protection, Primorje-Gorski Kotar County,
University of Rijeka, Rijeka, Croatia
e-mail: koraljka.vahtar-jurkovic@pgz.hr

K. R. Knežević
Kvantum d.o.o, Zagreb, Croatia
e-mail: ksenijaradic@gmail.com

M. Obad Šćitaroci et al. (eds.), *Cultural Urban Heritage*,
The Urban Book Series, https://doi.org/10.1007/978-3-030-10612-6_27

identified for use in the revitalisation and enhancement of historic gardens now and in the future.

Keywords Historic gardens · Historic parks · Revitalisation models · Enhancement criteria · (Heritage) Authenticity

27.1 Introduction

Gardens are the work of man and nature. They have always been considered creative work, and sometimes they reach the level of art. Gardens are not nature, although at first glance they may look natural. They are not nature because they are the result of man's work, interpretation, design and maintenance. Nature maintains itself, while gardens must be tended by man because they cannot survive without him.

Fragility, ephemeralness, growth and constant change are the characteristics of garden architecture. Plants are resilient and long lasting, and fragile and short lived at the same time, especially if we do not take care of them. They are living organisms that grow, age and die, and year after year, decade after decade, they change significantly. When are they the most beautiful, what is their original state, what is their initial matrix that we often call upon? Due to these specific features, gardens are remarkably different from other types of cultural heritage.

Gardens are often neglected; they lose their identity or disappear due to the natural process of decomposition or the lack of maintenance. In these cases, they need renewal, enhancement or revitalisation, depending on the level of degradation and what the idea of their future use is. Renewal would not be necessary if we continuously maintained and tended the gardens. They would age, but they would never grow old and disappear.

The question is how to renew a garden, how to renew and revitalise it if it has entirely disappeared or if only some residual parts are left, and how to enhance it if it is neglected? How can such a garden be modernised and adapted to the contemporary age and modern needs, for it is not always possible to keep gardens as museum specimens? It is difficult to achieve the sustainability of historic gardens and parks because they have never been financially sustainable, and maintenance costs have always been huge.

The stimulus for this research comes from the multitude of historic gardens in Croatia and the rest of the world which require constant maintenance and contemporary use. Despite legal and formal protection, gardens are often in an unsatisfactory condition. Numerous charters and declarations have been adopted, conservation methods are known and generally accepted, but the gardens are still rapidly becoming degraded or they are being renewed in a way that makes them lose the basic features of their identity. Therefore, we must find the right way to proceed, in other words to find acceptable revitalisation methods and enhancement criteria.

Our starting point is the fact that gardens belong to the cultural heritage, so in order to preserve them, we should act in accordance with the charters on cultural heritage protection and renewal methods while respecting the specific features typical of garden architecture.

The main research goal of this paper is to identify historical and current revitalisation models and enhancement criteria, which are the starting points for contemporary garden use and future models. The primary tasks are to identify the revitalisation models which respect the inherited identity features, ensure contemporaneity, allow contemporary intervention and enable the future development of historic gardens and parks.

Our hypothesis is that, for gardens as cultural heritage, approaches other than the use of conservation criteria can also be taken to find sustainable renewal models. In this way, we enrich the methodological framework, contribute to the diversity of heritage use, and increase the number of potential uses.

It is important to determine the appropriate permanent purpose (use) for all heritage sites, including historic gardens, because this will ensure long-term protection and economic sustainability. Determining the purpose depends on the value of the heritage and its ambient context, among other things. Renewal for the sake of renewal, without a sustainable purpose in the future, is not viable. Such renewals are certain to fail because they will not revitalise the heritage, although the heritage will remain physically renewed for a while.

When considering heritage renewal in general, which also includes historic gardens, the following terms are often considered: restoration, revitalisation and enhancement. *Restoration* refers to returning the heritage to its prior state (original state or one of its developmental stages), all the while respecting historical facts and documented sources. *Revitalisation* refers to the renewal of the heritage which opens up new possibilities for its use and provides the heritage with new life that is different from its prior (original) use or from its later established use. *Enhancement* refers to improving the former state in terms of maintenance, appearance, function and use, which does not require making huge and fundamental changes.

Restoration is not the subject of this paper since this topic has been methodologically well developed and elaborated and has been applied for years, based on which gardens around the globe have been renewed. This paper focuses on revitalisation, enhancement and reuse of historic gardens and parks. In many cases, the gardens have been abandoned and neglected and do not have an appropriate use (either the gardens themselves or the buildings in their vicinity) so repurposing them and using them in a different way from their original purpose is considered.

This paper focuses on historic gardens and parks, i.e. designed landscapes. The term “garden” will mostly be used in the paper, since the majority of the studied examples are smaller in size. Regarding typology, most of these gardens are the gardens of private buildings (castles, summer villas and villas) or are public town gardens and other public areas in the town.

Research was conducted on selected examples of typologically diverse cultural heritage, based on which historical and contemporary models were identified for use both today and in the future when revitalising and enhancing historic gardens.

27.2 Theoretical and Methodological Framework

The systematic development of the theory of protecting and renewing historic garden architecture began in 1971 at the first international colloquium on the protection and renewal of historic gardens. Two international institutions, ICOMOS and IFLA,[1] founded at that time the International Scientific Committee on Historic Gardens and Sites. In 1999, the name of the Committee was changed to the International Scientific Committee on Cultural Landscapes (ISCCL). Gardens are no longer the only focal point, since cultural landscapes are considered as well. Since then, historic gardens have been viewed as a part of the cultural landscape.[2] Although charters on historic gardens (Florence Charter, Carta italiana dei giardini storici) were adopted 37 years ago, they are still current and used when renewing historic gardens (Belle Arti e Paessagio 2013).

Historic gardens are being renewed because of the value of cultural heritage per se, an indicator of cultural achievement in terms of respecting inherited values. In addition, the renewal of historic gardens and parks increases the quality of landscape and the quality of life (Obad Šćitaroci 2018).

Researchers from different fields engaged in processes of cultural heritage renewal emphasise the importance of heritage management which primarily involves identifying optimal renewal models (Mc Manamon and Hatton 2000; Krellig and Rhode 2014; Messenger and Smith 2015). In this regard, the notion of respecting the authenticity of heritage is underlined as a key concept in cultural heritage renewal and management (Mc Manamon and Hatton 2000; Starn 2002).

Among recent studies exploring the subject of garden renewal, many papers analyse the use of contemporary materials in renewal procedures (Rhode 2012; Elliott 2010). Consequently, it is necessary to contextualise contemporary procedures in a certain model.

When planning the renewal of a historic garden, its integration into the surrounding area is crucial. Of equal importance is the implementation of an effective management plan and the need to involve and inform the local community about the value of the garden heritage in order to create a cultural connection between the heritage and the community and thus to start reusing the heritage (Sales 1995; Dreija 2012; Sales 2014).

When rethinking the renewal models, another essential starting point is to problematise, by means of detailed research, the significance of the heritage. Based on the significance of the garden, the starting point for the principles of renewal is set

---

[1]ICOMOS—International Council on Monuments and Sites, IFLA— International Federation of Landscape Architects.

[2]Cultural landscapes are considered "as complex systems where cultural relationships are developed within an ecological context, recognizing the mutual and reciprocal influence of nature and culture" (ICOMOS 1981). According to the UNESCO (2013) *Operational* Guidelines "cultural properties represent the combined works of nature and of man". There are three main categories: I. designed landscapes and created intentionally by man; mainly parks and gardens; II. organically evolved landscape, which may be a relict or continuing landscape; III. associative cultural landscape.

depending on the current circumstances, changes of purpose, and the need for the adaptation and development of the garden (Sales 2014; Bourke 2012; Jacques 2014).

Considering the ephemeralness and fragility of historic gardens as a specific type of cultural heritage whose purpose, renewal and use is always a challenge, scientists agree on one thing—that the theory of renewing historic gardens needs to be continuously upgraded with new methodology (Jacques 2014; Lambert and Lovie 2014). This paper offers a contribution here.

In this paper, rethinking the renewal of historic gardens as cultural heritage is based on the *Heritage Urbanism*[3] approach. Cultural heritage must be active because it contributes to the development of the local community and enhances people's living space. Revitalisation and enhancement of cultural heritage is placed in a spatial, urban planning and landscape context. Heritage is viewed together with its surrounding area (context), the effects of this area on the heritage and, vice versa, the effects of heritage on its surroundings (Obad Šćitaroci 2015). To renew a garden, it is essential to establish its identity features and conduct an assessment from different viewpoints: conservation, cultural and historical, architectural, urban planning, spatial and ambient, aesthetic, functional and ecological. This is the foundation for setting the criteria for contemporary interventions and renewal. Potential models for enhancement and/or revitalisation are determined by studying and comparing the historical and contemporary examples (Obad Šćitaroci and Bojanić Obad Šćitaroci 2017).

The research described in this paper is based on the aforementioned theoretical and methodological starting points and is conducted and verified with the help of many selected examples. The research results allow conclusions to be drawn on the features of identity of gardens, on basic criteria for the renewal of gardens and new interventions in their area, and for potential models applicable to the revitalisation and enhancement of historic gardens to be defined.

27.3 Reality of Gardens—The Examples Studied

Through prior long-term research of historic gardens and for the purpose of planning renewal projects, many examples from Croatia were studied which were compared with relevant international examples (Bojanić Obad Šćitaroci and Obad Šćitaroci 2004; Hajos 2007). For the purposes of this paper, new research was also conducted on the following historic gardens:

- Gardens of Renaissance summer villas in Dubrovnik (sixteenth to nineteenth century, Southern Adriatic)

[3]This is an approach that was explored and created as part of a research project under the title "Heritage Urbanism (HERU)—Urban and Spatial Models for the Revival and Enhancement of Cultural Heritage" conducted at the Faculty of Architecture of the University in Zagreb from 2014 to 2018 and funded by the Croatian Science Foundation (HERU-2032).

- Historicist gardens in Opatija (late nineteenth and early twentieth century, Northern Adriatic)
- Historicist spa gardens in Northern Croatia (late nineteenth and early twentieth century).

For every type of garden, identity features (identity factors) were studied, an assessment was conducted, and enhancement criteria and revitalisation models were established.

27.3.1 Gardens of Renaissance Summer Villas in Dubrovnik

The garden architecture of summer villas in Dubrovnik stems from the city's Renaissance urbanism marked by order, simplicity and functionality. The reason behind such a concept lies in the rational use of a small and narrow coastal belt of the former Republic of Ragusa (1358–1808) which covered an area of 1357 km^2 (Fig. 27.1). From the sixteenth century to the end of the nineteenth century, approximately 300 summer villas with gardens were built. About 150 of these are preserved today, most of which are in very poor state, either with no use or without permanent use. Fifty sites were explored in the area of Rijeka Dubrovačka (Obad Šćitaroci et al. 2017).[4]

Summer villas in the area of Dubrovnik (Figs. 27.2 and 27.3) were built outside the city in its agrarian areas, in accordance with the Roman concept of *otium et negotium*. They are characterised by being situated directly on the shore in most cases, surrounded by high boundary walls, and built of stone and on a stone (karst) surface. The villa and garden are closely connected, both by function and design.

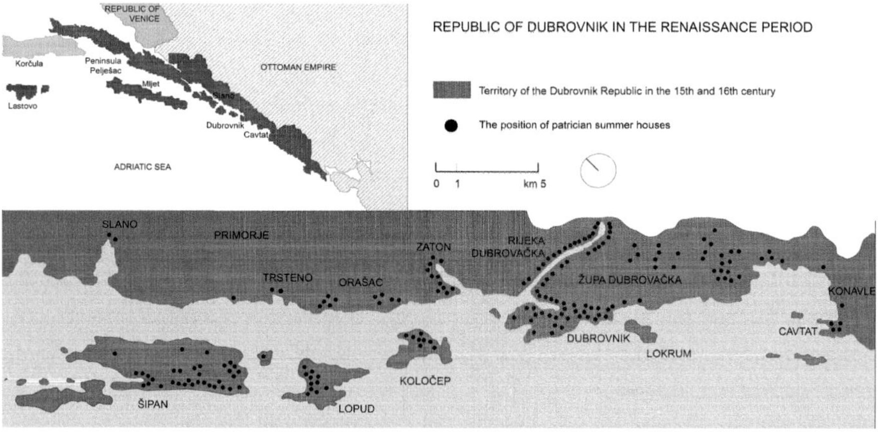

Fig. 27.1 Map of patrician summer houses in the Dubrovnik Republic in the fifteenth and sixteenth centuries

[4]This research was expanded and revised for the purposes of this paper.

Fig. 27.2 Remains of the Gučetić-Lazarević summer villa and garden, Rijeka Dubrovačka, Dubrovnik, 2016

Promenades or paths are laid out in the form of a grid bordered by small retaining walls that carry the stone columns for lattices or pergolas for grapevine. The space between the paths was bedded with soil where vegetation was planted, both native Mediterranean vegetation and vegetation brought from journeys abroad (lemons, oranges, pomegranate, rosemary, etc.).

An assessment of 50 gardens was conducted by using four main criteria: conservation; cultural and historical; architectural and structural; and spatial and ambient. There were also a total of 13 subcriteria.[5] A third of these gardens were renewed, third of them were transformed, and another third in ruins. Over 80% of them were from the fifteenth or sixteenth century, and nearly 40% have unique features.

The main criteria for the renewal and revitalisation of gardens in the Dubrovnik area are the preservation and reconstruction of their identity, the revitalisation of the

[5] Authenticity, the level of preservation and the value of the garden phases, rarity, age value, date of origin, cultural and historical value, aesthetic and artistic values, the state of the buildings, construction materials, the level of preservation of the exterior of the garden, preservation of the surrounding area, the level of preservation of the gardens and special urbanistic features.

Fig. 27.3 Sorkočević summer villa and garden, Rijeka Dubrovačka, Dubrovnik, 2017

ambience and the affirmation of their authenticity. For all the gardens which have kept their historical matrix and features, this means applying the criteria of integrity and authenticity and using the conservation renewal model. Many gardens are wholly destroyed, partitioned and cut off from the villa, so clear criteria are needed for potential contemporary interventions and adaptations which, despite contemporary needs, must affirm the authenticity and the heritage value of the Renaissance summer residences.

27.3.2 Historicist Gardens in Opatija—Public Gardens in the Function of Tourism

Opatija is a town of hotels and spas that developed in the second half of the nineteenth century and the beginning of the twentieth century. It is located on the North Adriatic Sea and has a favourable microclimate, so people have long visited Opatija for rest and recuperation. The town lies on a belt along the coast with hills in the background, and it has plenty of gardens and promenades. In Opatija, there are 10 public gardens, 13 gardens belonging to hotels and sanatoriums, several dozen villa gardens, 8 kilometres of coastal and inland promenades and many other public areas with vegetation (Vahtar-Jurković 2004, 2010) (Fig. 27.4).

Opatija is characterised by its urban planning concept, which is mostly preserved today and based on the unity of urbanism, architecture and landscape architecture established at the end of the nineteenth century at the time of the Austro-Hungarian Empire. The town's identity features therefore stem from its urban planning, architectural, cultural, historical, landscape and botanical values. Opatija's favourable climate and its historicist and Art Nouveau gardens and parks (Figs. 27.5 and 27.6),

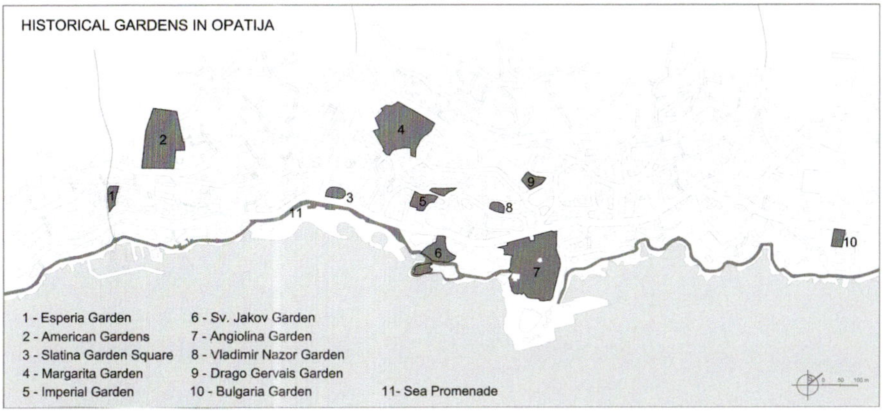

Fig. 27.4 Historical gardens in Opatija—Esperia Garden, American Gardens, Slatina Garden Square, Margarita Garden, Imperial Garden, St. Jakov Garden, Angiolina Garden, Bulgaria Garden, Drago Gervais Garden, Vladimir Nazor Garden, Sea Promenade

Fig. 27.5 Angiolina Garden, Opatija, 2007

and its coastal and forest promenades were pleasant places for walks, recreation, rest and recuperation. The town's gardens became its identity long ago, thanks to the lush natural vegetation (bay laurel woods) and plenty of exotic plants. Camellias, the symbol of Opatija, are the most common among the exotic plants.

Opatija in many ways is like Crikvenica, Veli Lošinj, Mali Lošinj and Rab also along the North Adriatic coast and can be compared to other Mediterranean tourist

Fig. 27.6 St. Jakov Garden, Opatija, 2007

destinations such as Portorož (Slovenia), Nice (France) and Sanremo and Rimini (Italy).

Some of the gardens have preserved their original features, some received modernist additions in the twentieth century, and some have no particular stylistic features. Managing such a situation requires appropriate criteria and allows for the application of different enhancement and revitalisation models. The main criteria should encompass not only individual gardens, but also the town as a whole where the gardens are an important part of its identity. This approach enables the application of different revitalisation and enhancement models and also allows for the introduction of new, appropriate features.

27.3.3 Historicist spa Gardens in Northern Croatia

Three thermal spas were studied in Northern Croatia: Lipik, Daruvar and Varaždinske Toplice. Daruvar and Varaždinske Toplice started to develop at the beginning of the nineteenth century on the foundations of the prior Roman baths, and Lipik began to develop at the end of the nineteenth century as a new thermal spa with no Roman tradition.[6] Lipik (Figs. 27.7 and 27.8) has not yet been restored following the war in

[6]In northern Croatia, there are several other examples of spa gardens—Stubičke Toplice, Krapinske Toplice and Topusko. They have lost their recognisable historical spa identity, so they are not

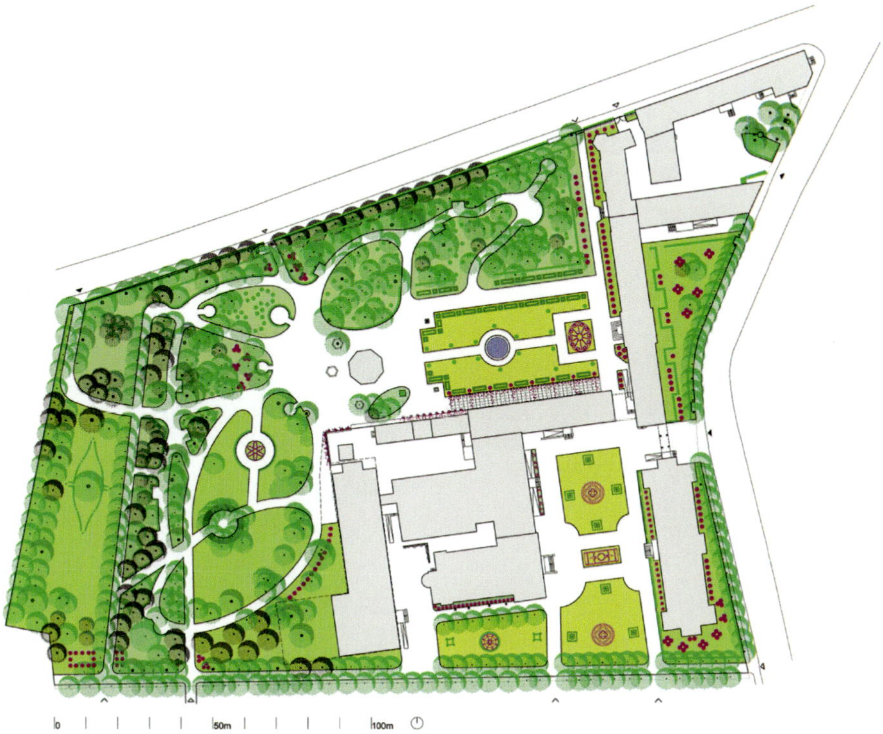

Fig. 27.7 Lipik, garden renewal plan, 2017, *Authors* Obad Šćitaroci M, Bojanić Obad Šćitaroci B and Radić Knežević K

the 1990s. The spa in Varaždinske Toplice (Fig. 27.9) is special with regard to its size and the well-preserved remains of Roman baths within the garden. The spa garden in Daruvar (Fig. 27.10) has great potential for modernisation alongside the affirmation of the historicist spa buildings (Bojanić Obad Šćitaroci and Obad Šćitaroci 2004; Obad Šćitaroci et al. 2014).

The distinctiveness of the spa gardens lies in the indivisibility and interlacing of the spa buildings and the gardens which together make a functionally designed whole. Without a garden, there is no spa. Gardens are not an addition, but a crucial part of the spa, an important part of the urban matrix and town environment. Enhancement criteria and revitalisation models for gardens and the entire spa complex stem from this symbiosis.

Spa gardens of continental Croatia are large in size (up to 10 ha), have no walls and are publicly accessible, and some of them have assumed the role of town square or promenade. Historicist parterres (flowers and topiaries) closely border the spa

---

discussed in detail in this paper, but they are still good examples for the application of enhancement criteria and revitalisation models.

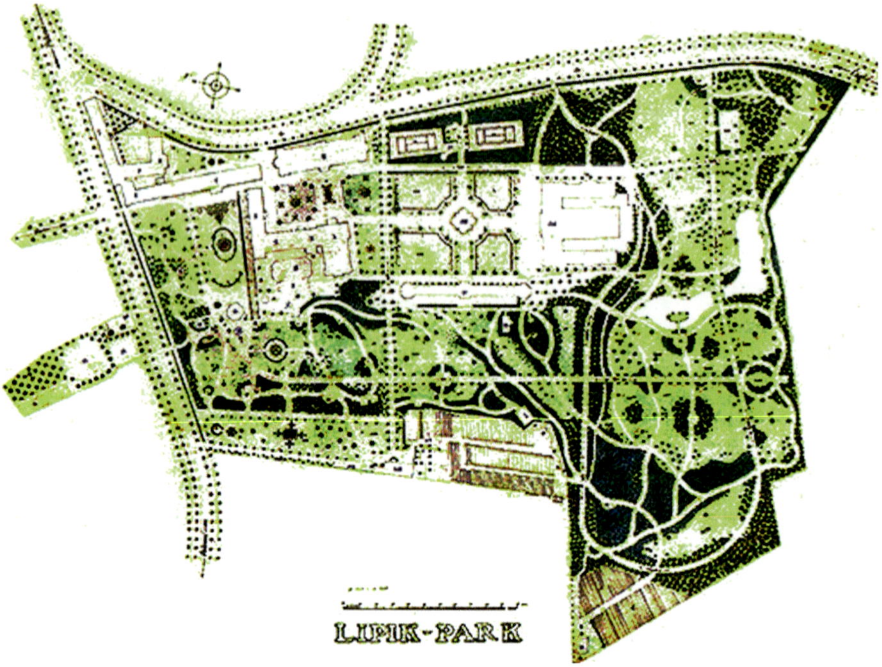

Fig. 27.8 Lipik, original garden plan, end of the nineteenth century (*Source* Šćitaroci private archive)

buildings, and the rest of the garden consists of romantic groves with tree-lined walkways which connect the buildings with other places within the garden.

Another special characteristic of spa gardens is the indivisibility of the buildings and the function of the spa, which implies the need for modernisation and additional contents aligned with the contemporary needs and technological requirements of a spa. Therefore, the current question of allowing new interventions in the spa requires criteria to be determined and appropriate enhancement and modernisation models applied whose consequences may be visible in the garden.

The main criterion for the enhancement and new interventions in spa gardens is the conservation and renewal of the historical (original) features and the revival of the prior ambience of the spa. This does not exclude the possibility of introducing new spa features in unfinished or less preserved parts of the garden.

Fig. 27.9 Julius spa garden in Daruvar, 2012

Fig. 27.10 Varaždinske Toplice, Roman baths in the garden, 2016

27.4 Garden Renewal Models—Appropriate and Acceptable Options

Garden renewal models are based on research that forms part of the Heritage Urbanism project and research on the selected examples covered in this paper. The models mentioned are applicable to the revitalisation and enhancement of historic gardens and parks (Obad Šćitaroci 2015; Obad Šćitaroci and Bojanić Obad Šćitaroci 2017, 2018).

By using the Heritage Urbanism method during the renewal of gardens and parks, research was conducted on the identity features, on assessing the value of the garden and its relationship with the surrounding urban environment or landscape, on setting the criteria for enhancement and new interventions and on determining the appropriate revitalisation models. It can be expected that new interventions in gardens and parks will enable both revitalisation and enhancement and will represent a high-quality modern contribution. Apart from the general criteria for new interventions in the gardens, special criteria for enhancement are set based on the specific features of each garden or park. Consequently, appropriate revitalisation models are selected or created. Each garden has its distinctive identity and a unique evaluation result, leading to a combination of basic and specific thematic models, thus creating a combined model.

27.4.1 Basic Models of Garden Renewal

Basic renewal models are the models used for all or most renewals. The conservation model is the primary model, followed by other models: urbanist, architectural, functional, aesthetic and ambient.

The *conservation model* implies the preservation of the existing state, taking into account the cultural, historical and other identity features, and renewal based on credible historical sources. This model is similar to restoration. It is a basic renewal model, but it cannot be the only model used for renewing historic gardens. Complementing this model with other appropriate models is expected when this is justifiable.

The *urbanist model* ensures that garden heritage is not considered or renewed in isolation, but in the context of its immediate and more distant environment. This allows for mutual influence and harmonisation, where the garden heritage becomes part of its surroundings and influences those surroundings (and vice versa, the surroundings influence the garden renewal). This model ensures the better, more appropriate development of a town, settlement and the landscape where the garden is located.

The *architectural model* implies carefully planning the renewal of the garden heritage, including the buildings next to the garden (summer residences, castles, villa, palaces, spa buildings) and buildings in the garden (pavilions, fountains, belvederes, rest places, etc.). Buildings and the garden itself should be considered a single organism, coordinated both functionally and by design.

The *functional model* rethinks garden renewal in the context of its prior functions, but in accordance with acceptable new functions and contemporary needs. If the heritage is not used appropriately and has no purpose, new life and long-term survival of the heritage cannot be ensured.

The *aesthetic model* is based on the thesis that gardens, much like other architectural structures, seek to achieve excellence and artistic distinctiveness based on stylistic and aesthetic values. This is why it is essential to use this model and aesthetic criteria when renewing garden heritage.

The *ambient model* considers the characteristics of context, which include not only the physical and material environment (urban, rural, landscape), but also the intangible influences (social, cultural, sociological, etc.) which can make garden heritage distinctive and can have a great effect on the sustainability of the renewed garden heritage. Applying the ambient model can contribute to the historical authenticity of gardens in the contemporary age.

After studying 50 sites in Rijeka Dubrovačka near Dubrovnik, it was discovered that combined models can be used in all cases where several models overlap (Fig. 27.11). The most applicable models are the conservation (38%) and ambient (26%) models. For permanently inhabited sites (48%), the functional model can be applied which enables contemporary garden use adapted to current needs. The urbanist model can be used in all cases where thorough urban rehabilitation of the entire built and unbuilt area of the summer residence is needed (100%).

By analysing the gardens of Opatija, it was found that the conservation model is predominantly used in cases where the original garden features are preserved (60%). The aesthetic model will be useful in the cases of modernist landscape additions in

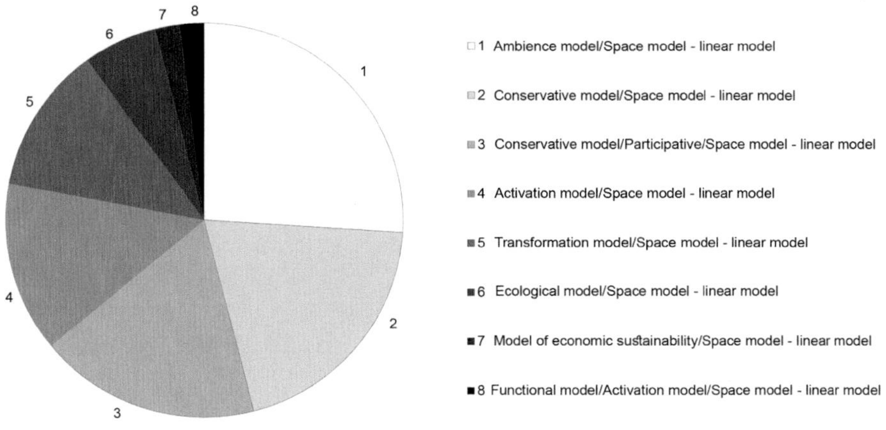

Fig. 27.11 Diagram—typological classification of potential restoration models of Dubrovnik villas and gardens in the Rijeka Dubrovačka area

the twentieth century (40%). For gardens with no distinctive stylistic features, two models are dominant: functional and urbanist.

Regarding spa gardens, apart from the conservation model, the urbanist model is the most prominent, while the application of the ambient model is crucial to revitalise the experience of the social scenery.

27.4.2 Thematic Models of Garden Revitalisation

Thematic or special models for the revitalisation of historic gardens and parks are based on a multidisciplinary approach which provides new possibilities in the contemporary age.

The *activation model* stimulates modern uses which will make the garden a place for events and activities. This is particularly important when the garden in question is not a public garden belonging to the town, but a private garden (originally belonging to a summer residence, a castle or a villa) which is repurposed and becomes a public or partly a public garden.

The *spatial model* is based on the thesis that individual historic gardens (small gardens in particular) cannot have much influence unless they are connected through a well-designed system. Therefore, it is recommended to link individual gardens by using one of the spatial models—*linear* (heritage trails), *networked* (connecting and overlapping different cultural and natural heritage), *dispersed* (with no visible connection) or some other model. Selecting spatial models depends on the character and size of the garden, its location within the town or landscape, the spatial idea of revitalisation at town or regional level, etc.

The *economic sustainability model* considers the long-term financial sustainability of a garden. This model is crucial for the revitalisation and enhancement of garden heritage when there is no permanent financial support available, so regular income must be ensured by repurposing the garden and using it appropriately.

The *contemporary interventions model* affirms the potential contemporary construction and contemporary interventions in garden heritage (just as in any other type of heritage), but with the preservation of the memory of space and heritage values. Innovations are encouraged, but only while respecting and affirming the heritage.

The *ecological model* is based on preserving, respecting and enhancing the natural factors, regardless of whether the garden heritage is located in the town or in the landscape. This model will be used more often for large parks where the emphasis is on the natural systems governed by the ecological, phytocoenological and other biotechnical factors.

The *transformation model* implies changing the function and use of the garden heritage compared to its original historical state. Re-establishing the original use is sometimes not possible, so an acceptable form of transformation is found.

The *participative model* is recommended for the enhancement and revitalisation of cultural heritage by the Council of Europe. This model includes the local community and all other stakeholders (bodies of authority, organisations, companies, citizens, etc.). It implies the participation and coordination of all state and regional bodies which must ensure the prerequisites for the effective implementation of renewal and the active use of the heritage as a national resource.

The linear spatial model can be applied to all the researched summer residence/villa gardens in the area of Rijeka Dubrovačka in order to create a tourist trail, thus providing a large number of heritage sites on a small stretch of land, with all of them situated along the road and the sea coast surrounding a deep, narrow bay. The economic sustainability model is applicable to almost all 50 summer residences considering that they are in poor state and have no future without large financial investments. The activation model is applicable in 14% of cases where the original characteristics of both the summer residence and the garden are significantly compromised. The ecological model can be used for 6% of summer residences situated at a river source which is a natural habitat for different flora and fauna. In 12% of sites, the transformation model can be employed because there are very few remains and great changes have occurred. The participative model can be adopted for 18% of sites where facilities and activities to serve the local community are more desirable.

For gardens in Opatija, the following special models can be used: activation, spatial, contemporary interventions (in a limited form), ecological and participative.

For the gardens spas, the models of economic sustainability, modern interventions (in a limited form) and the ecological model are applicable.

27.5 Conclusion

Compared to other forms of cultural heritage, garden architecture is particularly vulnerable due to its fragility and liability to change. Renewal is neither unambiguous nor standard, although the conservation method for the restoration of historic gardens and parks is clear. The research has focused on the revitalisation and enhancement of gardens and parks that are neglected or even abandoned and have lost their original use, so they are waiting their chance to start a new life with a new use. New use is the key to revitalisation and enhancement. The goal of this research was to identify the enhancement criteria and revitalisation models for garden heritage that will respect the features of historic identity and preserve authenticity, but also enable creative contemporary interventions and programmes to facilitate the future development of historic gardens and parks. This goal is accomplished by following the hypothesis that for gardens, just as for other forms of cultural heritage, apart from conservation criteria, different approaches can also be applied in finding sustainable renewal models. These other different approaches were the subject of the research described in this study.

The main research question was to identify appropriate and sustainable renewal models for historic gardens and parks, models that imply revitalising and enhancing/improving the current state. Models are identified by researching the historical and contemporary models of use and prior renewal, and by recognising features of identity and criteria for new interventions. The models identified were divided into two groups: (1) basic models which are used in most cases and (2) thematic or special models which are founded on a multidisciplinary view aiming to modernise heritage areas. These models do not offer concrete solutions, but at least provide a framework for finding one. A single model is rarely applied. Most often, several basic and thematic models that complement each other are applied, thus creating a combined model.

When renewing a garden, it is essential to create an ambience that provides an experience of historical authenticity. Historical appearance is impossible to recreate because gardens are constantly changing due to vegetation growth, but it is possible to create an ambience that brings the garden or a park closer to its original setting.

When a garden is well preserved and when archival cartographic and other sources are available, credible renewal can be successfully achieved. When there is no documentation available or when data are insufficient, and the existing state has significantly changed from the original or historical state, then revitalisation models may be applied and the *Heritage Urbanism* approach can be used to find an acceptable solution.

Acknowledgments The research is part of the scientific project "Heritage Urbanism—Urban and Spatial Planning Models for Revival and Enhancement of Cultural Heritage". It is financed by the Croatian Science Foundation [HRZZ-2032] and carried out at the University of Zagreb, Faculty of Architecture.

References

Belle Arti e Paessagio (2013) Carta italiana dei giardini storici http://www.beap.beniculturali.it/opencms/opencms/BASAE/sito-BASAE/mp/Uffici-musei-e-monumenti/Giardini-e-parchi-storici/Carta19903.html?id=8339&pagename=8339. Accessed 24 Feb 2018

Bojanić Obad Šćitaroci B, Obad Šćitaroci M (2004) Gradski perivoji Hrvatske u 19. Stoljeću/Public Parks of Croatian Towns in the 19th Century in the European Context. Šćitaroci/Faculty of Architecture University of Zagreb, Zagreb. ISBN 953-97121-3-0

Bourke M (2012) How will my garden grow? A philosophy for the restoration of historic gardens. J Gard Hist 3(1):49–54

Dreija K (2012) Historic gardens and parks: challenges of development in the context of relevant regulations, definitions and terminology. MOKSLAS—Liet Ateitis Sci—Future Lith 4(2):167–175. ISSN 2029-2252

Elliott B (2010) Changing fashions in the conservation and restoration of gardens in Great Britain. Bulletin du Centre de recherche du château de Versailles: Sociétés de cour en Europe, XVIe-XIXe siècle/European Court Societies, 16th to 19th Centuries. https://doi.org/10.4000/crcv.10764

Hajos G (ed) (2007) Stadtparks in der österreichischen Monarchie 1765–1918. Böhlao Verlag, Wien-Köln-Weimer

ICOMOS (1981) Historic gardens—the florence charter. https://www.icomos.org/images/DOCUMENTS/Charters/gardens_e.pdf. Accessed 24 Feb 2018

Jacques D (2014) The treatment of historic parks and gardens. J Archit Conserv 1(2):21–35

Krellig H, Rohde M (eds) (2014) Historic gardens and climate change: recommendations for preservation. Edition Leipzig, Leipzig

Lambert D, Lovie J (2014) All rosy in the garden? J Archit Conserv 12(3):83–106

Mc Manamon FP, Hatton A (eds) (2000) Cultural resource management in contemporary society. Routledge, London. ISBN 978-0-415-64241-5

Messenger PM, Smith GS (eds) (2015) Cultural heritage management: a global perspective. University Press of Florida, Gainesville. ISBN 978-0-8130-6085-9

Obad Šćitaroci M, Bojanić Obad Šćitaroci B, Radić K (2014) Spa garden in Daruvar—methods of renewal and reconstruction. YBL J Built Environ 2(2):5–16. ISSN 2063-997x

Obad Šćitaroci M (ed) (2015) Cultural heritage—possibilities for spatial and economic development—proceedings. Faculty of Architecture University of Zagreb, Zagreb. ISBN 978-953-8042-11-9 e-book

Obad Šćitaroci M, Bojanić Obad Šćitaroci B (eds) (2017a) Modeli revitalizacije i unaprjeđenja kulturnog naslijeđa - multidisciplinarni dijalog. Arhitektonski fakultet Sveučilišta u Zagrebu, Zagreb. ISBN 978-953-8042-30-0 e-book

Obad Šćitaroci M, Marić M, Medović M (2017b) Gardens in Rijeka Dubrovačka—Factors of identity and criteria for evaluation. Prostor 54:172–189. ISSN 1330-0652

Obad Šćitaroci M (2018) Heritage as an active space and spatial resource. In: Cocci Grifoni R et al (eds) Quality of life in urban landscapes. Springer, Cham, pp 341–349. ISBN 978-3-319-65581-9 eBook

Obad Šćitaroci M, Bojanić Obad Šćitaroci B (2018) Models of garden restoration as cultural heritage. In: Marić M (ed) Zbornik radova: Dr.sc. Bruno Šišić dubrovački krajobrazni arhitekt. Matica hrvatska, Dubrovnik, pp 111–121. ISBN: 978-953-7784-59-1

Rhode M (2012) La cura dei giardini storici - teoria e prassi. Olschki, Firenze

Sales J (1995) Garden restoration past and present. Gard Hist 23(1):1–9

Sales J (2014) Conserving historic gardens. J Archit Conserv 6(2):72–84

Starn R (2002) Authenticity and historic preservation: towards and authentic history. Hist Hum Sci 15(1):1–16. https://doi.org/10.1177/0952695102015001070

UNESCO (2013) Operational guidelines. http://landscapes.icomos.org/index.php/en/. Accessed 24 Feb 2018

Vahtar-Jurković K (2004) Opatija: Urbanistički razvoj i perivojno naslijeđe/Opatija: urban development and garden heritage. Glosa, Rijeka. ISBN 953-6081-38-5

Vahtar-Jurković K (2010) Opatijski gradski perivoji/the city parks of Opatija. Glosa/Faculty of Architecture University of Zagreb, Rijeka-Zagreb. ISBN 978-953-6081-71-4

Chapter 28
Revitalisation Models for Central European Country Houses

Boris Dundović, Mladen Obad Šćitaroci and József Sisa

Abstract A country house is a representative building which, apart from its residential function, also serves as a managerial focus for the wider estate. Throughout history, the representation, organisation, and management of a country house have served as complementary economic mechanisms that ensured that the seigniorial estate functioned as a (self-)sustaining system. After the dissolution of the Austro-Hungarian Monarchy in 1918, life in Central European country houses started to decline progressively. The ensuing turbulent national, social, and political situations on the territory of the former empire left the majority of the country houses to oblivion and decay. However, after a long hiatus, interest in this type of built heritage in Central Europe has significantly increased in the last few decades. In the present-day economy, recognising and implementing suitable models of active use for the manors of the region has become a pressing issue. This chapter begins with a presentation of the historical models of alterations, and deliberates on the adequacy of their implementation today. It proposes contemporary architectural and conservation models of revitalisation and includes the urban and spatial planning models which can contribute to the rehabilitation of wider cultural landscapes in rural parts of Central Europe.

Keywords Contemporary models · Cultural landscape · Economic sustainability · Management · Manor house

---

B. Dundović (✉)
Institute of History of Art, Building Archaeology and Restoration,
Vienna University of Technology, Vienna, Austria
e-mail: boris.dundovic@gmail.com

M. Obad Šćitaroci
Department for Urban and Spatial Planning and Landscape Architecture,
Faculty of Architecture, University of Zagreb, Zagreb, Croatia
e-mail: mos@arhitekt.hr

J. Sisa
Research Centre for Humanities, Institute for Art History,
Hungarian Academy of Sciences, Budapest, Hungary
e-mail: sisa.jozsef@btk.mta.hu

M. Obad Šćitaroci et al. (eds.), *Cultural Urban Heritage*,
The Urban Book Series, https://doi.org/10.1007/978-3-030-10612-6_28

28.1 Introduction

The history of architecture teaches us that only adaptable building types can successfully overcome the intensity of the economic, national, social, and political vectors bestowed upon them. Country houses are an architectural type that successfully evolved from Roman *villae rusticae* and mediaeval burgs, ultimately reaching its peak in the nineteenth century. The adaptability of country houses as a specific architectural type in the span of over fifteen centuries is evident in many subtypes that arose from specific functional, spatial, and economic needs (manors, curiae, mansions, villas, homesteads, summer houses, just to name a few). Yet, their core mechanism has never changed and is deeply embedded in the definition: a country house is a representative building which, apart from its residential function, primarily serves as a managerial focus for the wider estate.

In Central Europe, country houses are somewhat smaller in scale than their British or French counterparts, but the intensity of their spatial significance and cultural value for their regions and their communities remains the same. Central European country houses can be categorised into two major types: a *manor house* is an architecturally rich dwelling or a building complex which represents the administrative focus of a large estate, while a *curia* presents a simpler building of more modest appearance without architectural lavishness and serves as the centre of a smaller estate or a smaller part of a larger one (Obad Šćitaroci and Bojanić Obad Šćitaroci 2006). Additionally, an architecturally elaborated house coupled with a smaller estate is a type often called a *mansion*. The size and shape of a country house and its estate may have varied, but its core substance has remained the same.

The continuity of country house culture during the turbulent twentieth century highly depended on the givens of its cultural region. For instance, life in Central European country houses started to decline after the dissolution of the Austro-Hungarian Monarchy in 1918, resulting in their abandonment and progressive decay, especially during and after the Second World War. It took half a century for general interest in country houses to re-emerge, but after such a long-lasting gap, their common public narrative and symbolic value changed. As a result, country houses are today often pictured as architectural remnants of a culture forcibly interrupted by the tumultuous events of the twentieth century. Just a small number of country houses successfully maintained their role of functional nodes (Dundović et al. 2015) in their immediate landscape, whereas others were left to decay and are now perceived mostly as ruins, as the romantic archaeology of a vanished culture. It is important to surmount these melancholic narratives and re-introduce country houses in their core definition, while providing them with a new life and purpose.

This research ponders the functional and spatial potentiality of country houses of the former Lands of the Crown of Saint Stephen, with special focus on the country houses in present-day Croatia and Hungary. The two countries are comparatively relevant not solely because their country houses belonged to the same cultural region at the time of building, but also due to their markedly different approach to build heritage conservation and, more precisely, to country house revitalisation. The research

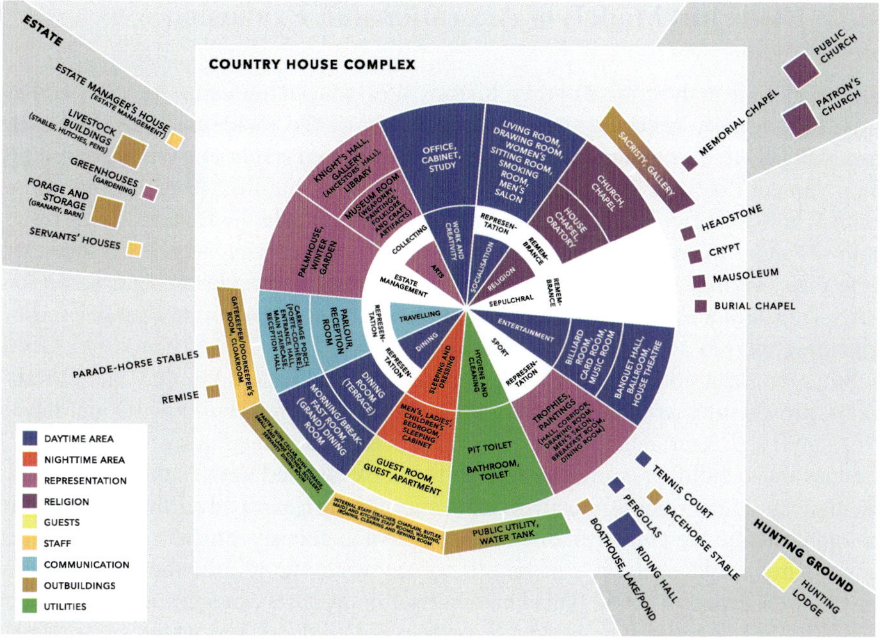

Fig. 28.1 An ideal functional scheme of a country house and its estate at the turn of the twentieth century. According to the *Source* Fekete (2002)

was prompted by the extremely poor state of country houses in Croatia and, on the other hand, by the upsurge of revitalised manors and mansions in the neighbouring Hungary in the past few decades.

While this research narrows its territorial reach to Croatia and Hungary, their country houses belonged to the greater tradition of country house building in Central Europe, resulting in characteristically similar physical features (Dundović et al. 2012) and spatial syntax (Fekete 2002) of both buildings and their estates (Fig. 28.1). The historical and contemporary models derived from this research are therefore applicable to the whole of Central Europe as a single cultural region.

The authors' principal aim is to deliberate the economically efficient potential of country houses as wider spatial complexes in need of models for active use. The focus of the research is not reserved only for the physical reconstruction of built heritage, but is mainly directed at models of functionally sufficient revitalisation that ultimately ensures economic sustainability, that is, a continuity of country-house life in the present-day economy. Both historical and contemporary revitalisation models are developed, presented, described, and evaluated according to the methodology of Heritage Urbanism.

28.2 Historical Models of Alteration and Expansion

Country houses in the course of their history rarely stayed the same. As living organisms, they usually reflected the changing needs of the owners, should that stem from the growth of the family, the increasingly elevated status of its head, or simply improvement due to a desire to have a more comfortable life. Just like any home, country houses were prone to constant change. This made them basically different from the other main traditional building type, churches, which tended to preserve more their original appearance and furnishings. This subchapter presents some formerly Transleithanian country houses where alterations and additions, for one reason or another, played a major role in their history.

Since construction had to be economical and no regulations tied the patrons' hands, pre-existing structures were often included in the new building. Count István Széchenyi, a Hungarian aristocrat especially interested in the theory and practice of country house building, summed it up in his book devoted to the subject in the following way: "In our country entirely new mansions are hardly ever built but, should the old home be the clumsiest and the least advantageously situated structure—no, only a simple, old wall—it will be the corner-stone, the trap of the new mansion for ever" (Széchenyi 1866). This characterisation may be exaggerated, but the fact remains that old country houses were seldom pulled down and whatever remained had to be re-used. And not only earlier homes but in some cases also other kinds of buildings were recycled into country houses. In Parádsasvár, earlier baths were incorporated into a house (Miklós Ybl, 1880–1882), in Horné Lefantovce (József Hubert, 1894) an old Pauline monastery was re-fashioned as a residence, along with the church now serving as the state hall. While remodelling, the builders usually (but not always!) took special care to disguise the old walls and create an entirely new-looking structure. Although this approach arose from the pragmatic use of existing structures instead of building them anew, these were the first attempts of the *integration model* in revitalisation, which kept the original building matter of the historical structure.

Remodelling sometimes seemed like reconstructing the former architecture and glory of an old building. In a way, such an operation anticipated what would later be called "restoration" in the nineteenth century. This was especially true of mediaeval structures and revived mediaeval styles, evident in efforts in the re-building of Bojnice Castle in an "authentic" mediaeval style (Sisa 2010). Here, the architect József Hubert had to restore, or rather freely recreate, mediaeval splendour in all the details, be it a sumptuous hall or just a small piece of furniture. Not surprisingly, contemporaries considered this operation a major project of building restoration. This *model of aesthetic enhancement*, along with the closely related *embellishment model*, is nowadays seen as outdated.

Otherwise, remodelling could be much more mundane and straightforward, and this was more typical. The purpose was to provide more space inside and a more modern appearance outside, regardless of the style of the pre-existing building. In the process, even relatively new mansions could be converted. This was the case of the country house in Rum in western Hungary. Here, the original structure, built by Adorján Bezerédj, displayed neo-Renaissance architecture. Yet with the change of the owners, Count Károly Kornis found the building too small and probably not sumptuous enough, neo-Renaissance being identifiable much more with the middle classes than with the aristocracy. Thus, between 1898 and 1901 he had the country house remodelled in what his contemporaries called a mix of "mansard" and "burgstyl" (castle style), more compatible with his status and as a reference to his ancestral home in Manastirea, Transylvania. The castle in Manastirea featured several towers and bastions, and so did the newly rebuilt country house in Rum. In the eyes of the contemporaries, the tower served as a symbol of noble ancestry and even had a heraldic significance. In this case, the aesthetic enhancement was a result of the *spatial expansion model*. This model was more clearly applied to the old neo-Classical house of the Zichy family in Belezna, a single-storey building that had to be enlarged by the architect János Bobula Jr. in such a way that the walls and the rooms remained more or less intact. To conserve the original appearance, the intervention was carried out with the technical ingenuity of the period. Therefore, in this case, we can acknowledge the *model of architectural innovation* as well.

In the nineteenth century, the technology of comfort developed substantially; in fact, this was the time when its full range appeared. However, due to the scarcity of documents and the inevitable destruction of original technical features, it is difficult to establish general facts. Taking into account the financial power of the patron and the general quest for comfort, it seems safe to assume that older country houses also incorporated a substantial amount of modern technology. This involved the installation of indoor plumbing. Water tanks and water pipes would provide running water as well as the possibility of installing bathrooms and flushing toilets. Several eighteenth-century stately homes were modernised along these lines by the descendants of the original builders or by their new owners, thus inaugurating the *models of architectural and infrastructural modernisation*.

At the other end of the social spectrum and on a larger spatial scale, nouveau-riche patrons also built fashionable homes, sometimes in addition to industrial sites (Sisa 2010). It was not uncommon for a factory and a country house to form part of a single complex; thus, the owners' source of income and their homes, financed from those incomes existed side by side. It was only in the last decades of the nineteenth century that the spatial and structural alterations gave way to a new paradigm in the enhancement of manor management and estate productivity: the *functional models*. The *economic production model*, applied to the mansion of József Törley, is a case in point. Törley, a champagne manufacturer and an emblematic figure of the country after the Austro-Hungarian Compromise, first built his champagne factory in 1888–1889 on the slopes of a hill in Budafok (now a suburb of Budapest). Incidentally, the site was identical to that of the by-then destroyed palace of Prince Eugene of Savoy. Some ten years later, he decided to build his mansion next to the factory. As a gesture of presence and of supervision, it looms over the factory. The choice of the style—French Gothic—was no accident; it obviously referred to the country where champagne was invented. Like a feudal lord, József Törley invariably employed a "house architect", Lajos Rezső Ray, and then, after the latter's death, his son, Vilmos Rezső Ray. The son was responsible for building the mansion in Budafok in the vicinity of the factory.

In sum, alterations and expansions were always an integral part of country house building and management of estates. They made the houses both more modern and more comfortable, and the estates more functional. Throughout history, Central European country houses were steadily enhanced according to the specific needs of the period, which ensured the continuity of their life in the ever-changing times.

28.3 Architectural and Conservation Models

After a long hiatus in the active life of country houses, marked by the ensuing turbulent national, social, and political situations in the territory of the former Austro-Hungarian Empire present during most of the second half of the twentieth century, interest in this type of built heritage in Central Europe has significantly grown over the last few decades. With the increased interest, the number of revitalisation endeavours has also increased. Therefore, in the present-day economy and tourism of Croatia and Hungary, recognising and implementing suitable and efficient models of active use for manors in the region has become a pressing issue.

In terms of the contemporary life of country houses and the surrounding estates as their immediate spatial complement, there are two major categories of architectural and conservation models for revitalisation: the *spatio-structural* and *functional models*. It is, however, important to underline that, in this categorisation, both architectural and conservation models do not hinder the development, but rather provide enhancement mechanisms which efficiently recognise the potential of economic sustainability while considering the identity factors such as the functional significance of a country house, its authentic features, and cultural value (Obad Šćitaroci et al. 2006).

28.3.1 Spatio-Structural Models

Numerous historical examples of models listed in the previous subchapter show that physical transformation regularly provided country houses with a new or enhanced active life. Apart from some of the models that are outdated (such as *embellishment* and *aesthetic enhancement*), most of them can be positively applied today.

In terms of conservation, spatio-structural models coincide with practical methods of the same terminology. The *restoration model* includes all the methods that replace the decayed or damaged elements while simultaneously preserving the original matter of a country house or its estate. It serves "the preservation of the original in the form in which it has come down to us, with its various layers and with its outstanding as well as its seemingly secondary or insignificant components" (Petzet 2004). Nádasdy Manor in Nádasdladány, Hungary, and Trakošćan Castle in Hrvatsko Zagorje, Croatia, are

Fig. 28.2 Nádasdy Manor in Nádasdladány, Hungary (left), and Trakošćan Castle in Hrvatsko Zagorje, Croatia (right). *Photos by* B. Dundović

both successful examples of the restoration model applied to the architecture and the surrounding landscape garden of a country house (Fig. 28.2). However, this specific model allows no significant transformation of the original structure, historical material, or physical features, and can therefore be suitable mostly to former country houses functioning as museums.

The *reconstruction model* is a more appropriate choice for country houses or their elements that are severely damaged or lost. It is suitable for architecture or landscapes that have irretrievably lost their original structure or features. It is a model that tries to preserve the original spatiality and atmosphere, although some or all of the elements lack the authenticity of matter. Such an example is Eltz Mansion in Vukovar, Croatia, which was severely damaged during the war in the 1990s but has been successfully reconstructed.

On the other hand, if a country house needs to accommodate a function that requires visible physical intervention or a structural transformation, the *model of physical transformation* makes for a more appropriate approach, as it relies on the creation of new concepts. An example of this model is the reconstruction of Janković Manor in Suhopolje, Croatia, which uses historical facts to reconstruct the estate layout and spatial concept of the edifices, while compensating for some of the missing outbuildings with contemporary structures. Contemporary architectural interventions complete the spatial needs of the new function—a hotel. Another similar example is Forgách Castle in Halič, Slovakia, now a hotel and spa centre, where the roof structure was reconstructed and modernised with a contemporary addition of a glass structure covering the inner courtyard. A physical transformation model can include: alteration, interpretation, retrofitting, modernisation, innovation, remodelling, architectural compensation, and contemporary intervention (architectural, structural, and/or infrastructural).

28.3.2 Functional Models

These models are the sets of roles country houses and their estates can assume and further intensify, along with the fundamental definition of the managerial focus of the estate. Mentioned in the subchapter above, we saw that Central European manors of the late nineteenth century had already started implementing several *economic models* applicable today: the *economic production model* when a manor house becomes the focus of production activity, the *business model* when it becomes a node of the service industry, and the *administrative model* when it is important to establish a managerial focus for several displaced instances of the same holding. Apart from the already mentioned Törley Estate in Budafok, Hungary, which today functions as a champagne and wine factory, there are several vineyard manors in Croatia that are also recognisable emblems for the products of their estates: for instance, the historical Belaj Manor and winery near Cerovlje in Istria (Fig. 28.3), the manor house in Bilje, which historically served as a hunting lodge of Prince Eugene of Savoy in Slavonia, or Turković Manor in Kutjevo, where the tradition of wine-making dates back to Roman times (Obad Šćitaroci and Bojanić Obad Šćitaroci 2001).

In relation to country houses, the repetition of historic functions does not imply redundancy but rather a synthesis of didactic information on historical layers and the creative potential of active use today. By simulating the historical mechanisms, we subject the vanished life of a country house to revision (Deleuze 1994, p. 212), thus exposing it to dynamic and productive processes of creative evolution, and ultimately intensifying it as an active participant in the present-day economy. This inherently creative repetition provides us with three main models: the *presentation model* and the *exhibition model*, which show former country-house life or its fragments, and the *revival model* which is the most authentic and most comprehensive of the three, but also economically the most inefficient. Veliki Tabor is a castle in Hrvatsko Zagorje, Croatia, where all three models are apparent: it presents the authentic original condition of the structure, exhibits the historical artefacts, and revives the scenes of former manorial life for visitors.

Fig. 28.3 Belaj Manor with its outbuildings and vineyards in Istria, Croatia. *Photo by* Belaj Winery

Other types of *tourism models* include the *therapeutic and medicinal model*, the *sport model*, the *leisure model*, and the *accommodation model*. For instance, the manor house in Balf, located near Sopron in Hungary, serves as a hotel for visitors of medicinal baths. The Janković Curia in Kapela Dvor near Virovitica in Croatia is, on the other hand, a heritage hotel adjacent to the transboundary biosphere reserve of the Drava, Mura and Danube rivers protected by UNESCO, offering recreational and learning activities.

With respect to their specific spatial syntax and, especially, the interdependence of interior and exterior space, country houses and their immediate surroundings can often be convenient for educational purposes. The *education*, *teaching*, and *academic models* are all complementary socio-functional models. One example where an elementary school located in a country house provides the historical structure with continuous care is Wenckheim Manor in Doboz near Békéscsaba, Hungary, with its facilities surrounded by a nature reserve. However, when deliberating upon a country house as an intensive node of knowledge dissemination, two contemporary models of utmost learning intensity need mentioning: the *research model* and the *science model*. A fine example of this approach is Bračak Manor near Zabok, Croatia, which was re-purposed and re-constructed as the energy centre and business incubator of the North-West Croatia Regional Energy Agency, consisting of offices, workshops, and presentation rooms for start-up companies in the field of energetics.

Although the majority of country houses in Hungary and Croatia are today re-constructed or re-stored due to individual efforts, the total number of successful results in manor revitalisation in Hungary significantly outweighs the Croatian efforts. However, there are two essential differences in the approach to manor revitalisation: (a) the Hungarian government implemented the National Mansion Programme, which presently includes 45 country houses and encourages "participatory planning in the project implementation process in the future in order to maximize the positive socio-economic impacts of cultural heritage" (Lazányi et al. 2016, p. 109); and (b) not all country houses in Hungary are viewed as historic monuments of larger cultural significance (for instance, only 5 of 45 country houses currently in the Programme are protected as historic monuments), which enables the efficient integration of contemporary intervention and historical layers to ensure the active revitalisation in present-day economic conditions (Nemzeti Kastély-és Várprogram 2016). The National Mansion Programme guarantees that authentic factors of identity will be conserved while a new active function is implemented, by including historical research and situation analysis, tourism impact assessment, international outlook, and benchmarking studies of cultural routes, clusters, feasibility, and funding (Fekete 2016). Without a doubt, country houses are functional nodes of a cultural region, but to acquire the necessary active intensity, they have to be deliberated upon as constituents of wider spatial systems (Dundović et al. 2015), which can be actualised by urban, landscape, and spatial planning models.

28.4 Urban, Landscape, and Spatial Planning Models

Country house complexes are factors of identity for a wider cultural landscape, and any proposed revitalisation model should consider the context of the region or area in which they emit their cultural significance. Croatian and Hungarian country houses are the result of spatial, natural, and cultural factors of the former Habsburg Monarchy and the Austro-Hungarian Empire which lasted until 1918, and their intraregional connections are thus "the basis for planning of the wider areas beyond national borders" (Dundović et al. 2015). Considering a country house as a focal node in the interconnected cultural network of a wider area, a number of model groups can be discussed: for instance, *urban planning models*, *landscape models*, *spatial models*, and *socio-cultural models*.

28.4.1 Urban Planning Models

Urban planning models provide tools for the planned programming of country house revitalisation in a region or a wider area. It is of great relevance whether a manor house is part of the urban fabric within, on the edge or outside the urban area of a settlement. Country houses are powerful urban generators, and for this reason it should be taken into consideration whether a settlement surrounding the country house (and its estate) historically existed or whether their urbanity is a present-day condition. For example, Janković Manor in Daruvar, Croatia, was an intensive generator of the urban condition for the town. It served as a managerial focus of several family estates in Central Slavonia but, as it was built adjacent to ancient Roman *thermae*, it encouraged the building of the town's thermal spa complex, which ultimately dictated its urban growth. Therefore, the country house with its surrounding garden is today located in the very centre of Daruvar. Urban planning models always need to consider the spatial relation of a country house to other structures of cultural functions and create a system of nodes connected with streets, pedestrian promenades, or even parks and gardens. However, due to their accessibility, manors which are part of urban structures are more adept at accepting active functions than those in the countryside.

28.4.2 The Landscape Model

The landscape model includes both a country-house complex or estate and its surrounding landscape. Two major questions arise when deciding upon appropriate tools of intervention: Is there a traffic connection and what are the specific elements of visual recognisability? The manorial landscape garden usually differs in structure and colour from the surrounding natural vegetation, making it easier to distinguish the country house as an authentic element of a landscape, even if it is not instantly visible. Therefore, the *landscape model* means that architectural intervention is always minimal, and attention is aimed primarily at the gardens, landscape gardens and forested area as places of recreation, sport, and leisure. Furthermore, the closely related *arboricultural model* involves the affirmation of landscape culture in the form of an arboretum. Along these lines, the country house can be the scene of garden and landscape exhibitions or a nursery garden for old varieties of plants. There are numerous historical country house arboretums with elaborated botanical concepts, such as the one surrounding Erdődy Manor in Vép, Hungary, with species imported from England, France, and Italy (Sisa 2007), or Hilleprand-Mailáth Manor in Donji Miholjac, Croatia, with 110 species of trees and shrubs (Obad Šćitaroci and Bojanić Obad Šćitaroci 2001). Gardens and landscape gardens are both historical and contemporary elements of country house revitalisation, and we often forget that they are an integral part of the manorial estate, as important as the manor house itself.

28.4.3 Spatial Models

Spatial models are intended for the wider spatial context of a country house. The relation of a country house estate to its surrounding area or region is activated primarily by traffic connections and the proximity of settlements. The spatial framework of the area also includes all the other neighbouring manorial estates, as they can be arranged in spatial or functional systems such as a manor network, manor ring, and manor sequence. The *manor network model* includes country houses of a spatial unit (for instance, Central Slavonia in Croatia or Northern Balaton in Hungary), of the same stylistic features (e.g. Baroque manors of Hrvatsko Zagorje in Croatia), of the same cultural route (e.g. along ancient wine roads), or according to means of transport (e.g. several manor houses near Zaprešić, Croatia, are easily visited within a day on foot), or because they belonged historically to the same owner family (e.g. manors and curiae of the Janković family from Lake Balaton in Hungary to Central Slavonia in Croatia). The *manor ring model* is a system of country houses that encompasses bigger towns or cities (e.g. the manor ring around Zagreb, Croatia). Accordingly, its linear variant is supported by the *manor sequence model*.

28.4.4 Socio-Cultural Models

Socio-cultural models include all means of collective social and cultural significance of a country house in a region or area, whether it is of local or regional relevance. This is a set of mostly immaterial models that exert greater economic impact by introducing a country house as a subjective signifier. For example, the *representation model* was historically reserved for the owner family (as a representation of their power) but today it is important for the local community culture to make the country house an emblematic and recognisable element of the cultural landscape. A local community often uses the *participatory model*, while corporations and industries rely on the *institutional model* (where the manor house becomes a visual signifier of an institution or a product). It is important to apply the *contingent model*, where all the country houses of a network work together to produce a succession of events in a certain period (for example, manor houses along Lake Balaton that provide tourists with a continuous succession of events in the summer months). Ultimately, due to ownership issues and other legal problems that often arise, it is important to use the *legislative model* (in the form of laws and acts provided by the state) to solve any legal problem blocking the further development.

28.5 Conclusion

It is usually understood that country houses, as any other types of monuments, "are designed to resist the forces of time in order to preserve and project a particular inflection of selected memories", which implies "an underlying idea that the event can be preserved in the material as long as the object persists" (Hale 2017). However, a country house (especially a Central European one) was historically a place of constant transformation, alteration, and reconsideration in order to accommodate new ideas and new concepts of active life on the estate. Since country houses are today also architectural documents of past times, we are additionally tasked to preserve their authenticity, integrity, and identity. Keeping this in mind, whichever of the presented historical or contemporary models we apply to a specific country house, the historic manor with its estate should never be a hindrance, but rather a node of potentiality, that is, a generator of local economic development.

Acknowledgements The research is part of the scientific project "Heritage Urbanism—Urban and Spatial Planning Models for Revival and Enhancement of Cultural Heritage". It is financed by the Croatian Science Foundation [HRZZ-2032] and carried out at the University of Zagreb, Faculty of Architecture.

References

Deleuze G (1994) Difference and repetition. Columbia University Press, New York

Dundović B, Obad Šćitaroci M, Bojanić Obad Šćitaroci B (2012) Prolegomenon to the comparison of stylistic features of Hungarian and Croatian manor houses in historicism. Prostor 20(2):352–367

Dundović B, Obad Šćitaroci M, Bojanić Obad Šćitaroci B (2015) Revitalisation of country houses along the route "Tracing the steps of the counts Jankovich": guidelines for nodes of functional intensities. In: Obad Šćitaroci M (ed) Cultural heritage—possibilities for spatial and economic development: proceedings. University of Zagreb, Faculty of Architecture, Zagreb, pp 114–119

Fekete CsJ (2002) Kastélyideál a századfordulón. Architectura Hungariae 4(1):13. http://arch.et.bme.hu/arch_old/korabbi_folyam/13/13fekete.html. Accessed 16 Mar 2018

Fekete CsJ (2016) A magyarországi kastélyprogramok bemutatása kitekintéssel az állami kastélyfenntartás külföldi gyakorlatára. http://www.terc.hu/documents/Kast%C3%A9lyprogram%20cikk.pdf. Accessed 16 March 2018

Hale M (2017) Radicalising architecture by redefining the monument. In: Boundas CV, Tentokali V (eds) Architectural and urban reflections after Deleuze and Guattari. Rowman & Littlefield International, London, pp 221–241

Lazányi O et al (2016) Multidisciplinary methodology development to measure socio-economic impacts generated by cultural heritage projects. In: Van Balen K, Vandesande A (eds) Heritage counts. Garant, Antwerp/Apeldoorn, pp 99–110

Nemzeti Kastély- és Várprogram (2016) Info jegyzet: képviselői információs szolgálat 26, 19 May 2016. http://www.parlament.hu/documents/10181/595001/Infojegyzet_2016_26_kastelyprogram.pdf. Accessed 22 Mar 2018

Obad Šćitaroci M, Bojanić Obad Šćitaroci B (2001) Manors and gardens in Croatia: Slavonia—from Zagreb to Vukovar. Šćitaroci, Zagreb

Obad Šćitaroci M, Bojanić Obad Šćitaroci B (2006) Two millennia long tradition of manors, burgs, villas and summer houses on the territory of Croatia. In: Lioce R (ed) Villas, stately homes and castles: compatible use, valorisation and creative management—experiences, tools and methods. Edizioni Lunargento, Venice, pp 135–151

Obad Šćitaroci M et al (2006) Criteria for valorisation of manors: addendum for selection of the most valuable Croatian manors. In: Lioce R (ed) Villas, stately homes and castles: compatible use, valorisation and creative management—experiences, tools and methods. Edizioni Lunargento, Venice, pp 175–189

Petzet M (2004) Principles of preservation: an introduction to the international charters for conservation and restoration 40 years after the Venice Charter. International charters for conservation and restoration: monuments & sites, vol I. ICOMOS, München, pp 7–29

Sisa J (2007) Kastélyépítészet és kastélykultúra Magyarországon: a historizmus kora. Vince Kiadó, Budapest

Sisa J (2010) The Hungarian country house 1840–1914. Acta historiae artium 51:139–205

Széchenyi I (1866) Pesti por és sár. Heckenast Gusztáv, Pest

Chapter 29
Models of Revitalisation and Enhancement of Cultural Heritage and Sustainable Use

Mladen Obad Šćitaroci and Bojana Bojanić Obad Šćitaroci

Abstract Models of revitalisation and enhancement of cultural heritage were studied through the Heritage Urbanism approach. A total of 17 models were identified and divided into three groups: Universal Heritage Models, Basic Heritage Models, and Thematic Models of the Heritage Approach. Appropriate models that can enhance and modernise heritage and preserve its identity features were chosen from these groups. In terms of dealing with heritage, the following models may be differentiated: conservation models, development models and models of use. Cultural heritage management is becoming a model to connect public, institutional, local and all users. Potential models for the enhancement and revitalisation of urban heritage and landscape heritage were studied. The study and identification of these models was based on multidisciplinary research and a multitude of selected case studies. The aim was to determine universal models which could contribute to finding sustainable and lasting new use for heritage. The research was based on the following hypotheses: heritage is not a burden but a potential for development and a strategic national resource; it is not sufficient to protect and conserve heritage, it must also be renewed and interpreted in a sustainable manner that will provide it with new life; heritage is a current challenge which can be resolved through its inclusion into the contemporary and future life of towns, settlements and communities; heritage is not to be perceived as a static object, but as a creative subject.

Keywords Cultural heritage · Enhancement models · Heritage re-use · Heritage urbanism · Heritage utilization · Revitalisation models · Sustainable heritage

M. Obad Šćitaroci (✉) · B. Bojanić Obad Šćitaroci
Faculty of Architecture, University of Zagreb, Zagreb, Croatia
e-mail: mos@arhitekt.hr; scitaroci@gmail.com

B. Bojanić Obad Šćitaroci
e-mail: bbojanic@arhitekt.hr; bobojanic@yahoo.com

M. Obad Šćitaroci et al. (eds.), *Cultural Urban Heritage*,
The Urban Book Series, https://doi.org/10.1007/978-3-030-10612-6_29

29.1 Introduction

Four years of research was devoted to the revitalisation and enhancement of cultural heritage by exploring universal models that could contribute to finding sustainable and lasting use of heritage. Many experts and researchers from different fields took part in the research, so it was both multidisciplinary and multinational. Cultural heritage experts and researchers in all countries face similar questions and attempt to solve comparable issues: how to prevent the deterioration of cultural heritage, how to preserve, renew and enhance it, how to modernise it and give it new life, and how to ensure its survival. The research primarily considered heritage that cannot be included on the UNESCO World Heritage List or on the list of national heritage. Focus was placed on "average" heritage that is important for the cultural identity of towns and municipalities, and which is the most prevalent type of heritage everywhere.

The stimulus for the research was the poor state of cultural heritage in Croatia, castles in particular. It soon became clear that other countries are also experiencing similar challenges regarding the maintenance, renewal, revitalisation and enhancement of cultural heritage, especially built heritage of all kinds and types. This topic was first researched at the beginning of the twenty-first century as part of the INTERREG IIIB EU programme (2004–2006) devoted to the revitalisation and enhancement of manors and villas (Obad Šćitaroci 2005, 2006). It became apparent then that European experts and researchers were faced with various challenges in rethinking and implementing procedures of renewal, revitalisation and enhancement of heritage. Another stimulus occurred in 2015 when 233 authors from 34 universities in 21 countries gathered at an international conference on the spatial and developmental possibilities of cultural heritage (Obad Šćitaroci 2015). After this conference, it became clear that the issues are global, although they are manifested locally. These two stimuli were enough to draw together interested experts from different countries in order to attempt to answer at least some of the questions.

29.1.1 Research Question and Aim

The research question of the HERU project[1] and the research task were to identify appropriate and sustainable models for the renewal and conservation of heritage which would respect the identity features of the heritage, but also to allow for contemporary interventions and new features by using appropriate criteria. The research explored the options available for cultural heritage enhancement through an integrated urbanist and spatial view in a multidisciplinary environment.

The focus of the research was on enhancing the methods of renewal and spatial protection of heritage to contribute to the development of the local community. The

---

[1]Urban and Spatial Models for Revival and Enhancement of Cultural Heritage (HERU-2032), 2014–2018, Faculty of Architecture, University of Zagreb.

aim was to raise awareness of cultural heritage as a strategic resource and to underline the possibility of capitalising on cultural heritage.

29.1.2 Starting Points and Theoretical Framework—Stimuli in Literature

The timeliness of the research topic is seen in the fact that it is unexplored, although there is a wealth of literature on heritage, on individual heritage-related topics and there have been numerous projects and actions related to the renewal and modernisation of the built heritage. However, there is a clear lack of research papers and literature attempting to address the following questions: how to transform passive heritage into active heritage; how to enhance and improve it; how to ensure its long-term survival; how to establish criteria for interventions in heritage; and how to set models for the revitalisation and enhancement of heritage.

Due to the absence of papers on models for the revitalisation and enhancement of heritage, literature that can indirectly help with this research was used. The literature was studied from two perspectives—theoretical considerations and examples of good practice (case studies). A theoretical view was necessary to consider the authenticity of the heritage with regard to changes occurring over time and contemporary needs and possibilities for maintenance. Many studies exist of specific examples, most often aiming to recognise features of identity. Methods have been less studied, and ever fewer studies exist on the criteria and models for the enhancement of heritage. In this context, it was decided to analyse examples which proved suitable for the identification of models in specific cases. By selecting and analysing many cases, models were identified which were consciously or intuitively used in the renewal of cultural heritage or in new interventions in it.

Different theoretical approaches opened the door to interdisciplinary views on heritage issues which directly stimulated or confirmed the need for the planned research.

Thomas (2018) and Whitehand (2015) discuss the relationship between urban morphology and urban conservation. While urban morphology strives to grasp the physical structure of towns and how it develops and changes over time, urban conservation tries to answer the question of how heritage space should be managed in the future. Traditional approaches to heritage protection focus on protecting individual buildings and their immediate surroundings. Contemporary approaches consider the historical dimension of the whole (broad) environment and are closely related to large-scale urban planning and architecture. This kind of broad view is a characteristic of the Heritage Urbanism approach, which is confirmation that its position related to heritage is correct and contemporary, as shown through the research presented in this book.

Economics and Heritage Conservation (De la Torre and Mason 1998) presents a multidisciplinary discussion on the evaluation of cultural heritage. Two different views, supported by arguments, clash. The first is the view of economists, and the second is that of culturalists. There are differences between economic and cultural values. Everyone agrees that it is impossible to view cultural heritage through traditional economic models where value is measured in terms of price and the market. Different values of cultural heritage are considered: economic value, aesthetic value, cultural value, political value, educational value. The concept of cultural capital explains the thinking on heritage in the economic context, since the value of cultural heritage should not be viewed through traditional economic categories.

New Activities for Cultural Heritage (Ceccarelli et al. 2017) shows the results of multidisciplinary research connected by four topics: architecture heritage, robotics, entrepreneurship and finance. Such a view opens up topics of contemporary synergistic activities in the frame of cultural heritage with technical and non-technical aspects. This technological approach sets new goals, systems and strategies which match the approaches presented in this book. Such an approach stimulates new research of cultural heritage.

In the ICOMOS Concept Note for the United Nations Agenda 2030 (Hosagrahar et al. 2016) cultural heritage is considered in relation to sustainable development with its many conditions, challenges and opportunities in the contemporary socio-economic, environmental and political context where each has an influence on heritage. Issues of global cultural heritage are identified in terms of urbanisation and the loss of local identity, unsuitable attitudes towards cultural heritage in urban planning and tourism policy, and the underestimation or lack of an ecological perspective in the urban planning of settlements. The Note emphasises that cultural heritage contributes to the sustainability of towns and settlements, it is a driver of economic development, contributes to social cohesion and to the promotion of sustainable urban and rural areas where it is pleasant to live. Such theses match the research starting points of this volume.

In the last fifteen years or so, a new field has opened up in the context of the need for the sustainable development of towns and spaces, based on contemporary conservation views presented at international conventions on conservation. More stimuli come from the activities and announcements of global and European organisations dedicated to cultural heritage (UNESCO, ICOMOS, ICCROM, English Heritage, WTO, and others).

The international research project of the European Union, SUIT—*Sustainable Development of Urban Historical Areas Through an Active Integration within Towns* (2000–2004) explored the sustainable development of historical areas in towns through their active integration.

A significant resource for studying cultural heritage and its role in urban planning is the book by the architect and urban planner Cohen (2001), *Urban Planning Conservation and Preservation*, in which the author emphasises that this type of planning is an entirely new area that needs to be included in contemporary urban planning.

Further impetus for research came from the following books published in recent years which offer different views on cultural heritage:

- *New Perspectives in Cultural Resource Management* (Mc Manamon 2017) where the authors offer different perspectives on the methods, policies, and procedures of historical and contemporary Cultural Resource Management;
- *The Cultural Landscape & Heritage Paradox* (Bloemers et al. 2010) which presents the methodology of cultural landscape research and management based on an interdisciplinary and transdisciplinary approach, and the principle of heritage *conservation through future development*;
- *Heritage for the Future—Realising the Economic and Social Potential of a Key Asset* (2004), analysing the economic and social developmental potential of European cultural heritage;
- *Management of Historic Centres* (Pickard 2001) where many examples (case studies) show models of managing the historic centres of European towns, and the conclusion emphasises the need to develop a sustainable management method for historic centres;
- *Landscape Interfaces: Cultural Heritage in Changing Landscapes* (Palang and Fry 2003) which presents cultural landscapes in Europe from different perspectives of research and management;
- *Reconnecting the City: The Historic Urban Landscape Approach and the Future of Urban Heritage* (Bandarin and van Oers 2015) which broadens UNESCO's current approach to the management of historic centre with a series of operative management procedures and various professional points of view.

The scientific colloquium "Models for Revitalisation and Enhancement of Cultural Heritage: A Multidisciplinary Dialogue" which presented preliminary research on the topic stated in the title was a solid foundation for researching models for the revitalisation and enhancement of cultural heritage (Obad Šcitaroci 2017a, b). In the proceedings of the colloquium, 66 reports by 79 authors were published focusing on models that could contribute to the enhancement of various types of heritage by offering a multidisciplinary perspective which had a crucial role in identifying the models presented in this chapter.

29.1.3 Research Method

The appropriate models for the revitalisation and enhancement of cultural heritage were identified in the examples studied. Many examples of heritage, primarily material heritage, but also intangible heritage, were used in the research. These examples were found in the following research topics: cultural landscape, archaeological heritage, fortified heritage, manors and castles, rural heritage, historic town tissue, small towns, villages, the modernist heritage (twentieth century), historic gardens and parks, public spaces, soundscape, heritage branding, etc.

Historical and contemporary/current approaches and attitudes towards heritage were studied in a large number of examples selected for research, and potential modules were identified. During the research, it was essential to note approaches that are still usable and/or still show a substantial effect on the enhancement of the heritage. In addition, it was important to note permanent principles in the research examples applicable in the past and in the present time, which means that they could also be applied in the future. Many principles are timeless, which means they are permanently applicable, regardless of when they were created. Models that were applied in the past and which can be applied in the future were defined based on the identified approaches.

29.2 Models of Revitalisation, Enhancement and Sustainable Use of Heritage

A number of the studied examples showed different ways of using urban cultural heritage, landscape cultural heritage, and intangible cultural heritage. The latest contemporary modes of use can be modernised or new models can be created based on contemporary needs and the contemporary understanding of heritage while respecting the identity features of heritage and the results of its evaluation (Obad Šćitaroci 2017a, b).

The models can be viewed as a focused way of using heritage. They are not templates meant to be copied. The ideas they contain can be used, but they need to be adapted to the context and to each specific type because every example of heritage is different. These models do not offer concrete solutions, but they do provide a framework for finding a solution. They encourage discussion and stimulate thinking. Some authors use the term "conceptual model" which offers fundamental principles and the basic functionality of the system which it represents (Van Balen and Vandeasande 2018). The selection of the model is influenced by the context, the factors of the heritage identity, the evaluated value of the heritage, and the criteria for new interventions in heritage sites.

Although multidisciplinary approaches to rethinking heritage differ as much as there are numerous types of tangible and intangible heritage, the enhancement models are a connection between them and are a constant. A large number of various heritage enhancement models without an "expiry date" were identified. Universal and lasting principles that were applied in the past, are still applied today, and can be applied in the future were also recognised. These models can therefore be seen as universal models of behaviour aiming at revitalising and enhancing any heritage.

The aim is to apply appropriate/acceptable models to specific heritage in a particular context in order to contribute to active and economically sustainable use of heritage and to find a permanent purpose for it. Appropriate models are found in historical and contemporary ways of handling heritage by identifying those that have a lasting value and future application, in other words those that can be re-interpreted for contemporary times and thus give new life to the heritage. Acceptable heritage enhancement and revitalisation models do not necessarily need to be profitable, which is a kind that is most often acceptable just to entrepreneurs/investors (Poljanec-Borić 2017).

As a result of the research, 17 models divided into three groups were identified: Universal Heritage Models, Basic Heritage Models, and Thematic Models of the Heritage Approach. By systemising these models, universal heritage models were recognised first. They are applicable in nearly all cases and are therefore universal. Basic models are generally present in any form of heritage renewal, enhancement and revitalisation. Thematic models (or at least some of them) can appear in combination with the basic ones, depending on the context, the character of the heritage, and the goals pursued.

According to the criterion of heritage management, the models were categorised into conservation models (dealing with conservation, maintenance and renewal), models of development (focused on research, revitalisation, adaptation, etc.) and models of heritage use (relating to culture, education, science, technology, tourism, etc.).

29.2.1 *Universal Heritage Models*

Universal models can be applied in almost all cases of heritage renewal, revitalisation and enhancement. The starting point in each case is protection and conservation, then establishing an urbanist and architectural concept, determining the appropriate use of the heritage to help it survive, and ensuring the legal security and economic sustainability of the heritage.

Protection and Conservation Model

The protection and conservation model implies identification, protection, evaluation, conservation and maintenance of the heritage; respect towards the cultural, historical and other features of the heritage identity during the renewal; and renewal based on credible historical sources. This is a basic model for cultural heritage renewal, but it cannot be used solely because this model alone cannot ensure the sustainability of the renewed cultural heritage. It must be modernised and complemented with other appropriate models which can enhance and revitalise the heritage.

Heritage protection is the foundation of this model. It prevents damage caused by inappropriate use or devastation. There are several relevant aspects of protection: administrative protection (based on the Act on the Protection and Preservation of Cultural Heritage); legal protection (during use and new architectural interventions); integral protection of the heritage and its surroundings (the connection of the heritage to its environment); and urbanist and spatial protection (integral protection built into spatial and urban planning).

In heritage renewal, the following approaches or methods can be applied: conservation (preservation of the existing state—in situ or ex situ); restoration/restitution (return to the original state); reconstruction (renewal based on the historical state); facsimile reconstruction; preserving the memory of heritage (presentation of an uncovered layer); contemporary interventions in heritage sites (when heritage is used for contemporary needs and has a new purpose); enhancement of neglected heritage.

Heritage Revitalisation Model
The revitalisation model is employed when heritage is no longer active, without a purpose or in a very poor state. According to the criterion relating to the scope and intensity of interventions, there are two types inherent in this model: *complete revitalisation* (when heritage has no use) and *partial revitalisation* (when some parts of the heritage site are (more) active, while others are not).

Revitalisation refers to physical renewal and the renewal of its contents. Physical renewal alone, regardless of how comprehensive it is, does not mean revitalisation. The result will be *passive* revitalisation which will not fulfil the desired goals. It is also necessary to renew the contents which will give new life to the heritage, thus ensuring *active* revitalisation. When justified and desired, various *activation models* are applied. They stimulate contemporary heritage use which can make the heritage site a place for events and social activities, even when the site was not originally a public area. Being open to the public and used by the public leads to activities that can ensure revitalisation.

Revitalisation can be *invasive* (when the basic features of cultural heritage are significantly changed or complemented) or *non-invasive* (when the identity features of the heritage are kept in their original state, with minimal contemporary interventions and technological modernisation). The latter will occur when the heritage in question is a monument which should be preserved in its original form and used as such. Revitalisation can be seen as an opportunity for improving the vitality of the heritage (e.g. streets, squares and public areas). *Identity re-affirmation* through a new urbanist and architectural concept will revive the forgotten identity features.

Heritage Enhancement Model
The enhancement model is used when heritage is partially active, has a purpose and is not in a very poor state. Heritage surely can and must be used better, so it needs enhancement and high-quality management, not only in terms of maintenance, but also in terms of activation programmes and new uses.

There are three type of enhancement to consider: regeneration, transformation and application. Enhancement can be implemented in different ways. New development with more significant changes or complementing the existing state are some of the options, as well as adaptation which could improve the functional state of the heritage. Marketing affirmation can aid heritage enhancement by attracting more visitors, which in turn encourages better maintenance.

Heritage Re-Use Models
Using heritage, which means that people accept it as part of their daily lives, guarantees the survival of the heritage. Using the cultural heritage stimulates community development. Since heritage most often cannot be used for its original or historical

purpose, it is necessary to think about new uses for the heritage and to seek acceptable *re-use* models. Models of heritage use must not go against preservation. They are expected to contribute to the sustainability of the cultural heritage. They can be viewed as a model for using urban heritage, landscape heritage and intangible heritage. There is a wide range of different scales and spaces to consider when rethinking heritage use—from buildings, housing complexes, parts of the town, the whole town, or even the whole region.

The heritage re-use model may consist of sub-models that differ depending on the type of heritage (tangible or intangible) and the goal (revitalisation, regeneration, re-use, modified re-use, development, renewal, revival, etc.). From the functional and organisational point of view, we can distinguish between the monofunctional and polyfunctional model, the traditional and contemporary (planning and management model) model, the individual and networked model, the transformation (a change of function and use) model, and the model of preserving the current state of the heritage. Functional models primarily consider renewal in the context of the re-use and enhancement of the previous functions which must be conducted in accordance with acceptable new functions and contemporary needs.

Economic Heritage Model—Model of Economic Sustainability

The economic model considers long-term financial sustainability during heritage use. This model is essential when there is no long-term financial support. The state, town or municipality budget frequently cannot ensure enough money for maintenance, so it is crucial to secure another financing model. If there is no continuous financial support from the budget, income must be ensured by finding new appropriate use for the heritage. Ignoring this model makes it impossible to achieve lasting sustainability and regular maintenance which are of great significance for the heritage.

The value of cultural heritage can be measured by the traditional economic criteria of price and market, but an evaluation—aesthetic, cultural, educational and others—is also necessary. Acceptable and objective value can be established only after evaluation.

It will be possible to revitalise and renew some heritage based on the traditional economic model (the one originally related to this heritage), but most heritage will require appropriate contemporary economic models of use. At the present time, cultural heritage management which is an integral part of economic planning is considered important.

Recently, heritage has been revitalised by applying the model of spatial branding. This model uses the method of managing the individual spatial identities of a place. This sort of management is accomplished by suitable architectural and urbanist interventions in space which might have a multiplying effect on the identity of this space, thus creating a potentially significant influence on the social, economic and spatial environment.

Legal Heritage Model

The legal model deals with issues relating to law and ownership, particularly issues regarding expropriation and legal relations between different owners. This model requires the preparation of implementation instruments and procedures at the state level. Without them, sustainable renewal and revitalisation of the heritage is not possible.

The legal model can be viewed from three different perspectives. The first aspect is related to technical laws and regulations related to buildings. However, in the case of built heritage, it is impossible to meet many of the contemporary technical and technological regulations, so renewal and revitalisation models must be adapted (to an acceptable level) in accordance with the features of the cultural heritage.

The second legal aspect is related to urban and rural built environments, i.e. urban and spatial planning where provisions for implementation are made at the level of the urban plan. These provisions set many conditions important for heritage renewal.

The third legal aspect stems from international conventions and declarations on the protection of cultural heritage which guide contemporary interventions at the global level.

29.2.2 Basic Heritage Models

Basic heritage renewal models are models that are difficult to avoid and are used in the majority of revitalisation and enhancement cases. These basic models are based on spatial and urban views on the heritage and its context, they are related to the transformation of built heritage, they adopt an integral approach, and link heritage and sustainable development.

Urban/Spatial Heritage Model

The urban/spatial model does not consider or renew heritage as an isolated entity, but rather in the context of its immediate and broader surroundings, and sometimes in the context of a part of the town or even a whole town or settlement. This way, heritage becomes part of its surroundings and has an influence on it, just as much as its surroundings have an influence on the future life of the heritage. Cultural heritage cannot survive outside its context—a settlement, a town or cultural landscape.

The urban/spatial model ensures heritage protection and renewal in two ways: first, by considering the specific heritage with its broader surroundings, and, second, with the help of master plans and spatial plans which set the criteria for heritage renewal, revitalisation and enhancement.

Master plans and spatial plans enable the spatial, functional and structural integration of heritage into the tissue and life of a town, settlement or landscape. Spatial and urban planning merge and connect into a whole all elements that appear in space—physical, natural and anthropogenic space, the communal and social infrastructure, roads and the movement system—thus creating a synthesis of heritage, both the existing and the planned heritage.

The urban/spatial model forms a connection between heritage and space, but also between different heritage sites. This approach creates connections and movement routes between heritage locations, ensures access to the heritage and affirms the heritage by including it into people's daily lives. It is crucial that heritage is not isolated and excluded but integrated into the town and landscape so people can notice, recognise and use it. This is the best way to preserve and enhance the cultural heritage.

Heritage Transformation Model

The transformation model implies a change in function, purpose and use of heritage in comparison with the original and historical state. Re-establishing the original or historical purpose is most often not possible, so an appropriate transformation model is selected that provides the heritage with a new purpose and contemporary use.

Transformation includes different approaches and/or methods which are applied to make contemporary interventions in historical space. These include functional transformation, urban transformation, structural transformation, ambient transformation, transformation of use, transformation of architecture/design, etc. However, it is of utmost importance to preserve the identity factors of the heritage and affirm the memory of space. The goal is to accomplish innovative contemporary revitalisation that will respect and affirm the heritage.

Heritage Integration Model

A fragmented, partial and sectoral approach to cultural heritage revitalisation does not yield good results. Such an approach cannot solve the spatial issues in towns, settlements and landscapes either. Therefore, an integrated approach to cultural heritage revitalisation, enhancement and management is considered a favourable solution.

The integrated approach model refers to planning, designing and managing cultural heritage. The possibilities of an integral approach are various. Integral local development is accomplished by connecting heritage with the local community so that people can recognise heritage as part of their identity and as a resource for improving their lives. The symbiosis between heritage and tourism can have a positive effect if part of the income from the use of heritage in tourism is returned to the heritage and the local community and invested for heritage enhancement. Establishing a dialogue and harmonising different factors in space can contribute to the development of an integration model.

In the case of archaeological heritage which is extensive and particularly vulnerable, an integral approach is relevant because it provides a lasting solution and enables the inclusion of in situ archaeological heritage into the life of towns, settlements and individual buildings or built structures.

In accordance with EU recommendations, an integrated approach should be embedded in the economic and social potential, in public policies and in the heritage of neighbouring and other countries. Ultimately, a more integrated approach needs to be taken with regard to the preservation and valorisation of heritage (Towards an integrated approach to cultural heritage for Europe, COM 2014/477).

Heritage Interaction Model—A Multidisciplinary Approach

The implementation of cultural heritage enhancement in the modern era is not possible without a multidisciplinary approach and without the cooperation of different national and/or public institutions. The issues of revitalisation and enhancement are not only related to conservation and culture, but go much further than that. The survival and enhancement of heritage is not possible without full and comprehensive consideration and management. A sectoral approach is unacceptable because it does not solve any issues. The renewal, enhancement and revitalisation of heritage can succeed only if different state departments (culture, spatial and regional development, fiscal policy, tourism, the economy, etc.) and different institutions and professions are included in the process, not by working individually but interacting together to reach a common goal—to make cultural heritage a sustainable national resource.

Cultural heritage management becomes a model for connecting the public, institutional, local and all users. The interaction model does not just include all data on heritage, but it primarily serves as a platform for sharing knowledge between practice and projects. The interaction model is *work in progress* which is continually developing and never stops.

The Council of Europe promotes participative heritage enhancement and the revitalisation model which involves local communities in a way that includes all stakeholders in the field (politics, associations, companies, individual manufacturers, citizens, etc.). This model implies networking and coordination between all national and regional resources which should ensure that all the prerequisites are met for the effective renewal and active use of heritage as a national resource.

Heritage Sustainable Development Model

The model of sustainable heritage development implies the long-term survival of heritage and its new life. The goal is to apply a model of sustainable development to renew all heritage and make it active. Renewing all the heritage includes not only the most valuable heritage, but also regional and local heritage sites which exist in large numbers. This sort of heritage in particular requires sustainable renewal models because it is not realistic to expect constant financial support from the state.

Although financial sustainability is important, there are other factors of sustainability, too, so in the context of cultural heritage, spatial, economic, environmental and social sustainability should also not be forgotten. Each of these aspects has its own criteria which must be coordinated and balanced. Entrepreneurial sustainability criteria are different from environmental and social criteria. Therefore, a sustainable development model must be considered together with other models and viewed in a multidisciplinary way in order to achieve the desired result and produce a visible effect.

It is recommended to link heritage sustainability and the creation of a sustainable model to the local community and the users of the renewed heritage. If all or most of the sustainability factors are achieved apart from the local community and users, there is a great chance that the heritage will not be revitalised sustainably because it will not be used.

29.2.3 Thematic Models of the Heritage Approach

The specialist point of view gives heritage a particular distinctiveness. Specialist views are used to create thematic models. Multidisciplinary and transdisciplinary views are a good starting point for profiling thematic heritage renewal models. Architectural, aesthetic, ecological, tourism, experiential and other models can offer interesting contemporary programmes, ideas and interpretations for the enhancement and revitalisation of heritage. Despite applying different approaches, the basic identity features and values of the heritage must never be brought into question.

System Model

Individual heritage (buildings and small structures) that can be viewed as a dispersed or a concentrated model cannot have a stronger activation and revitalisation effect if they are not connected through a designed system. Therefore, it is recommended to include individual heritage sites into one of the spatial thematic sub-models—*linear* (heritage trails), *networked* (connecting and overlapping various types of cultural and natural heritage), *dispersed* (connecting dispersed town or landscape heritage into a coherent and functional whole) or some other system.

Forming a connection is possible on the basis of a common theme, such as: manor hotels, hunting manors, manors turned into museums or restaurants, wine trails, etc. The theme is found in a region, micro-region, in the design and style features of the heritage, in historical or cultural routes, in families, etc. Territory can also be the criterion for connecting heritage into a logical and meaningful system, e.g. a heritage belt encircling a large town, heritage along the rivers or sea shores, island heritage, etc. A heritage system can be created by using one or more criteria to help find a theme for a particular area.

In the territorial or thematic principle, the system model is connected to the *Urban/Spatial Heritage Model* in order to integrate it into the spatial development of the local/regional community.

Architectural and Design Model

The renewal or re-use of an individual building or complexes which are considered cultural and built heritage is affected by the architectural idea/concept that stems from the character of the heritage and the criteria for its preservation and affirmation. At the same time, the architectural idea affects the revitalisation and the success of the use of the heritage. The main theme of the architectural/design model is *new in old*, i.e. contemporary architectural intervention in the cultural heritage building and heritage space. The eternal issue of the relationship between the new and the old is how to achieve a balanced relationship, how to harmonise the contemporary and historical design, and how to interpret heritage for the future.

The architectural and design idea cannot be put into strict models because this is individualised and comes from the creative process, inspired by the identity features and values of the heritage and complies with the conservation criteria. By studying a large number of examples of heritage, various concepts were discovered that marked the architectural and design model of heritage renewal.

Contemporary architectural intervention into heritage areas may follow the concept of contrast (emphasising new features in the heritage), mimicry (adapting to the heritage and its surroundings), contemporary interpretation of tradition and historical forms, or an urbanist and architectural redefinition, the concept of a house in a house, a museum presentation, the personalisation of urban places, urban characterisation, the matrix of cultural landscape and many other conceptual approaches.

Cultural Tourism Model
Cultural tourism, a fairly new form of tourism, has raised interest in heritage, which brings revenue to towns and anyone owning the heritage. This income is more indirect than direct, so the amount of money earned in the tourism industry allocated for heritage renewal (without which there is no cultural tourism) is still too small. This vicious circle has to be resolved in a sustainable manner as soon as possible so that the symbiosis between tourism and cultural heritage can survive in the long run.

Cultural tourism often occurs spontaneously, although we can and must plan it and create it. Renewal and revitalisation of all types and levels of heritage is a precondition for cultural tourism. By revitalising and affirming the tangible and intangible heritage in small and unknown places we strengthen economic activities as a result of the emergence and growth of tourism. Accordingly, there are various contemporary forms of tourist activities, such as creative and experiential tourism or thematic (focused on specific topics) tourism.

The cultural tourism model implies planning sustainable heritage tourism, enhancing historic architecture relevant for tourism, creating engaging tourist activities, evaluating cultural heritage for sustainable tourism, and conceptualising tangible cultural heritage in the tourism industry.

Experiential Model
The experiential model includes the aesthetic values of heritage and memory that are recognised in heritage. The experience is stronger if heritage is viewed in a new and unusual way and if experiments are made with the possibilities of heritage. The experiential qualities of soundscape, lightscape and walkscape are examples of new research on experience and the environment. The previous spatial aspects experienced when considering heritage and the urban or natural landscape are still applicable and contribute to the quality of the experienced cultural heritage model.

The sense of atmosphere is important not only for tourism. The atmosphere of cultural heritage is also relevant for daily life and for coexistence with the heritage. Experiencing heritage itself is often not enough, so it is necessary to provide heritage with contents that will enhance the visitor's experience and offer various levels of intensity of experience. Heritage keeps and awakens memories of people and events from the past. It helps profile the spatial identities of a place or an area. Regardless of how small the renewal procedures on lost and forgotten heritage are, they can still revitalise the atmosphere of the past that will stimulate other additional activities.

Ambient Authenticity Model
The ambient authenticity model respects the characteristics of the context which not only includes the physical and material environment (urban, rural, landscape) but also the intangible influences (social, cultural, sociological) that can make heritage special and can significantly affect the sustainability of the renewed heritage. The application of the ambient authenticity model can contribute to the historical authenticity of heritage.

The authenticity and quality of ambiance must be viewed from two perspectives—the aspect of heritage and the aspect of its surroundings. Cultural heritage that is renewed and revitalised has to maintain and enhance by means of renewal its authentic features in order to be credible and interesting. To increase the quality of the heritage, it is also necessary to ensure the ambient authenticity of the immediate and wider surroundings of the heritage. Without high-quality surroundings, the heritage loses its value and attractiveness. Therefore, the ambient authenticity model can contribute to heritage enhancement.

Landscape-Ecological Model

The starting point of the landscape-ecological model is preserving, respecting and enhancing the landscape and natural factors, regardless of whether the heritage is located in the town or in the landscape. This model will be applied more often in the cases of large cultural landscapes containing both built heritage and natural heritage where ecological, phytocoenological and other biotechnical aspects of the cultural landscape are predominant. In such cases, the concepts of a regional park or eco-museum can be applied.

The landscape-ecological model can also be used for the renewal and enhancement of large town parks and the renewal and revitalisation of historic gardens. In these cases, the landscape-ecological model will strengthen the urban area and support the ecological sustainability of historic gardens in an urban environment.

29.3 Conclusion

After the research conducted on a large number of case studies, 17 models were identified and divided into three groups—universal, basic, and thematic models. They can be applied when considering and developing cultural heritage renewal projects. These models and the research method used can serve as a stimulus for further research and for the updating and correcting of results. The models offered in this paper can be tested, verified and applied to all types of cultural heritage. Further verification and practical application will determine the sustainability of the individual models.

The identified models for the enhancement and revitalisation of cultural heritage answer the research question posed at the beginning of the research: what are the appropriate and sustainable models for the renewal and conservation of cultural heritage which would recognise and respect its identity features and which would serve as guidelines for contemporary interventions and for the introduction of new contents by using appropriate criteria?

The cultural heritage revitalisation and enhancement models will find practical application in the protection and renewal of cultural heritage, urban planning, architecture and landscape design, and they will also be useful for cultural heritage management and economic considerations relating to the heritage. The revitalisation and enhancement models will be helpful in making decisions regarding conservation and they will also stimulate the creation of a system of active protection. They do not offer concrete solutions, but they provide a framework for finding solutions. Selecting a

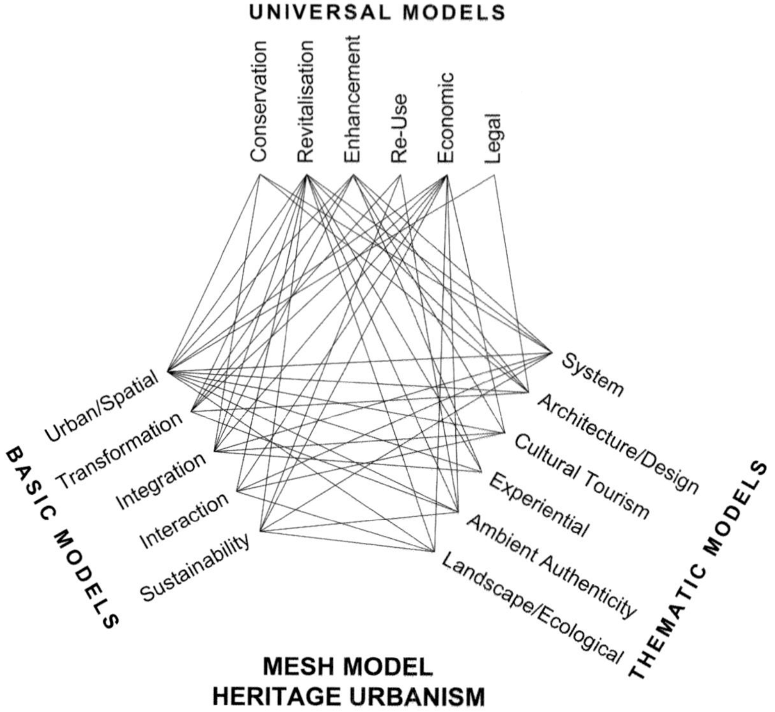

Fig. 29.1 Correlation diagram between universal, basic and thematic models of heritage urbanism approach

model is governed by the context, the heritage identity factors, the factors of influence, the value of the heritage and the criteria for new interventions on the heritage.

To revive the heritage, it is crucial to choose a use for the heritage that will ensure lasting active protection and economic sustainability in accordance with the value of the heritage. Renewed heritage does not equal revitalised heritage if it does not have a permanent purpose and if it is not functional and in harmony with its immediate and wider surroundings in terms of infrastructure and ambiance. Full revitalisation is achieved when users and the local community accept the renewed heritage and start using it.

Using only one renewal model is uncommon. Universal models are almost always used, and, most often, several different compatible basic and thematic models are applied together. They complement one another by creating a quality model mesh where quality arises from connecting different models (Fig. 29.1). Such a model is determined for each type of heritage, and in each case individually. Every example of heritage has its distinct identity and a unique evaluation result which together form the basis for a combination of basic and thematic models.

The research shows and proves that heritage is not a burden but a resource for development which needs to be activated and directed so that it becomes dynamic capital rather than dead capital. Renewal models promote understanding of cultural heritage as a strategic resource that can be capitalised. Preserving and renewing heritage is not enough, it must also be interpreted to be visible, not only to tourists but to the local community and its owners as well.

Cultural heritage management is often implemented and includes identification, protection, preservation, presentation, and interpretation of the heritage, as well as keeping it for future generations. Cultural heritage management is becoming a kind of *articulation model*—connecting the public, institutions, and local and all heritage users. It can be viewed as a continuously developing platform for sharing knowledge between practice and projects.

Appropriate models for specific heritage cases must be identified and selected from a large number of models, sub-models, types and approaches that will enhance the heritage and preserve its identity features, contribute to the active and sustainable use of the heritage, find lasting purposes for the heritage, and give it new life.

Heritage is often viewed as something that has survived, that is in a poor condition and is static. It is also seen as historical—it used to exist and live in the past, and it has to be protected today. Cultural heritage is frequently in disharmony with contemporary society. Heritage should not be seen as a problem; it should be included in the contemporary and future life of the community which owns it. The environment (a town, a village, a landscape) where the heritage is located is constantly changing. Besides, cultural heritage not only consists of built structures, but also of people, communities, ideologies and environment.

Are there any risks if artistic freedom is allowed in the renewal and interpretation of cultural heritage? Is it permitted to experiment with the heritage? Is the heritage a non-renewable resource? The research conducted and its results, i.e. the models identified, decrease the risk and thus provide an opportunity for experiment in accordance with the qualified evaluation and expert criteria based on the study of identity factors.

Today, efforts are being made to transform the static view of heritage into a creative process. Many towns in Europe have experienced economic growth and an increase in tourism thanks to the activation, revitalisation, renewal and re-use of the urban cultural heritage. The current issue is the relationship between cultural heritage and economic value because it is impossible to finance all cultural heritage as museum exhibits without a sustainable use. There must be a change of perspective: we must stop perceiving heritage as static objects and start viewing it as active creative subjects. A change of perspective will enable and facilitate a renewal of cultural heritage by applying the suggested revitalisation and enhancement models.

Acknowledgements The research is a part of the scientific project "Heritage Urbanism—Urban and Spatial Planning Models for Revival and Enhancement of Cultural Heritage". It is financed by the Croatian Science Foundation [HRZZ-2032] and carried out at the University of Zagreb, Faculty of Architecture.

References

Bandarin F, van Oers R (ed) (2015) Reconnecting the city the historic urban landscape approach and the future of urban heritage. Wiley Blackwell, Oxford. ISBN 978-1-118-38398-8

Bloemers T, Kars H, van der Valk A, Wijnen M (ed) (2010) The cultural landscape & heritage paradox. Amsterdam University Press, Amsterdam. ISBN 978 90 8964 155 7

Ceccarelli M, Cigola M, Recinto G (eds) (2017) New activities for cultural heritage. Springer, Cham. ISBN 978-3-319-67025-6 (Hardback), ISBN 978-3-319-67026-3 (E-Book)

Cohen N (2001) Urban planning conservation and preservation. Mc Graw-Hill, New York. ISBN-13: 978-0071375849

De la Torre M, Mason R (eds) (1998) Economics and heritage conservation. Getty Center, Los Angeles

Hosagrahar J, Soule J, Fusco Girard L, Potts A (2016) Cultural heritage, the UN sustainable development goals, and the new urban agenda—ICOMOS concept note for the United Nations Agenda 2030. International Council on Monuments and Sites (ICOMOS), 15 Feb 2016

Mc Manamon FP (ed) (2017) New perspectives in cultural resource management. Routledge, Abingdon-on-Thames. ISBN-13: 978-1138101128

Obad Šćitaroci M (ed) (2005) Contemporary use and creative management of manors, castles and villas, proceedings. European project "Villas, stately homes and castles: compatible use, valorisation and creative management"—Interreg IIIB Cadses. Faculty of Architecture University of Zagreb, Zagreb. ISBN 953-6229-40-4

Obad Šćitaroci M (ed) (2006) Manors and gardens—cultural heritage as a generator of economic development, proceedings. European project "Villas, stately homes and castles: compatible use, valorisation and creative management"—Interreg IIIB Cadses. Faculty of Architecture University of Zagreb, Zagreb. ISBN 953-6229-49-8

Obad Šćitaroci M (ed) (2015) Cultural heritage—possibilities for spatial and economic development. Faculty of Architecture University of Zagreb, Zagreb. ISBN 978-953-8042-10-2 paperback, ISBN 978-953-8042-11-9 e-book. www.arhitekt.hr/znanost/HERU/culturalheritage, tinyurl.com/heru2015-proceedings

Obad Šćitaroci M (ed) (2017a) Modeli revitalizacije i unaprjeđenja kulturnog naslijeđa - multidisciplinarni dijalog. Faculty of Architecture University of Zagreb, Zagreb. ISBN 978-953-8042-29-4 paperback, ISBN 978-953-8042-30-0 e-book http://www.arhitekt.unizg.hr/znanost/HERU/diseminacija/Diseminacija%20projekta%20i%20rezultata/HERU_zbornik_2017.pdf

Obad Šćitaroci M (2017b) Uvod u znanstveni kolokvij. In: Obad Šćitaroci M (ed) Modeli revitalizacije i unaprjeđenja kulturnog naslijeđa - multidisciplinarni dijalog. Faculty of Architecture University of Zagreb, 6–11. ISBN 978-953-8042-29-4 paperback, ISBN 978-953-8042-30-0 e-book

Palang H, Fry G (ed) (2003) Landscape interfaces: cultural heritage in changing landscapes. Kluwer Academic Publishers, Dodrecht/Boston/London. ISBN 978-94-017-0189-1

Pickard R (ed) (2001) Management of historic centers. Spon Press (Taylor&Francis Group), London/New York. ISBN 0-419-23290-7

Poljanec-Borić S (2017) Prikladni model razvojnog korištenja kulturne baštine. In: Obad Šćitaroci M (ed) Modeli revitalizacije i unaprjeđenja kulturnog naslijeđa - multidisciplinarni dijalog. Faculty of Architecture University of Zagreb, 18–22. ISBN 978-953-8042-29-4 paperback, ISBN 978-953-8042-30-0 e-book

Thomas RM (2018) Conservation, heritage and urban morphology—further thoughts. Urban Morphol 22(1):71–73

Van Balen K, Vandeasande A (2018) Innovative built heritage models. CRC Press/Balkema, Liden, The Netherlands. ISBN: 978-1-138-49861-7 hardback, ISBN 978-1-351-01479-3 e-book

Whitehand JWR (2015) Conservation, heritage and urban morphology. Urban Morphol 19(2):115–116

Whitehand JWR (2004) Heritage for the future—realising the economic and social potential of a key asset. The congress of local and regional authorities—local and regional action no. 6, Norwich UK—Proceedings, Council of Europe Publishing, Strasbourg

Whitehand JWR (2014) Towards an integrated approach to cultural heritage for Europe. Communication from the commission to the European Parliament, the Council, the European Economic and Social Committee and the Committee of the Regions, COM (2014) 477, Brussels, 22 July 2014

GPSR Compliance

The European Union's (EU) General Product Safety Regulation (GPSR) is a set of rules that requires consumer products to be safe and our obligations to ensure this.

If you have any concerns about our products, you can contact us on ProductSafety@springernature.com

In case Publisher is established outside the EU, the EU authorized representative is:

Springer Nature Customer Service Center GmbH
Europaplatz 3
69115 Heidelberg, Germany

Batch number: 10756373

Printed by Printforce, the Netherlands